Lignocellulosic Biorefineries

Cover Illustration:
Leaves: © Denis Tabler - Fotolia.com
Refinery: © christian42 - Fotolia.com

Lignocellulosic Biorefineries

Jean-Luc Wertz and Olivier Bédué

EPFL Press
A Swiss academic publisher distributed by CRC Press

CRC Press
Taylor & Francis Group

Taylor and Francis Group, LLC
6000 Broken Sound Parkway, NW, Suite 300,
Boca Raton, FL 33487

Distribution and Customer Service
orders@crcpress.com

www.crcpress.com

Library of Congress Cataloging-in-Publication Data
A catalog record for this book is available from the Library of Congress.

This book is published under the editorial direction of
Professor Hubert Girault (EPFL).

Great effort has been made to identify the owners of the figures used in this book and to obtain the appropriate permissions. For any specific concerns about this, please contact the publisher.

The Internet links provided in the reference sections are intended to help readers quickly access additional resources. There were last checked in April 2013; the permanence of these links cannot be guaranteed.

EPFL Press

The EPFL Press is the English-language imprint of the Foundation of the Presses polytechniques et universitaires romandes (PPUR). The PPUR publishes mainly works of teaching and research of the Ecole polytechnique fédérale de Lausanne (EPFL), of universities and other institutions of higher education.

Presses polytechniques et universitaires romandes, EPFL – Rolex Learning Center,
Post office box 119, CH-1015 Lausanne, Switzerland
E-mail: ppur@epfl.ch
Phone: 021 / 693 21 30
Fax: 021 / 693 40 27

www.epflpress.org

ISBN 978-2-940222-68-1 (EPFL Press)
ISBN 978-1-4665-7306-2 (CRC Press)

Printed in Spain

Foreword

Jean-Luc Wertz and Olivier Bédué have researched and written a timely work on the subject of lignocellulosic biomass, and how these materials can be put to use in a biorefinery. Lignocellulosic biomass is an important resource which already constitutes about 10% of the world's energy mix, although used predominantly for "cooking and heating." However, in the future it could supplement the fossil resources used in the chemical industry and for fuel.

The authors provide a complete overview of the topics related to the use of lignocellulosic biomass, except for the issue of procuring this resource. The authors offer a dynamic view of the biorefineries of tomorrow, ranging from photosynthesis, composition and characterization to using and deriving value from co-products, plus an assessment of technologies. They show that lignocellulosic biomass is a complex product with widely varied applications. Investments in research and development, which have been dedicated for about ten years now to biomass conversion, provide a good idea of its potential interest.

This very complete work will no doubt be among the basic publications referred to by those involved in making use of lignocellulosic biomass. It will help clarify the discussion of issues that are of interest to the public and also serve as an aid to public policy-making.

I am very pleased to see that the authors are former employees of Total. After contributing to a previous work, *Cellulose Science and Technology* (2010, EPFL Press), they now provide us with a wider vision of lignocellulose, the issues, and possibilities for the future. They are a good example of the expertise within our Group.

Jean-François Minster
Senior Vice President
Scientific Development
Total
March 2013

Table of Contents

Chapter 5
Enzymatic and non-Biological Degradation of Cellulose 189

Chapter 6
Hemicelluloses and Lignin, Other Key Constituents of Biomass 239

Preface

Environmental concerns and energy security are dominant themes for the world of today. Human activities, by burning fossil fuels such as oil, coal and natural gas, contribute both to the accumulation of greenhouse gas and climate change. Fossil resources will not last forever, and it is a certainty that, at some point in the future, there will not be enough to satisfy all our needs.

This book, *Lignocellulosic Biorefineries*, has been written to help bridge the gap between economic growth and environmental sustainability. Biorefineries convert biomass, a non-fossil renewable resource of biological origin, into bioenergy and biobased chemicals and materials. Photosynthetic organisms such as plants and bacteria convert about 100 billion metric tons of carbon from carbon dioxide into biomass per year. Biorefineries have been identified as a potential solution to reduce greenhouse gas emissions and to meet the growing demand for energy and for non-energy products. As discussed recently at the World Economic Forum in Davos, it is expected that by 2020 global biorefining will offer potential annual revenue of about US$ 300 billion along the entire biomass value chain.

First-generation biorefineries, which are already well implemented around the world, convert edible biomass such as corn, sugarcane and soybean. Lignocellulosic biorefineries, also called second-generation biorefineries, convert inedible plant biomass, made up primarily of cellulose, hemicelluloses and lignin. Some key advantages of lignocellulosic biorefineries include the fact that they do not compete with food production, provide much higher volumes, and could make use of agricultural and forestry residues and wastes.

The time is right for a comprehensive book that describes most important aspects of lignocellulosic biorefineries from the very complementary viewpoints of biology, biochemistry, chemistry and polymer science. Step by step, the book covers photosynthesis and biomass growth, first-generation biorefineries and the three major components of biomass before describing a wide variety of biomass processes, including pretreatments, biochemical conversion and thermochemical conversion, along with results from most recent case studies.

This book is intended to serve as a reference for policymakers, industry leaders, professors, teachers, scientists, engineers and students who have an interest in promoting the bioeconomy.

Acknowledgments

We express our gratitude to Mr. Jean-Michel BRUSSON from Total, who helped greatly to get funding for this book and brought accurate and useful advice to improve its content.

We also thank Mr. Jean-François MINSTER, Senior Vice President, Scientific Development, Total for accepting to write a foreword.

Jean Luc WERTZ dedicates this book to his family, and in particular to his wife Lydia and his two children, Vincent and Marie. He would also like to thank Prof. Michel PAQUOT for all their fruitful discussions and Julie ROÏZ for their interaction on biobased products.

Olivier BEDUE dedicates this book to his kids, Lea and Lucas.

Finally we thank Dr Frederick FENTER at the EPFL Press for his huge support and assistance during the publication of this book, as well as those publishers, persons, and scientific associations who granted permission to reproduce figures covered by copyright.

Jean-Luc Wertz and Olivier Bédué
Mai 2013

Chapter 1

Introduction

1.1 Climate change

Climate includes patterns of temperature, precipitation, wind and seasons. [1] Climate change not only refers to a change in the weather, it also refers to seasonal changes over a long period of time, such as the increased global average surface temperature over the 20th century and beyond. These climate patterns play an essential role in shaping natural ecosystems, and the human economies and cultures that are based on them.

Climate change can be due to natural variability and/or as a result of human activity. [1, 2] There is new and stronger evidence that most of the warming over the last 50 years is attributable to human activities (**Figure 1.1**). It is unlikely that the warming is to be entirely natural. Volcanic eruptions and variation in solar irradiance do not explain the warming in the latter half of the 20th century. From models of temperature changes caused by natural ***forcing***, [3] we should have observed a decrease in the global average temperature lately, whereas instead, we have observed an increase. [4]

Human contributions include increased levels of carbon dioxide (CO_2) and other ***greenhouse gases*** (***GHG***) in the Earth's atmosphere. One of the

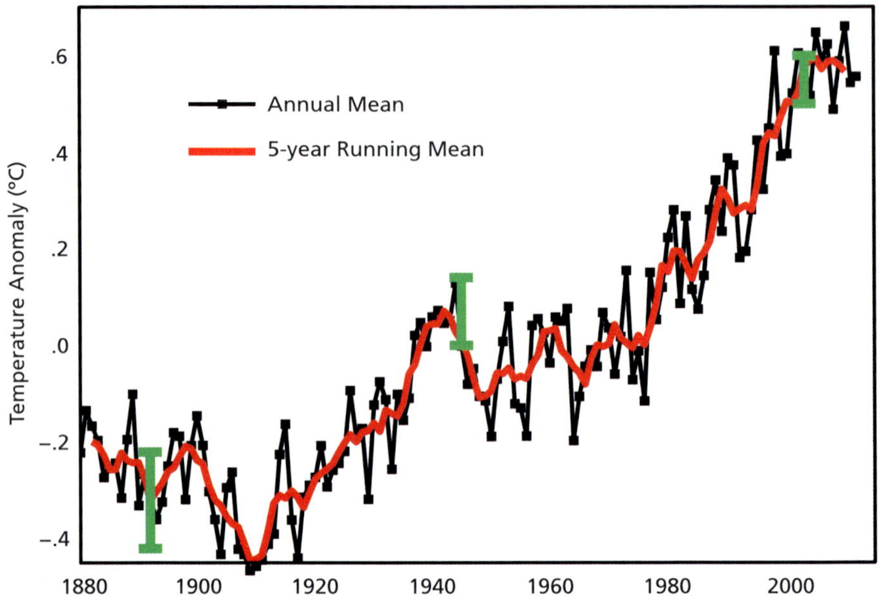

Figure 1.1 Line plot of global mean land-ocean temperature index, 1880 to 2011, with the base period 1951-1980. The dotted black line is the annual mean, and the solid red line is the 5-year mean. The green bars show uncertainty estimates. (This is an update of Fig. 1A in J. Hansen, M. Sato, R. Ruedy, K. Lo, D.W. Lea and M. Medina-Elizade, 2006. [5, 6])

most significant ways people contribute to greenhouse gases is by burning fossil fuels (oil, coal, natural gas). Each year, we release about 7 billion tonnes (t) of ancient stored carbon (1 tC = 3.66 tCO_2 equivalent) into the atmosphere by burning fossil fuels. [7, 8] It is estimated that natural processes can only absorb about half of that amount. Most of the burning occurs in running automobiles, in heating homes, powering factories, and in generating electricity. The clearing of land contributes as well. When trees are cut down in the context of deforestation, they are no longer available to absorb CO_2 from the atmosphere, and as they decay or burn, they actually release CO_2. Many scientists estimate that global carbon emissions need to be reduced by ~70%, a figure consistent with the goal of limiting the global average temperature increase to 2 °C (relative to pre-industrial levels), if the world is to avoid dangerous climactic changes. [7, 9]

The 20th century has been an era of petroleum that has provided humanity with a wide range of fuels and chemicals that have influenced most aspects of human life. About 90% of fossil oil is used in the production of energy (power, heat, and transport fuels), while ~10% is used to produce chemicals and materials. [10] In 2008, the total world energy consumption was ~5×10^{20} J with 78% derived from fossil fuels, 2.8% from nuclear power, 13% from traditional biomass, used primarily for cooking and heating, 3.2% from hydro-

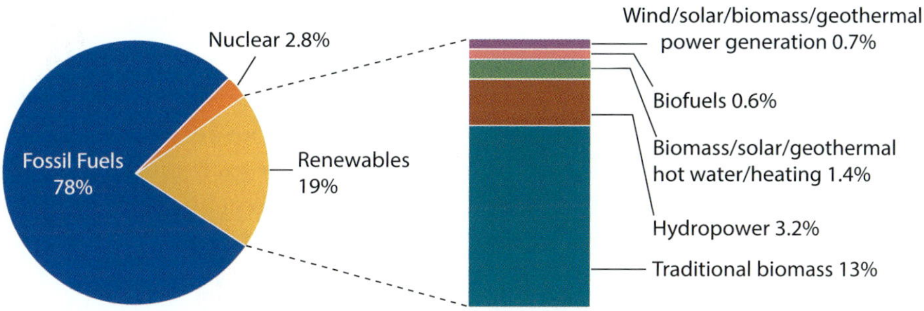

Figure 1.2 Renewable energy share of global final energy consumption, 2008. Copyright 2010 Deutsche Gesellschaft für technische Zussammenarbeit. [11]

power, 1.4% from biomass/solar/geothermal hot water/heating, 0.6% from biofuels, and 0.7% from wind/solar/biomass power generation, i.e. 19% from renewables (**Figure 1.2**). [11] Global renewable energy capacity grew at rates of 10-60% annually for many technologies during the five-year period from the end of 2004 through 2009.

A tremendous increase in energy demand is projected in the coming years with much of the demand coming from developing nations.

The United Nations Framework Convention on Climate Change (UNFCCC) meeting held in Copenhagen in December 2009 (the 15th Conference of the Parties, COP 15) named as its objective to develop a global response to the threats of climate change caused by human activities. [12] The Copenhagen Accord acknowledged that supporting developing countries' efforts to reduce emissions and adapt to the impacts of climate change will be essential to any new climate agreement. [13] To make these financial commitments a reality, it was agreed that the next two UNFCCC Conferences of the COP will need to establish a financial architecture to manage and distribute these funds.

The UNFCCC Conference, held in Cancun, Mexico, in November-December 2010 (COP 16), reached a series of agreements that includes measures to reduce greenhouse gas emissions, with the goals of limiting the global average temperature increase to 2 °C above pre-industrial levels, greater protections for forests, and a new UN-administrated climate fund for financing mitigation and adaptation activities in developing countries. [14] While the Cancun agreement does not set any binding targets, it lays the groundwork for a successor treaty to the Kyoto Protocol.

The 2011 COP 17 was hosted by Durban, South Africa in November-December 2011. The outcomes included a decision by Parties to adopt a universal legal agreement on climate change as soon as possible, and no later than 2015. [15]

1.1.1 Earth's energy budget and greenhouse effect

The ultimate source of energy that drives the climate change is radiation from the Sun. [16] About half of the radiation is in the visible part of the electromagnetic spectrum. The other half is mostly in the near-infrared part, with some in the ultraviolet part. Throughout the year each square meter of the Earth's spherical surface outside the atmosphere receives an average of 340 Watts of solar radiation, ~30% of which is immediately reflected back into space by clouds, by the atmosphere, and by Earth's surface (**Figure 1.3**). The remaining 240 Wm^{-2} is partly absorbed by the atmosphere, although most (163 Wm^{-2}) of it warms the earth's surface. The Earth's surface returns that heat to the atmosphere, partly as infrared radiation and partly as sensible heat and as water vapour (evapotranspiration) which releases its heat when it condenses higher up in the atmosphere. Under present conditions this exchange of energy between surface and atmosphere maintains a global mean temperature near the surface of 14 °C, decreasing rapidly with height.

For a stable climate, a balance is required between incoming solar radiation and the outgoing radiation emitted by the climate system. [16] Therefore, the climate system itself must radiate on average 240 Wm^{-2} back into space (**Figure 1.3**). Any physical object radiates energy of an amount and at wavelengths typical for the temperature of the object. For the Earth to radiate 240 Wm^{-2}, it should radiate at an effective emission temperature of −19 °C, with typical wavelengths in the infrared part of the spectrum. This is 33 °C lower than the average temperature of 14 °C at the Earth's surface. The explanation lies in the radiative properties of the atmosphere in the infrared wavelengths.

The atmosphere contains several trace gases (mainly water vapor, CO_2, methane, nitrous oxide, and fluorinated gases) that absorb and emit infrared radiation. [16] These so-called greenhouse gases absorb radiation, emitted by the Earth's surface, the atmosphere, and clouds, except in a transparent part of the spectrum called the atmospheric window (**Figure 1.3**). They emit, in turn, infrared radiation in all directions, including downward to the earth's surface. Thus, greenhouse gases trap heat within the atmosphere. This mechanism is called the natural greenhouse effect. The net result is an upward transfer of infrared radiation from warmer levels near the Earth's surface to colder levels at higher altitudes. The infrared radiation is effectively radiated back into space from an altitude with a temperature of −19 °C, in balance with the incoming radiation, whereas the Earth's surface is kept at the much higher temperature of 14 °C. The natural greenhouse effect is part of the energy balance of the Earth (**Figure 1.3**).

Clouds also play an important role in the Earth's energy balance and, in particular, in the natural greenhouse effect. Clouds absorb and emit infrared

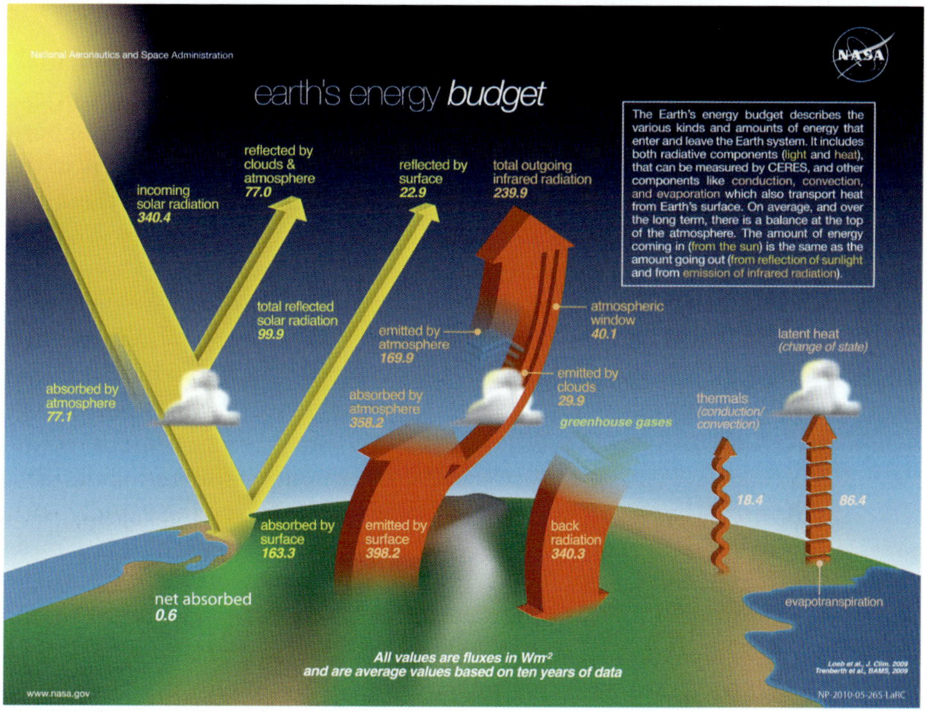

Figure 1.3 Earth's energy budget (values in Watts per m^2 (Wm^{-2}). Incoming solar radiation: 340; reflected by clouds and atmosphere: 77; reflected by surface: 23; absorbed by atmosphere: 77; absorbed by surface: 163; emitted by surface: 398; absorbed by atmosphere: 358; emitted by clouds: 30; emitted by atmosphere: 170; atmospheric window: 40; outgoing infrared radiation: 240; back radiation: 340; thermals (conduction/convection): 18; latent heat (change of state): 86. [17]

radiation and, thus, contribute to warming the Earth's surface, just as with greenhouse gases. On the other hand, most clouds are bright reflectors of solar radiation and tend to cool the climate system. The net average effect of the Earth's cloud cover in the present climate is a slight cooling. [16]

1.1.2 Carbon cycle

Carbon is the fourth most abundant element in the Universe, after hydrogen, helium, and oxygen, and is the building block of life on Earth.

The carbon cycle is critical to understanding climate change. Like water in the water cycle, carbon in the carbon cycle moves between various components of Earth by different processes. Understanding where carbon is stored and how it is captured and released is essential.

Introduction

On Earth, carbon cycles through the land, ocean, atmosphere, and the Earth's interior in a major biogeochemical cycle. [18] It is one of the most important cycles of the Earth and allows for carbon to be recycled and reused throughout the biosphere. [19] Most of Earth's carbon is stored in rocks. The rest is in the ocean, atmosphere, plants, soil, and fossil fuels.

Carbon flows between each reservoir in an exchange which has slow and fast components: the ***geological carbon cycle***, which operates over large time scales (millions of years), and the ***biological carbon cycle***, which operates at shorter time scales (days to thousands of years). [19, 20] Any change in the cycle that shifts carbon out of one reservoir puts more carbon in the other reservoirs. Changes that put carbon gases into the atmosphere result in warmer temperatures on Earth.

In the geological (slow) carbon cycle, carbon moves between rocks, soil, ocean and atmosphere. [18, 19] CO_2 in the atmosphere reacts with some minerals to form calcium carbonate. This mineral is then dissolved by rainwater and carried to the oceans. Once there, it can precipitate out of the ocean water, forming layers of sediment on the sea floor. As the Earth's plates move, these sediments are subducted underneath the continents. Under the great heat and pressure, calcium carbonate melts and reacts with other minerals, releasing CO_2. The CO_2 is then re-emitted into the atmosphere by volcanism.

In the biological (fast) carbon cycle, carbon moves from the atmosphere, through the Earth's ecosystems (the biosphere), and back to the atmosphere. [21, 22] The biological carbon cycle is largely the movement of carbon through life forms on Earth. Plants and phytoplankton are the main components of the cycle. [19]

The diagram of the fast carbon cycle between land, atmosphere and oceans is shown in **Figure 1.4.**

Reservoirs

The global carbon cycle can be thought to include six reservoirs of carbon interconnected by pathways of exchange: the atmosphere, the oceans, the fossil fuels, the Earth's crust, the land plants, and the soils. **Table 1.1** indicates how much carbon, in Gt of carbon equivalent (1Gt = 1 gigaton = 1 billion metric tons), is stored in the various reservoirs, and under which main form. [19, 23, 24]

Atmosphere

The concentration of CO_2 in the atmosphere is ~0.04%. [24] In the year 2000, the globally averaged concentration was ~0.0368%, or 368 ppm, equivalent to ~780 GtC. Most of the carbon in the atmosphere is CO_2, but small amounts

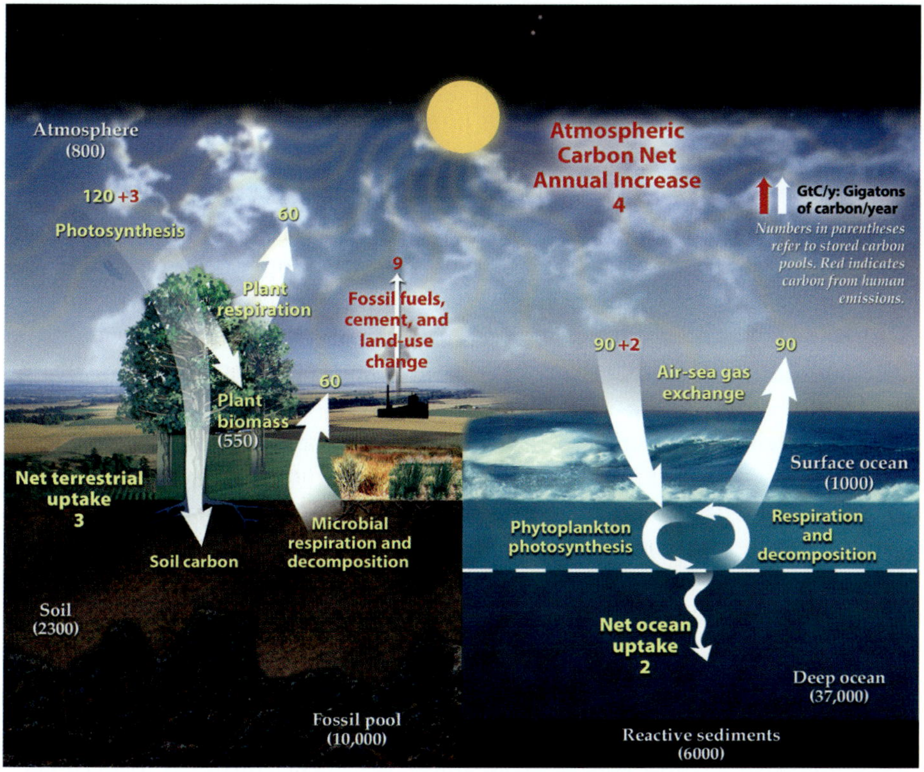

Figure 1.4 Fast carbon cycle. The diagram shows the movement of carbon between plants, soils, atmosphere and oceans. Yellow numbers are natural fluxes, and red are human contributions in gigatons of carbon (GtC, giga = 10^9) per year. White numbers indicate stored carbon in GtC. Human contributions include burning fossil fuels and clearing land. Of the 9 GtC per year released by people to the atmosphere, ~5 GtC have been taken up by land plants and oceans while ~4 GtC have stayed in the atmosphere. [19]

Table 1.1 Reservoirs of carbon throughout the Earth. [19, 25, 26]

Reservoir	Amount in billion of metric tons of carbon (GtC)	Main form of carbon
Atmosphere	800	CO_2
Ocean (surface; deep)	38,000 (1000; 37,000)	Dissolved CO_2
Earth's crust	>60 10^6	$CaCO_3$
Fossil fuels	10,000	Hydrocarbons
Land plants	550	Carbohydrates
Soils	2300	Dead organic matter and in living bacteria and fungi

of carbon exist in the forms of methane (CH_4), carbon monoxide (CO), and non-methane hydrocarbons. These trace gases are important because they modify the chemical and/or radiative properties of the Earth's atmosphere. Methane is present at ~1.7 ppm and is much less stable than CO_2. CO, which is present at ~0.1 ppm, is not a greenhouse gas, but its chemical reactivity affects the abundances of ozone and methane which are greenhouse gases. Non-methane hydrocarbons, which are present in even smaller concentrations, are also unstable.

Oceans

The total amount of dissolved inorganic carbon in the world's oceans is ~37,000 Gt, and the amount of organic carbon is ~1000 Gt. [24] The world's oceans contain ~50 times more carbon than the atmosphere. The amount of carbon contained in the living vegetation of terrestrial ecosystems is somewhat less than that present in the atmosphere. Most terrestrial carbon is stored in the vegetation and soils of the world's forests. The amount of carbon stored in recoverable reserves of coal, oil, and gas is estimated to be close to 10,000 GtC, larger than any other reservoir except the deep sea and the Earth's crust.

Land plants and soils

The amount of carbon contained in the living vegetation of terrestrial ecosystems (~550 GtC) is somewhat less than that present in the atmosphere. [24] Soils contain ~4 times that amount (2300 GtC). Most terrestrial carbon is stored in the vegetation and soils of the world's forests.

Fossil fuels

The amount of carbon stored in recoverable reserves of coal, oil and gas is estimated to be close to 10,000 GtC, larger than any reservoir except the oceans and the Earth's crust. [24] Until the industrial revolution, this reservoir was not a significant part of the short-term cycle of carbon.

Fluxes

Table 1.2 shows the magnitudes of the fluxes (in GtC per year) between one reservoir and another. [18, 19, 25, 26, 27]

Between land and atmosphere

Biology plays an important role in the movement of carbon through the processes of ***photosynthesis*** and ***respiration***. [19, 28] Life on Earth is based on the production of carbohydrates from solar energy and CO_2 (photosynthesis) and

Table 1.2 Fluxes of carbon between one reservoir and another. [18, 19, 25, 23, 27, 24]

Flux of carbon	Amount in GtC per year (4)
Photosynthesis	120
Plant respiration	60
Soil respiration	60
Plant to soils	60
Fossil fuel formation	0.0001
Fossil fuel burning	6.3
Land use change/deforestation	2
Atmosphere to ocean	107
Ocean to atmosphere	105
Carbonate formation	0.3
Weathering and erosion	0.6
Volcanoes	0.1
Net flux of carbon	
Fossil fuels	6
Land use change	2
Atmospheric increase	3
Oceanic uptake	2
Other terrestrial uptake	3

the metabolism (respiration) of those carbohydrates to produce the chemical energy that facilitates growth and reproduction. Through the process of photosynthesis, solar energy is captured and converted by plants and then stored in the chemical bonds of carbohydrates (**Equation 1.1**):

$$6\ CO_2 + 6\ H_2O + \text{energy} \rightarrow C_6H_{12}O_6 + 6\ O_2 \tag{1.1}$$

Plants and animals burn these carbohydrates through the process of respiration, the reverse of photosynthesis (**Equation 1.2**).

$$C_6H_{12}O_6 + 6\ O_2 \rightarrow 6\ CO_2 + 6\ H_2O + \text{energy} \tag{1.2}$$

Respiration releases the energy contained in carbohydrates for use in metabolism and renders the carbohydrate fuel back to CO_2. Together, respiration and decomposition (respiration that breaks down organic matter mostly by bacteria and fungi, with release of CO_2) return the biologically fixed carbon back to the atmosphere. The amount of carbon that moves through the biological cycle is 1000 times greater than the amount of carbon that moves through the geological cycle on an annual basis.

Photosynthesis and respiration also play an important role in the geological cycle of carbon. [19] The presence of land vegetation enhances the weathering of soil, leading to the long-term (but slow) uptake of CO_2 from the atmosphere, ultimately to produce carbonate. [18, 29, 30] In the oceans, some of the carbon taken up by phytoplankton to make shells of calcium carbonate settles to the bottom to form sediments. During times when photosynthesis exceeded respiration, organic matter slowly built up to form fossil deposits. All these processes represent a removal of CO_2 from the atmosphere and storage of carbon in geological sediments.

Between oceans and atmosphere

It is the amount of dissolved organic carbon that determines the atmospheric concentration of CO_2. [24] In the long term (millennia), the most important process determining the exchanges of carbon between the oceans and the atmosphere is the chemical equilibrium of dissolved CO_2, bicarbonate, and carbonate in the ocean.

Two additional processes keep the atmospheric CO_2 lower than it otherwise would be. [24] One process is referred to as the solubility pump and the other as the biological pump. The solubility pump is based on the fact that CO_2 is more soluble in cold waters. The biological pump also transfers surface carbon to the intermediate and deep ocean. Not all of the organic matter produced by phytoplankton is respired in the surface waters where it is produced; some sinks to deeper water. Together the two pumps keep the dissolved organic carbon concentration of the water surfaces ~10% lower than at depth. The concentration of CO_2 in the atmosphere would have been 720 ppm if both pumps were turned off.

Short-term exchanges of carbon between ocean and atmosphere result from the diffusion of CO_2 across the air-sea interface. [24] The diffusive exchanges transfer ~90 GtC per year across the air-sea interface in both directions.

Human role in the carbon cycle

In addition to the natural fluxes of carbon through the Earth's system, human activities, particularly fossil fuel burning and land use change (i.e. the net effects of deforestation and reforestation), are also releasing CO_2 into the atmosphere. [19] Fossil fuels presently account for 80-85% of total CO_2 emissions and land use change for 15–20%. [12] When burning fossil fuels, human

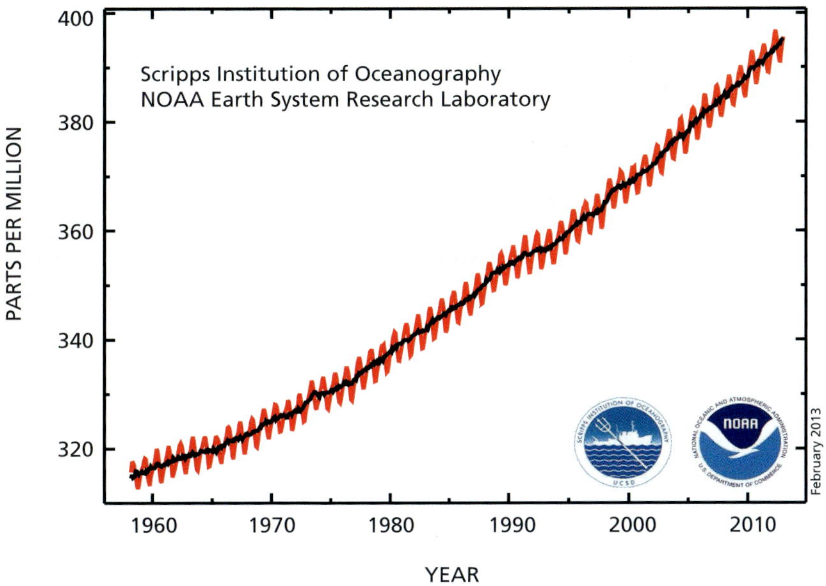

Figure 1.5 The Mauna Loa curve. The trend in atmospheric concentration for CO_2 in parts per million from 1958 to present. [34]

beings are effectively moving carbon more rapidly into the atmosphere than is being removed naturally through the sedimentation of carbon, causing atmospheric CO_2 to increase. Also, by clearing forests, human beings are transferring carbon from living biomass into the atmosphere. Therefore, the global carbon cycle is in strong disequilibrium. [31]

Total CO_2 emissions from human activities have grown at about 2% per year from 1800 to 2000. [12, 32, 33] However, fossil fuel emissions have accelerated since 2000 to grow at about 3.4% per year. Total CO_2 emissions are responsible for two-thirds of the growth of all greenhouse gas radiative forcing. The so-called Mauna Loa curve (**Figure 1.5**) shows that from 1958 to present, the concentration of CO_2 in the atmosphere has increased from ~317 ppm to over 392 ppm.

Not all of the CO_2 that has been emitted by human activities remains in the atmosphere. [18, 27] The oceans have absorbed some of it; as the CO_2 in the atmosphere increases, it drives diffusion of CO_2 into the oceans. However, balancing the global carbon budget requires additional carbon sinks. During the period ~1850–2000, there has been a 30% increase in the amount of carbon in the atmosphere (~280 ppm to 368 ppm, equivalent to 175 GtC). During this time, ~275 GtC have been released from the combustion of fossil fuels, and another ~155 GtC were released as a result of land-use changes (total of ~430 GtC released). Taking into account the fossil fuels emissions, the land-use net source, the atmospheric increase, and an oceanic uptake of ~140 GtC, the global carbon balance suggests that other terrestrial ecosystems have accumulated ~115 GtC during this time, at a steadily increasing rate.

Several scenarios could cause the land to take up more CO_2 than is released each year. [18, 35] For example, re-growth of forests since the massive deforestation in the Northern Hemisphere over the last century could account for some of the missing carbon, while changing climate could also contribute to greater uptake than release. The residual terrestrial sink illustrates the complexity of biogeochemical cycles, especially those in which living organisms play an important role. It is obviously important to understand the processes that control these sources and sinks in order to predict their behavior in the future.

Without CO_2 sinks, which remove and store CO_2 from the atmosphere, the total human CO_2 emissions since 1800 would have caused atmospheric CO_2 to increase from its pre-industrial value of 280 ppm to nearly 500 ppm. [12] However, the disequilibrium of the carbon cycle causes the vast human input of CO_2 to be repartitioned between the carbon stores in the atmosphere, land, and oceans. Consequently, land and oceans have taken up more than half of total CO_2 emissions since 1800 and the actual CO_2 accumulation in the atmosphere has raised the CO_2 concentration to only ~390 ppm. However, these natural sinks are vulnerable to climate and land use change: they are likely to weaken in the future because of several effects, including increasing ocean acidification, ocean circulation changes, and water, temperature and nutrient constraints on CO_2 uptake.

The fact that ***peak oil*** may be reached in the near future, resulting in severe oil price increases, and rising levels of atmospheric CO_2, resulting in climate change, have triggered a global development toward alternative renewable resources. [12] Consumer demand for green products and proactive companies have also become a driver for the development of alternative renewable resources.

Impact of biobased products

Biobased products are products derived partially or totally from biomass. [36] They include (1) bioenergy defined as biofuels, power and heat, and (2) biobased chemicals and materials.

Biofuels can contribute towards reducing emissions of CO_2, and reduce the world's dependence on non-renewable oil. [37, 38] This is because as biofuels burn, they release no more carbon dioxide than was absorbed during growth by the plants from which they derive, so that the total contribution of carbon dioxide to the atmosphere is zero over the timescale of the growth of the crop and the use of the fuel. Bioenergy is produced in a cycle that mimics the Earth's ecological cycles and minimizes pollution (**Figure 1.6**). [39] This contrasts with burning fossil fuels, which releases carbon dioxide that has been

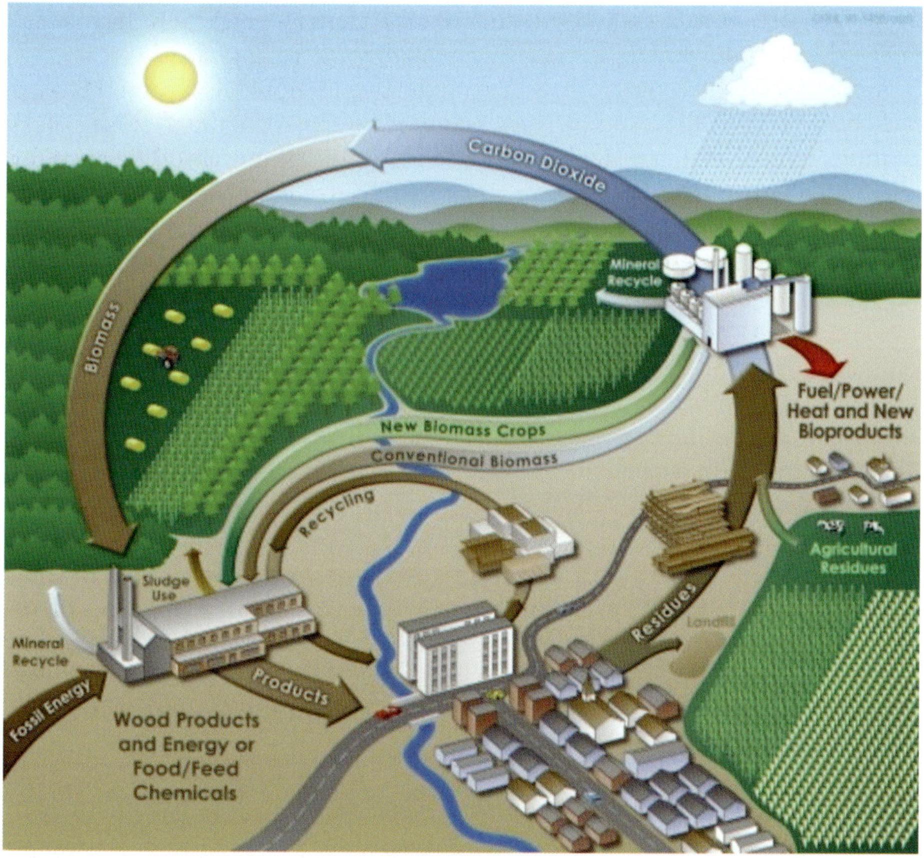

Figure 1.6 Bioenergy cycle. Most of the carbon (CO_2) to create bioenergy is taken from the atmosphere and later returned to the atmosphere. The nutrients to create it are taken from the soil and later returned to the soil. The residues from one part of the cycle form the inputs to the next part of the cycle. Reproduced by permission of Oak Ridge National Laboratory. [39]

captured for millions of years and therefore makes a net contribution of CO_2 to the atmosphere.

The production and use of biobased chemicals and materials can also make a significant impact on industry's CO_2 emissions. [40] Biomass can be seen as storehouses of solar energy. Plants' energy significance goes beyond pumping out carbon compounds to be refined into liquid biofuel. It can also lead to the manufacture of new sustainable chemicals and materials derived from new industrial plant-based feedstocks instead of traditional fossil-based feedstocks. Thus, innovation can contribute both towards reducing energy consumption and waste and achieving sustainable industrial and societal development.

1.2 Biomass and its conversion to products

1.2.1 Definition

Biomass can be defined as material of biological origin excluding material embedded in geologic formation and/or fossilized. [36] This definition refers to the short carbon cycle, i.e. the life cycle of biological materials (e.g. plants, algae, marine organisms, forestry, micro-organisms, animals, and biological waste from households, agriculture, animals, and food/feed production).

The Earth's biomass represents an enormous store of energy. It has been estimated that about one eighth of the total biomass produced annually would provide all of humanity's current demand for energy. [41, 42] Furthermore, since biomass can be regrown, it is a potentially renewable resource. One of the most attractive aspects of biomass energy is that it doesn't contribute to the enhanced greenhouse effect, provided that the biomass is harvested sustainably.

Biobased products are produced in a ***biorefinery*** that integrates biomass conversion processes. [43] The concept of a biorefinery is analogous to a petroleum refinery where a feedstock, crude oil, is converted into fuels and a wide range of valuable chemicals and materials. [44] In the case of a biorefinery, biomass is used as a feedstock to produce biofuels and a wide range of valuable chemicals and materials.

1.2.2 Biomass feedstock

Biobased products can be manufactured from various biomass feedstocks. [43, 45] Two categories of feedstock and products dominate: first and second generation. ***First-generation products*** are manufactured from edible biomass such as starch-rich or oily plants. ***Second-generation products*** utilize biomass consisting of the residual non-food parts of current crops or other non-food sources, such as perennial grasses or algae (products from algae are sometimes also referred to as third-generation products). These are generally considered as having a significantly higher potential to replace fossil-based products.

First-generation feedstock

Sugar / starch crops

The most common type of biorefinery today uses sugar- or starch-rich crops. [43] Sugar crops such as sugar cane, sugar beet, and sweet sorghum store the chemical energy as simple sugars (mono- and disaccharides), which can be easily extracted from the plant material for subsequent fermentation to bioethanol or biobased chemicals. Sugar cane is restricted to certain locations like Brazil due to weather and soil requirements.

Starch-rich crops, such as corn, wheat and cassava (manioc), store the energy as starch, a polysaccharide. Starch can be hydrolyzed enzymatically to deliver a sugar solution, which can subsequently be fermented and processed into biofuels and biobased chemicals. The processing of many starch-rich crops also delivers, as a side product, valuable animal feed that is rich in protein and energy.

The use of first-generation sugar/starch crops for production of biobased products (primarily bioethanol) has generated an important issue regarding land use change. According to the World Economic Forum, in 2010 there were about 400 operational first-generation biorefineries around the world. [43] Yet, despite the initial success of this biofuel feedstock, its future is debated because GHG emissions from land-use change are likely to exceed, or at least offset, much of the GHG savings obtained by using biofuels for transport. [46]

Vegetable oil/animal fat

The most common raw materials for biodiesel (fatty acid alkyl esters) are vegetable oils and animal fats (triglycerides). The process used is transesterification with an alcohol. Two categories of oil exist: pure plant oil (PPO) and waste vegetable oil (WVO). [43] Pure plant oil stems from dedicated crops such as rapeseed, soybean, palm and sunflower seeds. Animal fats are attractive feedstocks for biodiesel because their cost is lower than that of vegetal oil. [47] Use of waste vegetable oil or animal fats is an effective method of recycling daily wastes.

Sustainable and economic production of biodiesel from palm oil has proven to be a challenge. This is due to the significant land-use change and sustainability issues linked to pure plant oil production, and the high costs associated with the refinement of waste oil. [43]

Second-generation feedstock

Lignocellulosic biomass

Lignocellulosic biomass (or simply biomass) refers to inedible plant materials made up primarily of cellulose, hemicelluloses, and lignin. It represents the vast bulk of plant material. It includes:

- agricultural waste, such as straw, corn stover (leaves and stalks after harvest), corn cob (the hard cylindrical core that bears the kernels of an ear of corn), bagasse (dry dusty pulp that remains after juice is extracted from sugar cane), molasses (thick, dark syrup from the processing of sugar cane or sugar beet);

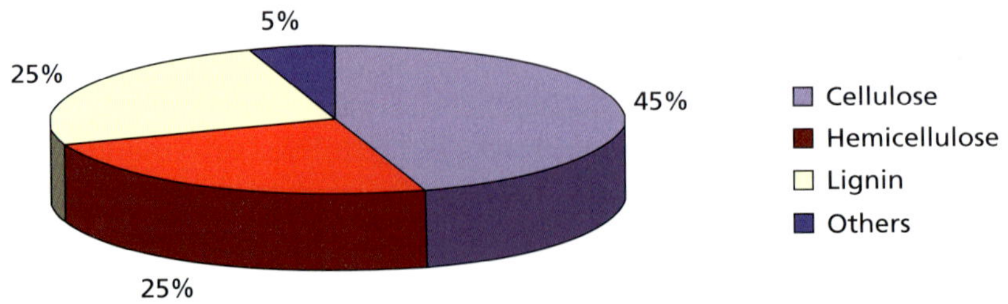

Figure 1.7 Typical composition of lignocellulosic biomass.

- forestry wastes, such as wood chips;
- a fraction of municipal and industrial (paper) wastes;
- fast-growing energy crops, such as miscanthus, switchgrass, short-rotation poplar or willow coppice.

These feedstocks exclude direct land-use change. Such biomass resources typically contain on dry weight basis 40–60% cellulose, 20–40% hemicelluloses, and 10–25% lignin (**Figure 1.7**).

Second-generation feedstock is very likely to be used industrially in the future for the production of fuels and other biobased products. However, it is more difficult than expected to convert lignocellulosic biomass into useful products.

Jatropha oil

The Jatropha curcas tree produces seeds containing 27–40% inedible oil, which can be converted to biodiesel. [43] An assessment of Jatropha curcas sustainability reveals a positive effect on the environment and GHG emissions, provided cultivation occurs on wasteland or degraded ground. Due to its labor-intensive production chain, it is thought that Jatropha is one of the drivers for rural development.

Microalgae

Microalgae are single-cell photosynthetic organisms known for their rapid growth and high-energy content (**Figure 1.8**). [48, 49, 50] Depending on the species, their sizes can range from a few μm to a few hundreds of μm.

Some algal strains are capable of doubling their mass several times per day. In some cases, more than half of that mass consists of lipids or triacylglycerides, the same material as in vegetable oils.

Microalgae have attracted considerable attention in recent years due to their potential value as renewable energy resources. [43] Focus has been on storage triglycerides, which can be used to synthesize biodiesel via transesterification. The remaining carbohydrate content can also be converted to bioethanol via fermentation.

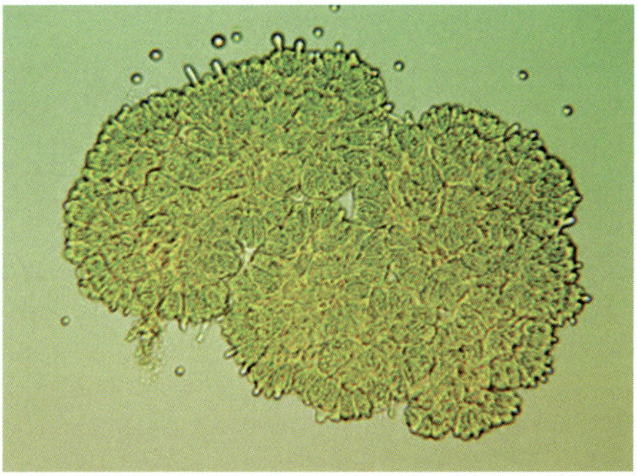

Figure 1.8 Photomicrograph of *Botryococcus* microalga, with oil droplets being released. [51]

The advantages of using algae-derived fuels are numerous. [43] First, they can provide between 10 and 100 times more oil per acre than other second-generation biofuel feedstock, and the resulting oil content of some micro-algae exceeds 80% of the dry weight of algae biomass, almost 20 times that of traditional feedstock. They need not compete with arable land. Furthermore, they are highly productive, quick to cultivate, and require only CO_2, sunlight, and water to grow.

Like plants, algae use the sunlight for the process of photosynthesis. They can be produced in two ways: ponds and photobioreactors (PBR). [52] Open ponds are the simplest of algae growing systems. An alternative to open ponds are close ponds where the control over the environment is much better than that for the open ponds. Close ponds cost more than the open ponds and considerably less than photobioreactors for similar areas of operation. A photobioreactor is a closed equipment which provides a controlled environment and enables high productivity of algae. [53]

However, several barriers remain to be overcome before the large-scale production of microalgae-derived biofuels can become a reality. [43] Producing low-cost microalgal biodiesel primarily requires improvements to algal biology through genetic and metabolic engineering, to achieve optimum characteristics, such as high growth rate, high lipid content, and ease of extraction. Advances in photobioreactor engineering are expected to lower the cost of production.

1.2.3 Conversion processes

Depending on the nature of the feedstock and the desired output, biorefineries employ a variety of conversion technologies. [43]

First-generation processes

First-generation liquid biofuels include bioethanol from sugar/starch and biodiesel from oil/fats. They are currently made mostly using conventional technology. [54] They have generated the "food versus fuel" issue that has claimed that the production of biofuel has diverted crops away from food to fuel.

First-generation chemicals and materials from sugar/starch include polymers, such as polylactic acid (PLA), and chemical building blocks, such as succinic acid and 1, 3 propanediol. [55, 56, 57, 58] Chemicals from vegetable oils include fatty acids and esters. [59] Bioethanol itself is a building block for the production of polymers, such as polyvinyl chloride (PVC), polyethylene (PE), polypropylene (PP) or polyethylene terephthalate (PET).

Transesterification and other conversion of triglycerides

Transesterification of plant oil or animal fats is a process by which triglycerides are reacted generally with methanol in the presence of a strong base catalyst to produce fatty acid methyl esters (***FAME***) as biodiesel and glycerol.

Besides the transesterification process, the Neste process entails direct hydrogenation of plant oil or animal fat triglycerides into the corresponding alkanes, thus removing oxygen from the oil. [60]

Fermentation

Sugar in sugar crops can be directly fermented to produce ethanol. The fermentation of sugar solutions from starch crops usually requires an enzymatic hydrolysis of starch with use of amylases to liberate the sugars from the plant material. The hydrolysis is followed by the microbial fermentation to produce ethanol.

Second-generation processes

Lignocellulosic biomass is not produced for food, and hence makes the "food versus fuel" debate much less of an issue.

Second-generation biofuels include cellulosic ethanol, BtL (biomass to liquid) diesel, and other BtL fuels. These biofuels are still generally in the demonstration plant stage, even for the most advanced.

Internationally, there is increasing interest in the use of cellulosic biomass for the production of second-generation chemicals and materials. Currently such chemicals and materials are generally in the research and development or demonstration stage.

There are two primary pathways for producing biobased products from lignocellulosic biomass: ***biochemical*** and ***thermochemical***. Biochemical conversion basically involves hydrolysis of the polysaccharides in the biomass, and fermentation of the resulting sugars into ethanol. Thermochemical conversion typically involves gasification or other thermal treatment of the biomass followed by catalytic synthesis or fermentation of the resulting gas or liquid into ethanol, diesel, and other fluids.

Fermentation

Before fermentation, lignocellulosic feedstock processing needs to separate the cellulose and hemicelluloses from the non-fermentable lignin. This is usually performed mechanically, followed by acid, alkali, and/or steam treatment. While the lignin is currently mostly burned to generate energy, cellulose and hemicelluloses are hydrolyzed enzymatically with use of cellulases to deliver sugar solutions for subsequent fermentation.

Gasification and Fischer-Tropsch process

Gasification of biomass converts carbonaceous materials into synthesis gas, namely H_2 and CO, known as syngas. Gasification is achieved at high temperatures in the presence of a limited amount of oxygen. The resulting syngas can then be used in the Fischer-Tropsch process to produce hydrocarbons.

The conversion of syngas via the Fischer-Tropsch process involves the catalytic conversion of syngas into liquid hydrocarbons ranging from methane to wax. A selective distribution of products is achievable via control of temperature, pressure, and the type of catalyst. When applied to biomass and biofuel production, the process is commonly referred to BtL.

Fast pyrolysis

Pyrolysis is the thermal decomposition of the biomass in the absence of oxygen to produce char, gas, and a liquid product (bio-oil) rich in oxygenated hydrocarbons. [61] It can be upgraded to lower the oxygen content and transported using the same infrastructure used by the oil industry. Bio-oil and upgraded oil can be used in applications ranging from value-added chemicals to transportation fuels.

Anaerobic digestion

Biodegradable waste or energy crops can be converted into a gaseous fuel called biogas, comprised primarily of methane and CO_2. Commercial processes typically run via ***anaerobic digestion*** or fermentation by ***anaerobic organisms***. [43, 62] This process is used as a renewable substitute for commercial natural gas.

1.3 Biorefineries

The IEA (International Energy Agency) Bioenergy Task 42 on biorefineries has defined *biorefining* as the sustainable processing of biomass into a spectrum of marketable products. The products intermediates and final products include

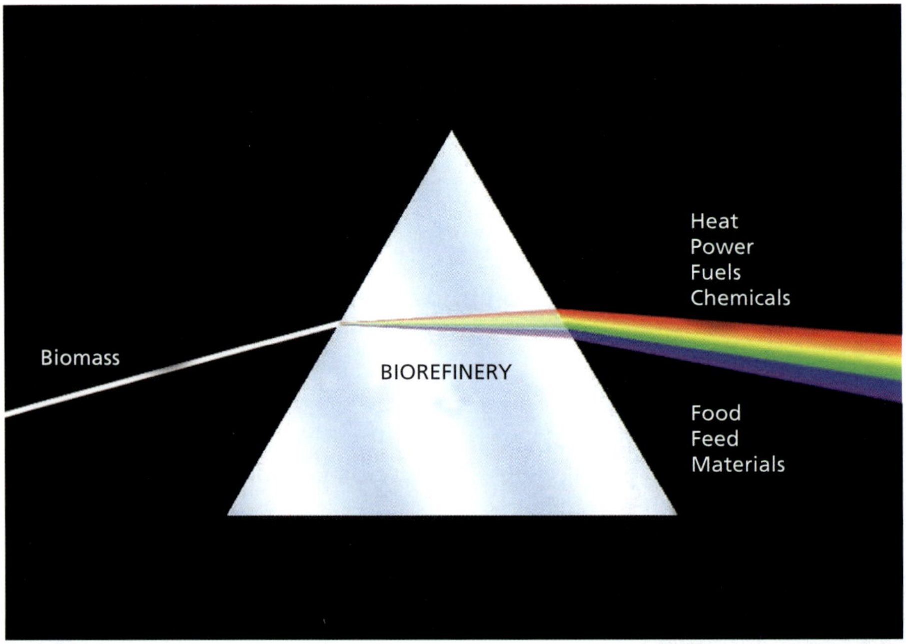

Figure 1.9 Biorefinery and its role in transformation of biomass. Reproduced by permission of IEA Bioenergy Task 42 Biorefinery. [65]

food, feed, chemicals, materials, and energy (defined as fuels, power and heat) (**Figure 1.9**). [63, 64]

The biorefinery concept is analogous to a petroleum refinery that produces multiple fuel and non-fuel products from petroleum (**Figure 1.10**). The traditional crude oil refinery converts oil into fuels, chemical ***building blocks*** for petrochemistry, and specialty chemicals, such as lubricants and solvents. Biomass refinery converts biomass into biofuels, chemical building blocks for agro-biochemistry, and specialty chemicals, such as biolubricants and biosolvents.

By producing multiple biobased products, a biorefinery takes advantages of biomass components and intermediaries, and maximizes the value that is derived from refining operations. [66]

One can distinguish between energy-driven biorefineries and product-driven (rather than energy-driven) biorefineries. [67] In product-driven biorefineries, the biomass is fractionized into multiple biobased products with maximal added value and minimal ecological impact, after which the process residues are used for power and/or heat production. In energy-driven biorefineries, the biomass is primarily used for the production of secondary energy carriers (biofuels, power, and heat), and process residues are sold as feed or, even better, are upgraded to added-value biobased products to optimize economics and ecologics of the full biomass supply chain.

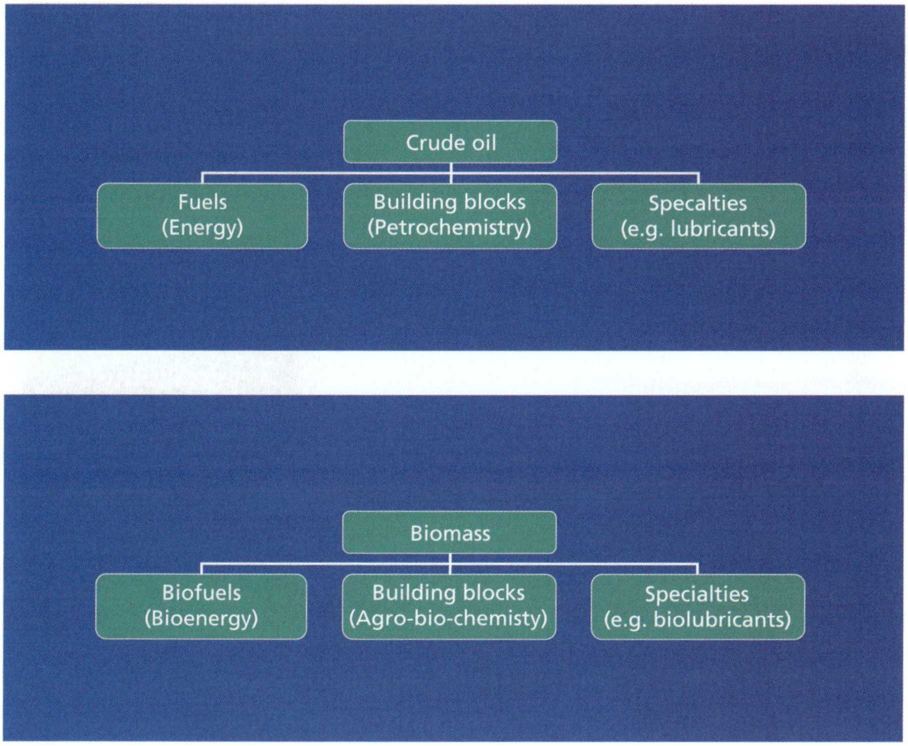

Figure 1.10 **Top** Oil refining into fuels, building blocks and specialties. **Bottom** Biomass refining into biofuels, building blocks and specialties.

Other biorefinery concepts sometimes used are: green biorefineries, whole crop biorefineries, lignocellulosic biorefineries, and marine biorefineries.

Biorefineries are most commonly classified as a function of their feedstocks: food plants (e.g. cereal grains) for ***first generation***, lignocellulosic materials for ***second generation***, algae for second or third generation.

1.4 Life cycle assessment (LCA)

As environmental awareness increases, industries and businesses are assessing how their activities affect the environment. [68] Many businesses have responded to this awareness by providing greener products and using greener processes. The environmental performance of products and processes has become a crucial issue. Many companies have found it advantageous to explore ways of moving beyond regulatory compliance by using pollution prevention strategies and environmental management systems to improve their environmental performance. One such tool is LCA. This concept considers the entire life cycle of a product, process, or activity (hereafter called "product").

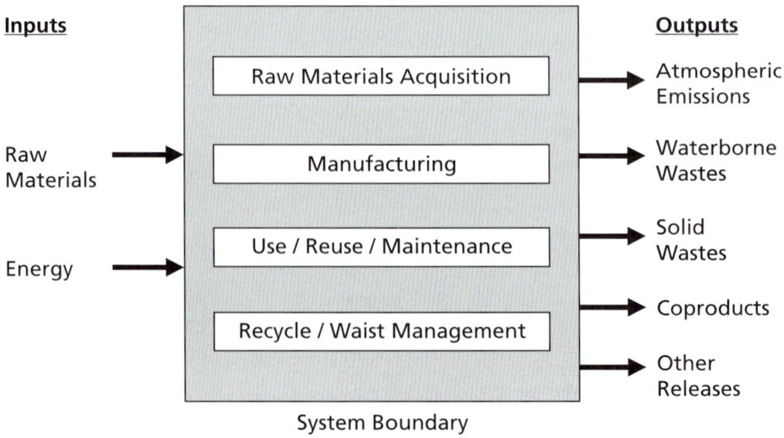

Figure 1.11 Life cycle stages including inputs/outputs. [68]

Life cycle analysis is a "cradle-to-grave" approach for assessing industrial systems. [68] "Cradle-to-grave" starts with the gathering of raw materials from the earth to create the product, and it ends when all materials are returned to the earth. LCA evaluates all stages of a product's life from the perspective that they are interdependent, meaning that one operation leads to the next. LCA enables the estimation of the cumulative environmental impacts resulting from all stages in the product life cycle. Therefore, LCA provides a comprehensive view of the environmental aspects of the product or process.

The term *life cycle* refers to the major activities in the course of the product's lifespan, from its manufacture, use, and maintenance, to its final disposal, including the raw material acquisition for its manufacture. [68] **Figure 1.11** illustrates the typical life cycle stages that can be considered in a LCA and the typical inputs/outputs measured.

Specifically, LCA is a four-component process (**Figure 1.12**) to assess the environmental aspects and potential impacts associated with a product by:

- Defining goal and scope.
- Compiling an inventory of relevant energy and material inputs and environmental releases.
- Evaluating the potential environmental impacts associated with identified inputs and releases.
- Interpreting the results to better inform decision-makers. [68, 69]

LCA is part of the ISO 14000 series of standards on environmental management. [70] The old ISOs 14040:1997, 14041:1999, 14042:2000 and 14043:2000 were replaced by the new ones, ISO 14040:2006 (Principles and Framework) and ISO 14044:2006 (Requirements and Guidelines). [71]

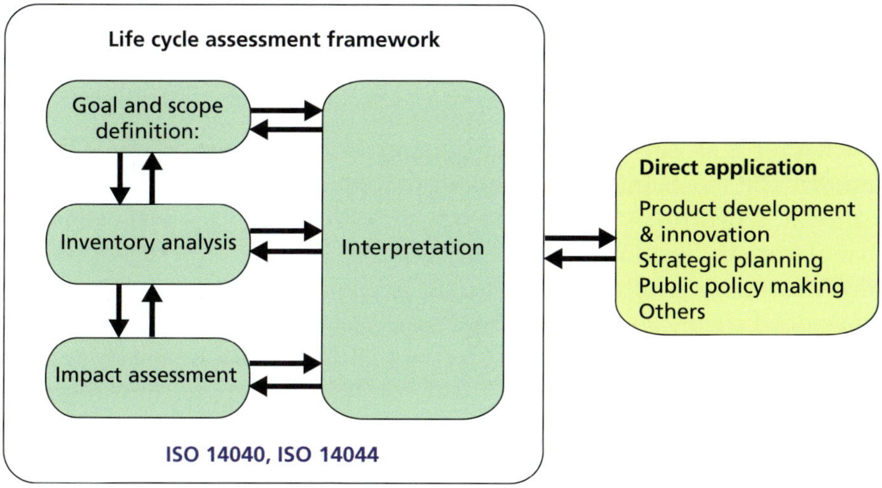

Figure 1.12 The four phases of an LCA. [72]

1.5 Structure of the book

Chapter 1 of the book starts with a brief review on climate change. Biomass is defined and the various biomass feedstocks for first- and second-generation biorefineries are described. The main biomass conversion processes are then briefly commented. The various concepts of biorefineries are presented.

Chapter 2 describes photosynthesis and biomass growth, focusing on the role of enzymes, light reaction mechanisms, Calvin cycle, and different metabolic pathways encountered in plants. Indeed, a biorefinery produces biobased products derived from photosynthesis.

Chapter 3 deals with an overview of first-generation biorefineries defined as using edible biomass feedstocks. First-generation biofuels include bioethanol, biodiesel, and biogas. Their production process is reviewed, and the global production of liquid biofuels is analyzed. Biobased chemicals and materials derived from sugar/starch plants and edible oil plants are described.

Chapter 4 presents cellulose, the predominant constituent of biomass. Its molecular, supramolecular, and morphological structure, its biosynthesis, and its properties, and applications are reviewed, using biology, chemistry and physics.

Chapter 5 deals with cellulose degradation. The enzymatic hydrolysis and the non-biological degradation are reviewed with a focus on useful technologies for biorefineries.

Chapter 6 presents hemicelluloses and lignin, the other key constituents of biomass. Their molecular structure, biosynthesis, and degradation are described.

Chapter 7 covers the pretreatments of lignocellulosic biomass in order to fractionate the material. The pretreatments include steam explosion, steam cracking, solvent fractionation, acid pretreatment, alkaline pretreatment, microwave, and other physical pretreatments.

Chapter 8 is devoted to the biochemical conversion of lignocellulosic biomass. The biochemical processes commonly involve, after a biomass pretreatment, a hydrolysis of the polysaccharides followed by a fermentation of ethanol from the resulting mixture of 6-carbon and 5-carbon sugars. The carbohydrates formed after hydrolysis and the lignins can also be valorized into value-added chemicals and materials. Case studies close the chapter.

Chapter 9 is about the thermochemical conversion of lignocellulosic biomass. Processes such as gasification and pyrolysis are able to convert biomass into a gas or liquid intermediate for further refining to valuable products. [73] Gasification typically involves high-temperature degradation of biomass with limited oxygen to produce syngas, a mixture of H_2 and CO, followed by a Fischer-Tropsch synthesis of linear hydrocarbons. Pyrolysis involves medium-temperature degradation of biomass in the absence of oxygen to produce liquids that can be refined in ways similar to crude petroleum. Case studies close the chapter.

Finally, **Chapter 10** gives perspectives regarding both the potential and the challenges of lignocellulosic biorefineries.

1.6 References

[1] Washington State, Department of Ecology, *Climate Change*, 2010 in www.ecy.wa.gov/climatechange/whatis.htm

[2] NASA, World Book, *Global Warming*, 2007 in http://www.nasa.gov/worldbook/global_warming_worldbook.html

[3] IPPC Fourth Assessment Report: Climate Change 2007 in http://www.ipcc.ch/publications_and_data/ar4/wg1/en/spmsspm-human-and.html

[4] United Nations Environment Programme (UNEP), UNEP/GRID-Arendal, 2005 in http://www.grida.no/_res/site/file/publications/vital-climate_change_update.pdf

[5] J. Hansen, M. Sato, R. Ruedy, K. Lo, D.W. Lea and M. Medina-Elizade, Global temperature change, *Proc. Natl. Acad. Sci., 103*, 4288, 2006 in http://pubs.giss.nasa.gov/docs/2006/2006_Hansen_etal_1.pdf

[6] NASA, Goddard Institute for Space Studies, GISS Surface Temperature Analysis, 2008 in http://data.giss.nasa.gov/gistemp/graphs_v3/

[7] M. Stein, *When technology fails: A Manual for Self-Reliance, Sustainability, and Surviving the Long Emergency*, Chelsea Green Publishing, 2008 in http://www.amazon.com/When-Technology-Fails-Revised-Expanded/dp/1933392452

[8] R. Hotinski, Princeton University, *Carbon Mitigation Initiative*, 2009 in http://cmi.princeton.edu/wedges/pdfs/teachers_guide.pdf

[9] EU Climate Change Expert Group 'EG Science' *The 2°C target*, 2008 http://ec.europa.eu/clima/policies/international/negotiations/future/docs/brochure_2c_en.pdf

[10] R. Hatti-Kaul, Lund University, *Biorefineries – A Path to Sustainability*, 2009 in www.scienceforum2009.nl/Portals/11/6Kaul-pres.pdf

[11] Renewable Energy Policy network for the 21st Century, *Renewables 2010, Global Status Report*, 2010.

[12] K. Richardson, W. Steffen, H. J. Schellnhuber, J. Alcamo, T. Barker, D. M. Kammen, R. Leemans, D. Liverman, M. Munasinghe, B. Osman-Elaha, N. Stern, and O. Waever, *Synthesis Report, Climate Change: Global Risks, Challenges & Decisions*, Copenhagen, 2009 in http://climatecongress.ku.dk/pdf/synthesisreport

[13] A. Ballesteros, *From Copenhagen to Cancun*, World Resources Institute, 2010 in http://www.wri.org/stories/2010/06/copenhagen-cancun-climate-finance

[14] http://news.mongabay.com/2010/1213-cancun_agreements.html

[15] United Nations, Framework Convention on Climate Change, 2012 in http://unfccc.int/meetings/durban_nov_2011/meeting/6245/php/view/reports.php

[16] IPCC Reports, Working Group 1, *The Scientific Basis* in http://www.ipcc.ch/ipccreports/tar/wg1/041.htm

[17] NASA, 2010 in http://science-edu.larc.nasa.gov/energy_budget/

[18] P. Przyborski and L. Remer, *The Carbon Cycle*, NASA, Earth Observatory, 2001 in http://earthobservatory.nasa.gov/Features/CarbonCycle/carbon_cycle2001.pdf

[19] H. Riebeek, *The Carbon Cycle*, NASA, Earth Observatory, 2011 in http://earthobservatory.nasa.gov/Features/CarbonCycle

[20] Global Greenhouse Warming, 2012 in http://www.global-greenhouse-warming.com/global-carbon-cycle.html

[21] http://wildbc.org/publications-resources/climatechange/Primer_Final.pdf

[22] http://www.vtaide.com/png/carbonCycle.htm

[23] Cornell University, *Modeling the Carbon Cycle*, 2005 in http://www.geo.cornell.edu/eas/education/course/descr/EAS302/302_06Lab11.pdf

[24] R.A. Houghton, *The Contemporary Carbon Cycle, Treatise of Geochemistry*, Vol 8, Elsevier, 2003 in http://www.lu.lv/ecotox/publikacijas-3-kursa-studentiem/sample8.pdf

[25] University of Colorado, Geological Sciences, Courses, GEOL 3520, *Carbon Cycle*, 2009 in http://www.colorado.edu/GeolSci/courses/GEOL3520/Carbon_cycle.pdf

[26] http://wildbc.org/publications-resources/climatechange/carbon-cycles-gr7.pdf

[27] http://www.ldeo.columbia.edu/edu/dees/V1003/lectures/global_carbon_cycle

[28] T. POWELL, Photosynthesis, Illuminating Alternatives to Solar Energy Research, Berkeley Science Review, 2009 in http://sciencereview.berkeley.edu/articles.php?issue=16&article=features_01_photosynthesis

[29] J.S. MOQUET, A. CRAVE, J. VIERS, P. SEYLER, E. ARMIJOS, L. BOURREL, E. CHAVARRI, C. LAGANE, A. LARAQUE, W.S. CASIMIRO, R. POMBOSA, L. NORIEGA, A. VERA and J.L. GUYOT, *Chem. Geol. 287*, 1, 2011 in http://www.citeulike.org/user/nilsmoosdorf/article/8671473

[30] ZMT Bremen in http://www.zmt-bremen.de/Kohlenstoff-_und_Naehrstoffkreislaeufe.html

[31] http://www.whrc.org/carbon/index.htm

[32] http://www.climatedata.info/Forcing/Emissions/Emissions/CO2.html

[33] http://unfccc.int/files/press/backgrounders/application/pdf/press_factsh_mitigation.pdf

[34] P. TANS, Trends in Atmospheric Carbon Dioxide – Mauna Loa, NOAA/ESRL, Available online at: http://www.esrl.noaa.gov/gmd/ccgg/trends

[35] M. PIDWIRNY, *The Encyclopedia of Earth, Carbon Cycle*, 2010 in http://www.eoearth.org/article/Carbon_cycle

[36] CEN/BT/WG 209 *Biobased Products – Standardization Gap Analysis*, Technical Report, Working Document, prCEN/TR 1, 2010 in http://www.sis.se/PDF/CEN-BT-WG209_N0118_Final_Report_of_CEN_BT_WG_209(2).pdf

[37] United Kingdom Parliament, Parliamentary Business, *How Can Biofuels Contribute Towards Defra's Policy Goals?*, 2003 in www.publications.parliament.uk/pa/cm200203/cmselect/cmenvfru/929/92905.htm

[38] http://www1.eere.energy.gov/biomass/abcs_biofuels.html

[39] Oak Ridge National Laboratory, *The Bioenergy Cycle: A Vision of the Future* in http://www.firstnationalpower.com/technology.html

[40] OECD, Industrial Biotechnology and Climate Change, 2011 in http://www.oecd.org/sti/biotechnologypolicies/49024032.pdf

[41] Nova: Science in the news, *Biomass – the growing energy resource*, Australian Academy of Science, 2009 in http://www.science.org.au/nova/039/039print.htm

[42] Massachusetts Agriculture in the Classroom, *Winter 2008 Newsletter*, 2008 in www.umass.edu/umext/mac/Newsletters/winter2008.htm

[43] World Economic Forum, *The Future of Industrial Biorefineries*, 2010 in http://www3.weforum.org/docs/WEF_FutureIndustrialBiorefineries_Report_2010.pdf and http://www.weforum.org/en/media/Latest%20News%20Releases/NR_Biorefineries

[44] D. GREER, *Creating Cellulosic Ethanol: Spinning Straw into Fuel*, CropChoice.com, 2005 in http://www.cropchoice.com/leadstryethanol052005.html

[45] JL WERTZ, O. BEDUE and J.P. MERCIER, *Cellulose Science and Technology*, EPFL Press, Lausanne, 2010

[46] European Federation for Transportation and Environment, 2011 in http://www.transportenvironment.org/what-we-do/what-science-says-0

[47] Extension, *Animal Fats for Biodiesel Production*, 2010, in http://www.extension.org/pages/Animal_Fats_for_Biodiesel_Production

[48] US Department of Energy, Energy Efficiency and Renewable Energy, *Biomass Program, Algal Bio*fuels, 2007, in http://www1.eere.energy.gov/biomass/pdfs/algalbiofuels.pdf

[49] M. MITRA and A. MELIS, *Optics Express 16*, 21807, 2008 in http://www.opticsinfobase.org/oe/abstract.cfm?uri=oe-16-26-21807

[50] S. ATSUMI, W. HIGASHIDE and J.C. LIAO, *Nat. Biotechnol. 27*, 1177, 2009, in http://www.nature.com/nbt/journal/v27/n12/full/nbt.1586.html

[51] European Biofuels Technology Platform, *Algae for Production of Biofuels*; 2010 in www.biofuelstp.eu/algae.html

[52] Oilgae, *Comprehensive Oilgae Report, Chapter XII - Algae Grown in Open Ponds, Closed Ponds & Photobioreactors*, 2010 in http://www.oilgae.com/ref/report/cha/12.html

[53] Oilgae, *Comprehensive Oilgae Report, Cultivation of Algae in Biophotoreactor*, 2010 in http://www.oilgae.com/algae/cult/pbr/pbr.html

[54] R. HATTI-KAUL, *Biorefineries – a path to sustainability*, Lund University, 2009 in www.scienceforum2009.nl/Portals/11/6Kaul-pres.pdf

[55] R. C. BROWN, Iowa State University, *The Future of Biorefining Agricultural Biomass*, 2005 in www.farmfoundation.org/news/articlefiles/949-robertc.brownpaper.pdf

[56] B. KAMM, P. R. GRUBER, and M. KAMM, Eds., *Biorefineries – Industrial Processes and Products. Statu Quo and Future Directions*. Vol. 1, Wiley-VCH, 2006 in www.wiley-vch.de/books/sample/3527310274_c01.pdf

[57] DuPont, The Miracles of Science, 2007 in http://www2.dupont.com/Renewably_Sourced_Materials/en_US/proc-buildingblocks.html

[58] http://www.bio-amber.com/index.html

[59] R.C. BROWN, *Issues in New Crops and New Uses*, J. Janick and A. Whipkey, Eds., ASHS Press, Alexandria VA, 2007, in www.hort.purdue.edu/newcrop/ncnu07/pdfs/brown30-38.pdf

[60] S.R SCHILL, *Heeding Hydrogenation*, Biodiesel Magazine, 2007 in http://www.biodieselmagazine.com/articles/1505/heeding-hydrogenation

[61] http://www.pnl.gov/main/publications/external/technical_reports/PNNL-18284.pdf

[62] S. Zinoviev, S. Arumugam, and S. Miertus, International Centre for Science and High Technology United Nations Industrial Development Organization, *Background Paper on Biofuel Production technologies*, 2007 in www.biofuels-dubrovnik.org/downloads/Background Paper biofuels CEE UNIDO.pdf

[63] IEA Bioenergy, T42, *Biorefineries: adding value to the sustainable utilisation of biomass*, 2009 in http://www.globalbioenergy.org/uploads/media/0909_IEA_Bioenergy_-_Biorefineries_adding_value_to_the_sustainable_utilisation_of_biomass.pdf

[64] IEA Bioenergy Task 42 on Biorefinery & EC FP6 IP BIOSYNERGY, *Adding Value to the Sustainable Utilisation of Biomass*, Biorefinery Training Course, Genth, 2009 in http://www.biosynergy.eu/fileadmin/biosynergy/user/docs/TrainingCourseDisseminationDOC.pdf

[65] IEA Bioenergy, *IEA Bioenergy Task 42 Biorefinery*, 2009 in www.biorefinery.nl/fileadmin/biorefinery/docs/Brochure_Totaal_definitief_HR_opt.pdf

[66] European Commission, *FP7 Cooperation Work Programme: Energy*, 2009 in ftp://ftp.cordis.europa.eu/pub/fp7/docs/wp/cooperation/energy/e_wp_200902_en.pdf

[67] CORDIS, *Biorefinery Course: Adding value to the sustainable utilisation of biomass*, 2009 in http://cordis.europa.eu/wire/index.cfm?fuseaction=article.Detail&rcn=19323

[68] Scientific Applications International Corporation (SAIC), *Life Cycle Assessment: Principles and Practice*, EPA/600/R-06/060, 2006 in http://www.epa.gov/nrmrl/lcaccess/pdfs/600r06060.pdf

[69] M. Patel, C. Bastiolo, L. Marini, and E. Wurdinger, *Life-cycle Assessment of Bio-based Polymers and Natural Fiber Composites*, Encyclopedia Biopolymers, Vol. 10, Wiley-VCH, 2003

[70] ISO, ISO 14000 essentials, 2011 in http://www.iso.org/iso/iso_14000_essentials

[71] ISO, ISO 14040 : 2006 in http://www.iso.org/iso/catalogue_detail?csnumber=37456

[72] A.L. Roes, *Ex-ante life cycle engineering, Application to nanotechnology and white biotechnology*, Thesis, Utrecht University, 2011, in http://igitur-archive.library.uu.nl/dissertations/2011-0112-200331/UUindex.html

[73] http://www1.eere.energy.gov/biomass/fy04/fundamental_biomass_gas.pdf

Chapter 2

Photosynthesis, the Ultimate Beginning for Biorefineries

2.1 Introduction

Photosynthesis is the main life process on Earth. [1] Through photosynthesis, land plants, algae and many species of bacteria, such as ***cyanobacteria***, convert light energy from the sun into chemical energy needed for most biological functions. More precisely, photosynthetic organisms capture and use solar energy to transform water and atmospheric carbon dioxide (CO_2) into molecular oxygen (O_2) and carbohydrates such as glucose (**Equation 2.1**) [2, 3]:

$$6\ CO_2 + 6\ H_2O + \text{energy} \rightarrow C_6H_{12}O_6 + 6\ O_2 \qquad (2.1)$$

Carbohydrates can be used later to supply the energy needs to the cell. Therefore, photosynthesis is a process in which solar energy is stored in the bonds of carbohydrates.

The amount of energy trapped by photosynthesis is immense, approximately 100 terawatts (1 TW = 10^{12} W), which is about six times larger than the power consumption of humanity (**Figure 2.1**). [4, 5] The energy produced

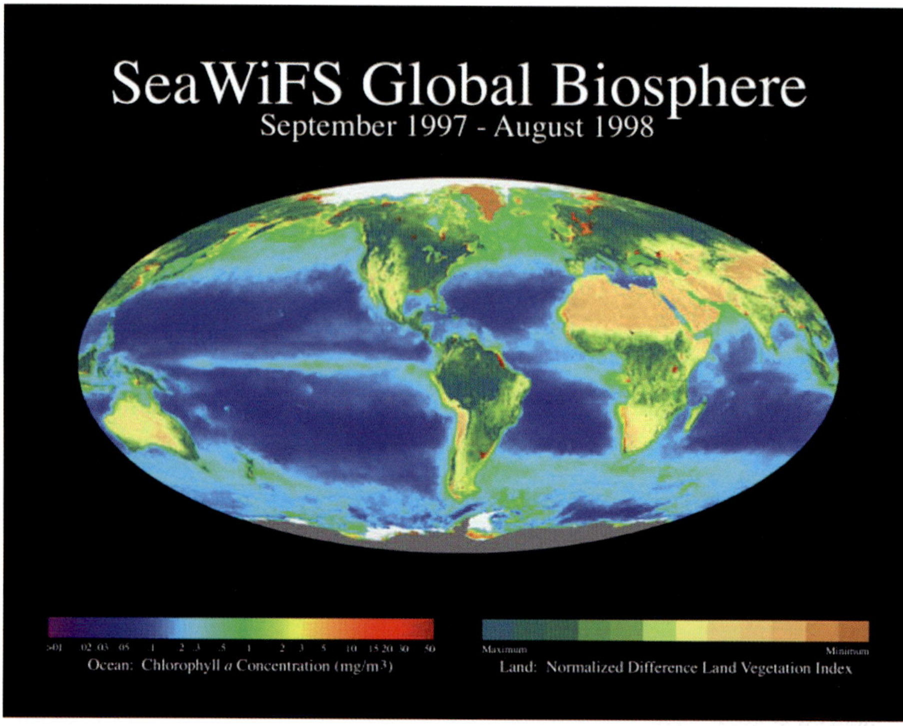

Figure 2.1 Global distribution of photosynthesis, including oceanic phytoplankton and land vegetation. **Ocean scale** (left): Chlorophyll a concentration (mg/m^3): from 0.01, dark blue to 50, purple. **Land scale** (right): Normalized Difference Land Vegetation Index: from maximum, blue to minimum, brown. Provided by the SeaWIFS Project, NASA/Goddard Space Flight Center and ORBIMAGE. [8]

by photosynthesis forms the basis of virtually all terrestrial and aquatic food chains. Consequently, photosynthesis is the ultimate source of the carbon in the organic molecules found in most organisms. [6] Photosynthetic organisms convert about 100 billion tonnes of carbon into biomass per year. [7]

Photosynthesis evolved early in the evolutionary history of life, when all forms of life on Earth were microorganisms and the atmosphere had much more CO_2. [7] The first photosynthetic organisms probably evolved about 3,500 million years ago, and used hydrogen or hydrogen sulfide as sources of electrons, rather than water. Cyanobacteria appeared later, around 3,000 million years ago, and changed the Earth for ever when they began to oxygenate the atmosphere, beginning about 2,400 million years ago. This new atmosphere allowed the evolution of complex life such as ***protists.*** Eventually, no later than a billion years ago, one of these protists formed a symbiotic relationship with a cyanobacterium, producing the ancestor of the plants and algae. The ***chloroplasts*** in modern plants are the descendants of these ancient symbiotic cyanobacteria.

Photosynthetic organisms are ***photoautotrophs***, which means that they are able to synthesize food directly from CO_2 using solar energy. [7] However, not all organisms that use light as a source of energy carry out photosynthesis, since ***photoheterotrophs*** use organic compounds, rather than CO_2, as a source of carbon. In photosynthetic organisms, photosynthesis releases oxygen. This is called oxygenic photosynthesis. Although there are some differences between oxygenic photosynthesis in plants, algae and cyanobacteria, the overall process is quite similar. However, there are some types of bacteria that carry out anoxygenic photosynthesis, which consumes CO_2 without releasing oxygen.

CO_2 is converted into carbohydrates in a process called carbon fixation. [7] Carbon fixation is a redox reaction, so photosynthesis has to supply both the energy to drive the process, and the electrons to convert CO_2 into carbohydrates, which is a reduction reaction. In essence, photosynthesis is the opposite of cellular respiration, which is carried out through ***glycolysis***, the ***Krebs cycle*** and the ***electron transport chain***. [9] In cellular respiration, carbohydrates are oxidized to produce CO_2, water and energy (**Equation 2.2**):

$$C_6H_{12}O_6 + 6\ O_2 \rightarrow 6\ CO_2 + 6\ H_2O + \text{energy} \tag{2.2}$$

In green plants and algae, photosynthesis takes place within chloroplasts, which are specialized cellular compartments. In photosynthetic bacteria, which lack membrane-bound compartments, the reactions take place directly in the cell. [10]

Photosynthesis occurs in two stages (**Figure 2.2**). [7] In the first stage, ***light-dependent reactions***, or light reactions, capture the solar energy and use

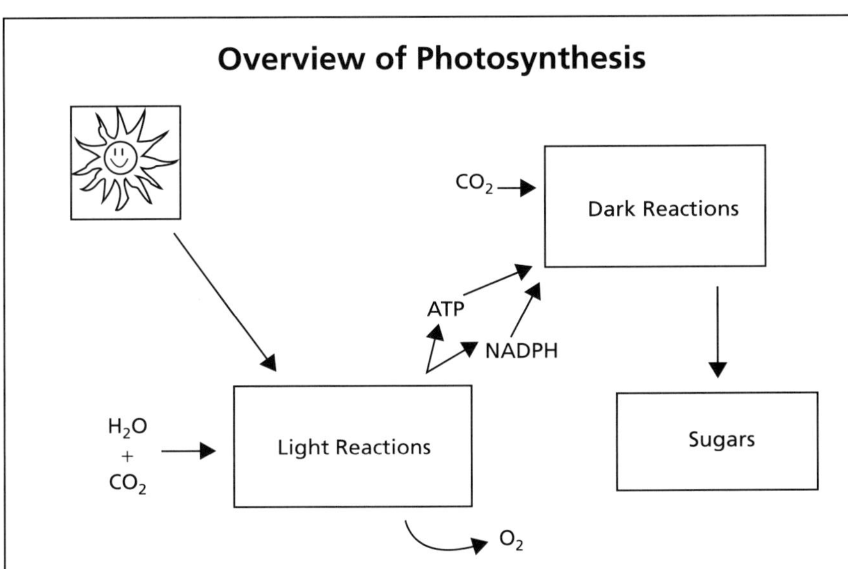

Figure 2.2 Overview of photosynthesis. [13]

it to make the energy-storage molecules ATP and NADPH. [11] Light reactions in land plants, green algae and cyanobacteria involve four major protein complexes, ***photosystem II*** (***PSII***), ***cytochrome*** b_6f, ***photosystem I*** (***PSI***) and ***ATP synthase***. During the second stage, the ***light-independent reactions***, or dark reactions, use these molecules to capture and reduce CO_2. [12] In dark reactions, the enzyme RuBisCO is responsible for starting the carbon fixation cycle.

2.2 Chloroplasts

2.2.1 Leaves

Leaves are the structures in plants that have evolved primarily to absorb sunlight and to convert this energy into carbohydrates. The cross-section through the blade of a typical leaf reveals 4 distinct tissue layers and a arrangement of veins, which are the vascular tissue of the leaf (**Figure 2.3**). [14]

1. Upper epidermis. This is a single layer of cells containing few or no chloroplasts. The cells permit most of the light that strikes them to pass through to the underlying cells. The upper surface is covered with a waxy cuticle.
2. Palisade layer (one of the two main types of ***mesophyll*** tissue). This consists of one or more layers of cylindrical cells oriented with their long axis

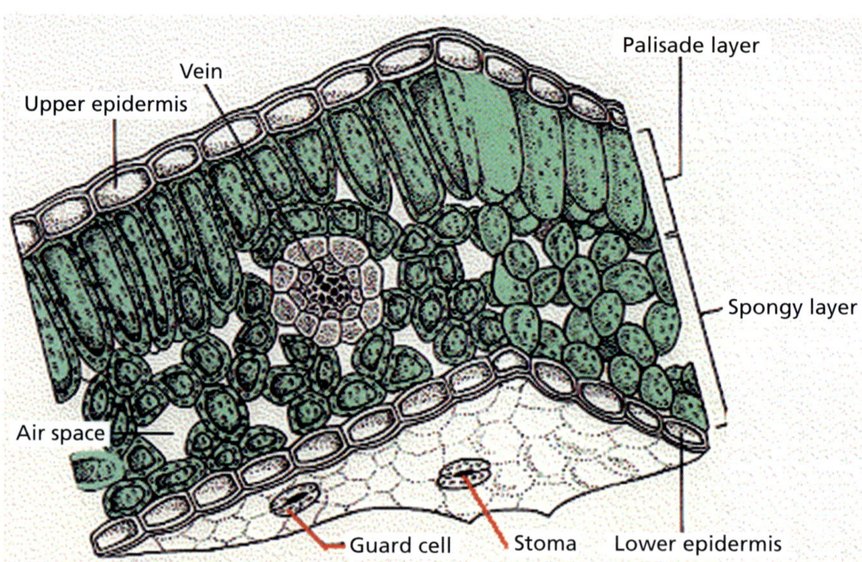

Figure 2.3 Cross-section through the blade of a typical leaf. Kimball's Biology Pages © John W. Kimball, distributed under a Creative Commons Attribution. [14]

perpendicular to the plane of the leaf. The cells are filled with numerous chloroplasts and carry on most of the photosynthesis in the leaf.

3. Spongy layer (one of the two main types of mesophyll tissue). Its cells are irregular in shape and loosely packed. They contain a few chloroplasts. During the day, these cells give off O_2 and water vapor to the intercellular chambers (air spaces) that surround them. They also pick up CO_2 from the intercellular chambers. The chambers are interconnected and eventually open to the outside through pores called ***stomata***. [15] Stomata allow gas exchange and water evaporation from inside the leaf. Because dehydration can be a serious problem, the stomata close when the plant is under water stress. When closed, CO_2 needed for the dark reactions cannot enter.
4. Lower epidermis. Typically, most of the stomata are located in the lower epidermis. Although most of the cells of the lower epidermis resemble those of the upper epidermis, each stoma is flanked by two cells called guard cells. These differ from the other cells of the lower epidermis not only in their shape but also in having chloroplasts. The guard cells regulate the opening and closing of the stomata.

The leaf also contains vascular bundles, which are known as veins (**Figure 2.4**).

Figure 2.4 Leaf veins. Kimball's Biology Pages © John W. Kimball, distributed under a Creative Commons Attribution. [14]

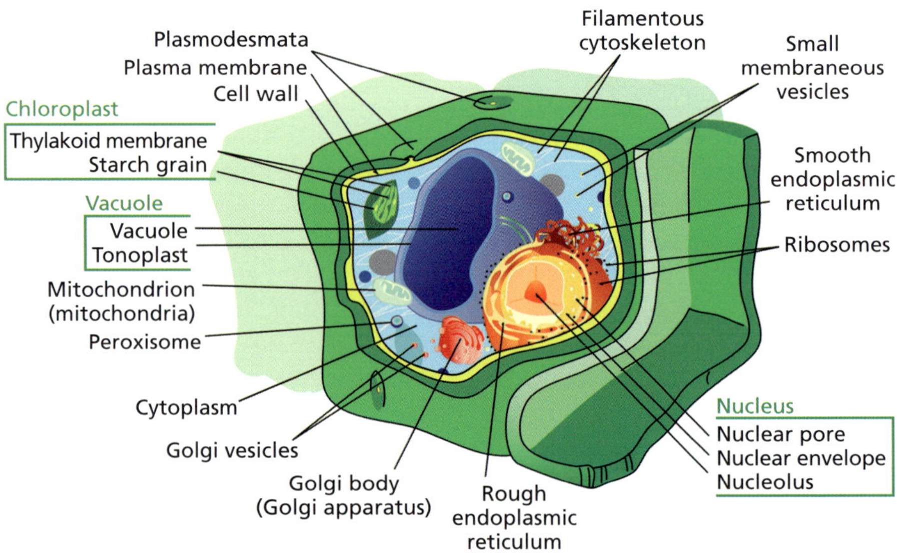

Figure 2.5 Plant cell structure. [16]

The veins are the circulatory system of the leaf. They are located in the mesophyll layer and are made up of xylem, which supplies water and minerals up from the roots, and phloem, which removes synthesized food.

Figure 2.5 shows a typical plant cell and its major ***organelles***. The nucleus contains the DNA of the cell. Respiration takes place in mitochondria. The vacuole is a fluid-filled bag. The major role of vacuoles is to maintain turgor pressure.

2.2.2 Chloroplasts and photosynthetic membranes

Chloroplasts are organelles known as ***plastids*** located in the cytoplasm of photosynthetic plant cells. They contain photosynthetic pigments. They are about 5 μm in diameter and 2-3 μm thick. [17, 18] A typical photosynthetic cell contains an average of 50 to 60 active chloroplasts (**Figure 2.6**). [19, 20]

Chloroplasts are composed of three types of membranes which create three separate compartments in each chloroplast (the intermembrane space, the stroma and the thylakoid space or lumen) (**Figure 2.7**). [18]

Chloroplasts are bounded by two membranes, which do not contain chlorophyll. [22, 23] Of these two membranes, the outer one is permeable to metabolites of small molecular weight; it contains proteins that form large aqueous channels. The inner membrane is the permeability barrier of the chloroplast; it contains transport proteins that regulate access to the stroma, an aqueous fluid enclosed in the inner membrane. The stroma is where the dark reac-

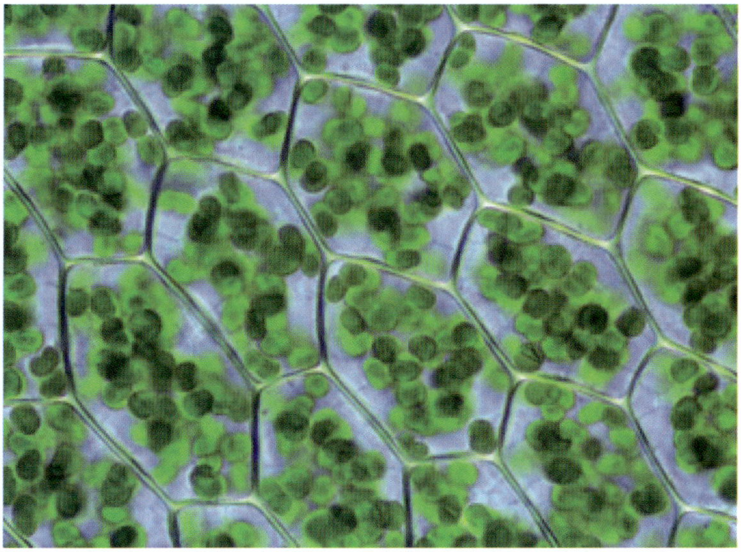

Figure 2.6 Chloroplasts visible in the cells of *Plagiomnium affine*, many-fruited Thyme-moss. Reproduced by permission of Kristian PETERS – Fabelfroh under the Creative Commons Attribution-Share Alike 3.0 Unported license. [21]

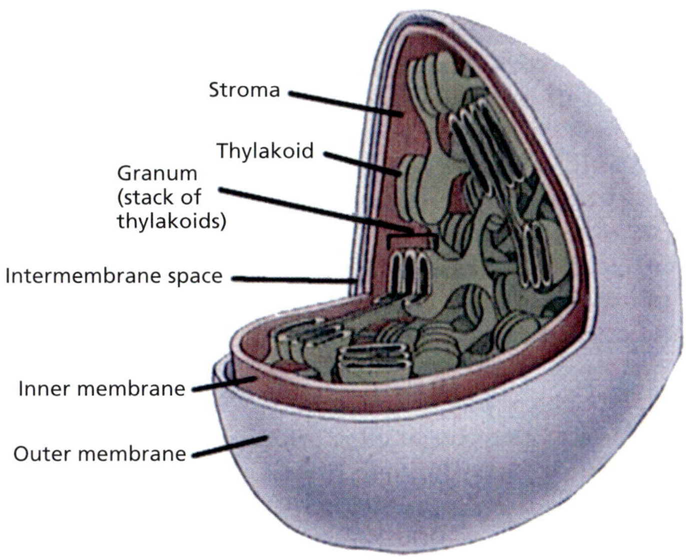

Figure 2.7 Chloroplast diagram illustrating the different components. [18] Chloroplasts are lens-shaped organelles that are ~5 micrometers in diameter. They are composed of three types of membranes which create three separate compartments in each chloroplast. The outer membrane is permeable to most ions and metabolites while the inner membrane contains transport molecules that regulate access to the stroma, a fluid body enclosed between the inner membrane and the thylakoid membrane. The thylakoid membrane is a system of interconnecting, disk-shaped compartments (thylakoids). The thylakoids are stacked like coins, these stacks are called the grana. Thylakoids enclose a single interconnected luminal space. Reproduced by permission of Pearson publishing.

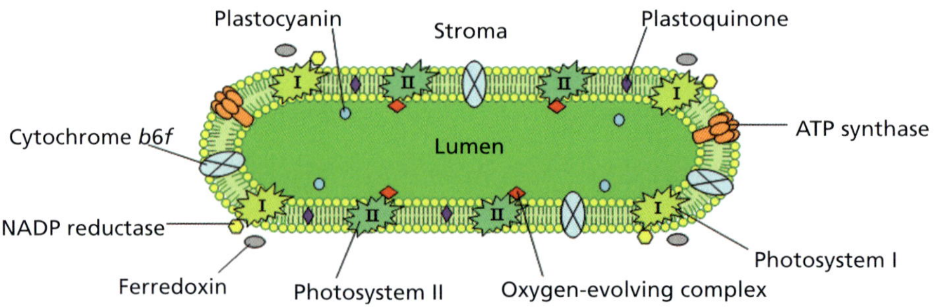

Figure 2.8 Thylakoid membrane with embedded protein complexes. [25]

tions of photosynthesis occur. The third membrane, the thylakoid membrane is the site of photosynthesis. [24] In each chloroplast, the ***thylakoid membrane*** is believed to constitute a single sheet that forms numerous small, flattened vesicles, the thylakoids, which commonly are arranged in stacks termed grana. The spaces between all the thylakoids constitute a single continuous compartment, the ***thylakoid lumen***. Four major photosynthetic protein complexes are embedded in the thylakoid membrane (**Figure 2.8**).

A chloroplast's primary function is to capture and store sunlight in the form of chemical energy. [18] The chloroplasts achieve this through the light-trapping pigments, such as ***chlorophyll***, and enzymes found in the thylakoid membranes. Some of the sunlight energy is used to break down water. The remaining energy is passed into the stroma where CO_2 is fixed to form carbohydrates.

2.3 Photosynthetic pigments and antenna complexes

2.3.1 Photosynthetic pigments

Pigments are molecules that absorb light. [2] When a photon of light strikes a photosynthetic pigment, an electron in an atom contained within the molecule becomes excited. Excited electrons move further from the nucleus of the atom. The excited molecule can pass the energy to another molecule or release it in the form of light or heat.

Chlorophyll *a* (Chl *a*) is the main photosynthetic pigment in all organisms except bacteria. Other pigments called accessory pigments absorb slightly different wavelengths of light. The combination of all of the pigments increases the range of colors that plants can use in photosynthesis. Accessory pigments include chlorophyll *b* (Chl *b*) and a group of pigments called ***carotenoids***. The most well known carotenoid is ***beta-carotene*** which absorbs different

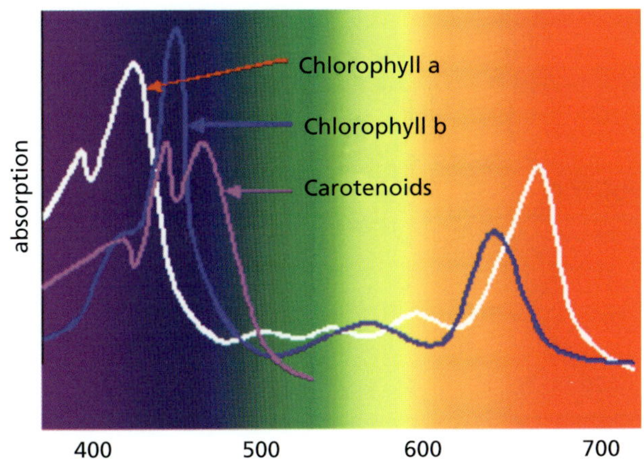

Figure 2.9 Absorption curves of chlorophyll *a*, chlorophyll *b* and carotenoids. Reproduced by permission of Michael J. Gregory. [2]

wavelengths than the chlorophylls. Carotenoids do not participate directly in photosynthetic reactions but are able to pass their energy to chlorophyll *a* (**Figure 2.9**).

Chlorophyll *a* and *b* both absorb the blue light and the red light best, resulting in an overall green appearance of most plants. [13]

Chlorophyll is composed of two parts; the first is a ***porphyrin*** ring with magnesium at its center, the second is a hydrophobic phytol tail (**Figure 2.10**). The ring has many delocalized electrons that are shared between several of the C, N, and H atoms; these delocalized electrons are very important for the function of chlorophyll. The tail is a 20 carbon chain that is highly hydrophobic and stabilizes the molecule in the hydrophobic core of the membrane where photosynthesis occurs.

Structurally, the only difference between chlorophyll a and b is the functional group indicated in green in the figure. The CH_3 group is present in chlorophyll *a* where chlorophyll *b* has a CHO group. However, functionally, the two structures are very different.

2.3.2 Antenna complexes

In most photosynthetic organisms, chlorophyll and carotenoid molecules are embedded within a protein barrel, forming protein-pigment complexes called ***light-harvesting complexes*** (***LHC***) or antenna complexes. [26, 27] LHCs are composed of hundreds of molecules of chlorophylls and accessory pigments. [13]

The functions of LHCs are:

- to capture the energy from photons;

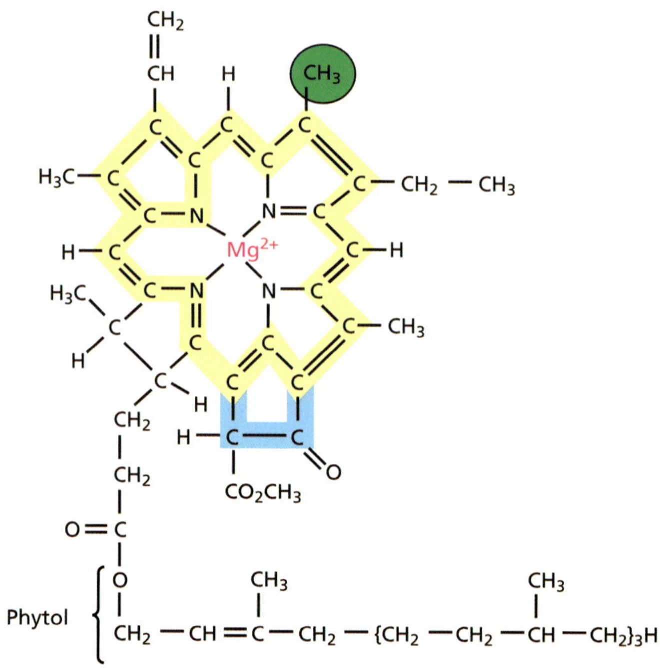

Figure 2.10 Structure of chlorophyll *a*. [13]

- to transfer the molecular excitation from one molecule to another;
- to deliver the molecular excitation to a reaction center, where photoinduced charge separation occurs.

Antenna complexes in purple bacteria

Anaerobic photosynthetic ***purple bacteria*** are great masters of harvesting light. [28] Nearly all the energy gained by the absorption of a photon is transferred on to the reaction center. Most of them possess two LHCs besides the reaction center, and these are integrated into the plasma membrane. [29, 30] The peripheral antenna complexes, termed LH-II, have absorption at shorter wavelengths than the internal antenna complex, termed LH-I, and the reaction center itself. In most purple bacteria, the two classes of LHCs are arranged concentrically. The inner concentric ring consists of LH-I, in contact with the reaction center; the outer rings consist of LH-II (**Figure 2.11**). [28]

The LHCs from purple bacteria are oligomers. The basic unit is a heterodimer consisting of two small protein subunits referred to as α- and β-apoproteins. The $\alpha\beta$-heterodimers of peripheral LHCs bind three bacteriochlorophyll *a* molecules (**Figure 2.12**).

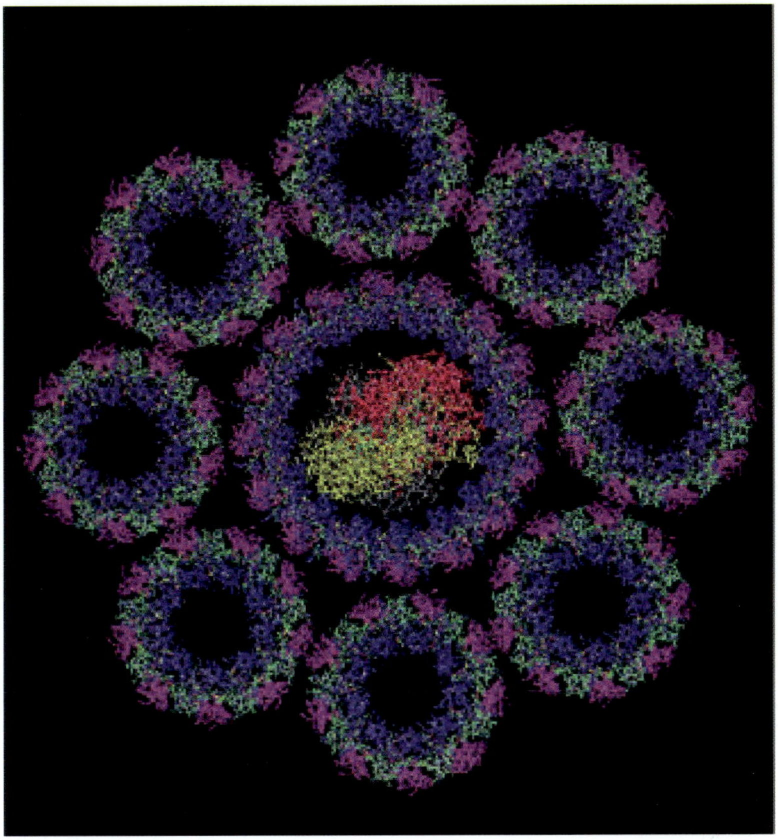

Figure 2.11 Concentric arrangement of peripheral LHCs (LH-II) and internal LHCs (LH-I) around the reaction center (RC), as shown by most purple bacteria. Reproduced by permission of the Theoretical and Computational Biophysics Group, University of Illinois at Urbana-Champaign. © 2002 Cambridge University Press. [28, 31]

Core antenna complexes

In addition to the peripheral antenna systems that vary between organisms, photosystems I and II from land plants, green algae and cyanobacteria, as well as the photosystems from heliobacteria and green sulfur bacteria, contain a core antenna system. [33] The core antenna is an intrinsic part of the photosystem. The core antenna protein domain flanks the reaction center core on both sides. The core antenna domain of PSI is fused with the reaction center domain and contains 90 chlorophylls, while there are two separate antenna proteins in PSII, containing together ~35 chlorophylls.

The core antenna system is much smaller in the anoxygenic type-I photosystems than in PSI. [33] The photosystem from heliobacteria may contain ~35 chlorophylls, while the photosystem from green sulfur bacteria has an even

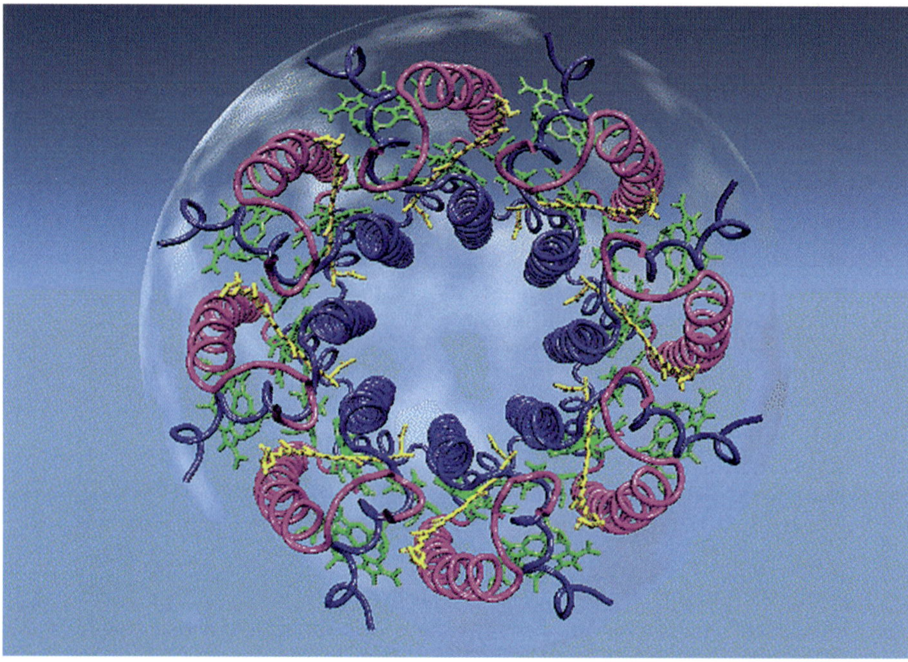

Figure 2.12 Crystal structure of the LH-2 octameric complex from the purple bacterium *Rhodospirillum molischianum*. View from above. The octamer of $\alpha\beta$ heterodimers is shaped as a hollow cylinder of 90 Å diameter formed by the helices of the two apoproteins, with the α-apoprotein inside and the β-apoprotein outside. Bacteriochlorophyll a, green; α apoprotein, blue; β apoprotein, magenta; carotenoid lycopene, yellow. Reproduced by permission of the Theoretical and Computational Biophysics Group, University of Illinois at Urbana-Champaign. © 1996, Elsevier. [29, 32]

smaller antenna size with ~10 chlorophylls. The purple bacteria reaction centers are the only photoreaction centers that do not contain a core antenna system. In addition to the core antenna system, all photosynthetic organisms have developed peripheral antenna systems (LHCs). In contrast to the reaction centers which have all evolved from a common ancestor, nature has developed different types of antenna complexes.

Peripheral antenna complexes of higher plants

Higher plants contain membrane integral antenna proteins, LHC-I (associated with photosystem I) and LHC-II (associated with photosystem II). [33] Both contain large amounts of these antenna complexes but especially LCH-II, which is the second most abundant protein on earth. In contrast to the core antenna system of PSI and PSII, which contain only chlorophyll *a*, the LHC-I and LHC-II complexes also contain chlorophyll *b*, which allows plants to absorb more light from the visible spectrum. The structure of LHC-II has been determined from spinach [34] and pea. [35] The LHC-II serves as the antenna to PSII and regulates the flow of energy within the photosynthetic apparatus. [36, 37] In the crystals and in the native membrane, LHC-II is a trimer; each monomer consists of three transmembrane helices. 20 LHC-II trimers are organized in an icosahedral particle.

Each monomeric LHC-II contains a single polypeptide, 8 molecules of chlorophyll *a*, 6 molecules of chlorophyll *b*, 4 carotenoids and a phospholipid. [34, 38]

The *Lhcb* (the letter b referring to PSII) gene family encodes several light-harvesting Chl a/b-binding (LHC) proteins associated with PSII. [39] Native LHC-II (major light-harvesting complex of higher plants) isolated from plant tissues consists of three isoforms, Lhcb1 to 3, which form homo- and heterotrimers. [36] All three isoforms are highly conserved among different species, suggesting different functional roles. LHC-II absorbs and transfers energy to the reaction center through the so-called minor LHCs (Lhcb4 to 6), which are similar to LHC-II in pigment content and gene sequence.

The major peripheral antenna complex of photosystem I in plants is LHC-I. The structure of the plant PSI-LHC-I supercomplex has been determined. The supercomplex consists of a PSI monomer which binds four LHC-I proteins (Lhca1 to Lhca4; the letter a referring to PSI). Each LHC-I protein consists of three transmembrane helices which bind up to 15 chlorophyll molecules.

2.4 Stages of photosynthesis

As shown in **Figure 2.2**, photosynthesis is a two-stage process. The first process is the light reactions, which include absorption of light, electron transport leading to NADPH, and generation of ATP. The second process is the dark reactions, which involve the conversion of CO_2 into carbohydrates (carbon fixation). [22]

Both stages of photosynthesis are coupled and controlled so as to produce the amount of carbohydrate required by the plant. The light reactions are catalyzed by proteins in the thylakoid membrane. The enzymes that convert CO_2 into carbohydrates are soluble components of the stroma.

2.5 Light reactions

The thylakoid membrane contains four major different complexes of integral membrane proteins. [38] The exact structure of these complexes differs from group to group (e.g., plant vs. alga) and even within a group (e.g., illuminated in air or underwater). But, generally, in land plants, green algae and cyanobacteria, one finds the photosystem II, the cytochrome b_6f, the photosystem I and the ATP synthase. The core components of the light reactions are found in two specially and functionally distinct units called photosystems. They are named photosystem I and photosystem II after the order of their discovery. The light reactions actually start with photosystem II.

2.5.1 Photosystem II (PSII)

Structure

General features

PSII is a large multisubunit complex of land plants, algae and cyanobacteria that acts as light-driven water:plastoquinone oxidoreductase. [40] Powered by light, this enzyme catalyzes the chemically and thermodynamically demanding reaction of water splitting. [41] In so doing, it releases O_2 and provides the reducing equivalents required for the conversion of CO_2 into carbohydrates. [42, 43]

Essentially, PSII includes (**Figure 2.13**):

- A ***reaction center*** (***RC***) core where light is converted into electrochemical potential energy and where the water-splitting reaction occurs. Water is oxidized to dioxygen at the ***oxygen evolving complex*** (***OEC***) which was revealed as a cluster of four Mn ions and a Ca ion surrounded by amino acid side chains. [44, 45, 46]
- Peripheral light-harvesting systems (Lhcb proteins).

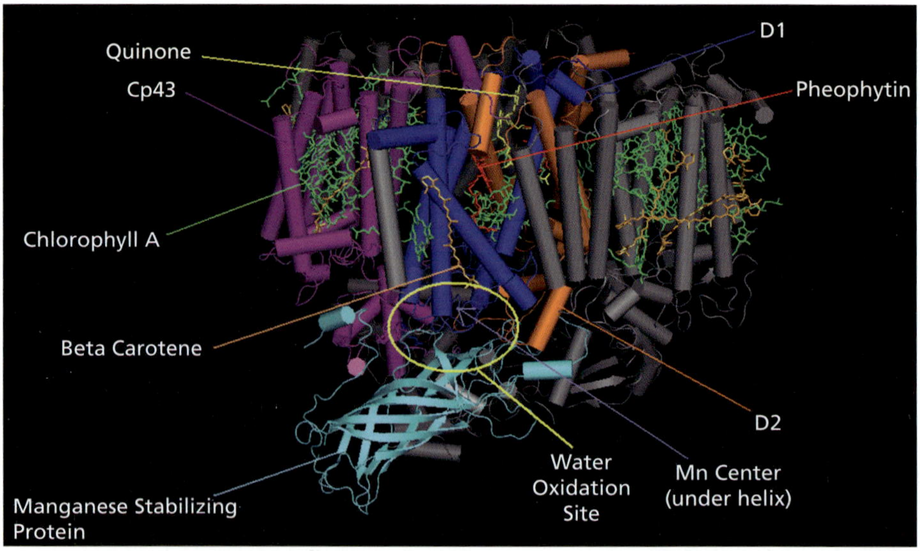

Figure 2.13 Cyanobacterial photosystem II, Monomer, PDB 2AXT. D1 and D2, reaction center proteins; Cp43, antenna chlorophyll *a* binding proteins. Reproduced by permission of Curtis NEVEU under the Creative Commons Attribution-Share Alike 3.0 Unported (CC BY-SA 3.0) license. [47]

In its complete form, PSII consists of many different proteins as well as a large number of cofactors including chlorophyll, pheophytin (chlorophyll lacking Mg^{++}), carotenoids, quinones (Q_A and Q_B, primary and secondary quinone acceptors), non-heme iron and manganese, which together trap solar energy to drive the water splitting reaction. [40, 48, 49]

A common feature of a wide range of membrane complexes is that they exist in oligomeric states in vivo despite the fact that when isolated as monomers they can carry out their enzymatic activities. [50] In oxygenic photosynthetic organisms, PSII normally functions as a dimer while PSI in cyanobacteria is trimeric.

The general architecture of PSII of higher plants is shown in **Figure 2.14**. [51]

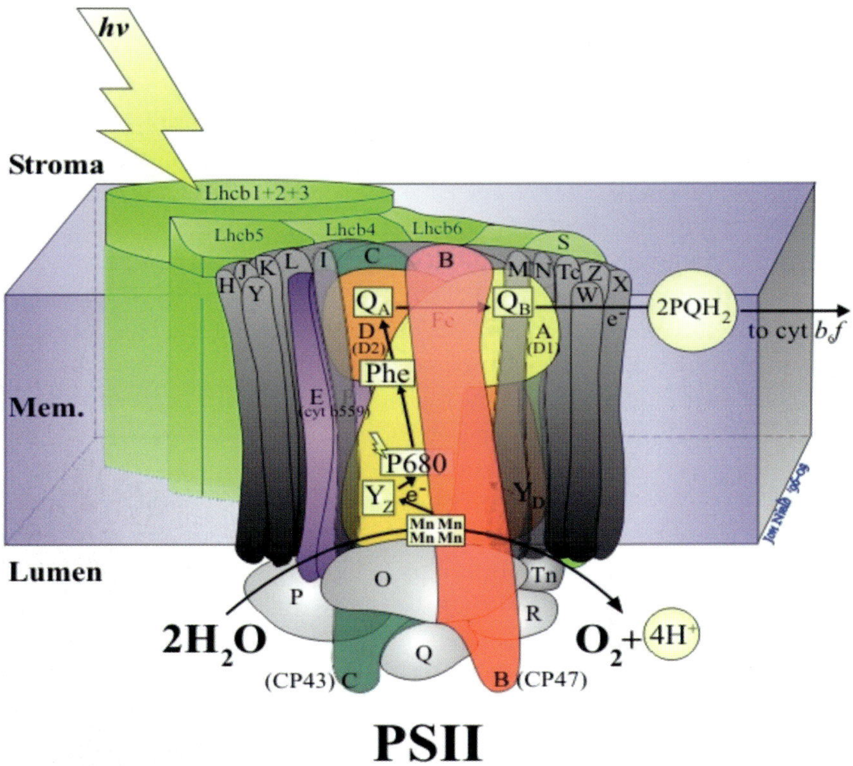

Figure 2.14 View of a higher plant PSII complex (based on a single reaction center). [51] P680,tetramer of chlorophyll *a* molecules; Mn, manganese cluster; Y_Z, tyrosine; Phe, pheophytin; Q_A, Q_B, quinones; PQH_2, plastoquinol; A to Z, core proteins encoded by the psba to psbz genes; A (D1), D (D2), reaction-center proteins; B (CP47), C (CP43), chlorophyll *a* binding proteins; E, F, cytochrome *b*559; Lhcb1 to 6, chlorophyll *a*/chlorophyll *b* binding proteins (CAB) 1 to 6 encoded by *cab* genes; one electron is transferred from P680 to Phe, then to Q_A and Q_B; four electrons are transferred from H_2O to P680 via Y_Z. Reproduced with the permission of Professor James Barber.

The RC core is composed of two homologous proteins known as D1 and D2, which are encoded by the *psbA* and *psbD* genes ("psb" for subunits of photosystem II). [41] These proteins each consist of five transmembrane α-helices. These two proteins are remarkably similar to the M and L subunits that constitute the RC of anaerobic purple bacteria. In both cases, they bind cofactors which facilitate primary charge separation. The D1 and D2 proteins bind one pheophytin molecule each and form the Q_A and Q_B binding sites of the acceptor side of PSII, respectively. [52] In addition, both D1 and D2 participate in the coordination of the P680 chlorophyll assembly (where P stands for pigment and 680 for its absorption maximum at 680 nm), a tetramer of equally spaced chlorophyll *a* molecules. [53] The Mn cluster on the donor side of the PSII is coordinated by amino acid residues in the D1 protein. The PSII RC core also contains a nonheme ferrous iron located on the acceptor side of the protein in close proximity to Q_A and Q_B. [54, 55] The nonheme iron is ligated by four histidine residues provided by D1 and D2. [56]

Closely associated with the D1 and D2 proteins are two chlorophyll *a*-binding proteins (CPs) called CP43 and CP47. [41] They are encoded by the *psbB* and *psbC* genes, respectively. They too are structurally homologous, each having six transmembrane α-helices. A striking feature of these two light-harvesting proteins is the very large extrinsic loops that join the luminal ends of transmembrane α-helices (V) and (VI). The Mn cluster is coordinated by amino acid residues in the C43 protein. [52] There are also a number of other low-molecular mass subunits (PsbE, PsbF, Psb H, PsbI, PsbJ, PsbK, PsbL, PsbM, PsbN, PsbT, PsbX, PsbY and PsbZ [57]) usually having a single transmembrane helix, which are rather featureless except for the PsbE and PsbF proteins that provide histidine ligands for the high-potential heme of ***cytochrome*** (Cyt) b559. [41]. Cytochrome *b*559 is a heterodimer composed of one α subunit (PsbE), one beta (PsbF) subunit, and a heme cofactor.

Finally, the PSII RC core complex has several extrinsic proteins attached to its luminal surface, the nature of which varies with different types of organisms, although the PsbO protein is ubiquitous to all types of oxyphototrophs. [41] In plants and green algae, PsbP and PsbQ associate with PsbO, whereas in red algae and cyanobacteria, they are replaced by the PsbU and PsbV proteins, where the latter is a cytochrome ($Cytc_{550}$). In all cases, they form a protein shield over the catalytic site of water splitting (OEC)

The PSII RC core complex is serviced by peripheral light-harvesting systems which vary among different organisms. [41] In plants and green algae, this antenna system is intrinsic and composed of Lhcb proteins that bind Chl *a* and Chl *b*, whereas in cyanobacteria and red algae, phycobilins arranged within the extrinsically located phycobilisomes carry out this function. There are usually ~200-300 light harvesting pigment molecules serving one PSII RC. In higher plants and green algae, there are six types of Lchb proteins that bind both chlorophyll *a* and *b*. [48] Most light-harvesting systems are made up of

Lhcb1-3 and are known as LHC-II. LHC-II is organized as trimers. The other chlorophyll a/b-binding proteins, Lhcb4, Lhcb5 and Lhcb6, also known as CP29, CP26 and CP24, respectively are found as monomers. [33, 58]

The general structure of the PSII supercomplex of higher plants can be described in terms of its major components. [58, 59, 60] In its center, there is a dimeric core complex (C_2; each monomer comprising the RC and the inner antenna proteins CP43 and CP47) which is flanked on either side by subsets (m) of minor antenna proteins CP29, CP26 and CP24, the whole being, in turn, flanked by variable numbers of LHC-II trimers that had been designated as strongly bounded (S), moderately bound (M) and loosely bound (L). [61, 62] A typical PSII supercomplex may be represented by $C_2m_2S_2[M_2L_2]$. The brackets enclose optional trimeric components.

Photosynthetic RC from anaerobic purple bacteria

The structure of the photosynthetic RC from the purple bacterium *Rhodopseudomonas viridis* was determined by Deisenhofer et al. in 1985. [63, 64] This accomplishment earned Deisenhofer and his colleagues the 1988 Nobel Prize in Chemistry. Structure, composition and function of this type of RC are similar to the core of the PSII in plants. The purple bacterium RC consists of three transmembrane polypeptides which are called H (heavy), M (medium, corresponding to D2) and L (light, corresponding to D1), bound to a c-type cytochrome on the external side of the membrane. [65] The core of the complex is formed by the subunits L and M, and their associated cofactors: four bacteriochlorophyll *b* molecules, two bacteriopheophytin *b* molecules, two quinones, one non-heme iron and one carotenoid. [66] Two of the bacteriochlorophyll molecules form the primary electron donor (special pair). Electrons are then transferred from the special pair to another pair of chlorophylls and from there to other prosthetic groups (pheophytins and quinones).

PSII core complex

The PSII core complex (PSIIcc) of spinach was resolved by electron crystallography at 8-Å resolution by Rhee et al. [67] in 1998. The one of the cyanobacterium *T. elongatus* was resolved by X-ray crystallography at 3.8-Å resolution by Zouni et al. [68] in 2001, at 3.7-Å resolution by Kamiya et al. [69] in 2003, at 3.5-Å resolution by Ferreira et al. [70] in 2004, and at 2.9-Å resolution by Guskov et al. [71] in 2009. [62] In situ, it occurs as a homodimer. Per monomer it comprises 20 proteins 3 of which are extrinsic, has more than 40 cofactors, a molecular mass of 320 kDa and dimensions of 205 Å (length), 110 Å (width) and 105 Å (depth). The X-ray data allow modeling of the transmembrane α-helices, the identification of cofactors and their assignments to binding sites in the protein scaffold. The crystal structure of PSII from *T. elongatus* at 2.9-Å resolution allowed the assignment of 20 protein subunits and modeling of 35 chlorophyll *a* molecules and 12 carotenoid molecules, 25 lipids and 1 chloride ion per monomer. [71] Assembly of cyanobacterial PSII is highly coordinated and proceeds through a number of distinct assembly intermediates. [72]

The crystal structure of monomeric PSIIcc from *T. elongatus* at 3.6-Å resolution was reported by Broser et al. in 2010. [73] The packing of the molecules within the unit cell is different compared with crystals of dimeric PSIIcc. In the monomer, 19 protein subunits, 35 chlorophylls, 2 pheophytins, the non-heme iron, Q_A, two heme groups, 11 carotenes, 22 lipids, 7 detergent molecules, and the Mn_4Ca cluster could be assigned analogous to the dimer.

Functional organization

One can distinguish three main areas of PSII in which reactions occur. [51]

1. In the antenna, all the pigments molecules are constantly absorbing light, and when light of a certain wavelength (<680 nm) is absorbed, the absorbed energy is transferred from one molecule to another until it is ultimately transferred to a chlorophyll of the reaction center, which absorbs light of longer wavelength than any of its neighbors. [13, 74]

2. In the acceptor side of the reaction center, the absorbed energy moves one of the P680 chlorophylls into the excited stage which is highly unstable. The chlorophyll *a* molecule is de-energized by ejecting an electron which is immediately taken up by pheophytin (the primary electron acceptor). Now there is a reduced pheophytin, which is a strong reducing agent, and an oxidized chlorophyll *a*, which is a strong oxidizing agent, this process is called charge separation ($P680^+$-$Pheo^-$). Subsequent electron transfer steps have been designed through evolution to prevent the primary charge separation from recombining. [75] However, to be able to absorb more solar energy and to carry out photosynthesis, the oxidized $P680^+$ (the primary oxidant) needs to be reduced. Thus nature has found a way to reduce $P680^+$ and still maintain the charge-separated state. $P680^+$ is reduced through an electron transfer from a nearby tyrosine residue (Tyr), resulting in the original P680 and a neutral tyrosine radical. The electron on the $Pheo^-$ is transferred to a plastoquinones molecule (QA) that is permanently bound to PSII. Although plastoquinones usually acts as a two-electron acceptor, it works as a one-electron acceptor at the QA-site. The electron on QA^- is then transferred to another plastoquinone molecule that is loosely bound at the QB-site. Plastoquinone at the QB-site works as a two-electron acceptor, becoming fully reduced and protonated after two photochemical turnovers of the reaction center. The full reduction of plastoquinone requires the addition of two electrons and two protons, creating PQH_2 (**Figure 2.15**).
   The reduced plastoquinone then de-binds from the reaction center and diffuses into the hydrophobic core of the membrane until it encounters a specific binding site on the cytochrome b_6f complex. After which, an oxidized plastoquinone moves to the QB-site and the process is repeated.

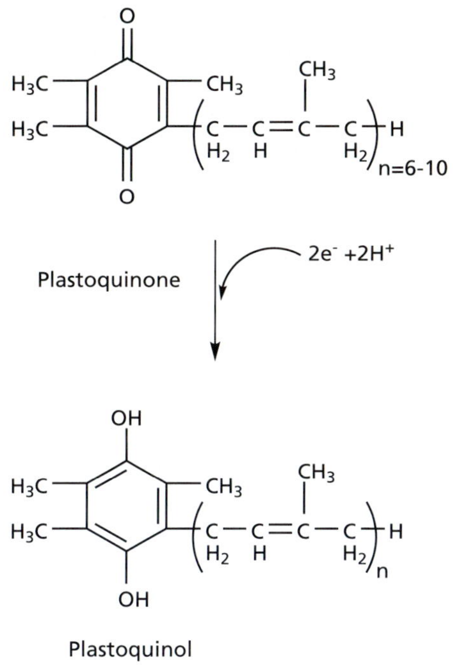

Figure 2.15 Plastoquinone to plastoquinol. [76]

Because the QB-site is near the outer aqueous phase, the protons added to plastoquinone during its reduction are taken from the outside of the membrane. P680+ gains then one electron coming from water oxidation.

3. In the donor side of the reaction center, the OEC catalyzes the water-oxidation reaction. Manganese, bound to water, calcium, chloride and oxygen (in manganese oxo bridges) rapidly reduces a local tyrosine (an intermediate reactant in the electron transfer from the Mn cluster to $P680^+$; Yz) which was oxidized as it reduced the oxidized $P680^+$. In the 1970's, Joliot and Kok worked out the mechanics of the four-step cyclic process required for turning over PSII during its water oxidation event. Each PSII reaction center works independently of one another and cycles through a set of five oxidation states, known as the S states ($S_0 \rightarrow S_4$). The current understanding of the mechanism of water oxidation proceeds through these states. **Figure 2.16** is a typical representation of the 4-step Joliot-Kok cycle. [77] Four incoming photons are required to complete the process.

The OEC starts at state S_0, and then with each incoming photon moves on to state S1, S2, S3 and finally S4. There are a total of five oxidation states of the OEC during the process. As the OEC cycles through this process, four electrons are removed from the bound water. Thus, the OEC couples the 4-electron photo-

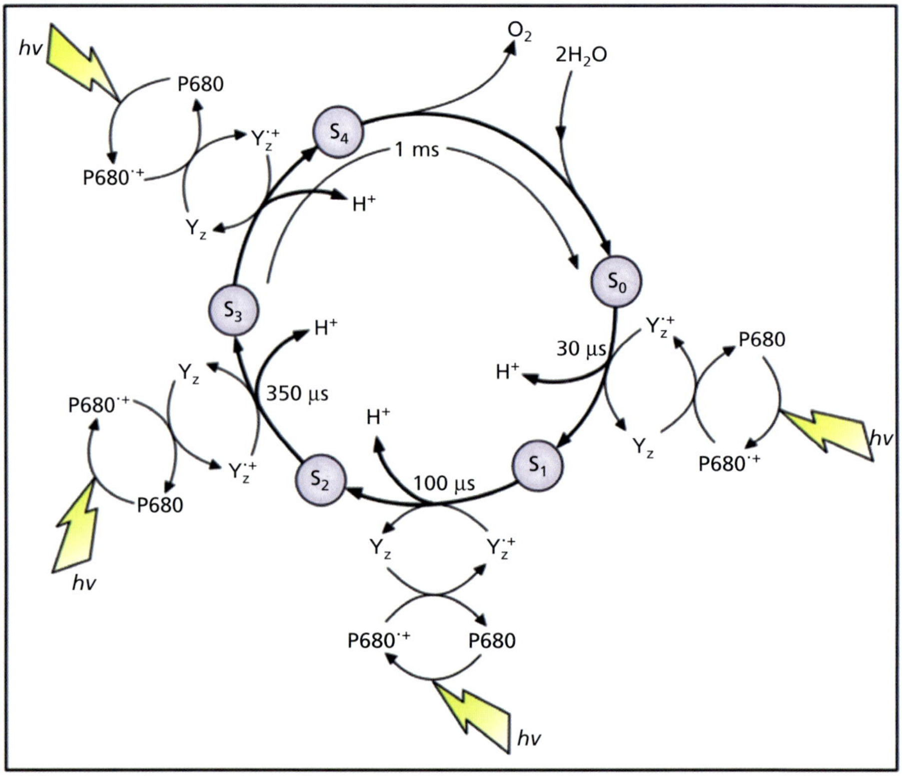

Figure 2.16 Four-step Joliot-Kok cycle. Yz, nearby Tyrosine residue. Reproduced with the permission of Professor James Barber. [45, 77]

chemistry of water oxidation with the one-electron photochemistry of the reaction center by sequentially storing oxidizing equivalents through five intermediate S-states, before one molecule of O_2 is released. Once the molecule of O_2 is evolved, the Mn cluster returns to its most reduced state. The deposition of protons during this process varies, but the figure depicts one proton released per photon absorbed.

The overall reaction of PSII is written below (**Equation 2.3**) [76]:

$$2H_2O + 4\ \text{photons} \rightarrow O_2 + 4\ H^+ + 4e^- \qquad (2.3)$$

2.5.2 Cytochrome b_6f complex

Structure

Cytochrome b_6f complex provides the electronic connection between PSII and PSI and generates a transmembrane electrochemical proton gradient for ATP synthesis (**Figure 2.17**). [78, 79] The electrons are transferred to PSI via plastocyanin (PC; copper-containing protein) or cytochrome c_6.

A 3.0-Å resolution crystal structure of the dimeric b_6f complex from the thermophilic cyanobacterium *Mastigocladus laminosus* reveals a large quinone exchange cavity, in which plastoquinone (PQ), a quinone-analog inhibitor, and a unique heme c_n, or heme *x*, are bound. [78] The core of the b_6f complex is similar to the analogous respiratory cytochrome bc_1 complex [81], but the domain arrangement outside the core is different and three prosthetic groups that are present in the b_6f complex are not found in the related bc_1 complex.

The b_6f complex is an integral membrane protein complex consisting of 8 subunits per monomer in the cyanobacterium *M. laminosus*, and 9 in plant chloroplasts. [79, 82] The additional subunit in the complex isolated from spinach is the enzyme ferredoxin:NADP-reductase (FNR). The *M. laminosus* monomer includes 13 transmembrane helices and 7 prosthetic groups

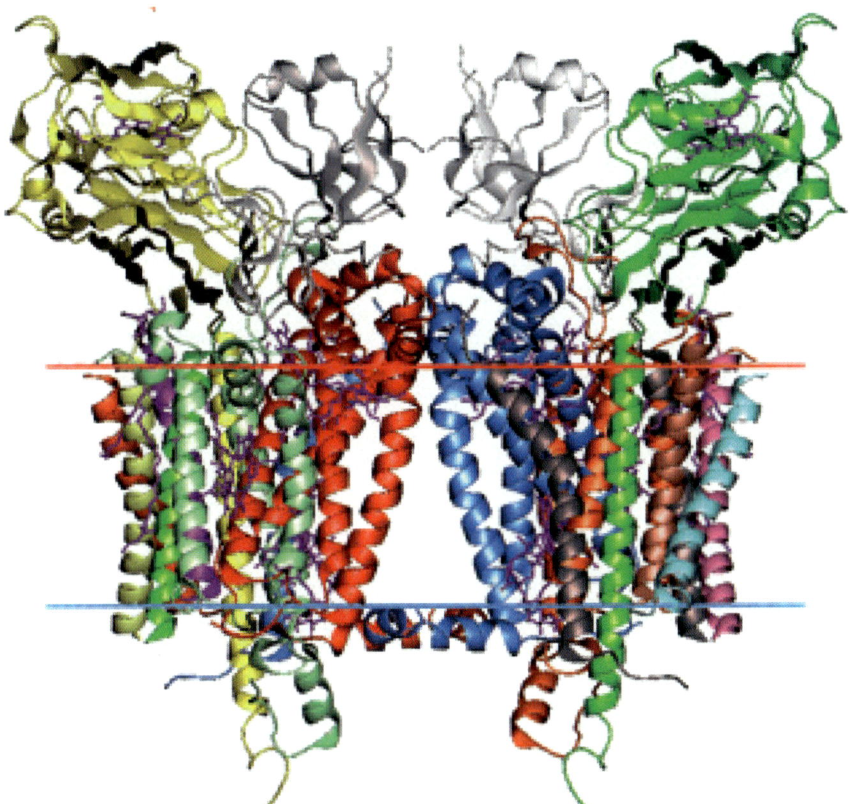

Figure 2.17 Crystal structure of the cytochrome b_6f complex from the single-celled green alga *Chlamydomonas reinhardtii* (PDB 1Q90). Boundaries of the lipid bilayer are shown by red and blue dots. Reproduced by permission of Andrei LOMIZE under the Creative Commons Attribution-Share Alike 3.0 Unported (CC BY-SA 3.0) license. [80]

(4 hemes, 1 [2Fe-2S] cluster, 1 chlorophyll a and 1 carotene). One of the four hemes is a unique heme c_n closely linked to heme b_n and with no amino acid side chains as axial ligands. [78, 79, 83] The two monomers of the dimer form a protein-free central cavity on each side of the transmembrane interface.

The b_6f complex from *M laminosus* contains four large protein subunits, including cytochrome f (PetA; "pet" for photosynthetic electron transport), cytochrome b_6 (PetB), the ***Rieske iron-sulfur protein*** (ISP) (PetC) and subunit IV (PetD), as well as four small hydrophobic subunits, PetG, PetL, PetL and PetN, leading to a dimer molecular weight of 217 kDa. The cytochrome b_6, which has four transmembrane α helices and contains two b-type hemes (hemes b_n and b_p on the electronegative (n) stromal and electropositive (p) luminal sides of the complex), as well as the unit IV, which has three transmembrane helices, are homologous to the N- and C-terminal halves of cytochrome *b* in the bc_1 complex. The Rieske iron-sulfur protein, consisting of an N-terminal single transmembrane α helix domain and a soluble extrinsic domain with a linker region connecting these two domains has a function similar to that of the ISP in the bc_1 complex. However, the *c*-type cytochrome f subunit, consisting of a single transmembrane helix, a heme, two extrinsic structural domains (a large heme-binding domain and a small domain) on the p side of the membrane, and an extended, bound water chain, is unrelated to cytochrome c_1 in the bc_1 complex. [78, 84, 85]

Functional organization

The cytochrome b_6f complex functions as a plastoquinol (PQH_2):plastocyanin/cytochrome *c6* oxidoreductase. [82] It catalyzes the oxidation of plastoquinol and the reduction of plastocyanin.

Within the b_6f complex, one electron is transferred from doubly reduced PQH_2 to a high-potential electron transfer chain, consisting of the ISP and cytochrome f on the p side of the membrane. [78] This results in the release of two protons to the aqueous lumen phase. Transfer of the second electron from PQH_2 across the complex through the two lower potential *b* hemes (or as anionic plastosemiquinone [86]) and the resulting proton uptake from the n side generate a proton electrochemical gradient across the membrane.

Electron transfer and proton translocation functions similar to those in the cytochrome b_6f complex are performed by the cytochrome bc_1 complex. [78] Lipophilic quinone species from the bulk membrane bilayer phase move between a site for reduction and proton uptake on the electronegative side of the membrane (Q_n site) and a site for oxidation and proton release on the electropositive side (Q_p site). The b_6f and bc_1 complex share the basic elements of this process, which is explained by a Q-cycle mechanism that couples electron transfer to generation of proton gradient. A structural framework for the Q-cycle is provided by the two large central cavities in crystal structures o f the dimeric bc_1 complex.

The following reactions (**Equations 2.4** to **2.12** and **Figure 2.18**) define the electron/proton transfer pathways in the b_6f complex in plants. [79] Like in a conventional bc_1 Q-cycle formulation, the n-side quinone is reduced through a low potential chain containing hemes b_p and b_n, b-type hemes on the p- and the n-sides of the complex.

$$PQH_2 + [2Fe\text{-}2S\ (ox)] \rightarrow PQ^{-\cdot} + [2Fe\text{-}2S\ red)] + 2H^+$$
$$[\text{p-side; } 2e^- \text{oxidation of } PQH_2] \quad (2.4)$$

$$[2Fe\text{-}2S\ (red)] + cyt\ f\,(ox) \rightarrow [2Fe\text{-}2S(ox)] + cyt\ f\,(red) \quad (2.5)$$

$$PQ^{-\cdot} + heme\ b_p\ (ox) \rightarrow PQ + heme\ b_p\ (red)\ [\text{low potential chain}] \quad (2.6)$$

$$PQ^{-\cdot} + O_2 \rightarrow PQ + O_2^{-\cdot}\ [\text{superoxide anion radical}] \quad (2.7)$$

$$2\ b_p\ (red) + (ox)/c_n\ (ox) \rightarrow 2\ b_p\ (ox) + b_n\ (red)/c_n\ (red) \quad (2.8)$$

$$b_n\ (red)/c_n\ (red) + 2H^+ + PQ \rightarrow b_n\ (ox)/c_n\ (ox) + PQH_2$$
$$[\text{n-side; } 2e^- \text{ reduction of PQ}] \quad (2.9)$$

$$2\ Fd\ (\text{red; from cyclic } e^- \text{ transport}) + FNR\ (ox) \rightarrow 2\ Fd\ (ox) +$$
$$FNR\ (red) \quad (2.10)$$

$$FNR\ (red) + b_n\ (ox)/c_n\ (ox) \rightarrow FNR\ (ox) + b_n\ (red)/c_n\ (red) \quad (2.11)$$

$$b_n\ (red)/c_n(red) + 2H^+ + PQ \rightarrow 2b_n(ox)/c_n(ox) + PQH_2 \quad (2.12)$$

51

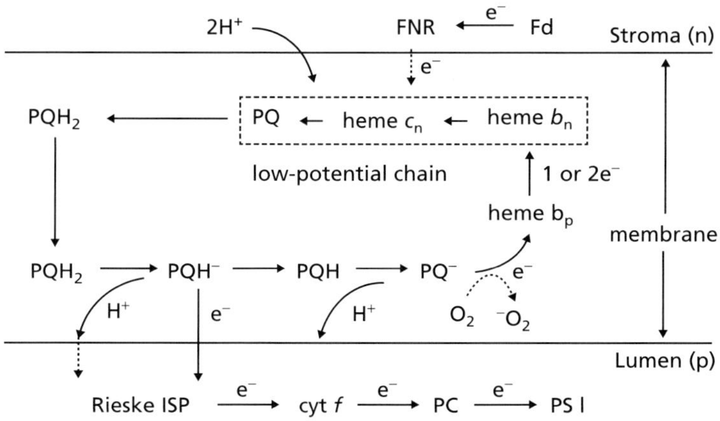

Figure 2.18 Model for the pathway of electron and proton transfer in the b_6f complex based on the Q cycle developed for the *bc*1 complex. Major differences in the b_6f complex are: the presence of the additional heme, *cn*, and electron input into b_6f through ferredoxin (Fd) and ferredoxin:NADP-reductase (FNR). PQH_2, dihydroplastoquinone; PQ, plastoquinone; PC, plastocyanin; ISP, iron-sulfur protein; Fd, ferredoxin; FNR, ferredoxin:NADP-reductase. [79]

The terminal electron acceptor on the b_6f complex is cytochrome f, which is oxidized by plastocyanin.

2.5.3 Photosystem I (PSI)

Structure

PSI is one of the most fascinating membrane protein complexes (**Figure 2.19**). [87] It comprises:

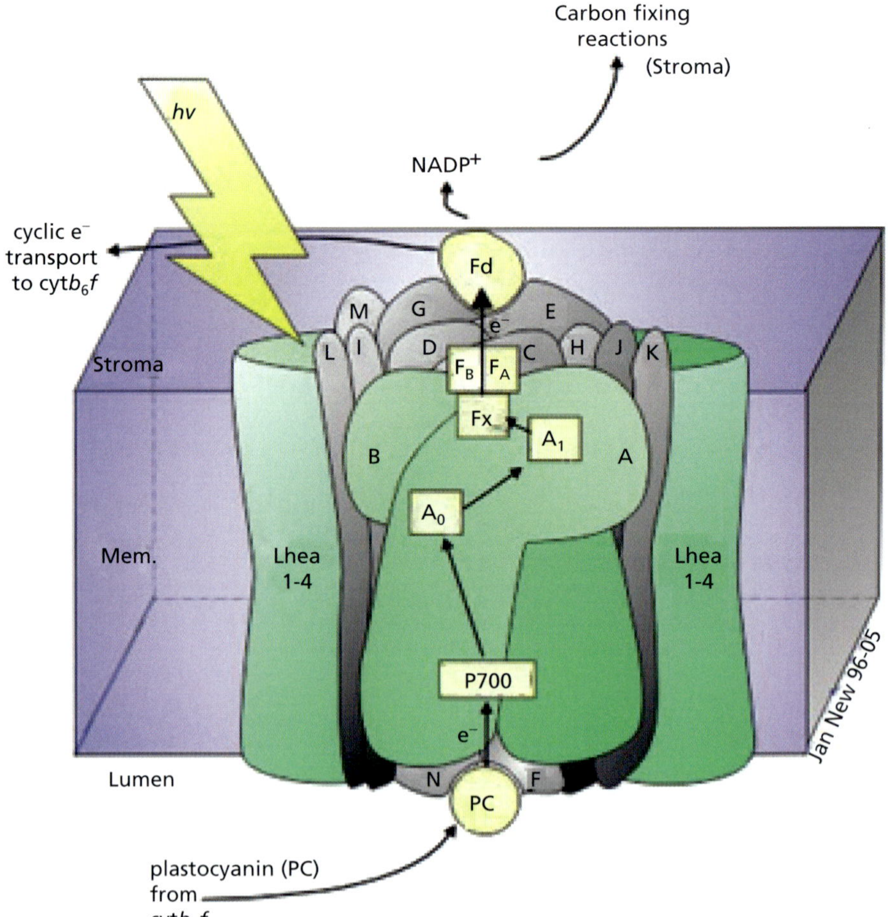

Figure 2.19 PSI structure and function. P700, chlorophyll *a* dimer; A_0, special molecule of chlorophyll; A_1, phylloquinone; Lhca1-4, PSI LHC proteins1-4; Fx, F_A, F_B, iron-sulfur center; Fd, ferredoxin; A-M, reaction center proteins. Reproduced with the permission of Professor James Barber. [89]

- A large antenna system (LHCI), consisting of chlorophyll *a*, chlorophyll *b* and carotenoids
- A reaction center which contains a chlorophyll *a* dimer called P700 (so named because its absorption spectrum peaks at 700 nm), a molecule (A) specialized in the capture of the P700 electron and different iron-sulfur centers, which play the role of electron transporters to the final PSI acceptor, ferredoxin (Fd). [88]

Its structure in a cyanobacterium has been completely elucidated. [90, 87] It probably closely resembles that of plants as well. In the thermophilic cyanobacterium *Thermosynechococcus elongatus*, PSI consists of 12 protein subunits. [91] A large number of cofactors (~100 chlorophyll *a* molecules, ~20 carotenoid molecules, 2 phylloquinone molecules and 3 Fe_4S_4 clusters) are coordinated with the protein subunits. In vivo, cyanobacterial PSI exists as a trimer consisting of 36 proteins to which 381 cofactors are non-covalently attached (**Figure 2.20**). [87]

PSI generates the most negative redox potential in nature (- 1V; **Figure 2.21**), and thus largely determines the global amount of enthalpy in living systems. Plant PSI is composed of a reaction center containing 12 subunits, termed each Psa followed by a capital letter ("Psa" being used for photosystem I reaction center components [92]), and a four subunit membranal light-harvesting antenna (LHCI). LHCI is exclusively found in land plants and green algae and consists of four different but homologous chlorophyll a/b binding proteins, termed Lhca1-4 (PSI LHC proteins1-4). [93]

The structure of plant PSI from pea at 3.4 Å resolution reveals 17 protein subunits, 168 chlorophylls, 2 phylloquinones, 3 Fe_4S_4 clusters and 5 carotenoids. [94]

Functional organization of photosystem I

The path taken by the electrons starts with the absorption of light by an antenna pigment and transfer of the energy to a P700 chlorophyll *a* at the PSI reaction center. [74] Absorption of energy by P700 causes the excitation of an electron and its transfer to A_0 (special chlorophyll *a* molecule), which is the primary electron acceptor of PSI. [95] The electron subsequently passes to A_1 (phylloquinone molecule) and then to an iron-sulfur center named F_x. From F_x the electron is transferred through two more iron-sulfur centers (F_A and F_B), which are bound by a peripheral protein at the stromal side of the membrane. The electron is finally transferred to ferredoxin (Fd), a small iron-sulfur protein that is external to the PSI complex on the stromal surface of PSI. The reduction of $NADP^+$ to form NADPH is catalyzed by a large enzyme called ferredoxin-$NADP^+$ reductase (FNR), which contains an FAD (flavin adenine dinucleotide) prosthetic group capable of accepting and transferring two electrons. An individual ferredoxin molecule can donate only one electron, so that two ferredoxins act together in the reduction (**Equation 2.13**):

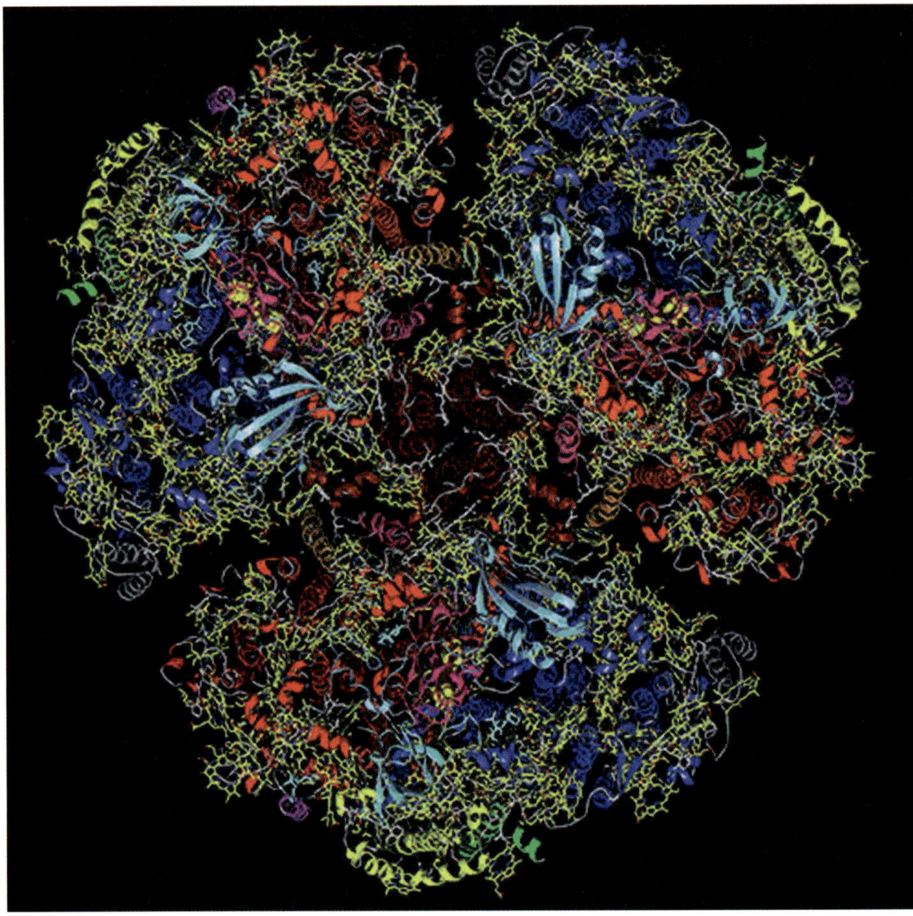

Figure 2.20 X-ray crystal structure of PSI trimer in the cyanobacterium *Thermosynechococcus elongatus* at 2.5 Å resolution. Reproduced by permission of Elsevier. [1, 91, 87]

$$2\ \mathrm{Fd_{red}} + \mathrm{H^+} + \mathrm{NADP^+} \xrightarrow{FNR} 2\ \mathrm{Fd_{ox}} + \mathrm{NADPH} \tag{2.13}$$

The removal of a proton from the stroma also adds to the proton gradient across the thylakoid membrane. The electron-deficient reaction-center pigment ($P700^+$) is reduced by an electron donated by plastocyanin. The overall reaction for PSI can be written as (**Equation 2.14**):

$$4\ \mathrm{e^-} + 2\ \mathrm{H^+} + 2\ \mathrm{NADP^+} \xrightarrow{4\,photons} 2\ \mathrm{NADPH} \tag{2.14}$$

The electron transport pathway from H_2O to $NADP^+$ is usually called the ***Z-scheme*** (**Figure 2.21**).

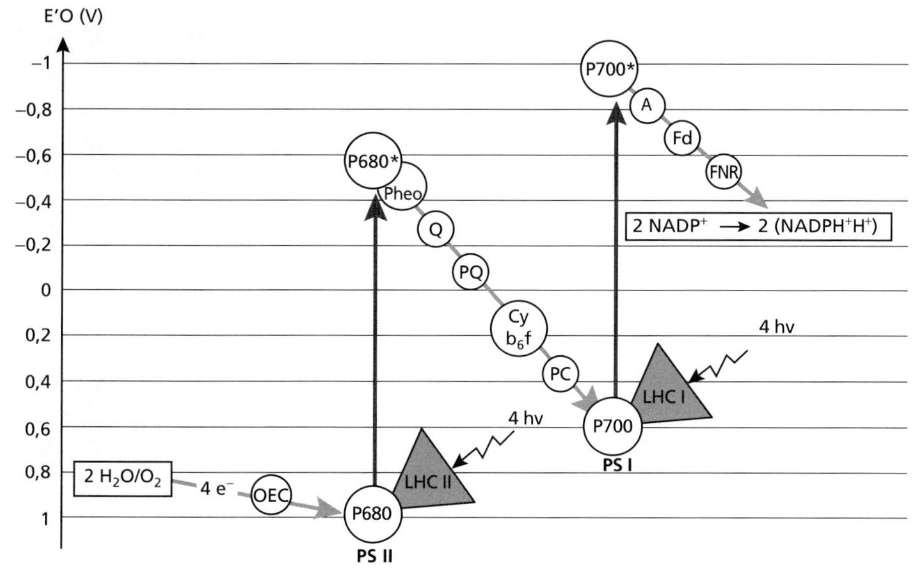

Figure 2.21 Z-scheme, acyclic electron transfer pathway. Drawn based on diagrams by Biologie et Multimédia – Université Pierre et Marie Curie – UFR des Sciences de la Vie. [96]

Whenever molecules gain or lose energy is involved in the scheme. [97] The Z-scheme is an energy diagram for electron transfer in the light reactions. It applies equally well to photosynthesis by plants, algae and cyanobacteria. The vertical energy scale shows each molecule's ability to transfer an electron to (i.e. to reduce) the next one from left to right. The molecules at the top can transfer electrons easily to the molecules below them. However, for electron transfer from those at the bottom to those above them, the reaction requires input of outside energy. Using this pathway, photosynthetic organisms transform light into electrical energy (electron flow) and hence into chemical energy as reduced NADPH and ATP. Although it is not a structural diagram, the Z-scheme gives the sequence of electron flow (oxidation and reduction). It is also called noncyclic ***photophosphorylation*** because it produces ATP in a one-way process (unlike cyclic photophosphorylation and pseudocyclic phosphorylation). [98]

The energy released as electrons pass down the gradient between PSII and PSI is harnessed by the cytochrome b_6f complex to pump protons against their concentration gradient from the stroma of the chloroplast into the interior of the thylakoid (lumen). [99] As their concentration increases inside, a strong diffusion gradient is set up.

In cyclic photophosphorylation, the electrons expelled by the energy of light absorbed by PSI pass, as normal, to ferredoxin (**Figure 2.22**).

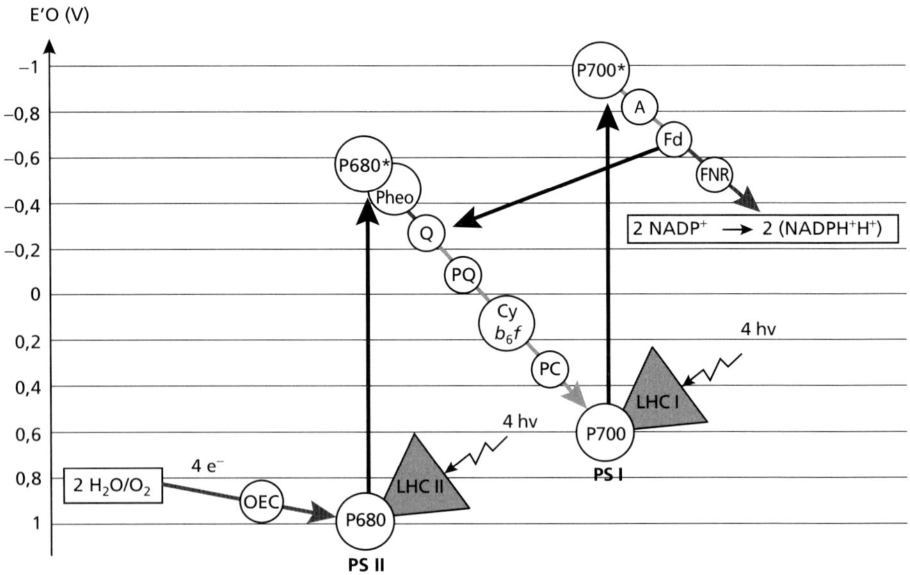

Figure 2.22 Cyclic photophosphorylation. Drawn based on diagrams from Biologie et Multimédia – Université Pierre et Marie Curie – UFR des Sciences de la Vie. [96]

But instead of going to make NADPH, they pass to plastoquinone and on back into the cytochrome b_6f complex. Here, the energy each electron liberates pumps two protons into the interior of the membrane. PSII does not take part in electron transfer. Therefore, there is no oxidation of water, neither reduction of $NADP^+$. This process is truly cyclic because no outside source of electrons is required.

2.5.4 ATP synthase

Structure

In eubacteria, chloroplasts and mitochondria, the synthesis of ATP (adenosine-5′-triphosphate; **Figure 2.23**) is carried out by a highly complex machine known as ATP synthase (or F-type ATPase or F-ATPase). [100]

F-ATPases are multisubunit membrane-bound complexes that couple the transmembrane proton motive force to the synthesis of ATP from ADP and orthophosphate (P_i) by a rotary mechanism (**Figure 2.24**). [101] They consist of two main structural domains, a globular catalytic domain known as F_1 and a membrane domain known as F_0, linked together by a central stalk and a peripheral stalk (or stator, the stationary part). During ATP synthesis, the central rotor turns about 150 times every second.

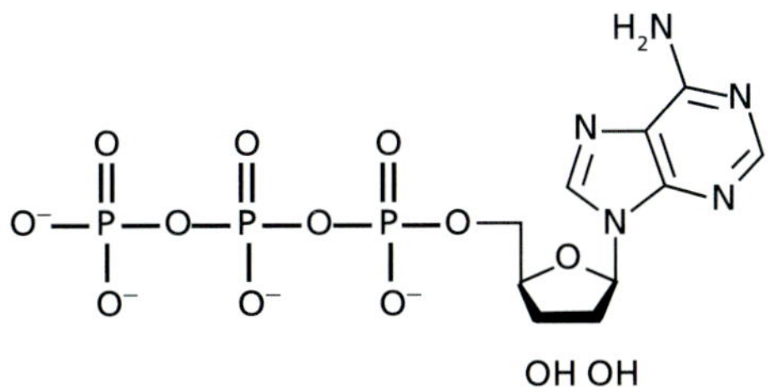

Figure 2.23 Adenosine-5′-triphosphate (ATP).

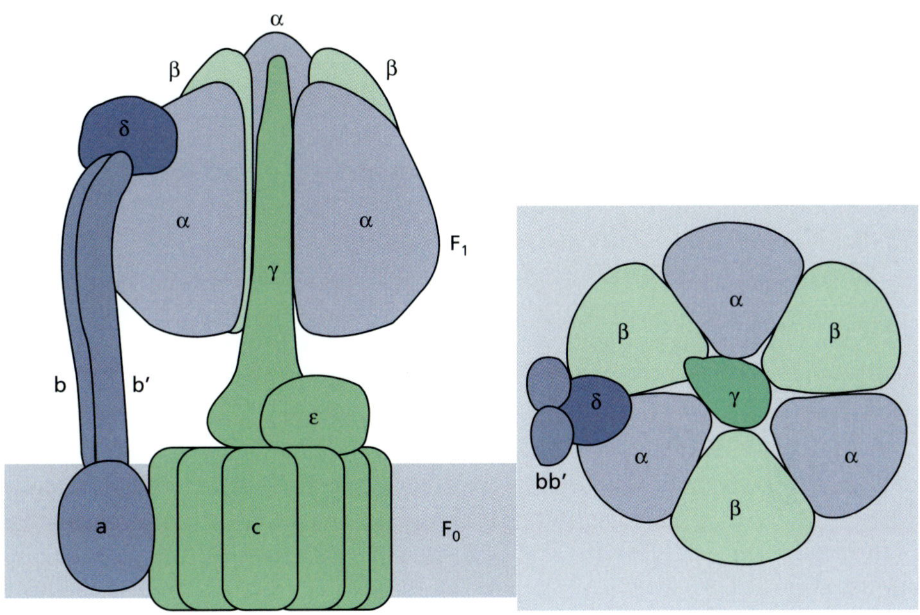

Figure 2.24 Structure of ATP synthase. Side view (left) and cross-section (right). Drawn based on diagrams by Biologie et Multimédia – Université Pierre et Marie Curie – UFR des Sciences de la Vie. [102]

The hydrophilic F_1 domain (which protrudes into the stroma and possesses the catalytic activity) comprises of five polypeptides ($\alpha, \beta, \gamma, \delta, \varepsilon$) in a stoichiometry $\alpha_3\beta_3\gamma\delta\varepsilon$. The characteristics of F_1 are:

- The α and β subunits are homologous to each other. Looking down to the membrane, they alternate around a ring.
- There are three nucleotide-binding catalytic sites, located at $\alpha\beta$ interfaces but predominantly involving residues of the β subunits.

- Each α subunit contains an ATP-binding site, but this site is inactive in catalysis.
- Mg^{2+} binds with the adenine nucleotides in both α and β subunits.

The membrane F_0 domain is made up of three hydrophobic polypeptides (subunits a, b, and c) in a stoichiometry ab_2c_{10}. F_0 forms the transmembrane pore or channel through which protons move down their gradient to drive ATP synthesis via a process called ***chemiosmosis***. [103]

Functional organization

The binding change mechanism for ATP synthesis has been proposed by Paul Boyer. He shared the Nobel Prize for this model, which accounts for three catalytic sites in F_1 (**Figure 2.25**). [104]

It is proposed that an irregularly shaped shaft linked to F_0 rotates relative to the three β subunits, which are arranged in a ring. The rotation is driven by the flow of protons through F_0. The conformation of each β subunit changes sequentially, as it interacts with the rotating shaft. Each of the three β subunits is in a different stage of the catalytic cycle at any time.

This model is supported by two major lines of evidence:

At first, the crystal structure of F_1 with the central stalk has been determined by John Walker, [106] who shared the Nobel Prize for that achievement. [107, 108] The γ subunit was found to include a bent helical loop that constitutes a shaft within the ring of α and β subunits. The three β subunits were found to differ in conformation and bound ligand: a first β subunit binds to a non-hydrolyzable ATP analog (assumed to be the tight conformation); a second β subunit binds to a molecule of ADP (assumed to be the loose conformation); the third β subunit has an empty active site (assumed to be the open conformation).

Secondly, the rotation of the γ shaft relative to the ring of α and β subunits was demonstrated by Noji et al. in 1987. [109] Their experiment provided evidence that ATP synthase, only 10 nm across, is the smallest known rotary

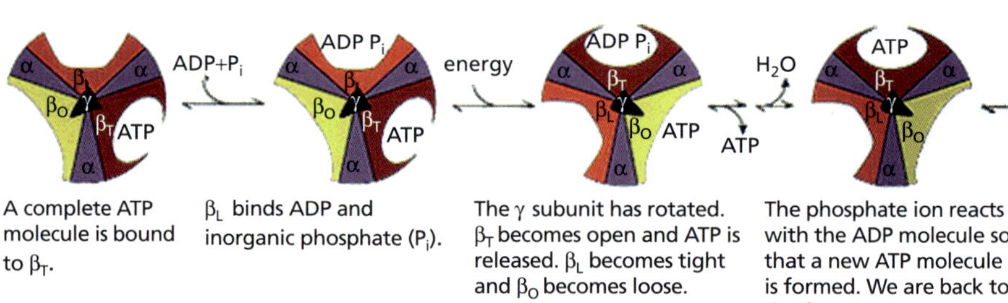

Figure 2.25 Binding change mechanism. Reproduced by permission of The Royal Swedish Academy of Sciences. [105]

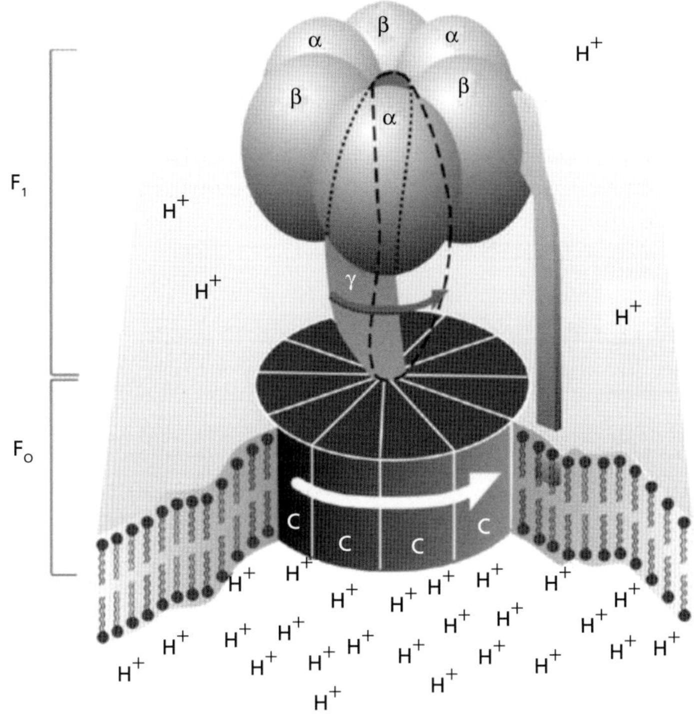

Figure 2.26 Simplified diagram of ATP synthase. The γ subunit rotates inside the cylinder formed by the six α and β subunits. Since the γ subunit is asymmetrical it compels the β subunits to undergo structural changes. This leads to the β subunits binding ATP and ADP with different strengths. Reproduced by permission of The Royal Swedish Academy of Sciences. [111]

motor spinning in all our cells 8000 rpm and burning the ATP fuel with nearly 100% efficiency. [110]

A simplified structure and function of ATP synthase is shown in **Figure 2.26**.

The F_0 domain, through which protons move, is located in the membrane, whereas the F_1 domain, which synthesizes ATP, is outside the membrane. When protons flow through the membrane via the disc of c subunits in the F_0 domain, the disc is forced to twist. The γ subunit is attached to the disc and consequently rotates with it. However, the six α and β subunits cannot rotate because they are locked in a fixed position by the b subunit, which is anchored in the membrane. Therefore, the γ subunit it rotates inside the cylinder formed by the six α and β subunits. Because the γ subunit is asymmetrical, it forces the β subunits to undergo structural changes leading to the β subunits binding ATP and ADP with different strengths. [112]

The overall process of the photosynthetic electron transport chain in chloroplasts is (**Equation 2.15**):

$$H_2O \rightarrow PSII \rightarrow plastoquinone \rightarrow cytb_6f \rightarrow plastocyanin \rightarrow PSI \rightarrow NADPH \quad (2.15)$$

The overall result of all light reactions is (**Equation 2.16**):

$$2\,H_2O + 2\,NADP^+ + 3\,ADP + 3P_i + 8\,photons \rightarrow O_2 + 2\,NADPH + 3\,ATP \quad (2.16)$$

2.6 Dark reactions

The dark reactions take place in the stroma of the chloroplast, and convert CO_2 to sugar. These reactions do not need light to occur, but need ATP and NADPH from the light-dependent stages. They involve a cycle called the ***Calvin cycle*** (or Calvin-Benson cycle) in which ATP and NADPH are used to make carbohydrates from CO_2 (**Figure 2.27**). [113] The incorporation of CO_2 into organic compounds is known as ***carbon fixation***.

The cycle comprises three steps:

1. **Carbon fixation.** A molecule of CO_2 combines with the 5-carbon compound ribulose 1,5-biphosphate (RuBP) to form a 6-carbon intermediate sugar via an enzyme called ***RuBisCO*** (ribulose-1,5-bisphosphate carboxylase/oxygenase). [23, 43, 114] The intermediate is unstable and immediately splits into two 3-carbon molecules of 3-phosphoglycerate (PGA). The reaction is shown in **Figure 2.28** and, in an overall form, in **Equation 2.17**.

$$6\,RuBP + 6\,CO_2 \rightarrow 12\,PGA \quad (2.17)$$

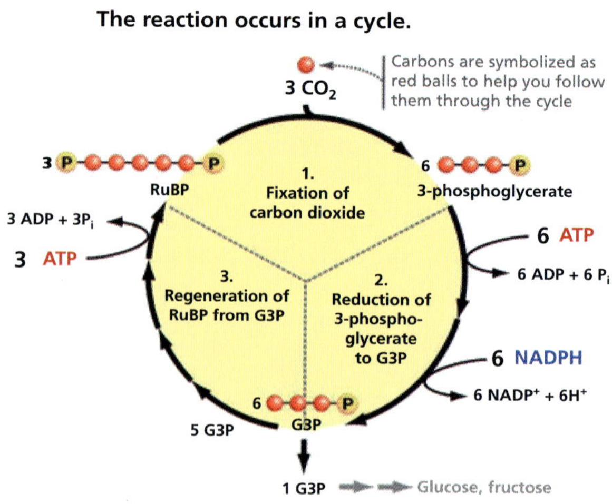

Figure 2.27 Calvin cycle. The cycle includes three phases: (1) fixation of CO_2 into 3-phosphoglycerate (PGA), (2) reduction of PGA to 3-phosphoglyceraldehyde (G3P), and (3) regeneration of ribulose-1,5-bisphosphate (RuBP) from G3P. Reproduced by permission of Pearson publishing [23]

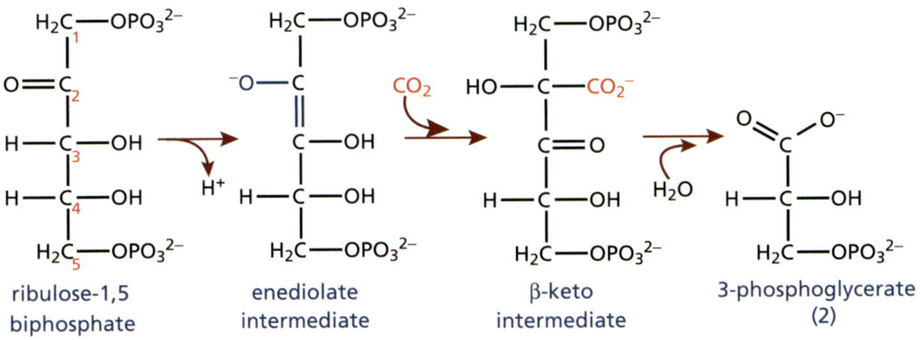

Figure 2.28 Combination of one molecule of RuBP with one molecule of CO_2 to form 2 molecules of PGA [115, 114]

The only enzyme known to break down CO_2 is RuBisCO (**Figure 2.29**). RuBisCO represents the beginning of the food chain. [116] Rubisco makes up 20-50 % of proteins in chloroplasts. [2] It acts very slowly, catalyzing three molecules per second. This compares to 1000 per second for typical enzymatic reactions. Large quantities are needed to compensate for its low speed. It may be the most abundant protein on earth.

In some bacteria, RuBisCO is composed of two identical subunits. In higher photosynthetic organisms, the enzyme is a hexadecamer consisting of eight large (L, 477 amino acids) catalytic and eight small (S, 123 amino acids) subunits in an L8S8 arrangement. [118]

2. **Carbon reduction.** PGA is reduced to glyceraldehyde 3-phosphate (G3P) in two steps. In the process, ATP and NADPH are used. From 6 CO_2, 12 G3P are obtained. Two of the 12 G3P are siphoned off from the cycle and used to make glucose or fructose (thus 6 carbons leave to make $C_6H_{12}O_6$), and the remaining 10 are used to regenerate RuBP. The overall reaction is shown in **Equation 2.18**

$$\begin{aligned}&12\ \text{PGA} + 12\ \text{ATP} + 12\ \text{NADPH} \rightarrow \\ &\quad 10\ \text{G3P} + 12\ \text{ADP} + 12\ \text{NADP}^+ + C_6H_{12}O_6\end{aligned} \tag{2.18}$$

3. **RuBP regeneration.** The 10 G3P are rearranged to make 6 RuBP. ATP is consumed for each RuBP formed (**Equation 2.19**). [119]

$$10\ \text{G3P} + 6\ \text{ATP} \rightarrow 6\ \text{RuBP} + 6\ \text{ADP} \tag{2.19}$$

Therefore, the overall reaction in the cycle (**Equation 2.20**) is:

$$\begin{aligned}&6\ CO_2 + 12\ \text{NADPH} + 18\ \text{ATP} + 12\ H^+ + 18\ H_2O \rightarrow \\ &\quad C_6H_{12}O_6 + 12\ \text{NADP}^+ + 18\ \text{ADP} + 18\ P_i + 6\ H_2O\end{aligned} \tag{2.20}$$

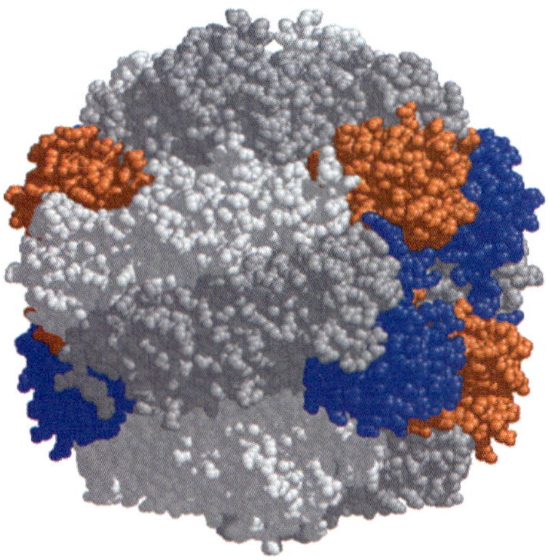

Figure 2.29 Spacefilling view of RuBisCO showing the arrangement of the large chain dimers (white and grey) and the small chains (blue and orange) [117]

This consumes all the products of the light reactions except O_2, and regenerates much of the reactants for the light reactions, thus giving the overall result for photosynthesis (**Equation 2.21**):

$$12\ H_2O + 6\ CO_2 + 48\ \text{photons} \rightarrow 6\ O_2 + C_6H_{12}O_6 + 6\ H_2O \qquad (2.21)$$

2.7 C3, C4 and CAM plants

2.7.1 Photorespiration

O_2 can compete with CO_2 for binding to RuBisCO, especially when CO_2 is low and O_2 is high. [115, 114] Reaction of O_2 with RuBP yields one molecule of PGA plus the 2-carbon compound 2-phosphoglycolate (**Figure 2.30, Equation 2.22**)

$$O_2 + \text{RuBP} \rightarrow 1\ \text{PGA} + 1\ \text{phosphoglycolate} \qquad (2.22)$$

This reaction, called ***photorespiration***, is the basis for the name RuBP Carboxylase/Oxygenase (RuBisCO).

The complex pathway that partly salvages carbon from 2-phosphoglycolate, via conversion to PGA, involves enzymes of chloroplasts, peroxisomes and mitochondria. This pathway recovers three quarters of the carbon from 2-phosphoglycolate as PGA while the rest is released as CO_2.

Photorespiration is a wasteful process, substantially reducing efficiency of carbon fixation. Major efforts have been made to modify the properties of

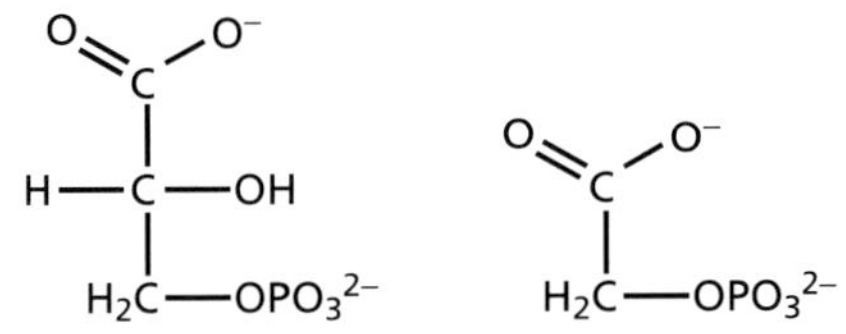

3-phosphoglycerate Phosphoglycolate

Figure 2.30 Products of photorespiration: 3-phosphoglycerate (PGA) and phosphoglycolate [115]

RuBisCO to eliminate the oxygenation reaction, especially using molecular genetics. [23] All the efforts so far indicate that the two reactions cannot be separated. Modifications in RuBisCO that reduce the oxygenase activity also reduce the carboxylase activity. Nature, however, has elaborated a system to avoid photorespiration, called the C4 photosynthetic pathway.

2.7.2 C3 vs C4 plants

In most plants, CO_2 is initially fixed via RuBisCo of the Calvin cycle into RuBP (also referred to as C3 pathway). [120] Because the product of this reaction is a C3 compound, such plants are called ***C3 plants***. Rice, wheat, and soybeans are among the C3 plants that are important in agriculture.

Some plants have developed a preliminary step to the Calvin cycle. [121] This preliminary step is known as the C4 pathway and the plants that fix carbon through this pathway are designated **C4 plants**. While most C-fixation begins with RuBP, the C4 pathway begins with a new molecule, phosphoenolpyruvate (PEP), a 3-carbon compound that is converted into oxaloacetic acid (OAA, a 4-carbon compound) when CO_2 is combined with PEP (**Figure 2.31**).

The OAA is converted into malic acid and then transported from the mesophyll cell into the ***bundle-sheath*** cell, where OAA is broken down into PEP and CO_2. The CO_2 then enters the Calvin cycle, with PEP returning to the mesophyll cell. The resulting sugars are now adjacent to the leaf veins and can be easily transported throughout the plant.

The capture of CO_2 by PEP is mediated by the enzyme PEP carboxylase, which has a stronger affinity for CO_2 than RuBisCo. When CO_2 levels decline below the threshold for RuBisCo, RuBP is catalyzed with O_2 instead of CO_2. C4 plants, which often grow close together, have had to adjust to decreased levels of CO_2 by artificially raising the CO_2 concentration in certain cells to prevent photorespiration. C4 plants evolved in the tropics and are adapted to higher temperatures than the C3 plants found at higher latitudes. Common C4 plants include corn, sugarcane, hemp, sorghum and crabgrass. [122]

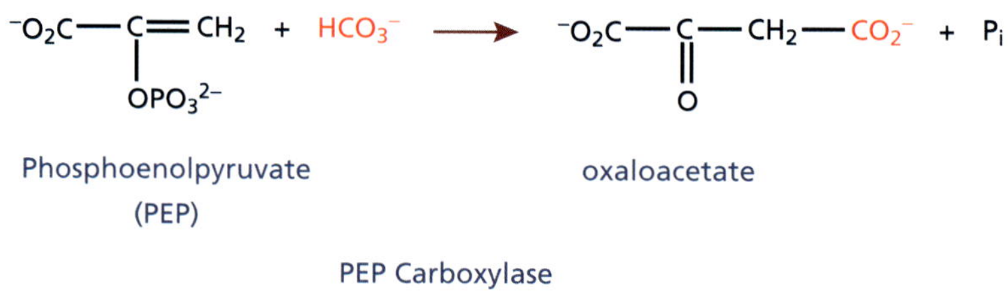

Figure 2.31 C4 plants: from 3-carbon PEP to 4-carbon OAA. [115]

The leaves of C4 plants are structured differently than those of C3 plants. [2] Photosynthesis occurs within the mesophyll cells in C3 plants, which form a dense layer on the upper surface of the leaf and a spongy layer on the lower surface (see **Figure 2.3**). Bundle-sheath cells surrounding the veins are not photosynthetic (**Figure 2.32**)

Dense layers of mesophyll cells surround the bundle-sheath cells of C4 plants. [2] Both the bundle-sheath cells and the rings of mesophyll cells are photosynthetic (**Figure 2.33**).

C4 plants have a pathway that allows them to fix CO_2 at lower concentration than C3 plants. [2] This gives them an advantage during hot, dry conditions because they can keep the stomata closed for longer periods of time to prevent dehydration before photorespirations occurs. The Calvin cycle occurs in the mesophyll cells of C3 plants but only in the bundle-sheath cells of C4 plants. The CO_2 concentration of the bundle-sheath cells of C4 plants is kept high because the mesophyll cells that surround them move CO_2 into the bundle sheath. Photorespiration in C4 plants is reduced when the stomata are closed because CO_2 movement maintains a higher CO_2 concentration in the bundle sheath than would otherwise be possible.

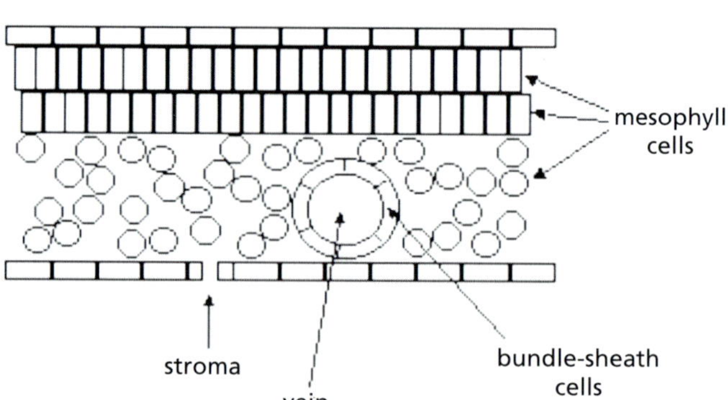

Figure 2.32 A cross-section of a C3 leaf. Reproduced by permission of Michael J. Gregory. [2]

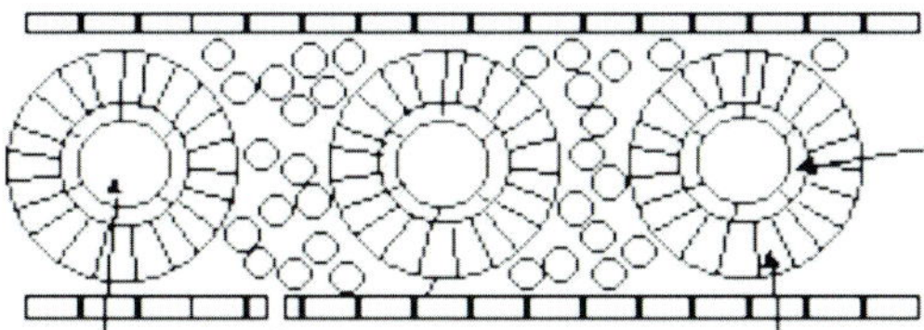

Figure 2.33 Cross-section of a C4 leaf. Reproduced by permission of Michael J. Gregory [2]

The C4 mechanism requires extra ATP but under hot, dry conditions C4 plants are two to three times more efficient than C3 plants. [2] At moderate conditions, C3 plants are at an advantage.

2.7.3 CAM plants

CAM (crassulacean acid metabolism) photosynthesis is found in most desert plants, particularly the succulents (plants that store water in thick, fleshy leaves). [2, 121] The Calvin cycle occurs in mesophyll cells of these plants but during the day when the stomata are closed. The stomata are open only at night, when it is relatively cool and humid. CO_2 fixation occurs when the stomata are open at night. CAM plants store the CO_2 mostly in the form of malic acid via carboxylation of phosphoenolpyruvate to oxaloacetate, which is then reduced to malate. CO_2 is released to the cells during the day. Thus, in contrast to C_4 metabolism, which physically separates the CO_2 fixation to PEP from the Calvin cycle, CAM temporally separates these two processes. CAM plants have a different leaf anatomy from C3 plants.

CAM plants are 5 to 7 times more efficient than C4 plants. [2]

2.7.4 Comparison of C3, C4 and CAM photosynthesis

C3, C4 and CAM photosynthesis is compared in **Table 2.1**. [123]

2.8 Photosynthetic efficiency

The photosynthetic efficiency is the fraction of light energy converted into chemical energy during photosynthesis.

Approximately 480 kilojoules (kJ) of free energy are stored in plant biomass for each mole of CO_2 fixed during photosynthesis. [124] Solar radiation striking the earth on an annual basis is equivalent to 178,000 terawatts (10^{12} W),

Table 2.1 Comparison of C3, C4 and CAM photosynthesis [123]

Property	C3 plants	C4 plants	CAM plants
Product of CO_2 fixation	3-carbon PGA	4-carbon OAA	4-carbon OAA
Chloroplasts	one type	dimorphic	one type
Theoretical energy requirement (CO_2:ATP:NADPH)	1:3:2	1:5:2	1:6.5:2-
Photosynthesis inhibited by O_2?	Yes	No	Yes
Photorespiration detectable?	Yes	Only in bundle sheath	Late afternoon
Photosynthesis rate (mg CO_2 fixed $dm^{-2}h^{-1}$)	15-30	40-80	(low)
Temperature optimum for photosynthesis (°C)	15-25	30-47	35
Dry matter production (bushels/acre)	Low (26, soybean; 30, wheat)	High (87, maize; 50, sorghum)	Low, variable

i.e. 15,000 that of current global energy consumption. Although photosynthetic energy capture is estimated to be ten times that of global annual energy consumption, only a small part of this solar radiation is used for photosynthesis. Approximately two thirds of the net global photosynthetic productivity worldwide is of terrestrial origin, while the remainder is produced mainly by phytoplankton (microalgae) in the oceans. Since biomass originates from plant and algal photosynthesis, both terrestrial plants and microalgae are appropriate targets for scientific studies relevant to biomass energy production.

Any analysis of biomass energy production must consider the potential efficiency of the processes involved. [124] Although photosynthesis is fundamental to the conversion of solar radiation into stored biomass energy, its theoretically achievable efficiency is limited both by the limited wavelength range applicable to photosynthesis, and the quantum requirements of the photosynthetic process. Only light within the wavelength range of 400 to 700 nm (photosynthetically active radiation, PAR) can be utilized by plants, effectively allowing only 45 % of total solar energy to be utilized for photosynthesis. Furthermore, fixation of one CO_2 molecule during photosynthesis, necessitates a ***quantum requirement*** of ten (or more), which results in a maximum utilization of only 25% of the PAR absorbed by the photosynthetic system (at least eight photons are required to split 2 water molecules to form 2 NADPH and

3 ATP, which can be used to fix one molecule of CO_2, i.e. ~1665 kJ of light energy are required to store ~480 kJ in the plant [125]). On the basis of these limitations, the theoretical maximum efficiency of solar energy conversion is approximately 11%. In practice, however, the magnitude of photosynthetic efficiency observed in the field, is further decreased by factors such as poor absorption of sunlight due to its reflection, respiration requirements of photosynthesis and the need for optimal solar radiation levels. The net result is an overall photosynthetic efficiency of between 3 and 6% of total solar radiation.

Recent examination of the potential yields has revealed that the maximum conversion efficiency of solar energy to biomass is 4.6 % for C3 photosynthesis at 30 °C and today's 380 ppm atmospheric CO_2, but 6 % for C4 photosynthesis. [126] This advantage over C3 will disappear as atmospheric CO_2 nears 700 ppm.

Intriguingly, it has been documented that light-absorbing molecules in some photosynthetic proteins capture and transfer energy according to quantum-mechanical probability laws instead of classical laws up to 180 °K. [127, 128, 129]

This contrasts with the long-held view that long-range quantum coherence between molecules cannot be sustained in complex biological systems, even at low temperatures. [127] Recent observations on light-harvesting proteins isolated from marine cryptophyte algae suggest that distant molecules within the photosynthetic proteins are wired together by quantum coherence for more efficient light-harvesting even at ambient temperature (**Figures 2.34** and **2.35**).

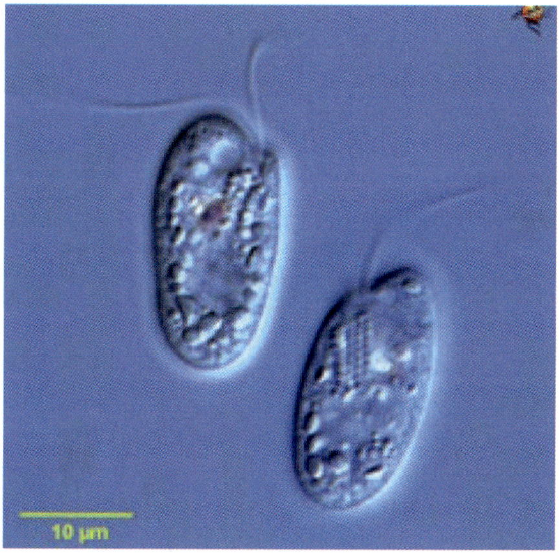

Figure 2.34 Marine microalgae, like the *Chroomonas* pictured here, employ quantum coherence to boost the efficiency of photosynthesis. Reproduced by permission of Provasoli-Guillard NCMA. [128]

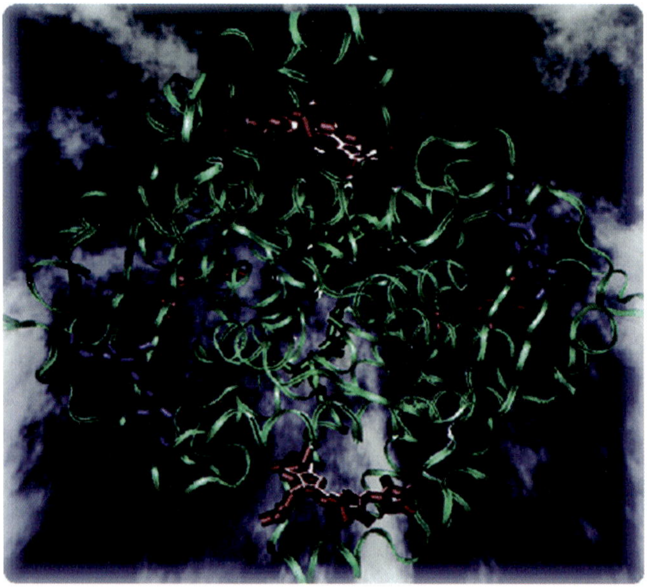

Figure 2.35 Some light-harvesting proteins use quantum coherence in energy transfer. Reproduced with permission from G.D. SCHOLES, *J. Phys. Chem. Lett. 1, 2*, 2010. Copyright (2012) American Chemical Society. [129]

Quantum coherence describes how more than one molecule interacts with the same energy from one incoming photon at the same time. Rather than the energy from a particular photon choosing one route to pass through the photosynthetic system, it travels through multiple channels simultaneously, allowing it to pick the quickest route.

2.9 References

[1] D. Boudreau, *Shedding Light on Photosynthesis*, Arizona State University, 2007 in http://researchstories.asu.edu/stories/shedding-light-photosynthesis-922

[2] M. Gregory, *The Biology Web, Photosynthesis*, Courses, General Biology 1, Clinton Community College, in http://faculty.clintoncc.suny.edu/faculty/michael.gregory/files/bio%20101/bio%20101%20lectures/photosynthesis/photosyn.htm

[3] T. Powell, *Photosynthesis, Illuminating Alternatives to Solar Energy Research*, Berkeley Science Review, 2009 in http://sciencereview.berkeley.edu/articles.php?issue=16&article=features_01_photosynthesis

[4] http://en.wikipedia.org/wiki/Photosynthesis

[5] Institute of Science in Society, *La récupération de l'énergie du soleil avec la photosynthèse artificielle*, 2009 in http://www.i-sis.org.uk/ArtificialPhotosynthesisFR.php

[6] M. McGinley and D. Swarthout, *Photosynthesis*, Encyclopedia of Earth, 2010 in http://www.eoearth.org/article/Photosynthesis

[7] Science Fair Projects and Experiments, *Photosynthesis, K12 Experiments & Background* Information, 2010 in http://www.juliantrubin.com/encyclopedia/botany/photosynthesis.html

[8] NASA Goddard Space Flight Center, Sea-viewing Wide Field-of-view Sensor (SeaWIFS) Project, in http://oceancolor.gsfc.nasa.gov/SeaWiFS/BACKGROUND/Gallery/biosphere.jpg

[9] http://chemwiki.ucdavis.edu/Biological_Chemistry/Photosynthesis/Photosynthesis_overview

[10] New World Encyclopedia Contributors, 2010, in http://www.newworldencyclopedia.org/entry/Photosynthesis

[11] G.E. Kaiser, 2001 in http://student.ccbcmd.edu/biotutorials/photosyn/ldr.html

[12] G.E. Kaiser, 2001 in http://student.ccbcmd.edu/biotutorials/photosyn/lindr.html

[13] South Carolina Department of Natural Resources, Biology 205, *Photosynthesis* in http://www.dnr.sc.gov/ael/personals/pjpb/lecture/lecture.html

[14] J.W. Kimball, Research Coordination Networks, Biology, *The Leaf*, 2010 in http://users.rcn.com/jkimball.ma.ultranet/BiologyPages/L/Leaf.html

[15] M.J. Farabee, Estrella Mountain Community College, Maricopa, 2010 in http://www.emc.maricopa.edu/faculty/farabee/biobk/biobookps.html

[16] http://en.wikipedia.org/wiki/Plant_cell

[17] F.B. Sampson and G. Marie Taylor, *Tuatara 11*, issue 3, 1963 in http://www.nzetc.org/tm/scholarly/tei-Bio11Tuat03-t1-body-d1.html

[18] SouthEastern Ohio Regional Freenet, http://www.seorf.ohiou.edu/~tstork/compass.rose/photosynthesis/Chloroplasts2.html

[19] S. Di Fiore, Q. Li, M.J. Leech, F. Schuster, N. Emans, R. Fischer and S. SCHILLBERG, *Plant Physiol. 129*, 1160, 2002 in http://www.plantphysiol.org/cgi/content/full/129/3/1160

[20] L. Bogorad, *Trends Biotechnol. 18*, 257, 2000 in http://www.ncbi.nlm.nih.gov/pubmed/10802561?dopt=Abstract

[21] http://en.wikipedia.org/wiki/Chloroplast

[22] H. Lodish, A. Berk, S.L. Zipurski, et al., *Molecular Cell Biology*, 4th Ed. Section 16.3 *Photosynthetic Sages and Light-Absorbing Pigments*, US National Library of Medicine, National Institutes of Health, 2000 in http://www.ncbi.nlm.nih.gov/bookshelf/br.fcgi?book=mcb&part=A4434

[23] University of Illinois at Chicago, *Photosynthesis* in http://www.uic.edu/classes/bios/bios100/lectures/ps01.htm

[24] School of Chemistry, University of Bristol, UK, Molecule of the Month, *Thylakoid membranes* in http://www.chm.bris.ac.uk/motm/oec/motmc.htm giving access to http://www.chm.bris.ac.uk/motm/oec/images/thkmemb.jpg

[25] http://en.wikipedia.org/wiki/Thylakoid

[26] C. BRAUCHLE, Lüdwig-Maximilians-Universität München, 2006 in http://www2.phys.chemie.uni-muenchen.de/braeuchle/index.php?id=12

[27] http://www.uniprot.org/keywords/42

[28] Theoretical and Computational Biophysics Group, University of Illinois at Urbana-Champaign, *Quantum Biology of the PSU*, 2006 in http://www.ks.uiuc.edu/Research/psu/psu.html

[29] J. KOEPKE, K. HU, C. MUENKE, K. SCHULTEN and H. MICHEL, Structure 4, 581, 1996 in http://www.ks.uiuc.edu/Publications/Papers/PDF/KOEP96/KOEP96.pdf

[30] Theoretical and Computational Biophysics Group, University of Illinois at Urbana-Champaign, Light Harvesting Complex-II, 2006 in http://www.ks.uiuc.edu/Research/bio_ener/LH_2

[31] X. HU, T. RITZ, A. DAMJANOVIC, F. AUTENRIETH and K. SCHULTEN, *Quarterly Reviews of Biophysics 35*, 1, 2002 in http://www.ks.uiuc.edu/Publications/Papers/PDF/HU2002/HU2002.pdf

[32] University of Illinois at Urbana-Champaign, Theoretical and Computational Biophysics Group, *Light Harvesting Complex II* in http://www.ks.uiuc.edu/Research/bio_ener/LH_2

[33] P. FROMME., *Photosynthetic Protein Complexes: A Structural Approach*, Wiley-VCH Verlag Gmbh, 2008 in http://www.amazon.fr/Photosynthetic-Protein-Complexes-Structural-Approach/dp/3527317309

[34] Z. LIU, H. YAN, K. WANG, T. KUANG, J. ZHANG, L. GUI, X. AN and W. CHANG, Nature 428, 287, 2004 in http://www.nature.com/nature/journal/v428/n6980/full/nature02373.html

[35] W. KUHLBRANDT, D.N. WANG and Y. FUJIYOSHI, *Nature 367*, 614, 1994 in http://www.nature.com/nature/journal/v367/n6464/abs/367614a0.html

[36] J. STANDFUSS and W. KUHLBRANDT, *J. Biol. Chem. 279*, 36884, 2004 in http://www.jbc.org/content/279/35/36884.full

[37] V.H.R. SCHMID, *Cell. Mol. Life Sci. 65*, 3619, 2008 in http://www.springerlink.com/content/4530814t3724r640

[38] J.W. KIMBALL, Research Coordination Networks, Biology 2009 in http://users.rcn.com/jkimball.ma.ultranet/BiologyPages/L/LightReactions.html

[39] H. TERAMOTO, T. ONO and J. MINAGAWA, *Plant Cell Physiol. 42*, 849, 2001 in http://pcp.oxfordjournals.org/content/42/8/849.abstract

[40] Wolfson Laboratories, Department of Biological Sciences, Imperial College London, 2003 in http://www.bio.ic.ac.uk/research/barber/PSIIreview/greenplants_start.html

[41] J. BARBER, *Photosystem II: an enzyme of global significance*, Novartis Medal Lecture, *Biochem. Soc. Trans. 34*, 619, 2006 in http://www.biochemsoctrans.org/bst/034/bst0340619.htm

[42] P. PREUSS, , *Spinach, or the Search for the Secret of Life as We Know It*, Science Beat, Berkeley Lab 2004 in http://www.lbl.gov/Science-Articles/Archive/sb/July-2004/2_spinach.html

[43] L. ORR and GOVINDJEE, *Photosynthesis and the Web: 2008*, Arizona State University, University of Illinois at Urbana-Champaign, 2007 in http://www.life.illinois.edu/govindjee/photoweb/subjects.html

[44] J. BARBER, *Inorg. Chem 47*, 1700, 2008 in http://www.cup.uni-muenchen.de/ac/kluefers/homepage/L/bac/oec_2008.pdf

[45] J.D. JOHNSON, The manganese-calcium oxide cluster of Photosystem II and its assimilation by cyanobacteria, Florida State University, 2006 in http://www.chm.bris.ac.uk/motm/oec/motm.htm

[46] Y. PUSHKAR, J. YANO, P. GLATZEL, J. MESSINGER, A. LEWIS, K. SAUER, U. BERGMANN and V. YACHANDRA, *J. Biol. Chem. 282*, 7198, 2007 in http://www.jbc.org/content/282/10/7198.full

[47] http://en.wikipedia.org/wiki/Photosystem_II

[48] J. NIELD, O. KRUSE, J. RUPRECHT, P. DA FONSECA, C. BUCHEL and J. BARBER, *J. Biol. Chem 275*, 27940, 2000 in http://www.jbc.org/content/275/36/27940.full.pdf

[49] J. NIELD, C. FUNK and J. BARBER, Phil. Trans. R. Soc. Lond. B 355, 1337, 2000 in http://www.ncbi.nlm.nih.gov/pmc/articles/PMC1692865/pdf/11127988.pdf

[50] O. KRUSE, D. ZHELEVA and J. BARBER, FEBS Letters 408, 276, 1997 in http://www.uni-bielefeld.de/biologie/Algenbiotechnologie/kruse/Literatur/Kruse%20et%20al.%201997.pdf

[51] Wolfson Laboratories, Department of Biological Sciences, Imperial College London, 2003 in http://www.bio.ic.ac.uk/research/barber/psIIimages/PSII.html

[52] N. KEREN, M. LIBERTON and H.B. PAKRASI, *J. Biol. Chem. 280*, 6548, 2005 in http://www.jbc.org/content/280/8/6548.full

[53] J. BARBER, *Bioelectrochemistry 55*, 135, 2002 in http://www.ncbi.nlm.nih.gov/pubmed/11786359

[54] J.P. MCEVOY and G.W. BRUDVIG, *Biochemistry 47*, 13394, 2008 in http://www.ncbi.nlm.nih.gov/pubmed/19053286

[55] S. SHCOLNICK and N. Keren, *Plant Physiol. 141*, 805, 2006 in http://www.plantphysiol.org/cgi/content/full/141/3/805#FIG1

[56] R. TAKAHASHI, A. BOUSSAC, M. SUGIURA and T. NOGUCHI, *Biochemistry 48*, 8994, 2009 in http://pubs.acs.org/doi/abs/10.1021/bi901195e

[57] J. BARBER and J. NIELD, *Phil. Trans. R. Soc. Lond. B 57*, 1329, 2002 in http://rstb.royalsocietypublishing.org/content/357/1426/1329.full.pdf

[58] S. DE BIANCHI, L. DALL'OSTO, G. TOGNON, T. MOROSINOTTO and R. BASSI, *Plant Cell 20*, 1012, 2008 in http://www.plantcell.org/cgi/content/full/20/4/1012

[59] C. BUCHEL, J. BARBER, G. ANANYEV, S. ESHAGHI, R. WATT and C. DISMUKES, *Proc. Natl. Acad. Sci. 96*, 14288, 1999 in http://www.ncbi.nlm.nih.gov/pmc/articles/PMC24429

[60] E.P. MORRIS, B. HANKAMER, D. ZHELEVA, G. FRISO and J. BARBER, *Structure 5*, 837, 1997 in http://www.ncbi.nlm.nih.gov/pubmed/9261075

[61] G.C. PAPAGEORGIOU, *Florescence of Photosynthetic Pigments in Vitro and in Vivo*, Chlorophyll a Fluorescence, Advances in Photosynthesis and Respiration, Vol. 19, 43, Springer, 2004 in http://www.springerlink.com/content/nj1374816u301365

[62] G.C. PAPAGEORGIOU and GOVINDJEE (Eds), *Chlorophyll a fluorescence: a signature of photosynthesis*, Springer 2004 in http://www.amazon.com/Chlorophyll-Fluorescence-Signature-Photosynthesis-Respiration/dp/140203217X

[63] J. DEISENHOFER, O. EPP, K. MIKI, R. HUBER and H. MICHEL, *Nature 318*, 681, 1985 in http://www.nature.com/nature/journal/v318/n6047/abs/318618a0.html

[64] T. MCANANEY, Stanford University in http://www.hhmi.org/research/investigators/figs/deisenhofer_nobel1.html

[65] H. MICHEL, O. EPP and J. DEISENHOFER, *Embo J. 5*, 2445, 1986 in http://www.ncbi.nlm.nih.gov/pmc/articles/PMC1167138/pdf/emboj00173-0040.pdf

[66] J. DEISENHOFER and H. MICHEL, *Embo J. 8*, 2149, 1989 in http://www.ncbi.nlm.nih.gov/sites/ppmc/articles/PMC401143/pdf/emboj00132-0032.pdf

[67] K.H. RHEE, E.P. MORRIS, J. BARBER and W. KUHLBRANDT, *Nature 396*, 283, 1998 in http://www.ncbi.nlm.nih.gov/pubmed/9834037

[68] A. ZOUNI, H.T. WITT, J. KERN, P. FROMME, N. KRAUSS, W. SAENGER and P. ORTH *Nature 409*, 739, 2001 in http://www.ncbi.nlm.nih.gov/pubmed/11217865

[69] N. KAMIYA and J.R. SHEN, *Proc. Natl. Acad. Sci 100*, 98, 2003 in http://www.ncbi.nlm.nih.gov/pmc/articles/PMC140893

[70] K.N. FERREIRA, T.M. IVERSON, K. MAGHLAOUI, J. BARBER and S. IWATA, *Science 303*, 1831, 2004 in http://www.sciencemag.org/content/303/5665/1831.abstract

[71] A. GUSKOV, J. KERN, A. GABDULKHAKOV, M. BROSER, A. ZOUNI and W. SAENGER, *Nat. Struct. Mol. Biol. 16*, 334, 2009 in http://www.ncbi.nlm.nih.gov/pubmed/19219048

[72] P.J. NIXON, F. MICHOUX, J. YU, M. BOEHM and J. KOMENDA, *Ann. Bot.*, doi: 10.1093/aob/mcq059, 2010 in http://aob.oxfordjournals.org/content/early/2010/03/25/aob.mcq059.full

[73] M. BROSER, A. GABDULKHAKOV, J. KERN, A. GUSKOV, F. MUH, W. SAENGER and A. ZOUNI, *J. Biol. Chem. 285*, 26255, 2010, in http://www.jbc.org/content/285/34/26255

[74] G. KARP, *Cell and Molecular Biology: Concepts and Experiments*, 6th edition, p. 218, 2009 in http://www.amazon.com/Cell-Molecular-Biology-Concepts-Experiments/dp/0470483377

[75] J. WHITMARSH and GOVINDJEE in Concepts in *Photobiology: Photosynthesis and Photomorphogenesis*, Ed. G.S. Singhal, G. Renger, K.D. Irrgang and Govindjee, Narosa Publ./New Delhi, and Kluwer Academic/Dordrecht in http://www.life.illinois.edu/govindjee/paper/gov.html#40

[76] B. MILES, Texas A&M University, 2003 in http://www.tamu.edu/classes/bmiles/lectures/photosystems.pdf

[77] J. BARBER, Int. J. Photoenergy. 6, 43, 2004 in http://downloads.hindawi.com/journals/ijp/2004/641529.pdf

[78] G. KURISU, H. ZHANG, J.L. SMITH and W.A. CRAMER, *Sciences 302*, 1009, 2003 in http://www.bio.purdue.edu/people/faculty/cramer/cramerlab/lab_PDFs/1009.pdf

[79] Cramer Lab, Perdue University, 2010 in http://www.bio.purdue.edu/people/faculty/cramer/cramerlab/cytbf.html

[80] http://en.wikipedia.org/wiki/Cytochrome_b_6f_complex

[81] A.S. SARIBAS, S. MANDACI and F. DALDAL, *J. Bacteriol. 181*, 5365, 1999 in http://www.ncbi.nlm.nih.gov/pmc/articles/PMC94043

[82] J. YAN and W.A. CRAMER 278, 20925, 2003 in http://www.bio.purdue.edu/people/faculty/cramer/cramerlab/lab_PDFs/JYISP.pdf

[83] D. STROEBEL, Y. CHOQUET, JL. POPOT and D. PICOT, *Nature 426*, 413, 2003 in http://www.nature.com/nature/journal/v426/n6965/full/nature02155.html

[84] S.E. MARTINEZ, D. HUANG, M. PONOMAREV, W.A. CRAMER and J.L. SMITH, *Protein Sci. 5*, 1081, 1996 in http://www.ncbi.nlm.nih.gov/pmc/articles/PMC2143431/pdf/8762139.pdf

[85] C.J. CARRELL, B.G. SCHLARB, D.S. BENDALL, C.J. HOWE, W.A. CRAMER and J.L. SMITH, Biochemistry 38, 9590, 1999 in http://www.bio.purdue.edu/people/faculty/cramer/cramerlab/lab_PDFs/phorm_struct.pdf

[86] P. JOLIOT and A. JOLIOT, *Proc. Natl. Acad. Sci. 91*, 1034, 1994 in http://www.pnas.org/content/91/3/1034.full.pdf

[87] I. GROTJOHANN and P. FROMME, *Photosynthesis Research 85*, 51, 2005 in http://www.ncbi.nlm.nih.gov/pubmed/15977059

[88] F. MOREAU and R. PRAT, *Structure et fonctionnement du PS I*, Université Pierre et Marie Curie, UFR des Sciences de la Vie, 2008 in http://www.snv.jussieu.fr/bmedia/Photosynthese-cours/11-PSI.htm

[89] Wolfson Laboratories, Department of Biological Sciences, Imperial College London, 2003 in http://www.bio.ic.ac.uk/research/barber/psIIimages/PSI.html

[90] J.W. KIMBALL, Research Coordination Networks, Biology 2009 in http://users.rcn.com/jkimball.ma.ultranet/BiologyPages/L/LightReactions.html

[91] P. FROMME, *Structure and Function of Photosystem I* in http://www.public.asu.edu/~pfromme/Research/web-Photosystem%20I.pdf

[92] R.B. HALLICK, *Plant Mol. Biol. Rep. 7*, 266, 1989 in http://www.springerlink.com/content/hq7415j2521546n7

[93] European Synchrotron Research Facility, *The Structure of Plant Photosystem I*, 2009 in http://www.esrf.eu/UsersAndScience/Publications/Highlights/2003/MX/MX02

[94] A. AMUNTS, O. DRORY and N. NELSON, *Nature 447*, 58, 2007 in http://www.nature.com/nature/journal/v447/n7140/full/nature05687.html and http://www.tau.ac.il/lifesci/departments/biochem/members/nelson/documents/nelson36.pdf

[95] R. GOVINDJEE, *The Z-Scheme Digram of Photosynthesis* in http://www.life.illinois.edu/govindjee/ZSchemeG.html

[96] F. MOREAU and R. PRAT, *La Photosynthèse, Le Schéma en Z*, Université Pierre et Marie Curie, UFR des Sciences de la Vie, 2008 in http://www.snv.jussieu.fr/bmedia/Photosynthese-cours/12-schemaz.htm

[97] GOVINDJEE and R. GOVINDJEE, *Photosynthesis and the "Z"-scheme*, 2000 in http://www.life.illinois.edu/govindjee/textzsch.htm

[98] J.W. KIMBALL, Research Coordination Networks, Biology, *Photosynthesis: The role of Light*, 2009 in http://users.rcn.com/jkimball.ma.ultranet/BiologyPages/L/LightReactions.html

[99] J.W. KIMBALL, Research Coordination Networks, Biology, *Cyclic Photophosphorylation*, http://users.rcn.com/jkimball.ma.ultranet/BiologyPages/L/LightReactions.html#CyclicPhotophosphorylation

[100] J. WALKER, *ATP Synthase*, Mitochondrial Biology Unit, Medical Research Council, UK in http://www.mrc-mbu.cam.ac.uk/research/atp-synthase

[101] V.K. DICKSON, J.A. SILVESTER, I.M. FEARNLEY, A.G.W. LESLIE and J.E. WALKER, *Embo J. 25*, 2911, 2006 in http://www.nature.com/emboj/journal/v25/n12/full/7601177a.html

[102] F. MOREAU and R. PRAT, *La Photosynthèse, Structure et Fonctionnement de l'ATPsynthase*, Université Pierre et Marie Curie, UFR des Sciences de la Vie, 2008 in http://www.snv.jussieu.fr/bmedia/Photosynthese-cours/15-ATPsynthase.htm

[103] R.H. GARRETT & C.M. GRISHAM, Biochemistry 2nd Ed., Chapter 21, *Electron Transport and oxidative phosphorylation*, http://web.virginia.edu/Heidi/chapter21/chp21.htm

[104] J.J. DIWAN, *Biochemistry of metabolism, F1F0 ATP Synthase*, 2007 in http://rpi.edu/dept/bcbp/molbiochem/MBWeb/mb1/part2/f1fo.htm

[105] Nobelprize.org, 2013 in http://www.nobelprize.org/nobel_prizes/chemistry/laureates/1997/illpres/boyer_walker.html

[106] J.P. ABRAHAMS, A.G.W. LESLIE, R. LUTTER and J.E. WALKER, *Nature 370*, 621, 1994 in http://www.nature.com/nature/journal/v370/n6491/abs/370621a0.html

[107] C. GIBBONS, M.G. MONTGOMERY, A.G.W. LESLIE and J.E. WALKER, *Nat. Struct. Biol. 7*, 1055, 2000 in http://www.ncbi.nlm.nih.gov/pubmed/11062563?dopt=Abstract

[108] PDB, 1E79, *Bovine F-1 ATPase inhibited by DCCD (dicyclohexylcarbodiimide)* in http://www.pdb.org/pdb/explore.do?structureId=1E79

[109] H. NOJI, R. YASUDA, M. YOSHIDA and K. KINOSITA JR, *Nature 386*, 299, 1997 in http://www.nature.com/nature/journal/v386/n6622/abs/386299a0.html

[110] PicoTwist, Single Molecule Biophysics in http://www.picotwist.com/index.php?content=smb

[111] The Official Web Site of the Nobel Prize, Press Release: The 1997 Nobel Prize in Chemistry in http://nobelprize.org/nobel_prizes/chemistry/laureates/1997/press.html

[112] G.B. KAUFFMAN and L.M. KAUFFMAN, *Chem. Educator 4*, 28, 1999, in http://www.springerlink.com/content/xpp23m6641174393

[113] J.S. CARTER, Clermont College, University of Cincinnati, *Photosynthesis*, 2004 in http://biology.clc.uc.edu/Courses/Bio104/photosyn.htm

[114] I. ANDERSSON, *J. Exp. Bot. 59*, 1555, 2008 in http://jxb.oxfordjournals.org/content/59/7/1555.full.pdf

[115] J.J. DIWAN, *Biochemistry of Metabolism, Calvin Cycle – Photosynthetic Carbon Reactions*, 2008 in http://www.rpi.edu/dept/bcbp/molbiochem/MBWeb/mb2/part1/dark.htm

[116] R. BERGMANN and G. MURPHY, *Ribulose-1,5-bisphosphate carboxylase*, University of Hamburg, 2010 in http://www.biologie.uni-hamburg.de/lehre/bza/1rxo/e1rxoe.htm

[117] http://en.wikipedia.org/wiki/RuBisCO

[118] R. Bergmann and G. Murphy, University of Hamburg, *Ribulose-1,5-bisphosphate carboxylase from Spinacia oleracea*, 2006 in http://www.biologie.uni-hamburg.de/lehre/bza/1rxo/e1rxom.htm

[119] Auburn University, Biology 1020, Chapter 10: *Photosynthesis* in http://www.auburn.edu/academic/classes/biol/1020/singh/lecturenotes/chapter10.doc

[120] A. Stern, University of Massachusetts Amherst, *Origin of the Organic Soup*, http://www.bio.umass.edu/biology/conn.river/photosyn.html

[121] P. Von Sengbush, *C3, C4 and CAM, Regulation of The Activity of Photosynthesis*, University of Hamburg, 2003 in http://www.biologie.uni-hamburg.de/b-online/e24/24b.htm

[122] Spiritus-Temporis.com, *C4 carbon fixation*, 2005 in http://www.spiritus-temporis.com/c4-carbon-fixation

[123] S.G. Saupe, *Autotrophism : Carbon Reactions*, College of St. Benedict/St. John's University, 2009, http://employees.csbsju.edu/ssaupe/biol327/lecture/photosyn-carbon.htm

[124] FAO, *Renewable biological systems for alternative sustainable energy production*, Agricultural Services Bulletins, 1997 in http://www.fao.org/docrep/w7241e/w7241e05.htm#1.2%20photosynthesis%20and%20biomass

[125] G.R. Kantharaj, *Photosynthesis*, Lecture Notes for Masters and PreUniversity Students, http://preuniversity.grkraj.org/html/7_PHOTOSYNTHESIS.htm

[126] X.G. Zhu, S.P. Long and D.R. Ort, Curr. Opin. Biotechnol. 19, 153, 2008 in http://etmd.nal.usda.gov/bitstream/10113/36097/1/IND44280883.pdf

[127] E. Collini, C.Y. Wong, K.E. Wilk, P.M.G. Curmi, P. Brumer and G.D. Scholes, *Nature 463*, 644, 2010 in http://www.nature.com/nature/journal/v463/n7281/abs/nature08811.html

[128] D. Biello, Scientific American, February 3, 2010 in http://www.scientificamerican.com/article.cfm?id=shining-a-light-on-plants-quantum-secret

[129] G.D. Scholes, *J. Phys. Chem. Lett. 1, 2*, 2010 in http://pubs.acs.org/doi/abs/10.1021/jz900062f

Chapter 3

Features of First Generation Biorefineries

3.1 Introduction

This chapter is aimed at providing the technological and production features of first-generation biorefineries for facilitating the understanding and development of second-generation biorefineries, knowing that first generation refers to feedstocks derived primarily from food plants.

The chapter is divided into three sections: first-generation biofeedstocks, first-generation biofuels, and first-generation biobased chemicals/materials.

First-generation biofeedstocks have been introduced in Chapter 1. They mainly comprise sugar/starch crops and vegetable oil. In this section, the main feedstocks – sugarcane for sugar source, cereal grains for starch source, and vegetable oil – are described, their quantities produced worldwide are given, and their main chemical features are analyzed.

First-generation biofuels include bioethanol, ETBE (ethyl tert-butyl ether, an oxygenated gasoline additive), biodiesel, and biogas. The production of bioethanol by fermentation of sugars, the synthesis of biodiesel from triglycerides, and the formation of biogas by anaerobic digestion of organic wastes are described. A general review of the global production of liquid biofuels and an analysis of the available feedstocks per country are also given.

The third section deals with first-generation biobased chemicals/materials. One can distinguish between (1) sugar-based products derived from sugar/starch plants, (2) oil-based products derived from plant oils, animal fats, and recycled greases, and (3) bioethanol-based products, such as polyethylene and PVC.

3.2 First-generation biofeedstocks

3.2.1 Sugar feedstocks

Introduction: saccharose

High amounts of saccharose or table sugar (sometimes called sucrose) are present in sugar cane and sugar beet. Saccharose is a non-reducing disaccharide. The molecule does not possess an anomeric center. It is composed of an α-linked D-glucose unit and a β-linked D-fructose (**Figure 3.1**).

Saccharose is a good substrate for fermentation. Saccharose is the major carbon source used by the yeast *Saccharomyces cerevisiae* (**Figure 3.2**) during production of baker's yeast, fuel ethanol, and several distilled beverages. [1] It is generally accepted that sucrose fermentation proceeds through extracellular hydrolysis of the sugar, mediated by the periplasmic enzyme invertase, producing glucose and fructose that are transported into the cells and metabolized to ethanol and CO_2.

Sugarcane

In Brazil, the sugarcane-based ethanol substitutes half of the gasoline that would be used if it did not exist, and its cost is competitive even without the subsidies that initially launched the program. [3] That has been accomplished in 30 years; the Brazilian Ethanol Program was launched in the 1970s to reduce the dependence on oil imports.

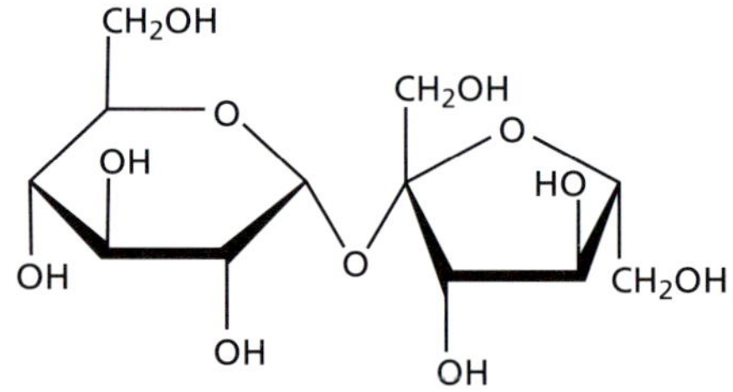

Figure 3.1 Structure of saccharose.

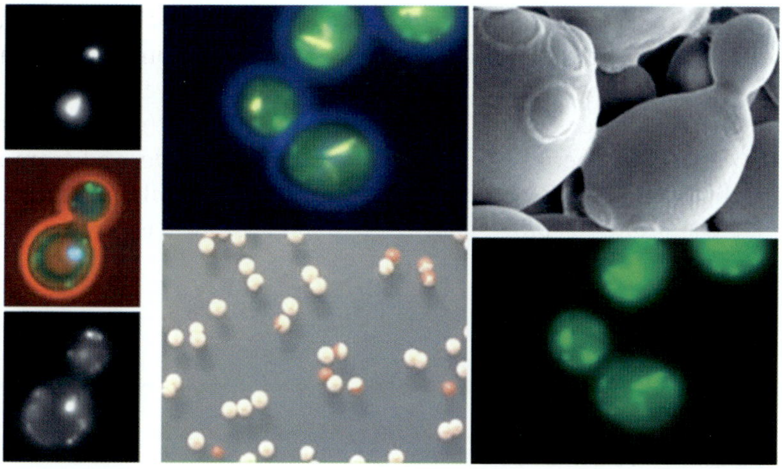

Figure 3.2 The yeast, *Saccharomyces cerevisiae*. Yeasts are unicellular fungi. The cells are about the same size as a red blood cell (10 μm diameter) and about 10x larger in diameter than a bacterium such as Staphylococcus. Reproduced by permission of Professor Mick TUITE, Kent fungal Group, University of Kent, UK. [2]

Sugarcane is a semi-perennial plant with C4-type photosynthetic cycle, family Gramineae. The aerial part of the plant is essentially formed by stalks, containing saccharose, and by tips and leaves, which form the sugarcane straw, as shown in **Figure 3.3**. These components altogether sum ~35 tons of dry material per hectare.

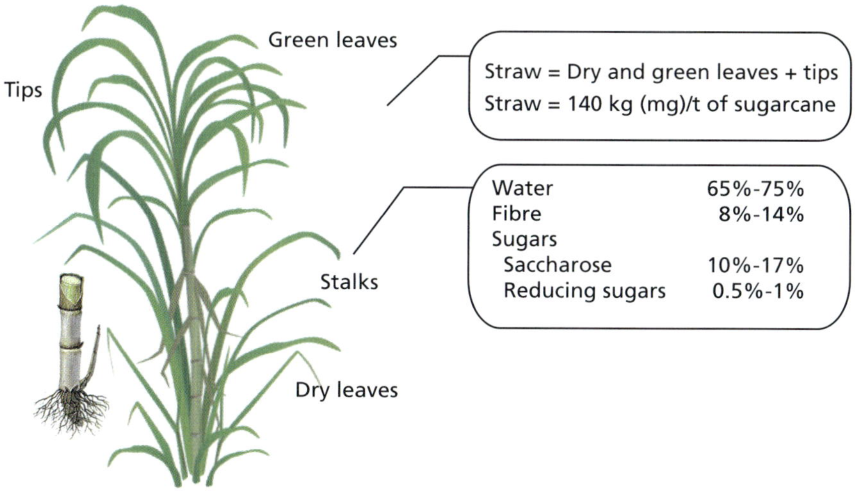

Figure 3.3 Typical sugarcane biomass structure. Source: Seabra (2008). [3]

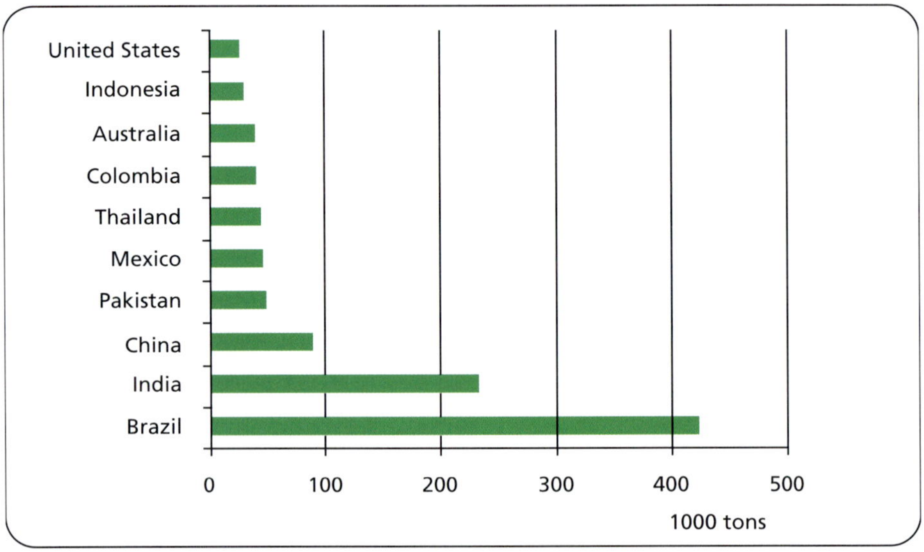

Figure 3.4 Leading sugarcane producing countries in 2005. Source: FAO (2007). [3]

Sugarcane is the one of the most important commercial crops all over the world. It occupies more than 20 million hectares in which nearly 1,300 million tons were produced in 2006/2007. Brazil stands out as the leading producer with a cropland area of ~7 million hectares, representing close to 42% of total production (**Figure 3.4**).

3.2.2 Starch feedstocks

Introduction: starch

High amounts of starch are found in cereal grains, potatoes, and legumes (beans). Starch is the principal polysaccharide used in plants to store glucose for later use as energy. It consists of two structurally different α-linked D-glucans: ***amylose*** and ***amylopectin***. Amylose is a linear α-1,4-linked D-glucopyranose polymer. It is the α-analogue of cellulose (**Figure 3.5**). The difference in linkage conformation gives rise to very different three-dimensional molecular structures. Native amylose is crystallized in the form of a double helix, each single strand having a sixfold symmetry (six monomers per turn of the helix), and cellulose, generally in a twofold helix. [4, 5]

Amylopectin is also an α-linked D-glucopyranose polymer, but it is branched. The glycosidic linkages too are 1,4, except at the branch point where they are 1,6 (**Figure 3.6**). Amylopectin, like amylose, has typically a helicoidal structure. The 1,4 linked sections can be seen to coil into a helical shape, and the 1,6 linkage forms a helical branch away from the main section. [6]

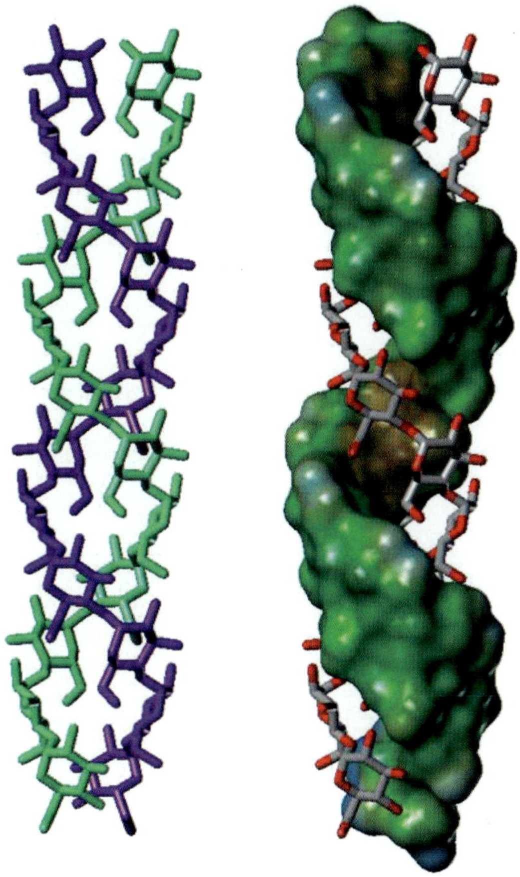

Figure 3.5 Molecular drawing for the double helix found in starch. *Starch Structure and Morphology*, Serge PÉREZ & Anne IMBERTY, http://cermav.cnrs.fr/glyco3d/lessons/starch/ [5]

The ratios of amylose and amylopectin can vary depending on the source (standard corn, waxy maize, rice, potato…), but amylase represents typically 15–35% of global starch amount.

Starch molecules arrange themselves in the plant in semi-crystalline granules. Each plant species has a unique starch granular size: rice starch is relatively small (~2 μm), while potato starches have larger granules (up to 100 μm).

Starch is easily hydrolyzable to glucose by human and animal enzymes to produce glucose, which is readily metabolized by yeast to produce ethanol at high yields.

Starch polymers for non-food non-plastic applications represent a mature sector with limited growth. [8] In Europe in 2007, 17% of the total consumption of starch is used for bioethanol production, 33% is dedicated to other non-food purposes, and 50% is used as food and feed (**Table 3.1**).

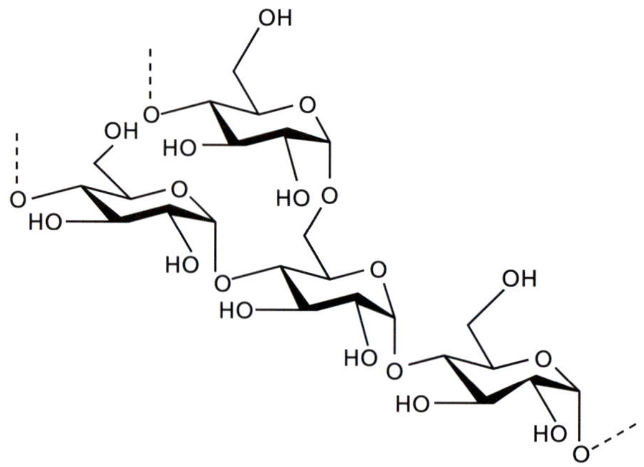

Figure 3.6 Amylopectin with α-1,4-linkages, except at the branch points where they are α-1,6. [7]

Table 3.1 Overview of starch use for food and non-food purposes in Europe in 2007.

Sector	Consumption (10 [6] tonnes)
Confectionary & drinks	2.9
Processed food	2.6
Feed	0.1
Total food /feed	**5.6**
Corrugating & paper making	2.6
Pharmaceuticals & chemicals	0.7
Other non-food	0.4
Total non-food (without starch for ethanol used as fuel)	**3.7**
Fuel ethanol	1.9
Total	**11.2**

Within the non-food sector that excludes biofuels, starch for non-plastic applications represents more than 90%, while starch plastics represent less than 10%. By far the most important non-plastic application of starch is corrugated board and paper making, representing a demand of nearly 3 Mt, which is approximately 70% of the total demand for starch used in Europe for non-food purposes (excluding bioethanol). Here starch and starch derivatives are used primarily as sizing agents.

Table 3.2 Starch content and theoretical ethanol yield of cereal grains.

Feed grain	Starch (%)	Theoretical ethanol yield[1)] (L/kg)
Corn	72 (65-76)	0.52
Wheat	77 (66-82)	0.55
Barley	57 (55-74)	0.41
Sorghum	72 (68-80)	0.52
Oat	58 (45-69)	0.42
Rice	79 (74-85)	0.57

1) Theoretical yield of ethanol is 0.51 kg ethanol/kg glucose. Upon hydrolysis, 1 kg starch produces 1.11 kg glucose.

Cereal grains

Cereal grains are by far the main starchy biofeedstocks for producing ethanol. Cereal grains include corn, wheat, barley, sorghum, oat and rice. They are used mostly for food and feed. [9] However, because of their high starch contents, they are also good feedstocks for conversion to biofuels and other biobased products. Bioethanol is the only biofuel that has been produced commercially in large quantity from these feedstocks. Starch contents and theoretical ethanol yield of relevant cereal grains are given in **Table 3.2**.

The world's farmers produced more grain in 2011 than ever before. [10] Estimates from the U.S. Department of Agriculture show the global grain harvest coming in at 2,295 million tons (**Figure 3.7**).

Nearly half the calories consumed around the world come directly from grain, with grain-fed animal products making up part of the remainder. Three grains dominate the world harvest: wheat (a C3 plant), rice (a C3 plant), which are primarily eaten directly as food, and corn (a C4 plant), which is largely used as a feedgrain for livestock. Wheat was the largest of the world's grain harvests until the mid-1990s. Then corn production surged ahead in response to growing demand for grain-fed animal products and, more recently, for fuel ethanol.

3.2.3 Vegetable oil

Triglycerides are the main constituents of vegetable oils and animal fats (**Figure 3.8**). The minor ingredients include mono- and diglycerides, free fatty acids, phosphatides, sterols, fatty alcohols, fat-soluble vitamins, and other substances. [11]

Triglyceride is a triester derived from glycerol and three fatty acids. Triglyceride molecules differ in the nature of the fatty acid bound to glycerol. Car-

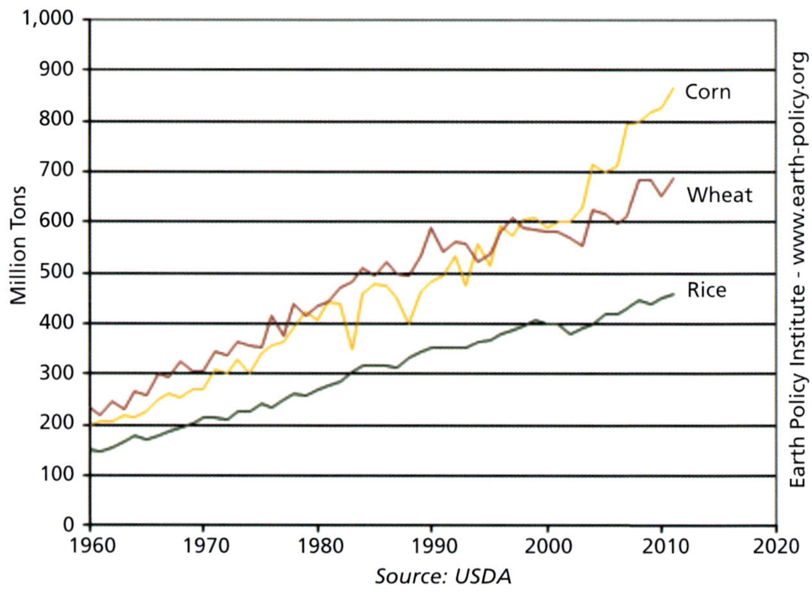

Figure 3.7 World corn, wheat and rice production, 1960-2011. Source: Earth Policy Institute - www.earth-policy.org

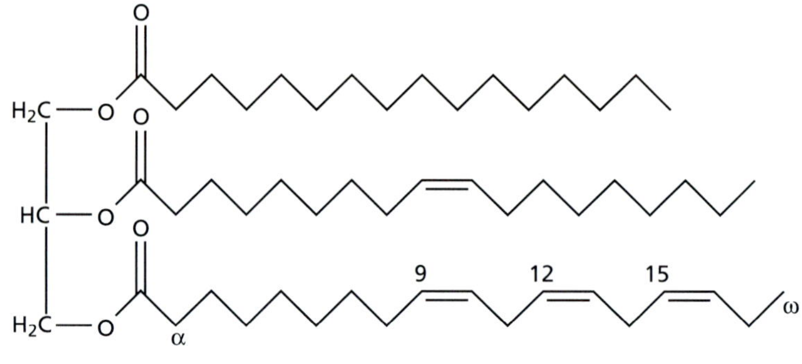

Figure 3.8 Example of a triglyceride. [12]

boxylic acids with chain length in the range of 6 to 24 carbon atoms (C6 to C24) are traditionally called *fatty acids.* [13] Fatty acids also vary according to their saturation, i.e. the number of double bonds: saturated fatty acids with no double bond, such as myristic (C14:0), lauric (C12:0), palmitic (C16:0) and stearic (C18:0) acids, and unsaturated fatty acids containing one or more double bonds, such as oleic (one double bond, C18:1), linoleic (two double bonds, C18:2) and linolenic (three double bonds, C18:3) acids. The position of the double bond can vary along the carbon chain.

Table 3.3 World vegetable oil consumption (in million metric tons per year). [14]

Oil	2000	2002	2004	2007	2008
Palm	23.3	25.8	30.5	40.8	40.2
Soybeans	26.0	30.5	31.6	38.4	37.5
Rapeseed	13.1	11.6	15.5	18.0	18.4
Sunflower seed	8.6	8.3	8.6	10.1	9.0
Peanut	4.2	4.4	5.0	5.0	5.0
Cottonseed	3.6	3.5	4.6	4.9	4.8
Palm kernel	3.3	3.4	3.6	4.8	4.8
Coconut	2.7	3.2	3.4	3.3	3.4
Olive	2.5	2.5	2.7	3.0	2.9
Total	87.2	93.2	105.6	128.2	125.8

Fats and oils are practically always mixtures of different fatty acids in varying proportions. Hence the degree of saturation of the different fats and oils depends upon their various fatty acids content.

There are more than 350 identified oil-bearing crops, among which only soybean, palm, sunflower, safflower, cottonseed, rapeseed, and peanut oil are considered as potential feedstocks for biodiesel. [14] **Table 3.3** shows the world vegetable consumption between 2000 and 2008, with palm oil and soybean oil contributing to the majority of the consumption.

The prominent feedstock for biodiesel in Malaysia and Indonesia is palm oil, in Europe it is rapeseed oil, and an emerging source in India is jatropha oil. Soybeans are commonly used in the United States for food products, which has led to soybean biodiesel becoming the primary source for biodiesel in that country.

3.3 First-generation biofuels

3.3.1 Bioethanol

Bioethanol can be used to replace standard gasoline in adequate engines, or as an additive to reduce fossil fuel usage and improve combustion. Nowadays, standard motorizations can afford up to 10–15% ethanol in gasoline, and even higher in Brazil.

Bioethanol is an alcohol made by ***fermenting*** the sugar components of sugar and starch plants like sugarcane, sugar beet, corn, and wheat. The degree of complexity of biomass conversion to bioethanol by fermentation depends on the nature of the feedstock used. In this section, we will cover the two most representative cases: sugarcane ethanol from sugar feedstock and corn ethanol from starch feedstock.

Sugarcane ethanol

The least complex technology is to use plants containing simple sugars, which can be fermented directly to ethanol by yeast or other microorganisms. [15] Brazil is the world's largest producer of bioethanol from sugar cane.

The initial processing stages for sugarcane ethanol are basically the same as for sugar production, as shown in **Figure 3.9**. [3]

Once in the mill, sugarcane is generally washed and sent to the preparation and extraction phases. [3] Extraction is made by roll-mills that separate the sugarcane juice containing saccharose from the bagasse, which is sent to the mill's power plant to be used as a fuel. In some new units implemented

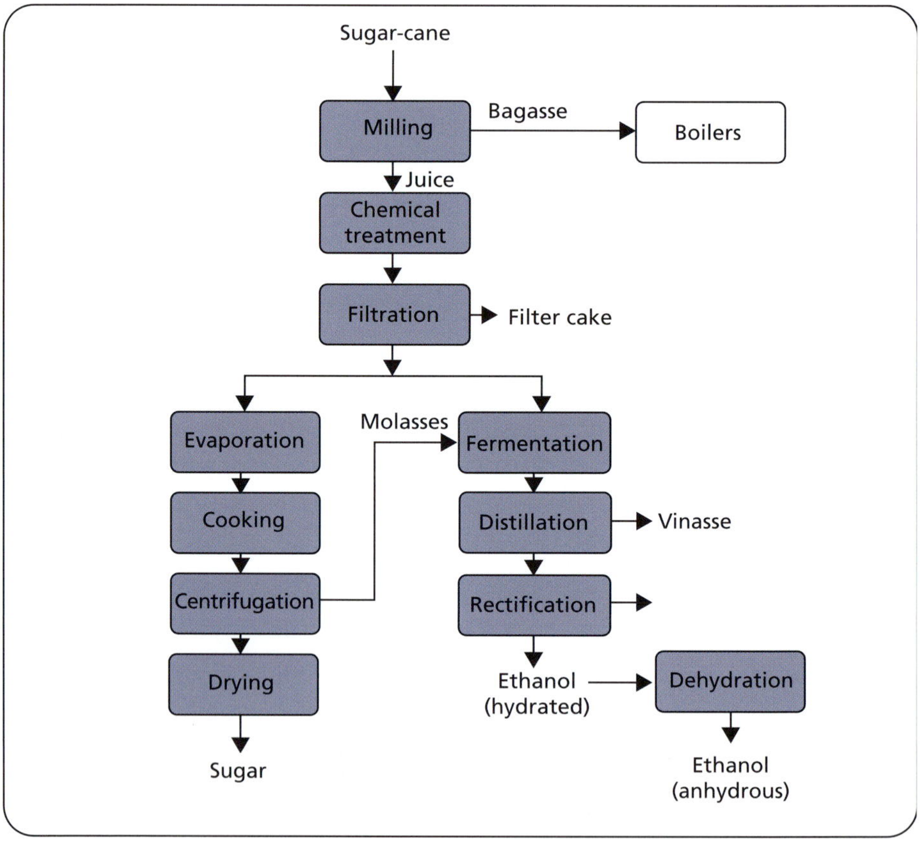

Figure 3.9 Sugar and sugarcane-based bioethanol production flowchart. Source: Seabra (2008). [3]

in Brazil, extraction by diffusion is being adopted. Produced in the mill or diffuser, the juice containing sugars can then be used in sugar or bioethanol production.

In sugar production, the juice is initially screened and chemically treated for coagulation and precipitation of impurities, which are eliminated through decanting. [3] The filter cake, used as fertilizer, is generated by recovering sugar out of the decanted slurry by means of rotary vacuum filters. The treated juice is then concentrated in evaporators and crystallized. In sugar and bioethanol production, only part of the saccharose available in the sugar cane is crystallized, and the residual solution with high sugar content (molasses) can be used in the process to recover more sugar. The molasses do not return to the sugar manufacturing process but can be used as an input for bioethanol production through fermentation.

In practice, sugarcane ethanol production uses either the sugarcane juice alone or a mix of juice and molasses as raw materials for fermentation, the latter being more frequent in Brazil. [3] In sugarcane-juice ethanol, the first stages of the manufacturing process, from sugar cane receipt to juice treatment, are similar to the sugar manufacturing process. After treatment, the juice is evaporated to balance its sugars concentration and, in some cases, it is mixed with molasses, generating sugarcane mash, a sugary solution that is ready for fermentation.

The mash is sent to fermentation reactors where yeasts (*Saccharoromyces cerevisiae*) are added to it and fermented for a period ranging from 8 to 12 hours, generating wine (fermented mash, with ethanol concentration from 7 to 10%). [3] Then, after fermentation, yeasts are recovered and treated for new use, while the wine is sent to distillation columns.

Corn ethanol

The other common technology for bioethanol production is to use starchy plants as feedstock. The commercial production of sugars from starch is the earliest example of modern industrial enzyme technology. In the United States, bioethanol is derived today almost completely from the starch contained in corn.

The steps for bioethanol production from corn include the following (**Figure 3.10**):

- **Milling:** The feedstock is passed through mills that grind them into flour. The flour is sieved to extract the pure starch dry milling or mixed with water, ground in the mill, and sieved alternatively to extract the pure starch after which the starch is dried in the open sun (wet milling). Due to less investments and running costs: dry milling process represents ~90% of existing corn-ethanol plants in USA. [16]
- **Liquefaction:** The meal of pure starch is mixed with water and α-amylases and passed through cookers where the starch is broken

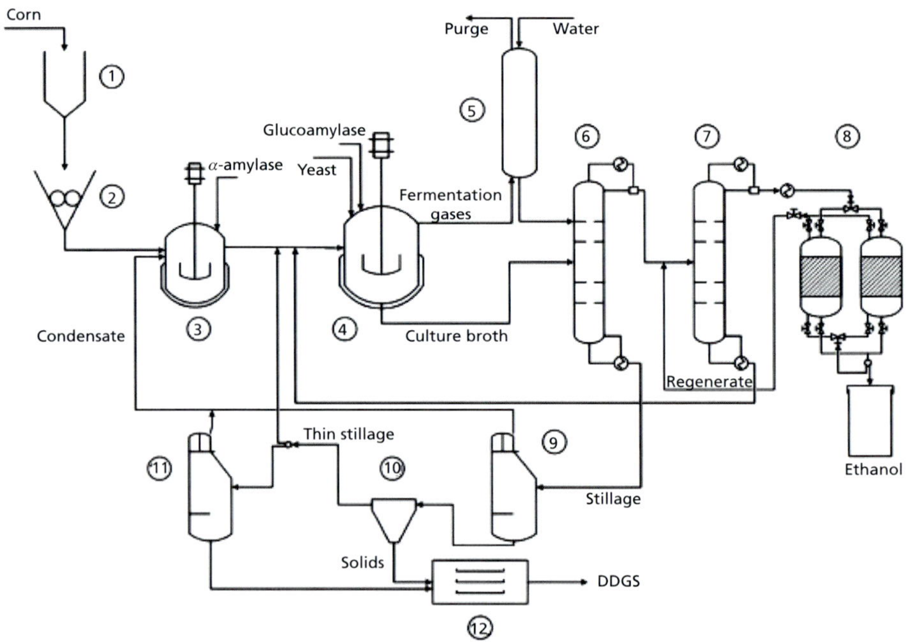

Figure 3.10 Simplified flowsheet for ethanol production from corn. 1. Washing tank. 2. Crusher. 3. Liquefaction reactor. 4. SSF (simultaneous saccharification and fermentation) reactor. 5. Ethanol absorber. 6. Concentration column. 7. Rectification column. 8. Molecular sieves. 9. First evaporator train. 10. Centrifuge. 11. Second evaporator train. 12. Dryer. DDGS, Dry Distillers Grains with Solubles. Copyright Elsevier 2008. [15]

down to dextrins with 5–10 glucose units and liquefied. Heat is necessarily applied to enable the liquefaction process in cookers with a high temperature stage (120–150 °C) and a lower temperature holding period (90 °C). The high temperatures reduce bacteria levels in the mash.

- **Saccharification:** The mash from the cookers is then cooled and the secondary enzyme (glucoamylase) added to convert the liquefied starch to fermentable glucose, a process called saccharification.
- **Fermentation:** Yeast is then added to the mash to ferment the sugars to ethanol and CO_2. If a continuous process is used, the fermentation mash is allowed to flow through several fermenters until the mash is fully fermented and then leaves the final tank. In a batch fermentation process, the mash stays in one fermenter for about 48 hours before the distillation process is started. In a SSF process, saccharification and fermentation are simultaneous.
- **Distillation and denaturing.**

3.3.2. ETBE (Ethyl Tert-Butyl Ether)

Oxygenated additives have been developed for ~30 years now to replace lead-containing products in gasoline. They increase octane rating of the fuel by increasing compression ratio, promote cleaner combustion and, therefore, reduce hazardous emissions.

Both alcohols (especially bioethanol) and ethers can be used as oxygenated additives. The most well known, obtained from fossil fuel resources, is MTBE (methyl tert-butyl ether). ETBE (ethyl tert-butyl ether) is now more and more commonly used and can replace advantageously MTBE. As it is obtained by catalytic reaction between bioethanol and isobutylene (**Figure 3.11**), it is more expensive than MTBE, but at least it can be considered partially renewable.

Due to high solubility in water and slow biodegradation, ETBE like MTBE necessitates specific storage and transportation equipments and infrastructures (pipe-lines, tanks, etc,).

The recent new European Directive (2009/30/EC) has increased oxygen content in gasoline up to 3.7wt%, enabling ethanol and ethanol derivatives usage to be multiplicated by 4 up to 2020 (**Figure 3.12**).

3.3.3 Biodiesel

Biodiesel is generally defined as a mixture of fatty acid alkyl methyl esters (***FAME***) made from vegetable oils, animal fats, or recycled greases. [19] It is produced mainly through transesterification, in which fat or oil (triglycerides) is reacted with an alcohol (methanol for FAME, but ethanol or other alcohols are also used). The reaction is conducted in the presence of a strong base catalyst to form biodiesel and glycerol (**Figure 3.13**). An excess of methanol is required to obtain a high degree of conversion. [20]

The conventional catalysts in natural oil transesterification processes are selected among bases such as alkaline or alkaline earth hydroxides or alkoxides. [20] However, transesterification could also be performed using acid catalysts or using metallic base catalysts or oxides of tin, magnesium, or zinc. All these catalysts act as homogeneous catalysts and need to be removed from the products after the methanolysis step.

In conventional industrial biodiesel processes, vegetable oil methanolysis is achieved using a homogeneous catalyst system operated in batch or con-

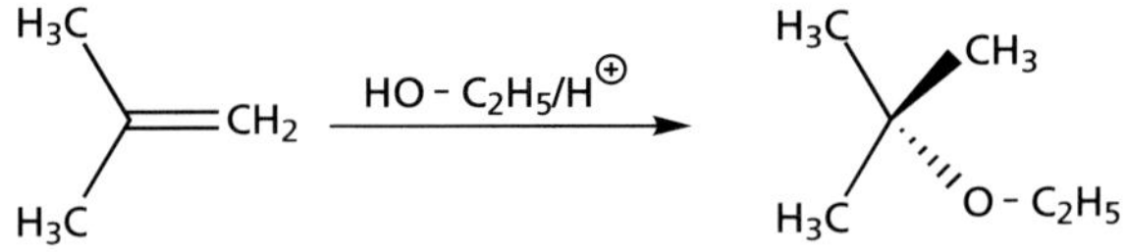

Figure 3.11 ETBE synthesis scheme. [17]

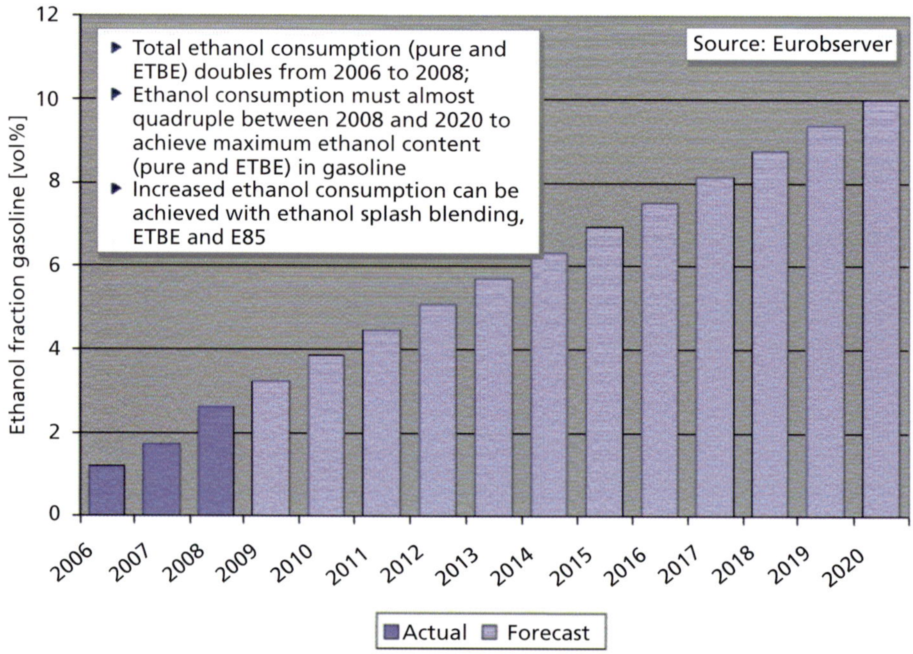

Figure 3.12 Ethanol and derivatives % forecast 2008-2020. Source: Eurobserver. [18]

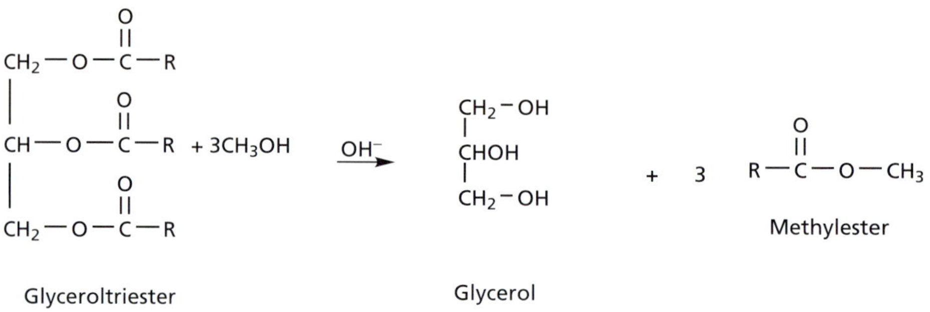

Figure 3.13 Transesterification reaction for production of first-generation biodiesel (FAME).

tinuous mode (**Figure 3.14**). Sodium hydroxide or sodium methylate is often used as catalyst. Sodium is recovered after the transesterification reaction as sodium glycerate, sodium methylate, or sodium soaps in the glycerol phase. An acidic neutralization step is required to neutralize these salts. In that case, glycerol is obtained as an aqueous solution containing sodium chloride. When sodium hydroxide is used as catalyst, side reactions forming sodium soaps generally occur. This type of reaction is also observed when sodium methylate

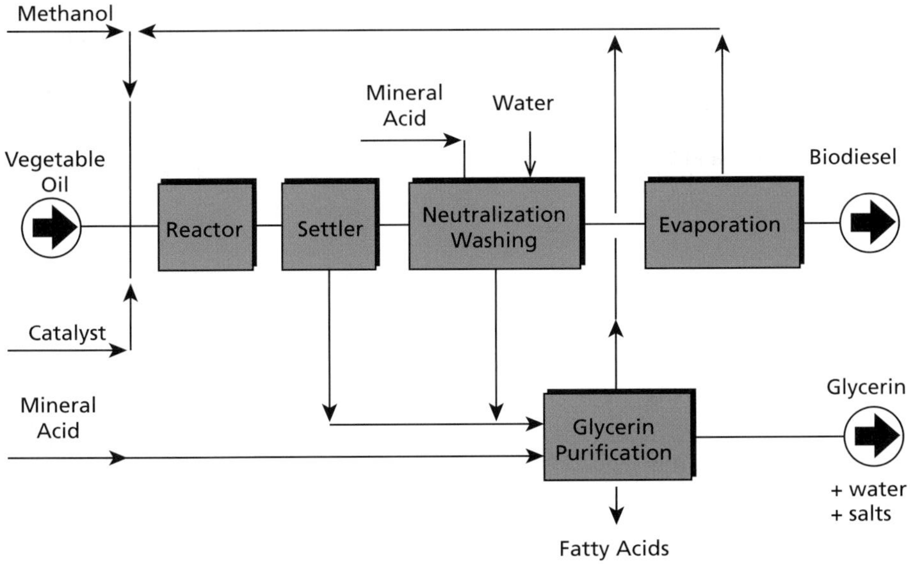

Figure 3.14 Global scheme for a typical continuous homogeneous catalyzed process for biodiesel production through transesterification. Reproduced by permission of Elsevier. [20]

is employed and traces of water are present. The sodium soaps are soluble in the glycerol phase and must be isolated after neutralization by decantation as fatty acids.

Much effort has been expended in search of solid acid or basic catalysts that could be used in a heterogeneous catalyzed process. [20] In this new continuous process, the transesterification reaction is promoted by a completely heterogeneous catalyst. This catalyst consists of a mixed oxide of zinc and aluminium, which promotes the transesterification reaction without catalyst loss. The reaction is performed at higher temperature and pressure than homogeneous catalysis processes, with an excess of methanol. This excess is removed by evaporation and recycled to the process with fresh ethanol. The desired chemical conversion, required to produce biodiesel at the European specifications, is reached with two successive stages of reaction and glycerol separation in order to shift the equilibrium of methanolysis (**Figure 3.15**).

Besides the transesterification process, the Finnish company Neste Oil has developed a proprietary process that produces a premium diesel from animal fats or vegetable oils, called NExBTL. [21] The Neste process entails direct hydrogenation of plant oil or animal fat triglycerides into the corresponding alkane, thus removing oxygen from the oil. Unlike yellow transesterified biodiesel, NExBTL is a clear, colorless paraffin. Neste's proprietary process involves catalysts, high pressure, high temperature, and hydrogenation. By-products include propane and gasoline.

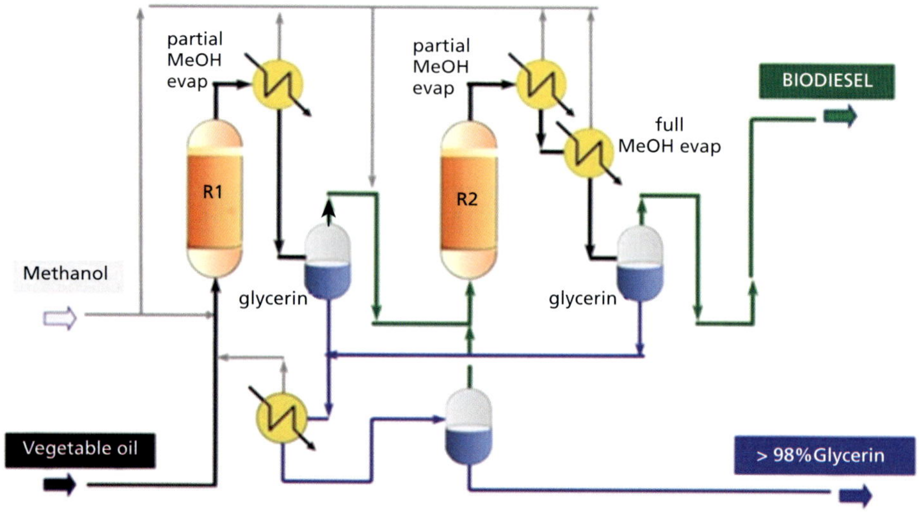

Figure 3.15 Simplified flow sheet of a heterogeneous process for bioethanol production through transesterification. Reproduced by permission of Elsevier. [20]

As diesel motorizations are not much sensitive to fuel composition, biodiesel can be used pure or blended with petroleum diesel in various proportions. Most states encourage use of biodiesel by tax reductions.

As for the production of ethanol from starch/sugar plants, the production of biodiesel from feedstocks linked with deforestation is a subject of controversy, together with water pollution problems for plant growth or farming. In that sense, use of algae or waste residues for biodiesel production is becoming a growing topic of interest.

3.3.4 Biogas

Biogas is a mixture of primarily biomethane (CH_4) and CO_2. It is produced by anaerobic digestion of organic wastes, such as sewage, manure, food wastes, and landfill (**Figure 3.16**). [22] Anaerobic digestion technology is well established, hence biogas is generally considered as a first generation biofuel.

Anaerobic digestion includes four major process stages: (1) hydrolysis, where the large organic molecules are broken down into simple sugars, fatty acids and amino acids; (2) acidogenesis, where the dissolved compounds from hydrolysis are consumed by acidogenic (fermentative) bacteria and broken down into even simpler organic compounds; (3) acetogenesis, where simple molecules formed through the acidogenesis phase are further digested by acetogens to produce largely acetic acid as well as CO_2 and H_2; and (4) methanogenesis, where methanogens utilize the intermediate products of the preceding stages and convert them into primarily CH_4 and

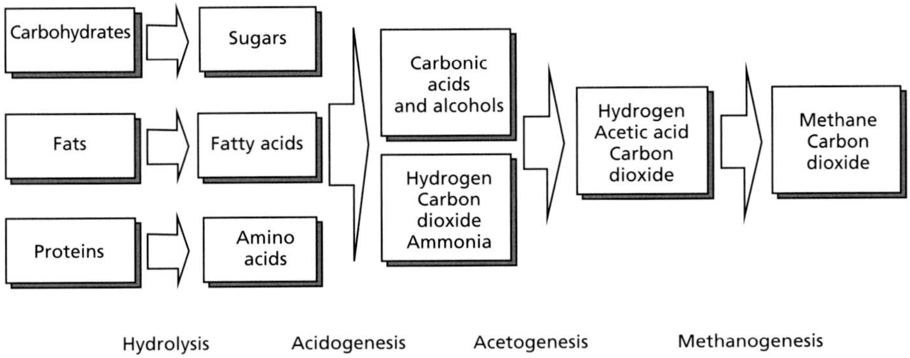

Figure 3.16 The four key process stages of anaerobic digestion: hydrolysis, acidogenesis, acetogenesis, and methanogenesis. Different micro-organisms are responsible for the processes in each stage. [23]

CO_2. [24, 25] The chemical reaction for the overall process can be written as follows (**Equation 3.1**):

$$C_6H_{12}O_6 \rightarrow 3\ CO_2 + 3\ CH_4 \tag{3.1}$$

After removal of contaminants, biomethane is the same as natural gas and can be used for heating or for electricity generation, or as a transport fuel in the form of Liquid Natural Gas (LNG) or Compressed Natural Gas (CNG). [22, 26]

3.3.5 Global production of liquid biofuels

Overview

Global biofuels production grew by only 0.7% in 2011, the smallest increase since 2000 (**Figures 3.17** and **3.18**). [27, 28] Increased output in North America was offset by declines in South & Central America and in Europe. Biodiesel accounts for just 27.5% of global biodiesels output, but accounted for all of the growth in global biofuels output. Global ethanol output declined by 1.4%.

Bioethanol

In June 2012, the Global Renewable Fuels Alliance (GRFA), in collaboration with F.O. Licht, forecasted global fuel ethanol production to hit 85.2 billion liters in 2012. [29] The GRFA predicted a 1% growth in ethanol output in 2012, up from the 84.5 billion litres produced in 2011.

According to GRFA, the United States and Brazil continue to be the largest producers of ethanol with production continuing at a steady pace in 2012. [29] Production levels in Africa remain relatively low. In the European region, pro-

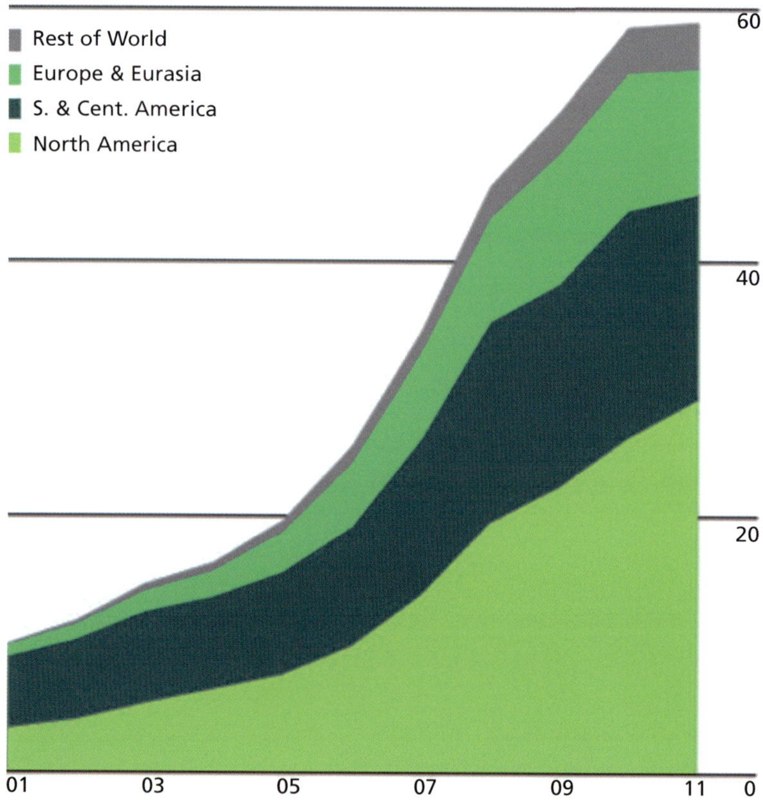

Figure 3.17 World biofuels production in million tons of oil equivalent (mtoe).Source: BP. [27]

duction continues to grow at a robust pace. In 2011, Europe saw its production grow by 4% over 2010. In 2012, Europe's production increase will more than double last year's growth rate. Europe is expected to produce 4.9 billion litres of ethanol, which is an 11% increase over 2011 (**Table 3.4**).

Biodiesel

In March 2012, F.O. Licht projected world biodiesel production in 2012 at 18.9 million tons, which is slightly up from the 18.8 million tons in 2011 (**Table 3.5**). [30]

Biodiesel production in ten leading countries and in the world in 2011 is shown in **Table 3.6**. [31]

Biogas

European primary biogas energy output rose to 8.3 mtoe in 2009, which is 346.8 ktoe more than in 2008 (**Table 3.7**). [32] This figure only includes the output intended for recovery and excludes biogas flares.

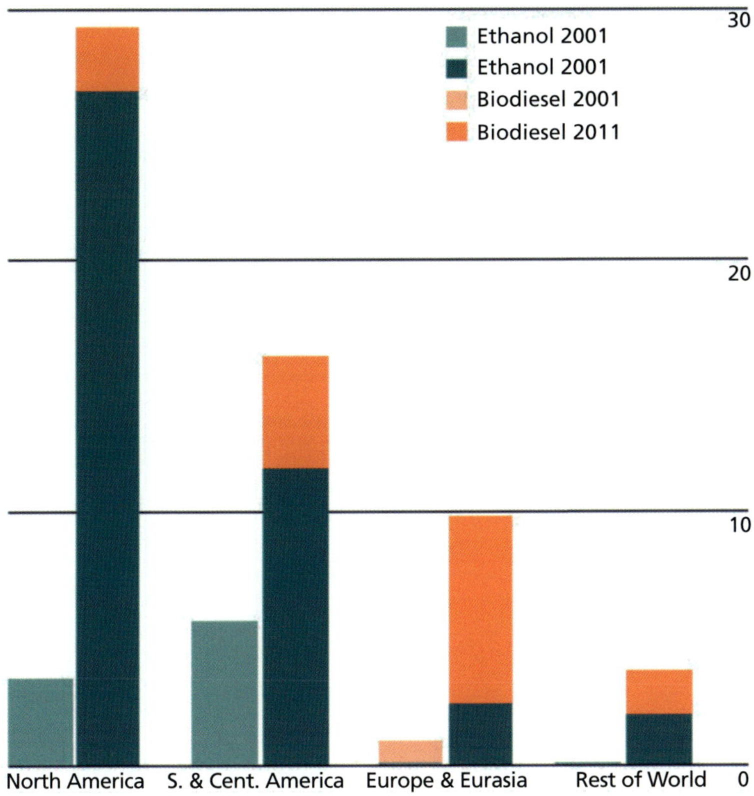

Figure 3.18 World biofuels production in 2011 compared to 2001, million tons of oil equivalent (mtoe). Source: BP. [27]

Table 3.4 World ethanol fuel production in million liters (source: F.O. Licht). [29]

	2006	2007	2008	2009	2010	2011	2012 (projections)
Europe	1,627	1,882	2,855	3,645	4,254	4,429	4,973
Africa	0	55	65	100	130	150	235
Americas	35,685	45,546	60,402	66,416	77,548	76,402	76,065
Asia/ Pacific	1,940	2,142	2,753	2,927	3,115	3,520	3,965
World	39,252	49,625	66,075	73,088	85,047	84,501	85,238

According to the American Biogas Council, in 2011 there were just over 160 anaerobic digesters on farms and ~1500 more operating at wastewater treatment plants in the United States. [33, 34]

Table 3.5 World biodiesel fuel production in million tons (source: F.O. Licht). [30]

	2011	2012 (projections)
Europe	8.1	7.9
South America	5.1	5.8
North & Central America	3.0	2.7
Asia	2.4	2.3
World	18.8	18.9

Table 3.6 Biodiesel production in leading countries and the world, 2011. [31]

	Production (million tons)
United States	2.80
Germany	2.78
Argentina	2.43
Brazil	1.40
France	1.20
Indonesia	0.63
Spain	0.52
Italy	0.52
Thailand	0.52
Netherlands	0.39
World	18.84

Table 3.7 Primary biogas energy output in leading EU countries and in Europe (in ktoe). [32]

	2008	2009
Germany	4229.5	4213.4
United Kingdom	1625.4	1723.9
France	453.1	526.3
Italy	410.0	444.3
Netherlands	225.7	267.9
Spain	203.2	183.7
European Union	7999.3	8346.0

3.4 Biobased chemicals and materials derived from food plants

3.4.1 Overview

Although transportation fuels are, and will be in the foreseeable future, the main output from a biorefinery, chemicals and materials derived from biomass are also products with significant economic potential. [35] Already commercially available biobased chemicals and materials include polymers, surfactants, lubricants, solvents, building block molecules, and composites.

Starch- and sugar-based products

Much attention has been paid to the starch- and sugar-based biorefineries. Starch-based biorefinery produces starch and carbohydrate derivatives from it, but also small amounts of oil, protein, and fiber. [35] Commercially significant chemicals and materials derived from starch are currently limited to nutritive sweetener and lactic acid used in the production of PLA. A number of carbohydrate derivatives are possible.

A joint report from the U.S. Department of Energy (DOE) identified twelve building block chemicals that can be produced from sugars via biological or chemical conversions. [36] These include three diacids (succinic, fumaric, and malic acids), seven carboxylic acids with additional functionality (2-5-furan dicarboxylic acid, 3-hydroxy propionic acid, aspartic acid, glucaric acid, glutamic acid, itaconic acid, and levulinic acid), one ketone (3-hydroxy-butyrolactone), and three polyols (glycerol, sorbitol, and xylitol/arabinitol). 1,3-Propanediol, a precursor in the production of plastics, has also been identified as an attractive building block molecule from carbohydrate.

Furthermore, the use of bioethanol-based ethylene as raw material for chemical products, especially for polyethylene, polypropylene, polyethylene terephthalate, and polyvinylchloride offers to the market an alternative to fossil-based commodity plastics. [37, 38]

Oil- and fat-based products

Other kinds of biobased products are derived from plant oils, animal fats, or recycled greases. For such an oleochemical biorefinery, the primary products are oil (triglyceride) and meal, the latter of which contains significant quantities of protein and fiber and some residual oil. [35] The oil can be either hydrolyzed to fatty acids and glycerol or converted into methyl (or ethyl) esters and glycerol by transesterification (FAME). The fatty acids and esters are potential platform chemicals for the production of many derivative chemicals used in high value products. Technologies already exist for upgrading of the glycerol by-product to 1,3-propanediol or acrylic acid. [39, 40]

3.4.2 Biobased polymers and plastics

Definitions and classification

Biobased polymers belong to the larger family of biopolymers. The meaning of the prefix "bio-" in biopolymer can refer either to "biobased" or "biodegradable" or "biocompatible." [41] Bioplastics are plastics manufactured using partially or entirely biopolymers. Plastics themselves refer to materials (usually polymers) that can be molded or shaped, commonly by heat and pressure.

Partially or entirely biobased plastics are obtained from natural polymers such as chitin, cellulose, starch, and proteins, which are extracted from biomass. These polymers can be mixed with synthetic polymers and/or additives, such as plasticizers, compatibilizers, and/or chemically modified to make them behave like plastics. This also includes polymers industrially produced from biobased precursors such as lactic acid or ethanol.

These biobased plastics may or may not be biodegradable or biocompatible. One development sector here is the replacement of traditional fossil-based materials by "green carbon"-based materials.

Partially or entirely biodegradable plastics are composed of either fully biobased polymers (like starch or modified starch), or partially biobased materials (like mixes of starch or natural fibers/particles with fossil-based plastics), or even fully fossil-based polymers (polyesters). In this case the targeted application concerns mainly materials that will have a visible impact on the environment (plastic bags and cups), such as disposable products.

In the last years, it has become increasingly clear that a very broad range of plastics can be produced fully or partially from biomass and that these plastics can be tailored to be fully or partially biodegradable (**Figure 3.19**). [42-44]

Current market and forecasts

Shen et al. [42] in 2009 estimated the global capacity of emerging biobased plastics at 0.36 Mt by the end of 2007. As shown in **Figure 3.20**, the worldwide capacity of biobased plastics, according to company announcements, will increase from 0.36 Mt in 2007 to 2.32 in 2013 and to 3.45 Mt in 2020. [42] This is equivalent to average annual growth rates of 37% between 2007 and 2013 and 6% between 2013 and 2020. The announced capacities in 2020 are 1.30 Mt starch plastics, 0.83 Mt PLA, 0.61 Mt biobased ethylene, 0.44 Mt PHA, 0.21 Mt biobased epichlorohydrin and 0.06 Mt other biobased plastics. These figures should be compared with a global plastics production capacity of over 200 Mt.

A study carried out in 2011 by the University of Applied Sciences and Arts in Hanover for the trade association European Bioplastics predicted that global bioplastics production capacity will reach 1.7 Mt in 2015, more than doubling 2010 capacity levels (**Figures 3.21** and **3.22**). [45] The traditional biodegradable and compostable polymers are expected to see only modest growth, while

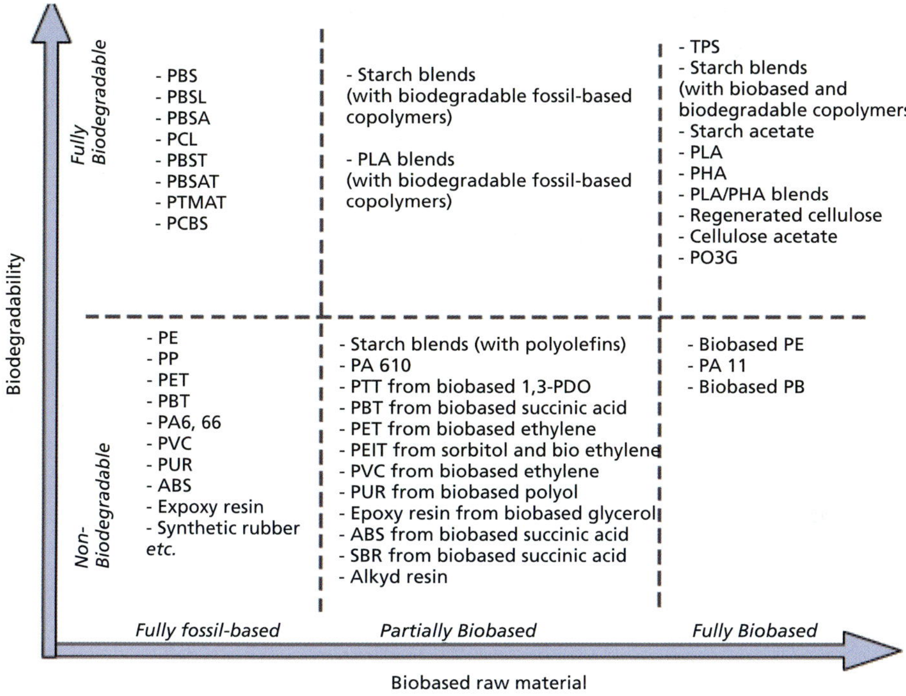

Figure 3.19 Classification of current and emerging plastics, depending on source and biodegradability. PBS, polybutyrate succinate, PBSL, polybutyrate succinate-co-lactate; PBSA, polybutyrate succinate-co-adipate; PCL, polycaprolactone; PBST, polybutylene succinate terephthalate; PBSTAT, polybutylene succinate-co-adipate terephthalate; PTMAT, polytetramethylene adipate terephthalate; PCBS, poly(L-cystine bisamide-g-sulfadiazine); PE, polyethylene, PP, polypropylene; PET, polyethylene terephthalate; PBT, polybutyrate terephthalate, PA, polyamide; PVC, polyvinyl chloride; PUR, polyurethane; ABS, acrylonitrile butadiene styrene; PLA, polylactic acid; PTT, polytrimethylene terephthalate; PEIT, polyethylene-co-isosorbite terephthalate; SBR, styrene-butadiene rubber; TPS, thermoplastic starch; PHA, polyhydroxyalkanoates; PO3G, polytrimethylene ether glycol; PB, polybutylene. Reproduced by permission of Martin PATEL/Utrecht University. [42, 44]

durable bioplastics, which accounted for 7% of the global bioplastics production capacity in 2009 and for 40% in 2010, are expected to reach 60% in 2015.

Table 3.8 indicates industry's estimate of future markets for biobased plastics in the European Union. The level of growth strongly depends on the presence – or not – of supportive EU policy measures. [46]

In Europe, biobased biodegradable plastics are predicted to grow strongly in applications such as waste and shopping bags, tableware, and mulch for agriculture (**Table 3.9**). [44, 47]

By comparison, market data for biobased lubricants and composites are shown in **Table 3.10**.

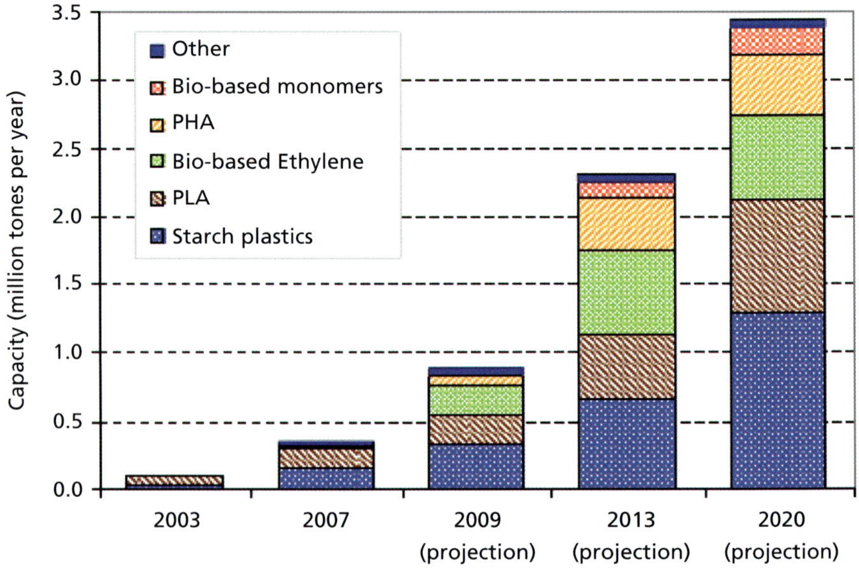

Figure 3.20 Worldwide capacity of biobased plastics until 2020. Reproduced by permission of Martin PATEL/Utrecht University. [42]

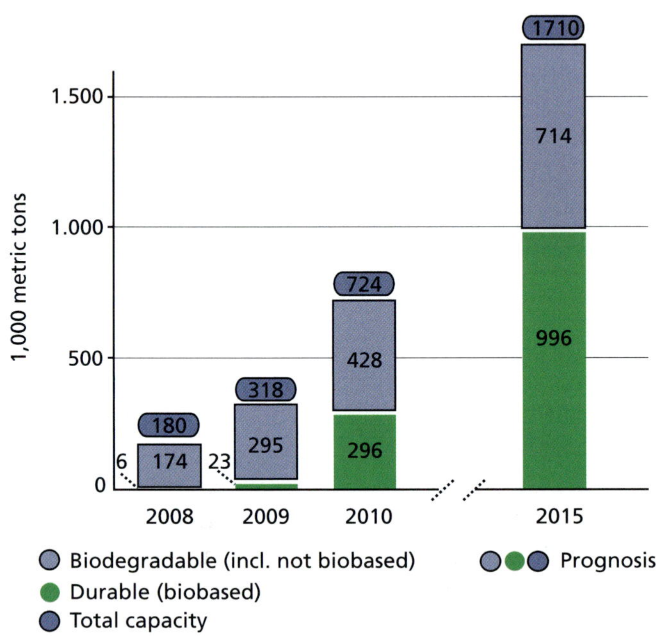

Figure 3.21 Global production capacity of bioplastics. Source: European Bioplastics/ University of Applied Sciences and Arts Hannover (Status May 2011). [45]

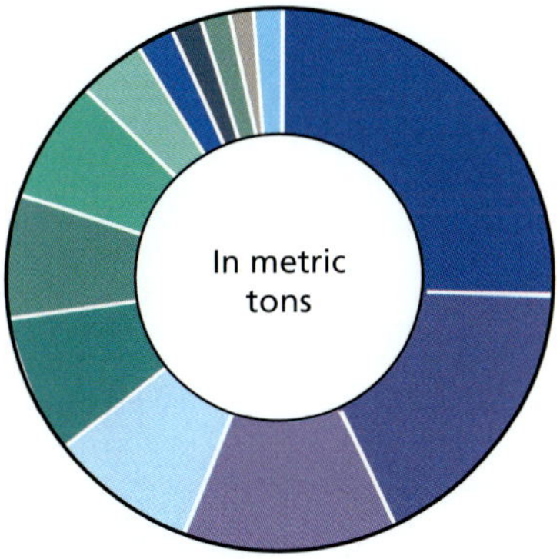

Type	Production capacity	Share
Bio-PE	450,000	26%
Bio-PET	290,000	17%
PLA	216,000	13%
PHA	147,100	9%
Biodegradable Polyesters	143,500	8%
Biodegradable Starch Blends	124,800	7%
Bio-PVC	120,000	7%
Bio-PA	75,000	5%
Regenerated Cellulose[1]	36,000	2%
PLA-Blends	35,000	2%
Bio-PP	30,000	2%
Bio-PC	20,000	1%
Others	22,300	1%
Total	1,709,700	100%

[1] *only hydrated cellulose foils*

Figure 3.22 Global biopolymers production capacity 2015 (by type). Source: European Bioplastics/University of Applied Sciences and Arts Hanover. [45]

Table 3.8 Biobased plastics: market situation 2010 and market forecast 2020. [46]

Application area	EU market volume total 2010 (t)	EU market volume 2020 without EU policy measures (t)	EU market volume 2020 with EU policy measures (t)
Short-life/ disposable	110,000	310,000	1,280,000
Durable	150,000[1)]	459,000	1,270,000
Total	260,000	769,000	2,550,000

1) including polyolefins (2012)

In the following sections devoted to first-generation biobased/biodegradable plastics and biobased/non biodegradable plastics, this chapter does not provide a complete overview of first-generation bioplastics. For the biobased/biodegradable category, polylactic acid (PLA), starch, and polyhydroxyalkanoates (PHAs) have been selected; and for the biobased/non-biodegradable category, green polyamides, green polyurethanes, green polyolefins and green PVC, green polyethylene terephthalate (PET), and Gaïalene derived from starch have been selected. Some other biobased plastics and their biocomponents will be discussed in Chapters 8, 9, and 10.

Table 3.9 Biobased biodegradable plastics: market situation 2010 and market forecast 2020. [46, 47]

Application type	Market volume EU total 2010 (t)	Market volume EU "bio" 2010 (t)	Market volume EU total 2020 (t)	Market volume EU "bio" 2020 (t)[1)]
Waste & shopping bags	1,000,000	30,000	480,000	260,000
Tableware	220,000	3,000	220,000	33,000
Mulch for agriculture	147,000	2,000	147,000	40,000
Total	1,367,000	35,000	847,000	333,000

1) with EU policy measures

Table 3.10 Biobased lubricants and composites: market situation 2010 and market forecast 2020. [46]

Biobased product	EU market volume 2010 total (t)	EU market volume 2010 "Bio" (t)	EU market volume "Bio" 2020 low legislation level (t)	EU market volume "Bio" 2020 high legislation level (t)
Lubricants	4,400,000	137,000	277,000	420,000
Composites	2,400,000	315,000	nd	830,000[1)]

1) EU total composite market 2020: 3,000,000 t

Biobased and biodegradable plastics

Polylactic acid (PLA)

PLA is one of the important biodegradable biobased polymers. Although there are several ways to obtain high molecular weight PLA, most industrial processes use polymerization of cyclic lactide intermediates obtained from prepolymers of lactic acid enantiomers (**Figure 3.23**).

The first step in PLA synthesis is the obtention of lactic acid (mix of S (or L) and R (or D) isomers) by bacterial fermentation of sugar from sugar/starch crops. Several bacterial systems can be used and produce different ranges of R/S ratios. [49] Homolactic fermentation is mainly industrially used, transforming hexoses into lactic acid with a yield of ~90%. Lactic acid is widely synthesized in the world (~200 kt/year) for application in food or cosmetics.

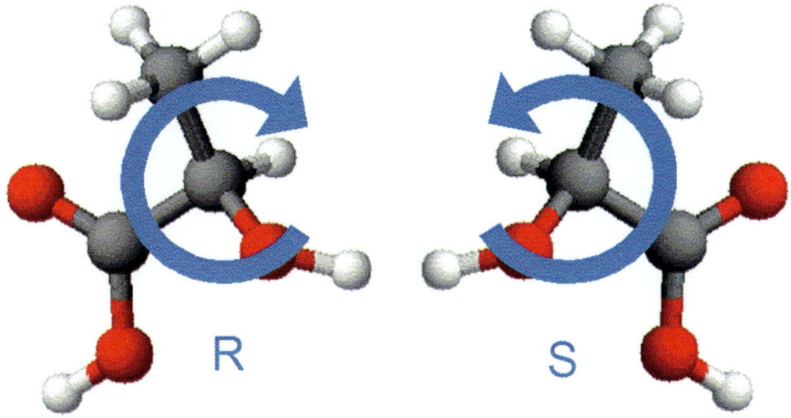

Figure 3.23 The two optical isomers of lactic acid: R-(or D)-lactic acid and S- (or L)-lactic acid [48]

High molecular weight polymers cannot be obtained directly by polyaddition because of large amounts of water released by the reaction. Prepolymers are, therefore, depolymerized and cyclized with the help of catalytic systems to produce intermediate cyclic lactides (dimer) that are then condensed through ring-opening polymerization (**Figure 3.24**). Another route for high molecular PLA is azeotropic dehydrative polycondensation of lactic acid. [50]

As three different lactide forms (**Figure 3.25**) coexist: L-lactide (also called (S,S)-lactide), D-lactide (also called (R,R)-lactide) and meso-lactide (also called (S,R)-lactide), and different catalysts are used for polymerization, different PLA polymer structures and tacticities can be obtained with quite various properties for final materials:

- Atactic/amorphous PLA (A-PLA) is obtained from mixtures of R and S lactide or from meso-lactide. It is transparent, with low tenacities, has no melting point and can be used for film extrusion.
- Homopolymers (L-PLA or D-PLA) are isotactic and show crystallinity with melting temperatures of ~160–180 °C. Industrially L-PLA is mainly used (e.g. in the NatureWorks process) after purification of L-lactide by distillation. These polymers can be used for extrusion or injection molding.
- Copolymers composed of isotactic stereoblocks can be obtained from racemic lactide (equimolar mixture of L and D forms). The PURAC Company (Thailand) delivers both pure L or D lactide to customers for selective copolymerization.

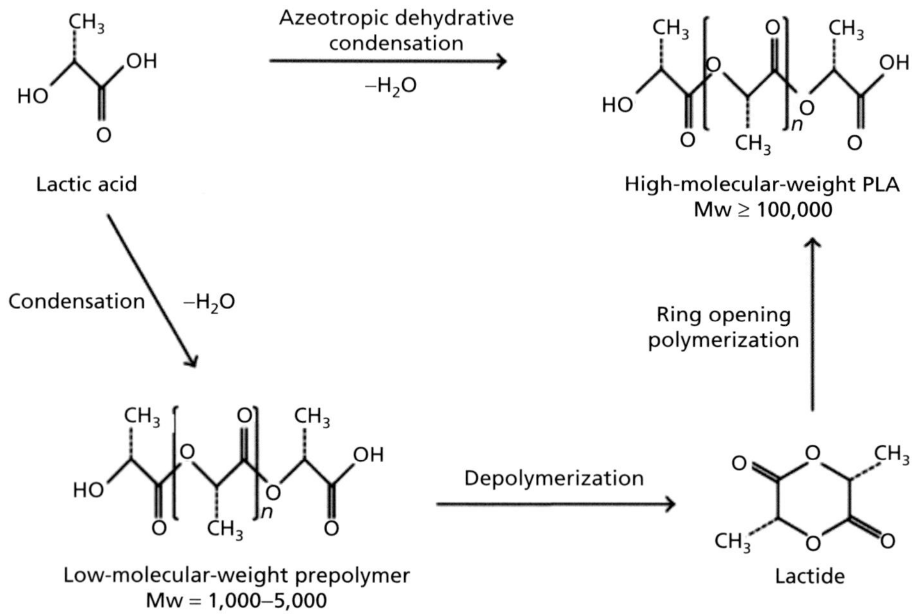

Figure 3.24 Production routes of lactides and PLA. [51]

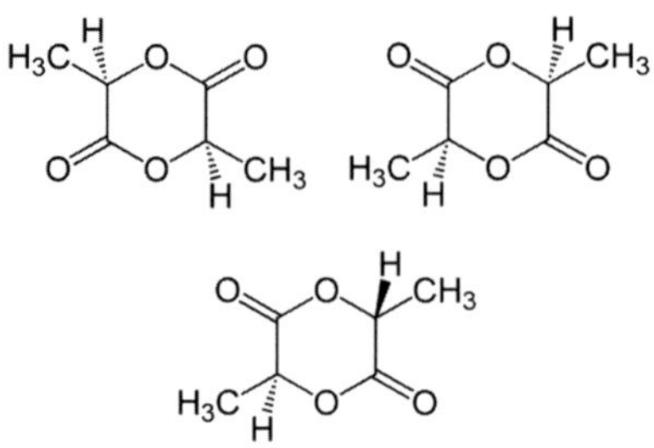

Figure 3.25 The 3 lactide forms: (R,R)-lactide or D-lactide (left above), (S,S)-lactide or L-lactide (right above) and (S,R)-lactide or meso-lactide (bottom). [52]

Polymerization mechanisms and stereochemistry control are critical for controlling PLA properties: research work on catalysts and conditions of ring opening polymerization is currently intensive. [53, 54]

Typical commercial grades of PLA are compatible with standard equipments, such as extrusion, injection molding, and calendaring, and can substitute PP, PS, or PET. Main applications are now in packaging (cups, bags, pots, bottles) and fibers (**Table 3.11**).

Table 3.11 Some relevant properties of PLA compared with PS and PP. [55]

	PP	PS	PLA
Crystallinity	70	0	Up to 80
Melting point (°C)	170	–	170–180
T_g (°C)	–15	95	57-60
E-modulus (GPa)	1.4	3.3	3.7
Tensile strength (MPa)	50	55	54

PLA is brittle, sensitive to heat, and quite permeable to water vapor, which is a handicap considering water bottling applications. Melt flow is also low, hindering foaming or melt blowing. Solutions are being developed in these areas by blending with additives or polymers.

Today, NatureWorks, a wholly owned subsidiary of Cargill, is the largest producer of PLA with a 140 kt/y capacity production facility (**Figure 3.26**). [56]

In 2007 Purac decided to enlarge lactide production capacity to 75 kt/y and developed specific partnerships with Sulzer, an engineering company, to sell both L-and D-lactides to customers and polymerization process for PLA (**Figure 3.27**).

Today Purac and Sulzer claim new grades of PLA, called "stereo-complex," with very high melt temperature (230 °C). They are produced in static mixer

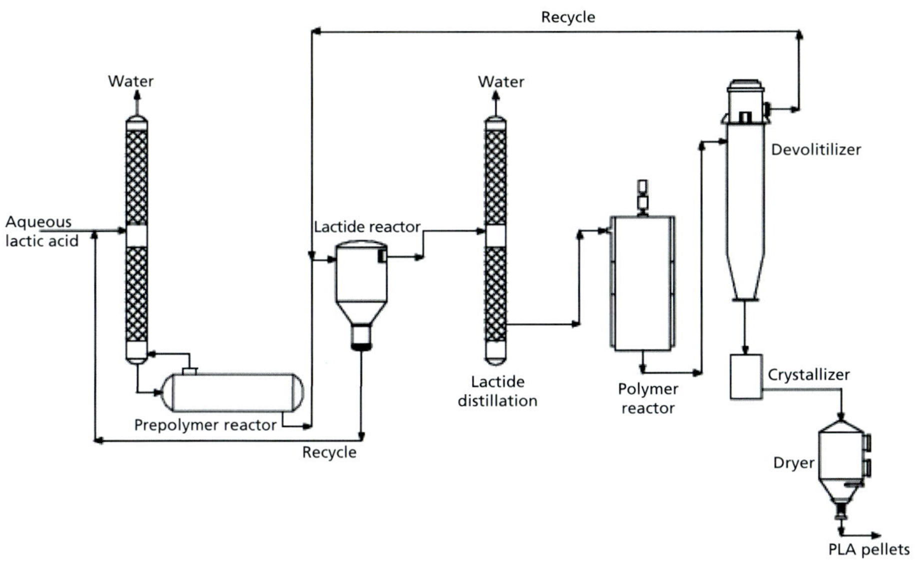

Figure 3.26 NatureWorks process for PLA production. [57]

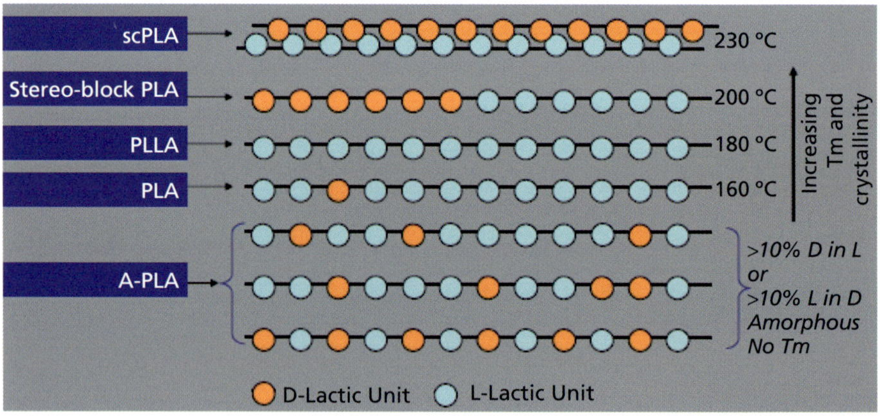

Figure 3.27 Different grades of PLA and their properties. A-PLA, amorphous PLA; PLLA, poly L-lactic acid; scPLA, stereo-complex PLA. [58]

units, PDLA acting as a catalyzer and complexing agent for PLA crystallization. This could be the second generation of PLA (scPLA), although one can notice that with the increase of crystallinity and thermal stability, the material shall become less biodegradable.

Futerro, a joint venture between Galactic and Total, agreements between Arkema/Synbra and Purac, the creation of entities or technologies like Loopla, Biocor, and Plarco for recycling PLA into lactic acid are the sign that the PLA adventure is only just beginning.

Starch derivatives and blends

In its natural form, starch is not thermoplastic and dissolves readily in water. But it can be plasticized/modified and blended/compatibilized with other polymers to get products with various properties and applications. Depending on the affinity of starch with other components of the mixture, several structure types can be shown, including nodules, co-continuous phase, or fully homogeneous mix. Disruption of granular structure of starch, addition of polymer, plasticizers, and additives are generally carried out in a continuous process using twin-screw extruders (also called reactive extrusion). Some ageing or fermentation steps can be added (**Figure 3.28**).

Most starch plastics and blends are produced for applications in packaging: total amount produced being around 150-200kt/y. Main producers are Novamont, IT (Mater-Bi®), Rodenburg, NL (Solanyl®), Biotec, DE (Bioplast®), Limagrain, FR (Biolice®), Biop, DE (BioPar®), PaperFoam, NL (Paperfoam®), Livan, CN (Livan®), Plantic, AU (Plantic®), Cereplast, USA (Cereplast®) and Biograde, AU/CN (Biograde®).

Novamont is clearly the leader in the sector of biodegradable materials from starch. The Mater-Bi trade name covers tailor made materials for a wide

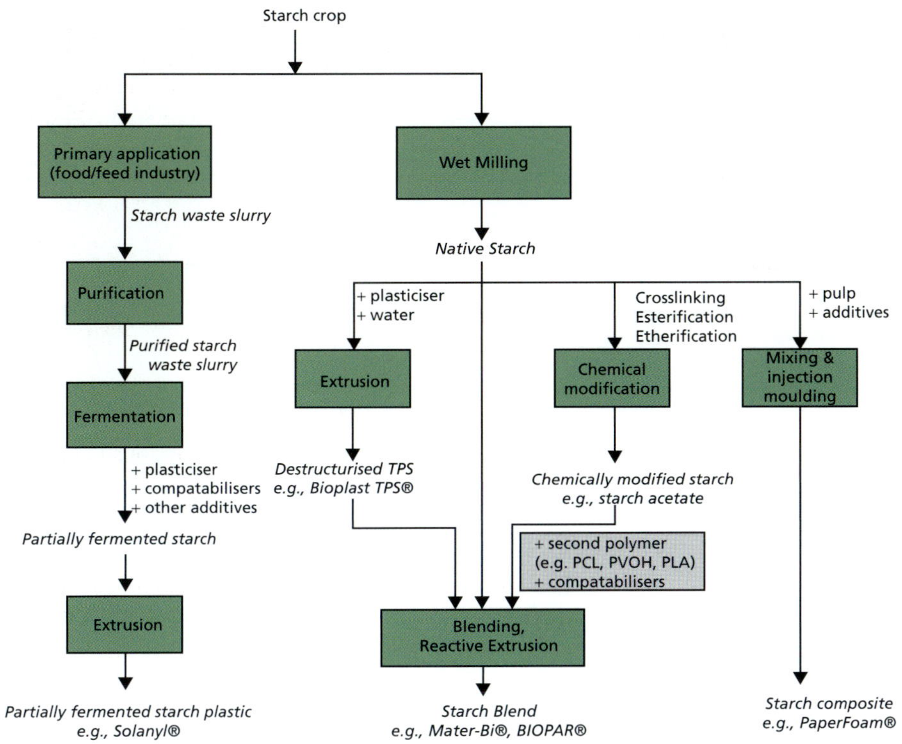

Figure 3.28 Plastics production technologies from starch. Reproduced by permission of Martin PATEL/Utrecht University. [41]

range of industrial applications. Now with 100 patents in the field of starch modifications and blends, Novamont has oriented its R&D efforts toward a biorefinery concept for developing more and more sophisticated biobased products (**Figure 3.29**).

Polyhydroxyalkanoates (PHAs)

Being produced by bacterial fermentation, PHAs are polyesters well positioned to be named *biobased polymers* (**Figure 3.30**).

PHAs are synthesized by a wide range of different microorganisms from carbohydrates, sugars, biomass residues, etc. Properties vary significantly depending on chemical structure, and research is underway to increase production/extraction yields and find copolymers with interesting properties (**Table 3.12**).

Main commercial grades are PHB (also called P3HB) and copolymer PHBV (also called P3HB-3HV). PHAs are considered to be the broadest biobased polymer because they are their own class and can have many different chemical structures. [61] Ranging from stiff and brittle plastics to rubber-

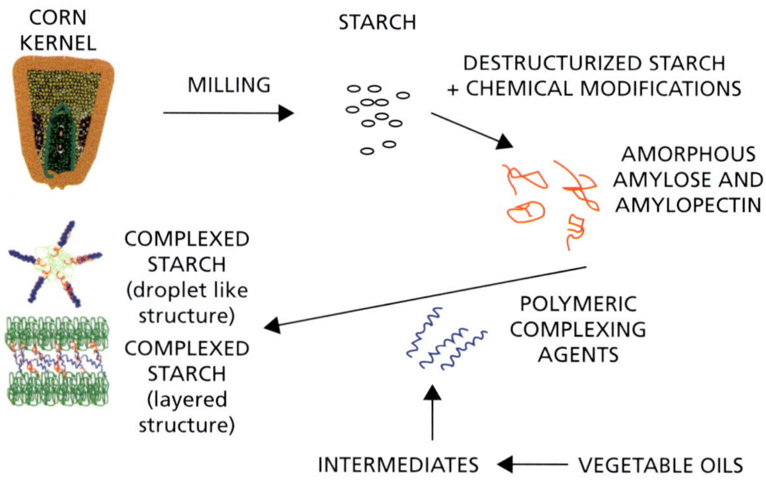

Figure 3.29 NOVAMONT's starch/vegetable oils technology. Reproduced by permission of NOVAMONT. [59]

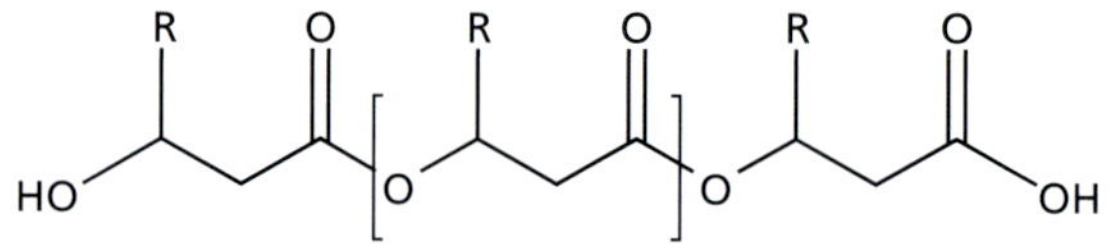

Figure 3.30 General formula of PHAs: in poly(3-hydroxybutyrate) (PHB), R = CH_3; in poly(3-hydroxybutyrate-co-3-hydroxyvalerate (PHBV), R = CH_3 or C_2H_5. [60]

like materials, these polymers have a wide range of properties. They also can exhibit properties similar to many synthetic polymers. However, they are more expensive than PLA. They are generally used in packaging.

Global production of PHAs is still confidential (20–30 kt/y), despite announcements made during the last decade. For more than 10 years now, Metabolix has been announcing ~50,000 t/y capacity installations in Great Britain but has been postponing regularly the availability of commercial grades. In 2012, Metabolix announced that the Archer Daniels Midland Company (ADM) has given notice of termination of the Telles, LLC joint venture for PHA bioplastics. [62] In 2007 Meridian, Inc. bought Procter & Gamble technology for PHAs, but since then, nothing more has been noticed. It seems that interesting developments are happening in Asia where, for example, the Chinese company Tianan is very active on PHBV.

Biobased and not biodegradable plastics

Green polyamides (PA)

Biobased plastics are about 30% of Arkema technical polymers. The French company exploits its over 60 years of knowledge in castor oil chemistry. Castor

Table 3.12 Selected microorganisms with typical PHAs synthesized. [52]

Microorganisms	PHA synthesized	PHA chemical structure
Ralstonia eutropha	PHB [poly(3-hydroxybutyrate)]	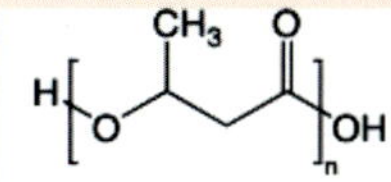
	PHBV [poly(3-hydroxybutyrate-co-3-hydroxyvalerate)]	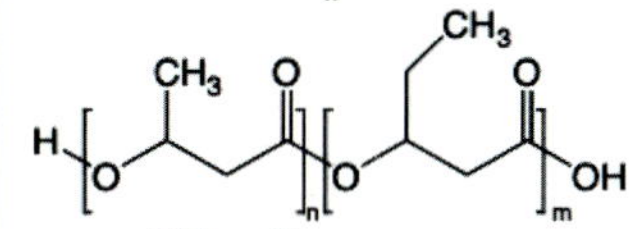
	P3HB4HB [poly(3-hydroxybutyrate-co-4-hydroxybutyrate)]	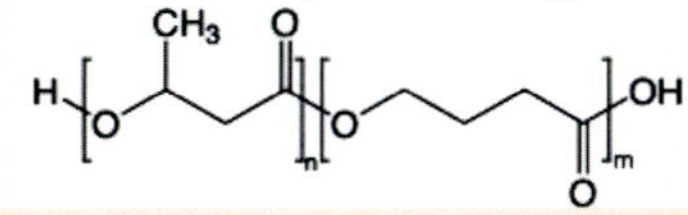
Aeromonas hydrophila	PHBHHx [poly(3-hydroxybutyrate-co-3-hydroxyhexanoate)]	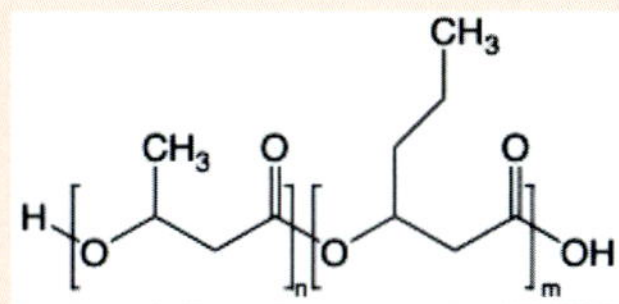
Pseudomonas putida	PHOHHx [poly(3-hydroxyoctanoate-co-3-hydroxyhexanoate)]	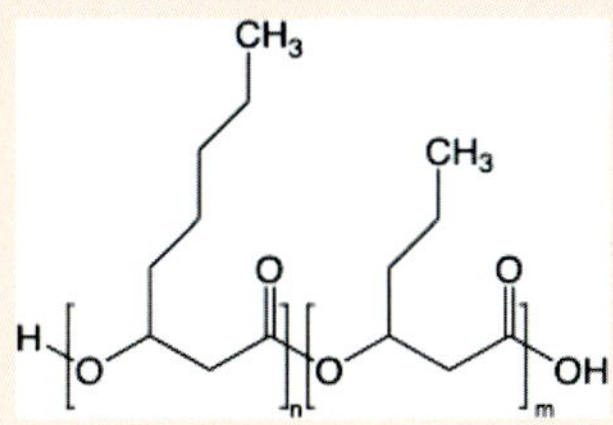

oil is extracted from the bean of the castor plant (*Ricinus Communis*) and is mainly composed of triglycerides of ricinoleic acid (an 18 carbon acid having a double bond in the 9–10 position and a hydroxyl group on the 12th carbon) (**Figure 3.31**). After saponification, the fatty acid is used for synthesis of various compounds including bioplastics. After several reaction steps, the acid is transformed into amino-11-undecanoic acid leading to polyamide 11 or Rilsan®.

Pebax® is a polyether-block-polyamide (PEBA) compound with elastomeric properties (**Figure 3.32**). It can be used as an additive or in blends for technical applications such as ski shoes. Now Arkema claims up to 100% biosourcing on both Rilsan® and Pebax® ranges.

Green polyurethanes (PU)

Many major chemical companies are developing the concept of biobased polyols for production of polyurethanes. The applications are diverse, including

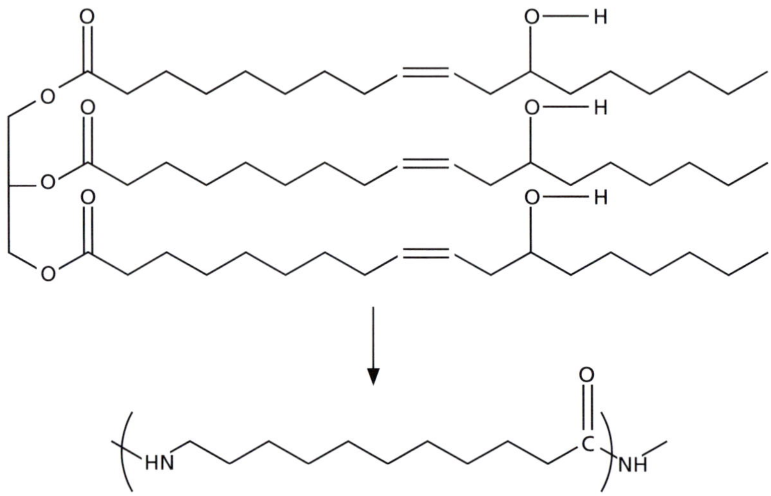

Figure 3.31 From castor oil to polyamide 11 (Rilsan®).

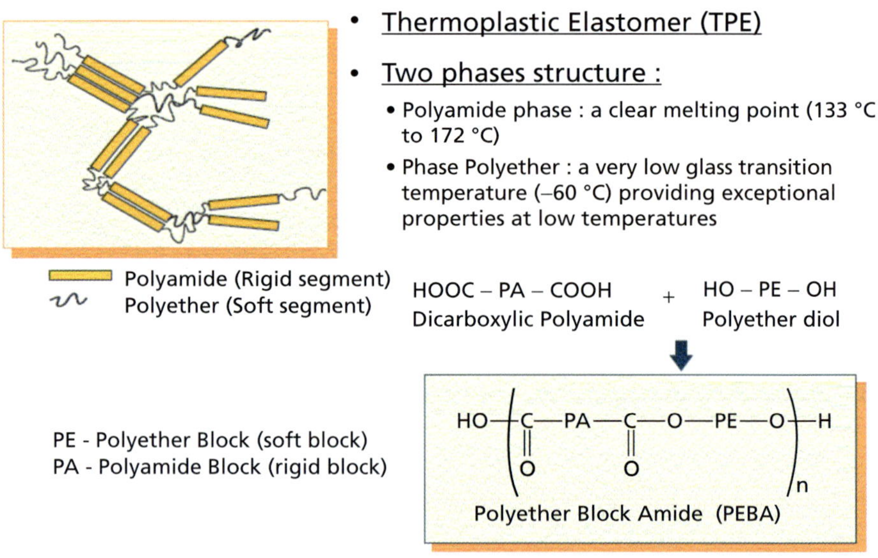

Figure 3.32 Pebax® structure and properties (courtesy from ARKEMA).

resilient foams for car cushioning and thermoplastic polyurethanes (TPU) for molding applications. Natural resources can be of many kinds (starch, soy bean, animal fats, etc.).

DuPont's integrated corn-based biorefinery produces bio-1,3-propanediol (Bio-PDO®) from corn by fermentation/separation steps through a joint venture with Tate & Lyle. DuPont claims a 40% energy savings and a 20% GHG emission reduction, compared to standard production from petroleum. [63] Bio-PDO® is then polymerized into several polyols for polyurethane production.

Cargill BiOH® polyols are based on soybean oil and used for the production of flexible foams. In contrast to castor oil, soybean oil is mainly composed of polyunsaturated fatty acid derivatives (triglycerides of linoleic acid) and contains no free hydroxyl groups on the chain. Therefore, these hydroxyl groups have to be incorporated by ozonolysis or epoxidation. Cargill says its soybean oil-based polyols, under the brand BiOH, can allow its customers to create products with typical levels of renewable content from 5% to 20%. [64]

Green polyethylene (PE), polypropylene (PP) and polyvinyl chloride (PVC)

In 2010, Procter & Gamble unveiled a new long-term environmental vision concerning sustainability, which includes using 100% renewable or recycled materials for all products and packaging. The partnership with the Brazilian company Braskem ensures exclusive use by Procter & Gamble of green PE produced from sugar cane for the packaging of cosmetic products.

Operational in July 2010, Braskem's production unit is the first in the world to produce 100% ethylene from a renewable resource (200 kt/year). [65] Ethylene is produced from ethanol through catalytic dehydration, with the production being fully transformed into green PE. The process yields to both high density PE (HDPE) and low density PE (LDPE). Recently, Braskem developed a partnership with Lyondellbasell for a new LDPE production process and with Novozymes for a green PP production process.

Sugar cane resource should also be used in Brazil for a green PVC production unit planned by Solvay Indupa.

Green polyethylene terephthalate (PET)

The Coca-Cola Company (Dasani®) and Danone (Volvic®) have recently claimed new biobased PET bottles for water.

In this case, the PET used for bottle manufacturing is partially recycled and partially biosourced. Ethylene glycol or monoethylene glycol (MEG) needed for PET manufacturing is obtained from sugar cane ethanol via ethylene. The green carbon content of the final product is typically ~10–30%. The Japanese Company Toyota Tsusho has recently claimed to establish the first integrated supply chain of bio-PET in the world. [66]

Gaïalene derived from starch

In end of 2010, the French company Roquette, specialized in starch and starch derivatives, launched a product called GAÏALENE®: semi-crystalline thermoplastic resins produced from starch using an advanced grafting technology and designed for replacing standard polyolefins, PS, ABS, or PVC. [67]

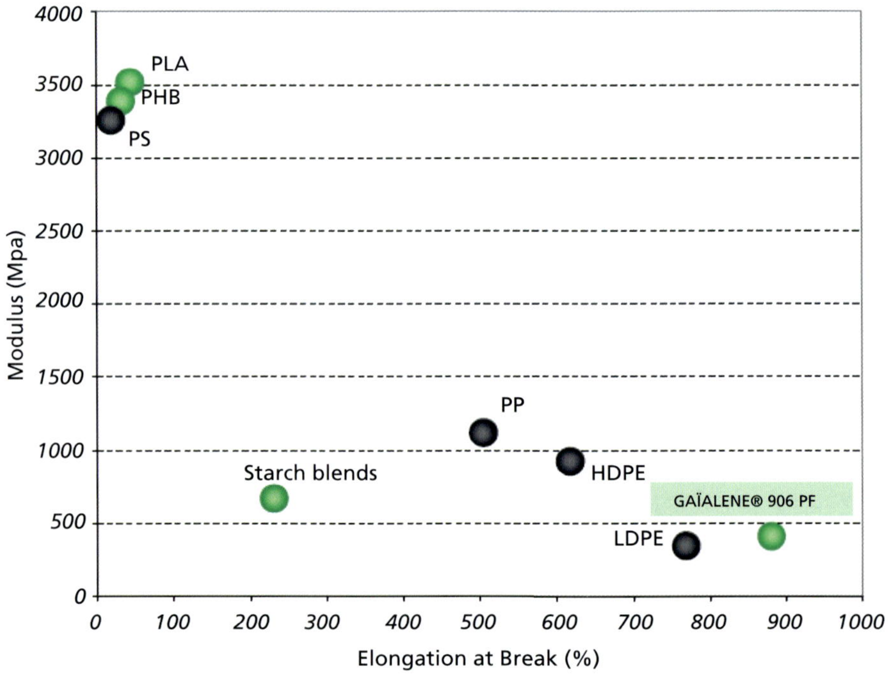

Figure 3.33 GAÏALENE® mechanical properties compared to other resins used for films (Courtesy of Roquette).

GAÏALENE® resins are clearly branded as biobased but are not biodegradable. The first grades of these plant-based resins are 52% biobased, can be processed on same equipments as usual by injection molding, blow film extrusion, extrusion blow molding, or compounding – but at relatively low temperatures (160–180°C) (see properties **Figure 3.33** and **Table 3.13**). These resins are fully recyclable. Roquette has now a capacity of 25,000 tons per year for these plant-based resins.

Table 3.13 GAÏALENE® typical mechanical properties and densities (Courtesy of Roquette).

	GAÏALENE® resins
Density	1.09–1.12
MVR (190 °C/10 kg)[1)]	40–110 cm^3/10 min
Flexion modulus	400–1300 MPa
Strain at break	70–900%
HDT B (0.45 MPa)[2)]	40–90 °C
Biobased content	52%

1) MVR, melt volume-flow rate 2) HDT, heat deflection temperature

3.5 Environmental performance

3.5.1 Life cycle assessments applied to first-generation biofuels

A study of life cycle assessments (LCA) applied to first-generation biofuels used in France was carried out for ADEME in 2010. [68] It provides not only an update in the biofuels' energy balances and greenhouse gas emissions, but also an evaluation of three less studied impact indicators (eutrophication, photo-oxidation, and human toxicity).

It addresses the following pathways:

- Ethanol from wheat, corn, sugar beet, and sugar cane, as well as corresponding ETBE.
- Vegetable oil methyl esters (VOME) from rapeseed, sunflower, soybean, and palm, methyl esters from animal fat (AFME) and used cooking oil (WVOME, waste vegetable oil methyl ester).
- Pure vegetable oil (PVO) from rapeseed.

The study highlights the extreme complexity of energy and environmental balance evaluation process of first-generation biofuels "from well to wheel."

Generally, without taking into account the effects of land-use changes, biofuels produced in France (biodiesel and ethanol) display energy balance and emissions of greenhouse gases more favorable than those of reference fossil fuels (diesel and gasoline). Comparing energy and greenhouse gas emissions balances enables to group biofuels in three categories (**Figure 3.34**):

- Pathways with lowest benefits (reduction of less than 25% for non-renewable energy consumption and less than 50% for GHG emissions): ETBE from wheat, corn, and beet.
- Pathways with average benefits (reduction from 49–85% for non-renewable energy consumption and 47–77% for GHG emissions): ethanol from wheat, corn, beet, sugarcane, ETBE from sugarcane, biodiesels from colza, sunflower, palm, soy, pure vegetable oil (colza).
- Pathways with best benefits (reduction greater than 80% for non-renewable energy consumption and 90% for GHG emissions): biodiesels from waste vegetal oil and animal fat.

The study also underlines the impact of land-use changes that can be discriminative. When the development of energy crops leads, directly or not, to deforestation or disappearance of grasslands, humid zones, or primary forests, biofuel GHG balance can become negative.

Several other parameters can have significant impacts on the GHG emission balances of biofuel pathways, such as nitrous oxide emissions, which depend on local conditions (agronomic practices, soils, climate, and meteorology).

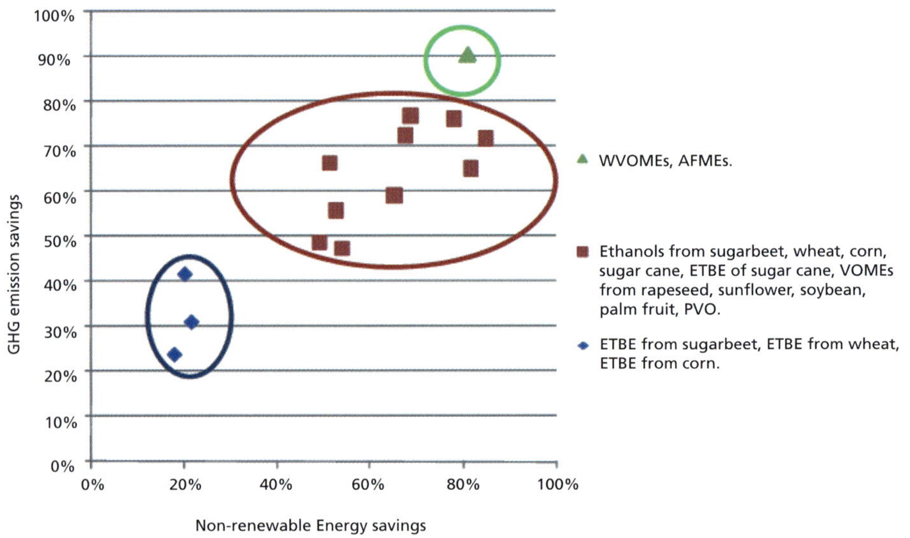

Figure 3.34 Energy and GHG emissions balances (exclusive of land-use changes) of first-generation biofuels compared to fossil fuels; For abbreviations, see text. [56]

Finally, the complexity of environmental impacts and the difficulty of modelling them complicates the reading of results from the three complementary indicators.

3.5.2 Sustainability of biofuels in Europe

The EU regulatory framework sets ambitious targets for the market uptake of biofuels in the transport sector by 2020. [69] Of particular relevance are the Directives on the "Promotion of the Use of Energy from Renewable Sources" (2009/28/EC) RED and the so-called "Fuel Quality Directive" (2009/30/EC) FQD. Directive 2009/28/EC includes a 10% target of energy from renewable sources in transport by 2020.

The framework sets environmental sustainability criteria for both biofuels and bio-liquids, with a threshold of 35% savings of GHG emissions and excludes the use of specific land-use categories, such as primary forests, wetlands, peatlands and areas with high biodiversity.

During 2009 and 2010, the European Commission (EC) worked intensively to give a better understanding of ***indirect land use change*** (ILUC) effects from increased first-generation biofuels use. If crops grown on existing arable land are used to make biofuels instead of food, this will likely cause ILUC because of the necessity to replace the food. Due to changes in the carbon stock of the soil and the biomass, indirect land-use change has consequences in the greenhouse gas (GHG) balance of a biofuel.

A new system, developed by the EC's Joint Research Center and adopted in June 2010 by the European Commisssion, encourages industry, goverments, and ONGs to set up voluntary certification schemes for all types of biofuels. It will help to ensure that all biofuels (including those imported into the EU) are sustainable and deliver high GHG savings (at least 35%) when compared to fossil fuels. This excludes the use of specific land categories, such as primary forests, wetlands, peatlands, and areas with high biodiversity.

EU law also requires the assessment of GHG emissions from both direct and indirect land-use changes (ILUC), N_2O emissions, soil carbon stock changes, definition of degraded lands, and updating and implementation of new biomass pathways. [69-71] Indirect land use change, which is now heatedly debated, occurs when the production of crops for biofuels in a given land pushes the previous activity to another location. If crops grown on existing arable land are used to make biofuels instead of food, this will likely cause ILUC because of the necessity to replace the food. Due to changes in the carbon stock of the soil and the biomass, indirect land-use change has consequences in the greenhouse gas (GHG) balance of a biofuel.

The sustainability assessment required by the EU law must include topics which may have positive or negative impacts on biofuel policies in tropical countries, including: air and soil quality, water footprint, food prices, and farm incomes.

An assessment of current and future production of second-generation biofuels is also required. Although second-generation biofuels are believed to offer sustained GHG advantages while reducing pressure on natural resources, conversion is both more complicated and more expensive.

End 2012, a draft EU proposal to eliminate food-crop subsidies for biofuels production and cap crop-based biofuels to 5% of EU transportation fuels was under consideration. [72] At present, biofuels represent 4.5% of the EU's transportation fuel consumption, essentially coming from food crops, so that food crop-based fuels would be capped at not much above their current levels. The EU was suggesting the establishment of quadruple-counting for advanced, non-food biofuels feedstocks, such as algae or waste residues. [73] For food-crops, this can mean a shift from biofuels to higher-value biobased products, which are generally unsubsidized and unmandated and offer good returns on investment for selected crops, such as maize. To the extent that biofuels producers and their existing plants can support different feedstocks, such as renewable sugars made from wastes, or waste-based fats, oils, and greases, the impact would be small. Another possibility for them would be to look at complementary technologies such as a shift from producing ethanol to the production of isobutanol or n-butanol for chemical markets.

3.5.3 Life cycle assessments of biobased polymers

LCA studies indicate that biobased polymers offer important environmental benefits not only for today but also for the future. [74]

Of all biobased plastics (pellets) studied, starch polymers are considered to perform best in environmental terms under the current state of the art, albeit with some differences among the various types of starch polymers. [74] Most LCA studies come to the conclusion that starch polymers (pellets and end products) are more beneficial in environmental terms than their petrochemical counterparts. Compared with starch polymers, the environmental benefits seem to be smaller for PLA. For PHA, the environmental advantage currently seems to be very small compared with conventional polymers.

In spite of uncertainties linked to e.g. impact assessment, it can be concluded that the results for the use of fossil energy resources and GHG emissions are already more favorable for most biobased polymers. [74] Overall, the available LCAs and environmental assessments strongly support the further development of biobased polymers.

3.6 References

[1] A.S. BATISTA, L.C. MILETTI and B.U. STAMBUK, *J. Mol. Microbiol. Biotechnol.* 8, 26, 2004 in http://www.ncbi.nlm.nih.gov/pubmed/15741738

[2] M.TUITE, *Yeast Prions and Chaperones* , Picture of Yeast Saccharomyces Cerisiae, 2010 in www.kent.ac.uk/bio/tuite/research.html

[3] BNDES and CGEE, *Sugarcane-based Bioethanol Energy for Sustainable Development*, Rio de Janeiro, BNDES, 2008 in http://www.sugarcanebioethanol.org/en/download/bioetanol.pdf

[4] S. PEREZ, M. KOUWIJZER, K. MAZEAU and S.B. ENGELSEN, *Modelling Polysaccharides: Present Status and Challenges*, CERMAV in http://glyco3d.cermav.cnrs.fr/glyco3d/lessons/modeling/index.html

[5] CERMAV in http://glyco3d.cermav.cnrs.fr/glyco3d/lessons/starch/page.php.73.html

[6] R. STEANE, BioTopics.co.uk, *The Amylopectin Molecule* in http://www.biotopics.co.uk/JmolApplet/amylopectinjdisplay.html

[7] http://en.wikipedia.org/wiki/Amylopectin

[8] L. SHEN, J. HAUFE and M.K. PATEL, *Product overview and market projections of emerging bio-based plastics, Universiteit* Utrecht, 2009 in http://nws.chem.uu.nl/publica/Publicaties%202009/PROBIP2009%20Final%20June%202009.pdf

[9] C. DRAPCHO, J. NGHIEM and T. WALKER, *Biofuels Engineering Process Technology*, The McGraw-Hill Companies, Inc., 2008 in http://www.mhprofessional.com/downloads/products/0071487492/DrapchoCh4.pdf

[10] J. LARSEN, *Grain Harvest*, Earth Policy Institute, 2012 in http://www.earth-policy.org/indicators/C54

[11] FEDIOL *Oil characteristics, Chemical composition*, 2012 in http://www.fediol.eu/web/chemical%20composition/1011306087/list1187970121/f1.html

[12] http://en.wikipedia.org/wiki/Triglyceride

[13] Fediol, *Oil characteristics, Fatty Acids*, 2012 in http://www.fediol.eu/web/fatty%20acids/1011306087/list1187970122/f1.html

[14] R.B. GUPTA and A. DEMIRBAS, Gasoline, Diesel and Ethanol Biofuels from Grasses and Plants, Cambridge University Press, 2010

[15] J.A. QUINTERO, M.I. MONTOYA, O.J. SANCHEZ, O.H. GIRALDO and C.A. CARDONA, Energy 33, 385, 2008 in http://biotecnologia.ucaldas.edu.co/publicaciones/Article%20EtOHprdn%20from%20cane%20and%20corn%20-%20compar%20analysis%20for%20Colombian%20case.pdf

[16] Informa Economics "A study assessing the opportunities and potential of corn-based products and technologies" (August 2009)

[17] ETBE synthesis from ethanol and isobutylene, WIKIPEDIA, 2011 in http://en.wikipedia.org/wiki/Ethyl_tert-butyl_ether

[18] J. BOELHOUWER A view on the mtbe and etbe market, World Biofuels Markets Conference 2010

[19] D. I. BRANSBY, *Cellulosic Biofuels Technologies*, Auburn University, 2007 in www.adeca.alabama.gov/C17/ATF/Document%20Library/Cellulosic%20Biofuel%20Technologies.pdf

[20] L. BOURNAY, D. CASANAVE, B. DELFORT, G. HILLION and J.A. CHODORGE, *New heterogeneous process for biodiesel production: A way to improve the quality and the value of the crude glycerin produced by biodiesel plants,* Catal Today 106, 190, 2005 in http://www.virtualmaze.com/sample/Biofuels%20Info/BioFuels/biofuelCatalysis.pdf and http://www.biorefinery.ws/en/bio_1stgen.php

[21] S.R SCHILL, *Heeding Hydrogenation*, Biodiesel Magazine, 2007 in http://www.biodieselmagazine.com/articles/1505/heeding-hydrogenation

[22] European Biofuels, Technology Platform, *Biomethane for use as a transport fuel*, 2009 in www.biofuelstp.eu/biogas.html

[23] J.H. PATIL, M.A LOURDU,.A. RAJ, C.C. GAVIMATH and B.M. MALI, *Int. J. Adv. Biotechnol. Res. 2*, 257, 2011 in http://bipublication.com/files/IJABRv2i2201105.pdf

[24] Industrial & Environmental Concepts, Inc., *Stages of Anaerobic Digestion Part 1*, 2010 in http://floatingcovers.wordpress.com/2010/05/06/stages-of-anaerobic-digestion-part-1

[25] S.M. KEDDY, *Modernizing Solid Waste Management in Rural Macedonia, Anaerobic Digestion of the Organic Fraction of Municipal Solid Waste*, 2006 in http://www.ewbuf.org/docs/ISWMP-AB.pdf

[26] U.S. Department of Energy, Energy Efficiency and Renewable Energy, Tribal Energy Program, Guide to Tribal Energy Development, *Anaerobic Digestion*, 2008 in http://www1.eere.energy.gov/tribalenergy/guide/printable_versions/biomass_biopower.html

[27] BP, *Statistical Review of World Energy 2012, Review by energy type, Renewable energy, Biofuels*, 2012 in http://www.bp.com/sectiongenericarticle800.do?categoryId=9037217&contentId=7068633

[28] BP, BP Statistical Review of World Energy June 2012 in http://www.bp.com/assets/bp_internet/globalbp/globalbp_uk_english/reports_and_publications/statistical_energy_review_2011/STAGING/local_assets/pdf/statistical_review_of_world_energy_full_report_2012.pdf

[29] Global Renewable Fuels Alliance, 2012 in http://www.globalrfa.com/pr_062612.php

[30] Toepfer International, 2012 in http://www.acti.de/en/4527.php

[31] Earth Policy Institute in http://www.earth-policy.org/datacenter/pdf/book_fpep_ch4_all.pdf

[32] EurObserv'ER *Biogas barometer, Journal des Energies Renouvelables*, 2010 in http://www.energies-renouvelables.org/observ-er/stat_baro/observ/baro200b.pdf

[33] K. Fehrenbacher, Apple's fuel cells will be powerd by landfills, GigaOM, 2012 in http://gigaom.com/cleantech/apples-fuel-cells-will-be-powered-by-landfills

[34] SlideShare, *3rd Annual Biogas USA West Conference*, San Francisco, USA, 2011 in http://fr.slideshare.net/MelanieBotting/biogas-usa-west-brochure-lr

[35] R. C. Brown, *The Future of Biorefining Agricultural Biomass*, Iowa State University, 2005 in www.farmfoundation.org/news/articlefiles/949-robertc.brownpaper.pdf

[36] T. Werpy and G. Peterson, Eds, *Top Value Added Chemicals from Biomass Volume 1, Results of Screening for Potential Candidates from Sugars and Synthesis Gas*, 2004 in http://www1.eere.energy.gov/biomass/pdfs/35523.pdf

[37] Biopact, *The bioeconomy at work: Solvay to produce green PVC from sugarcane ethanol*, 2007 in http://news.mongabay.com/bioenergy/2007/12/bioeconomy-at-work-solvay-to-produce.html

[38] A. Morschbaber, *Polymers Review 49*, 79, 2009 in www.informaworld.com/smpp/content~content=a910573893&db=all

[39] DuPONT Company, *Corn-Derived 1,3-Propanediol The Pure and Petroleum-Free Glycol Choice for the Cosmetic & Personal Care Industry* in http://www2.dupont.com/Sorona/en_US/uses_apps/tatelyle/tate_lyle.html

[40] C. Collette, *An Introduction to R&D*, Arkema Investors day, 2007 http://www.arkema.com/pdf/EN/finance/r_and_d.pdf

[41] FprCEN/TR 15932 *Plastics – Recommendations for terminology and characterisation of Bioplastics*, 2009

[42] L. Shen, J. Haufe and M.K. Patel, *Product overview and market projection of emerging bio-based plastics*, University of Utrecht, 2009 in http://nws.chem.uu.nl/publica/Publicaties%202009/PROBIP2009%20Final%20June%202009.pdf

[43] L. Shen, E. Worrell and M. Patel, *Biofuels, Bioprod. Bioref.* 4, 25, 2010 in http://onlinelibrary.wiley.com/doi/10.1002/bbb.189/pdf

[44] Best in Packaging, *The Promising Potential of Biobased Plastics (part 2)*, 2009 in http://bestinpackaging.com/2009/11/24/the-promising-potential-of-bio-based-plastics-part-2

[45] C. Smith, *Why green plastics are here to stay*, Compounding World, p. 45, June 2012 in http://www.chem.umn.edu/csp/pdfs/CWJune2012%20article.pdf

[46] ERRMA, *EU Public/Private Innovation Partnership "Building the Bio-Economy by 2020"* as part of "Innovation Union", A flagship initiative within the Europe 2020 strategy, 2010 in http://www.errma.com/index.php?option=com_docman&task=cat_view&gid=24&Itemid=24

[47] F.D. Innocenti, A Public Private cooperation for bio-based products, 2010 in http://ec.europa.eu/enterprise/sectors/chemicals/files/hlg/francesco_degli_innocenti_en.ppt#12

[48] http://fr.wikipedia.org/wiki/Chiralit%C3%A9_(chimie)

[49] GALACTIC, *History of Lactic Acid* , 2011 in http://www.lactic.com/index.php/lacticacid

[50] J. Ren, Ed., *Biodegradable Poly (Lactic Acid) Synthesis, Modification, Processing and Applications*, Tsinghua University Press and Springer, 2010

[51] Chae Hwan Hong, Si Hwan Kim, Ji-Yeon Seo and Do Suck Han, *ISRN Polymer Science 2012*, Article ID 938261, 2012 in http://www.isrn.com/journals/ps/2012/938261

[52] http://en.wikipedia.org/wiki/Lactide

[53] R.H Platel, R.M.Hodgson and C.K.Williams, *Biocompatible Initiators for Lactides Polymerization*, Polymer Reviews 48, 11, 2008

[54] K.A. Porter, *Ring opening polymerization of lactide for the synthesis of poly(lactic acid)*, 2006 http://chemistry.illinois.edu/research/organic/seminar_extracts/2005_2006/06_Porter.pdf

[55] P. Van Mourik, J. Van Dam and S. Picken, *Materials Science in Design and Engineering*, VSSD: Science and Technology, Chapter 8, *Materials of biological origin*, 2012, http://mail.vssd.nl/hlf/m017ch08.pdf

[56] Sustainable Biomaterials Collaborative, *Biobased Materials and Sustainability*, 2012 in http://www.sustainablebiomaterials.org/faqs.biobased.php

[57] E.T.H. VINK, D.A. GLASSNER, J.J. KOLSTAD, R.J. WOOLEY and R.P. O'CONNOR, *Ind. Biotechnol.* 3, 58, 2007 in http://www.natureworksllc.com/~/media/The_Ingeo_Journey/EcoProfile_LCA/EcoProfile/NTR_Eco_Profile_Industrial_Biotechnology_032007_pdf.pdf

[58] R. HAAN, *Unique Purac Lactides for improved PLA production and properties*, Bioplastic Processing & Properties, Loughborough University, UK, 2009

[59] C. BASTIOLI and L. CAPUZZI, Novamont, *Biobased Chemicals and 3° Generation Biorefinery*, ISOPSC 2012 in http://www.unisa.it/uploads/7629/1.3_-_capuzzi.pdf

[60] Polyferm-Canada, *Polyhydroxyalcanoates* in http://polyfermcanada.com/

[61] Ontario Bioauto Council, *Biopolymers* in http://www.bioautocouncil.com/News/File.aspx?6567b330-6901-4f38-9b8d-0be2f5da5f28

[62] Metabolix, *Metabolix Announces Termination of Telles Joint Venture*, 2012 in http://ir.metabolix.com/releasedetail.cfm?releaseid=639044

[63] DuPont, *Fact Sheet, The Bio-PDO™ Sustainability Story* in http://www.c2es.org/docUploads/Fact%20Sheet%20-%20Bio-PDO%20LCA.doc

[64] ICIS.com, *ACC CPI: PU players going green*, 2012 in http://www.icis.com/Articles/2012/09/21/9597480/acc-cpi-pu-players-going-green.html

[65] Braskem Industrial Complex in http://www.chemicals-technology.com/projects/braskem-ethanol/braskem-ethanol2.html

[66] G-Value, News release , *First supply chain of Bio-PET in the world*, 2010 in http://www.toyota-tsusho.com/english/press/uploadfiles/Final_E_22-54103_1013-1.pdf

[67] Roquette, *Gaïalene – an innovative range of plant-based plastics* at K'2010, 2010 in http://www.roquette.com/delia-CMS/t1/article_id-6225/topic_id-2056/gaialene-r-an-innovative-range-of-plant-based-plastics-at-k-2010.html

[68] ADEME, *Life Cycle Analyses Applied to First Generation Biofuels Used in France*, 2010 in http://www2.ademe.fr/servlet/getDoc?sort=-1&cid=96&m=3&id=70548&ref=12441&nocache=yes&p1=111

[69] European Commission, *Biofuels and Bioenergy: assessing the impacts and availability* in http://iet.jrc.ec.europa.eu/sites/default/files/documents/information_sheets/en/biofuels_and_bioenergy_en.pdf

[70] European Commission, Joint Research Centre, Institute for Environment and Sustainability, *The Biofuels Debate - Indirect Land Use Change*, 2012 in http://ies.jrc.ec.europa.eu/our-activities/support-for-eu-policies/indirect-land-use-change.html

[71] R. EDWARDS, D. MULLIGAN and L. MARELLI, JRC Scientific and Technical Reports, *Indirect Land Use Change from Increased Biofuels Demand*, 2010 in http://ec.europa.

eu/energy/renewables/consultations/doc/public_consultation_iluc/study_4_iluc_modelling_comparison.pdf

[72] J. LANE, *Is the EU Abandoning Biofuels?* Renewable Energy World.com, 2012 in http://www.renewableenergyworld.com/rea/news/article/2012/09/is-the-eu-abandoning-biofuels

[73] Checkbiotech, *Draft EU proposal on crop-based biofuels puts advanced biofuels into spotlight*, 2012 in http://bioenergy.checkbiotech.org/news/draft_eu_proposal_crop_based_biofuels_puts_advanced_biofuels_spotlight

[74] M. PATEL, C. BASTIOLI, L. MARINI and E. WURDINGER, *Life-cycle Assessment of Bio-based Polymers and Natural Fiber Composites*, Encyclopedia Biopolymers, Vol. 10, Wiley-VCH, 2003

Chapter 4

Cellulose, the Predominant Constituent of Biomass

4.1 Introduction

Lignocellulosic biomass, or simply biomass, is chiefly composed of cellulose, hemicelluloses and lignin.

Cellulose will be reviewed in the present chapter, its hydrolysis in Chapter 5, and hemicelluloses and lignin in Chapter 6. For each biopolymer, its structure, its biosynthesis, its structure and function in plant cell walls, its degradation, and its properties and applications will be described.

Three structural levels will be analyzed for cellulose:

- the molecular level;
- the ***supramolecular*** level, which involves the arrangement of molecules into crystals;
- the ***morphological*** (or ultrastructural) level, which involves the organization of the crystals into ***microfibrils***, layers, cell walls, tissues, or other cellulose morphologies.

Then, the biosynthesis of cellulose will be related to the structure of native cellulose. Finally, the properties and applications of cellulose will be described in relation with its three structural levels.

4.2 Structure

4.2.1 Molecular structure

The cellulose molecule is a linear polymer of D-***anhydroglucopyranose units*** linked together by β-1,4-***glucosidic bonds*** (**Figure 4.1**). In other words, it is a β-1,4-D-***glucan***. The pyranose rings are in the energetically favorable [4] C_1 chair conformation.

The conventional numbering of the carbon atoms in the ring is shown in **Figure 4.1**, with two attached oxygen atoms at C1, hydroxyl substituents at C2 and C3, one attached oxygen atom at C4, and one hydroxymethyl group at C5. The positions 1 and 4 are involved in the inter-unit linkage. O1 is the oxygen atom of the glycosidic bond, O5 that of the ring, O2 and O3 those of the secondary alcohols and O6 that of the primary alcohol. C1 is an acetal centre along the whole chain except for the right-hand end, where it is a hemiacetal centre with inherent reducing properties. Thus, cellulose, as all 1,4-linked glucans, has one ***reducing end*** containing an unsubstituted hemiacetal, and one non-reducing end containing an additional hydroxyl group at C4.

Cyclization of D-glucose leads to the existence of two cyclic modifications: the six-membered form called ***pyranose***, and the five-membered form called ***furanose***. [1] Pyranose involves the aldehydic function and the oxygen at C5 (O5) of the acyclic aldohexose. Furanose involves the aldehydic function and O4. The five-membered rings of D-glucose are unstable relative to the six-membered rings. Cyclization with hemiacetal ring formation generates a new asymmetric carbon atom at C1, the ***anomeric center***, and therefore two diastereoisomers (non-mirror-image stereoisomers), called ***anomers***, designated

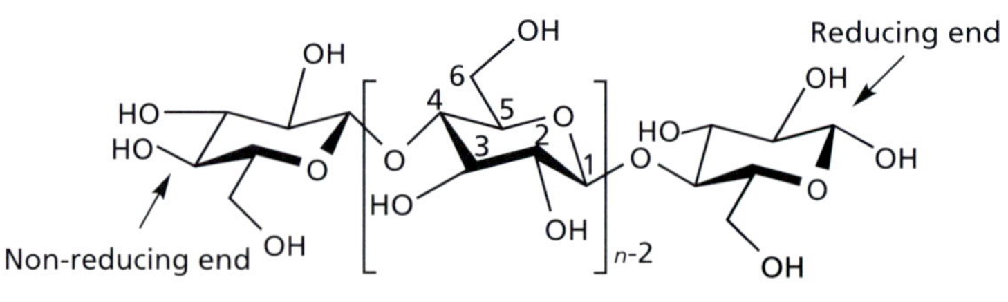

Figure 4.1 Molecular structure of cellulose, showing the numbering of the carbon atoms, the reducing end in red with a hemiacetal, and the non-reducing end in green with a free hydroxyl at C4.

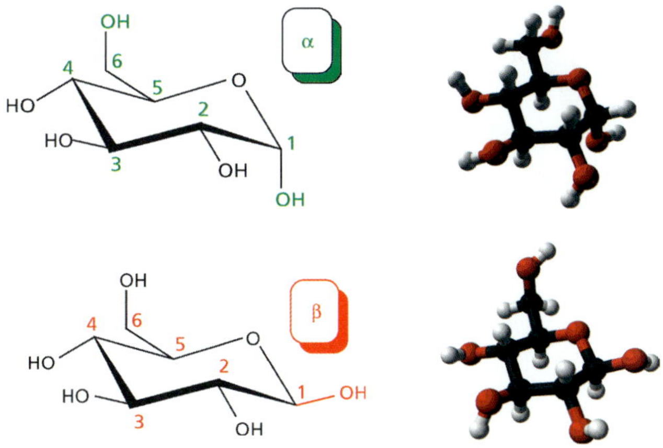

Figure 4.2 Haworth projections and ball-and-stick 3D models of α-D-glucopyranose (top) and β-D-glucopyranose (bottom).

α and β. The six-membered pyranose rings are in the [4] C_1 conformation, i.e. a chair shape (C) with C4 and C1 out of the reference plane containing the four other ring atoms (the superscript being the atom projecting through the side of the plane from which the ring numbering appears clockwise). For α-D-glucose, which is the most common crystalline modification of glucose, the hydroxyl group at the anomeric centre is **axial**, i.e. perpendicular to the mean plane of the ring, while for β-D-glucose, it is **equatorial**, i.e. in the mean plane of the ring (**Figure 4.2**).

Glycosides are formed from the reaction of sugars and alcohols ROH at the anomeric center. [1] In these molecules, the anomeric carbon atom of the carbohydrate part (***glycone***) is linked by an oxygen atom to the alcohol part R (***aglycone***), thereby defining ***glycosidic bonds***. The formation of glycosides can also be seen as the replacement of the hydrogen atom of the anomeric hydroxyl group by an alkyl (or equivalent) group. In the particular case where the alcohol is another sugar, the glycosides are ***disaccharides***. The character α or β of the anomeric configuration of the glycosidic linkage results from the anomeric form of the monosaccharide unit linked at C1. The β-character, like in cellulose, involves the formation of equatorial linkages. The α-character involves the formation of axial linkages.

In most crystal structures of cellulose, the molecule has a twofold helical (two monomers per turn of the helix) ***conformation*** [2], meaning that adjacent glucose units are oriented with their mean planes at an angle of 180° to each other (**Figure 4.3**). This conformation gives each molecule a flat ribbon-like structure that is stabilized by intramolecular hydrogen bonds.

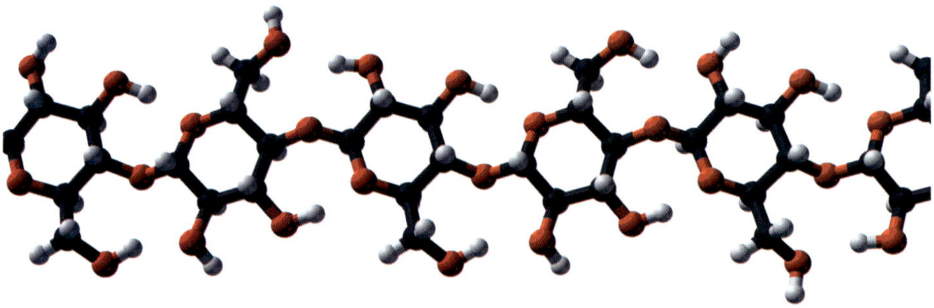

Figure 4.3 Conformational model of the cellulose chain in most crystal structures.

4.2.2 Supramolecular structure

Crystallinity of cellulose materials

The first X-ray diffraction patterns of native cellulose were taken in 1913. [3] These patterns brought the first scientific demonstration of the existence of highly ordered, crystalline regions. The intra- and intermolecular hydrogen-bonding system, as well as the chemical composition and conformation of cellulose chains, are responsible for their tendency to form highly ordered, crystalline domains.

The Bragg reflections from native and manmade celluloses appear broad and diffuse compared to those from well-developed single crystals. This observation is typical of semicrystalline polymers.

The relative amount of crystalline polymer in native cellulose varies widely with the source of the sample, as shown by the difference in sharpness of the X-ray diagrams (**Figure 4.4**).

Native celluloses have usually a higher crystallinity than manmade celluloses (**Table 4.1**) Degrees of crystallinity can reach very high values, particularly in *Valonia* algae and in animal ***tunicates***.

Crystal polymorphs and coordinate system

Several different crystalline lattices have been identified for pure celluloses. They are called cellulose I, II, III, and IV. Each ***polymorph*** is defined by its unit cell parameters. [8]

With only a few exceptions, all native celluloses consist of cellulose I. The unique crystalline lattice for almost all native celluloses was suggested from their similar X-ray diffraction patterns. This common pattern comes from a structure that is referred to as *cellulose I*. However, with the development of new investigation techniques, it has appeared that cellulose I is actually a composite of two crystalline ***allomorphs***, labeled Iα and Iβ. It has been shown that the Iα phase is metastable and can be converted into the thermodynamically more stable Iβ phase by annealing.

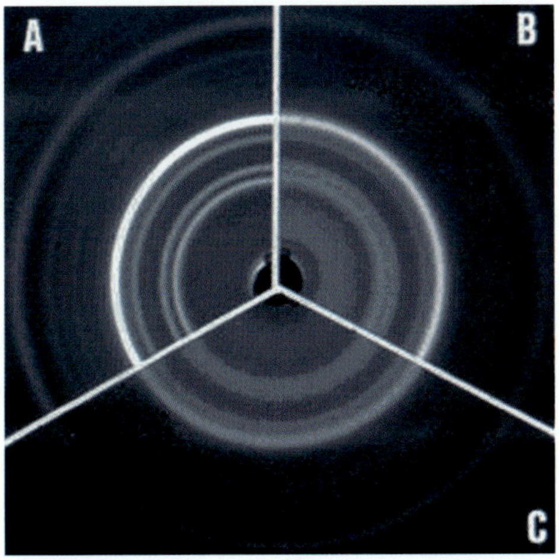

Figure 4.4 X-ray diffraction diagrams of cellulose materials: (A) tunicate cellulose; (B) cotton; (C) plant parenchyma. [4] Courtesy of Serge PEREZ, CERMAV.

Table 4.1 X-ray crystallinity range of native and manmade celluloses. [5, 6, 7]

Cellulose substrate	Crystallinity range (%)
Higher plants	50–75
Valonia	*90*
Tunicates	80
Bacteria	40–63
Manmade fibers	25–40

Celluloses that have been either dissolved and precipitated (***regeneration***), or treated with a concentrated alkaline swelling agent (~20%) and washed with water (***mercerization***) consist of cellulose II. Cellulose II is thermodynamically more stable than cellulose I, therefore the transformation from cellulose I to cellulose II is irreversible.

Cellulose III is formed by swelling cellulose I or II with amines or liquid ammonia, and subsequently removing the swelling agent anhydrously. Cellulose IV is classically formed by annealing cellulose III in glycerol.

The unit cell, the basis building block of a crystal, repeated infinitely in three dimensions, is characterized by (1) three vectors (a, b and c) that form the edges of a parallelepiped; and (2) three angles between the vectors (α, the angle between b and c; β, the angle between a and c; γ, the angle between a and b).

Two types of unit cells are found in cellulose: the ***monoclinic*** structure characterized by $a \neq b \neq c$ and $\alpha = \beta = 90° \neq \gamma$, with two chain segments per unit cell, and the ***triclinic*** structure characterized by $a \neq b \neq c$ and $\alpha \neq \beta \neq \gamma$, with one chain segment per unit cell.

The *a*, *b*, and *c* crystallographic axes used throughout this book satisfy the right-hand coordinate system (upper direction positive for *z* coordinates), with *c* as the cellulose chain axis and as the unique axis perpendicular to the two other in the monoclinic system. [9]

The directionality of a cellulose chain with respect to the *c* axis is given in **Figure 4.5**. [9] The chain sense is defined as ***parallel up*** when the *z* coordinate of O5 is greater than that of C5. In this case, the reducing end of the chain is oriented in the same direction with respect to the *c* axis. The reverse direction is defined as ***parallel down***. If the chains running side by side alternate in direction, with half the reducing ends at each side of the crystal, the chain packing is ***antiparallel***.

In this chapter, only the crystal structures of cellulose I and II will be reviewed, these two crystalline arrangements of cellulose being by far the most common.

Cellulose I

In their work published in 1929 and in 1937, Meyer, Mark, and Misch [10] proposed a unit cell of cellulose I that serves as early reference still today. This

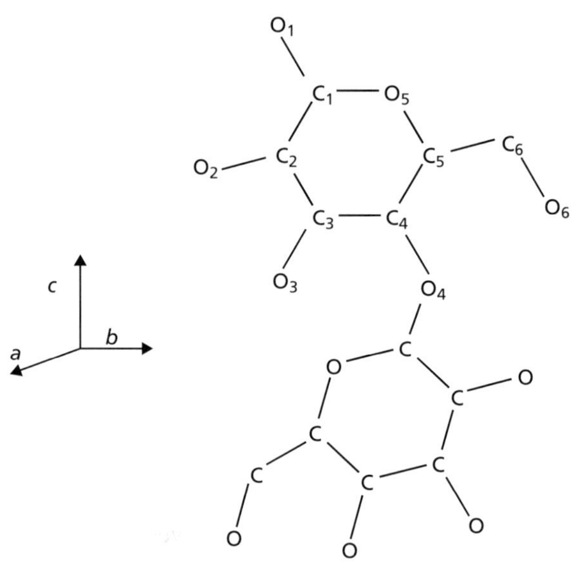

Figure 4.5 Definition of the directionality of the cellobiose repeat unit with respect to the crystallographic axes. [9]

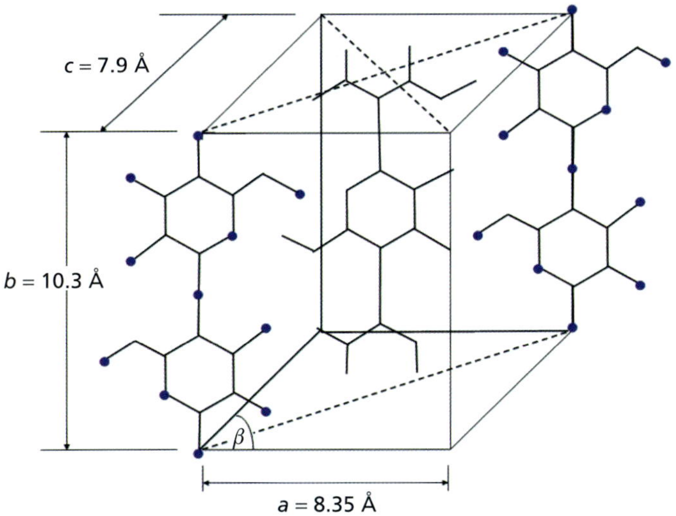

Figure 4.6 Original Meyer and Misch model for cellulose I, with the designation of the axes used in 1937. [10, 11]

unique crystalline lattice for all native celluloses was suggested from similitude in their X-ray diffraction patterns. In their model, the ***unit cell*** was monoclinic ($a = 7.9$ Å, $b = 8.35$ Å, $c = 10.3$ Å, $\gamma = 96°$) with two antiparallel cellobiose units, one at the origin (corner) and one at the center of the cell, oriented along the chain axis (**Figure 4.6**).

The alternating glucose units in the chains were rotated through 180°, involving that the two chains (the ***origin chain*** and the ***center chain***) of the cell were located on two independent two-fold screw axes with cellobiose as true ***repeating unit*** or ***asymmetric unit***. Furthermore, the two cellobiose segments were staggered by a quarter of the c axis (2.5 Å). In terms of lattice symmetry, the two chains of the cell were organized in a $P2_1$ space group ($P =$ ***primitive lattice***; $2_1 =$ two-fold screw axis), requiring adjacent glucose units in the same chain to be identical.

In 1974, Gardner and Blackwell, [12] from one side, and Sarko and Muggli, [13] on the other side, proposed a two-chain monoclinic unit cell, but with parallel chains having the same direction, as opposed to the antiparallel Meyer and Misch model. The direction of the chains was parallel down in the description given by Gardner and Blackwell, and parallel up in that given by Sarko and Muggli.

According to these last authors, intermolecular hydrogen bonds are formed between either adjacent origin chains or adjacent center chains of the unit cells, giving rise to hydrogen-bonded ***origin*** or ***center sheets*** along the (200) ***crystal planes***. There is no hydrogen-bonding between the origin and center chains. A less obvious force holding the structure together would be hydrophobic attraction perpendicular to the sheets. [8]

In parallel, it was found that among the native celluloses, algal (e.g. *Valonia*) and bacterial celluloses, on one hand, and celluloses from higher plants (e.g. ramie), on the other hand, give different diffraction patterns and infrared spectra. [8]

In 1984, Attala and VanderHart [14, 15] concluded from solid state [13] C NMR investigations that cellulose I was a system consisting of two distinct crystal phases, Iα and Iβ. The relative amount of each was dependent on native cellulose source. Algal and bacterial celluloses are rich in Iα, whereas celluloses from higher plants and tunicates are rich in Iβ. This biphasic character is in line with the observed difference in the diffraction patterns and infrared spectra. It was shown that the Iα phase is metastable and can be converted into the thermodynamically more stable Iβ phase by annealing in diluted aqueous NaOH or various organic solvents. [16, 17, 18]

In 1991, Sugiyama et al. [19] characterized the crystal structure of the Iα and Iβ phases from electron diffraction investigations on algal cellulose. Cellulose Iα has a triclinic unit cell containing one chain (*P*1 space group), whereas cellulose Iβ has a two-chain monoclinic cell (*P*2_1 space group) (**Figure 4.7**). The two phases can coexist within the same crystalline microfibril.

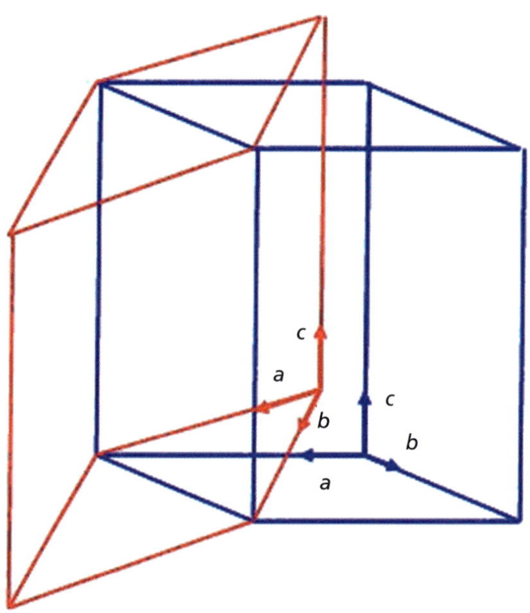

Figure 4.7 Relative orientation of the one-chain triclinic unit cell (cellulose Iα) with respect to the two-chain monoclinic unit cell (cellulose Iβ). Reproduced by permission of American Chemical Society. [19]

Table 4.2 Unit cell parameters of Iα and Iβ phases determined from electron diffraction [19], and synchrotron X-ray and neutron diffraction. [20, 21]

	Iα (triclinic)		Iβ (monoclinic)	
	Electron	X-ray and neutron	Electron	X-ray and neutron
a (Å)	6.74	6.717	8.01	7.784
b (Å)	5.93	5.962	8.17	8.201
c (Å)	10.36	10.400	10.36	10.380
α (deg)	117	118.08	90	90
β (deg)	113	114.80	90	90
γ (deg)	81	80.37	97.3	96.5

The one-chain triclinic structure requires all the chains to be packed parallel within the crystals but does not require adjacent glucose units in the same chain to be identical. In such a structure, a cellobiosyl residue is the crystallographic-independent repeat unit. By contrast, in cellulose Iβ, there are two independent chains, but in each of these, a glucosyl residue is the crystallographic-independent repeat unit. The cellulose molecules in cellulose Iα are organized in sheets along the (110) planes where the cellobiose segments are at the same level along the chain axis. The unit cell parameters of the two phases determined by Sugiyama et al., [19] and later confirmed by Nishiyama et al., [20, 21] are given in **Table 4.2.** The a value determined by electron diffraction is known to be overestimated since the electron beam has a tendency to swell the a parameter of cellulose. [21]

The main difference between the Iα and Iβ phases is a displacement of adjacent chains by a quarter of the c-axis period to give either a diagonally shifted (Iα) or a staggered (Iβ) arrangement of the cellobiose repeat units. [19]

In 1997, Koyama et al. [9] demonstrated the parallel-up packing of cellulose Iα and Iβ by an elegant electron microscopy study combining chemical staining of reducing ends and tilting microdiffraction. Polymerization by the cellulose synthase enzyme takes place at the nonreducing end of the growing chains.

In addition to electron microscopy, atomic force microscopy (AFM) has been used to study the surface structure of native cellulose microcrystals. [22, 23, 24] The AFM work using microcrystals from *Valonia* has revealed that it is possible to image the cellobiose repeat along the chain, and thus to identify the triclinic (Iα) structure on the surface.

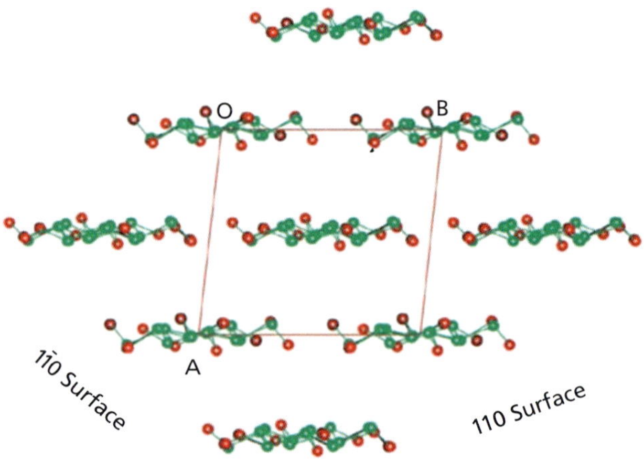

Figure 4.8 A segment of the cellulose Iβ crystal structure, showing the relationship of the cellulose chains to the *a* and *b* axes (origin O) and to the (1$\bar{1}$0) and (110) surfaces of a square microfibril. The intermolecular hydrogen bonds are in directions parallel to the *b* axis. Reproduced by permission of American Chemical Society. Copyright © 2002, American Chemical Society. [2, 20]

In 2002, Nishiyama, Langan, and Chanzy [20] determined in detail the crystal structure and the hydrogen-bonding system in cellulose Iβ from synchrotron X-ray and neutron fiber diffraction (**Figure 4.8**).

Fibrous oriented samples were prepared by aligning cellulose microcrystals from tunicate. The positions of hydrogen atoms involved in hydrogen bonding were determined from intensity differences between neutron fiber diffraction diagrams of intra-crystalline deuterated and hydrogenated samples. The resulting monoclinic structure (see **Table 4.8**) consists of two parallel chains (the origin and center chains) having slightly different conformations and organized in sheets packed in a parallel-up fashion, with all hydroxymethyl groups in the *tg* conformation and with different hydrogen bonding. [It should be noted that the ***conformation of the hydroxymethyl group*** is defined by two letters, the first referring to the torsion angle χ (O5-C5-C6-O6), i.e. the relative orientation of the C6-O6 bond to the C5-O5 bond about the C5-C6 bond, and the second to the torsion angle χ' (C4-C5-C6-O6); thus, an ideal *tg* (trans-gauche) conformation corresponds to the set of angles 180°, –60°]. Thus, the asymmetric unit contains two glycosyl units in the origin and center chains.

In 2003, Nishiyama, Sugiyama, Chanzy, and Langan [21] determined the crystal structure and hydrogen-bonding system in cellulose Iα from synchrotron X-ray and neutron fiber diffraction. Fibrous oriented samples were prepared by aligning cellulose microcrystals from the alga *Glaucocystis*, and their X-ray diffraction diagrams were recorded. The positions of hydrogen atoms involved in hydrogen-bonding were determined from neutron diffraction using hydrogenated and deuterated samples. The resulting structure is a one-chain triclinic unit (see **Table 4.2**) cell, with all glycosidic linkages identical and all hydroxymethyl groups in the *tg* conformation. However, adjacent pyranose rings in the same chain alternate in conformation, resulting in a cellobiose repeat and an asymmetric unit that contained two adjacent glycosyl units in the same chain. The chains organize in sheets in a parallel up fashion.

In conclusion to the section on cellulose I, **Figure 4.9** shows the projections of the crystal structures of cellulose Iα and Iβ down the chain axis, perpendicular to the chain axis and in the plane of the sheets, and perpendicular to the sheets. [21]

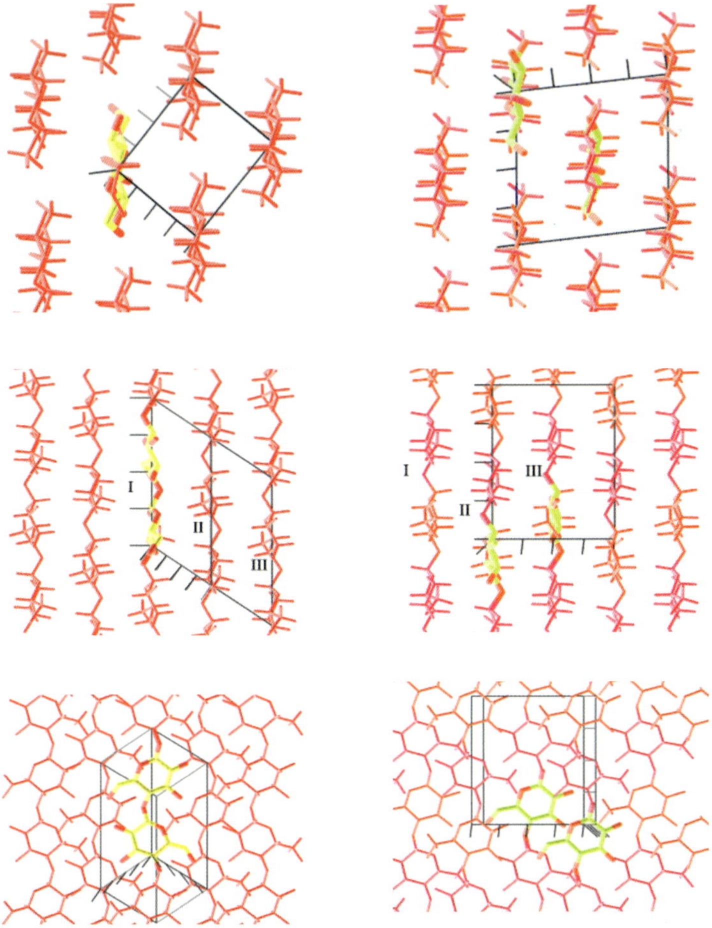

Figure 4.9 Projections of the crystal structures of cellulose Iα (left) and Iβ (right) down the chain axes (***top***), perpendicular to the chain axis and in the plane of the hydrogen-bonded sheets (***middle***), and perpendicular to the hydrogen-bonded sheets (***bottom***). The cellulose chains are represented by red skeletal models. The asymmetric unit of each phase is also represented in thicker lines with carbons in yellow. The unit cell of each phase is shown in black. Reproduced by permission of American Chemical Society. Copyright © 2003, American Chemical Society. [21]

The projections down the chain axis are strikingly similar. In the direction perpendicular to the chain axis and in the plane of the hydrogen-bonded sheets, the projections differ in the relative displacement of the sheets along the chain axis, as illustrated by the three adjacent sheets I, II, and III in **Figure 4.9** (middle). In both Iα and Iβ, the second sheet II is shifted in the up direction by ~$c/4$ relative to the first sheet I. The third sheet III is similarly shifted with respect to II by ~$c/4$ in Iα, but in Iβ it is shifted by ~$c/4$ in the down direction.

The most likely mechanism for solid-state conversion of cellulose Iα to Iβ is the relative slippage of the cellulose chains. This movement does not require the disruption of the hydrogen-bonded sheets but slippage by ~$c/2$ at the interface of sheets of type II and III in **Figure 4.9** (middle). [21]

The conversion of cellulose Iα to Iβ takes place at temperatures higher than 220–230 °C. [19, 21] It has been proposed that the inner cohesion of ambient cellulose crystals is due to van der Waals and/or C–H···O interactions between hydrogen-bonded sheets. Despite the apparent strength of these interactions at room temperature, they are clearly lessened by heating the crystals; the lattice is virtually constant in directions within the hydrogen-bonded sheets, but expands in the stacking direction. It has been suggested that librations of cellulose chains about axes lying within the sheets is the driving force for sheet separation and lubrication during conversion of cellulose Iα to Iβ.

Cellulose II

Celluloses that have been either dissolved and precipitated (regeneration) or treated with a concentrated alkaline solution (~20%) and washed with water (mercerization) consist of cellulose II. Cellulose II is thermodynamically more stable than cellulose I. Therefore, transformation from cellulose I into cellulose II is irreversible.

In 1929, Andress [25] proposed a unit cell for mercerized cellulose (cellulose II). The cell was monoclinic ($P2_1$ space group) with two antiparallel cellobiose segments staggered by about $c/4$. The chains were aligned on the 2-fold screw axes of the cell.

A major question in the study of cellulose II was the parallel or antiparallel packing of adjacent chains. Since the beginning of the investigations, the chain polarity has been proposed to be antiparallel to take into account crystallization from solutions of disorganized chains. As the packing has been determined to be parallel for cellulose I, the difficult question has arisen of how parallel cellulose I is converted into antiparallel cellulose II by mercerization, without dissolution of cellulose. A plausible mechanism has been suggested for this parallel to antiparallel conversion. Revol et al. [26] were the first to record a single crystal diffraction diagram on one section of a *Valonia* microfibril. The statistical antiparallelism of cellulose microfibrils has been proven in cross-section of *Valonia* cell wall. [27] It has been deduced from electron diffraction analysis of *Valonia* cell walls. [28, 29] It has been further clearly demonstrated by dark-field microscopy, when the polarity of each microfibril is clearly identified. [30] In agreement with this statistical model, the interdigitation mechanism leading to mercerization, proposed by Sarko et al., [31] becomes realistic.

In 1976, Kolpak and Blackwell, [32] on one hand, and Stipanovic and Sarko, [33] on the other hand, determined the crystal structure of regenerated cellulose (Fortisan) from X-ray fiber diffraction. The proposed structure consists of two antiparallel chains aligned along the 2-fold axes of the monoclinic ($P2_1$) cell.

In each proposal, both chains have equivalent backbone conformations but different conformations of their hydroxymethyl groups, i.e. *gt* for the corner chain and *tg* for the center chain. These hydroxymethyl positions and the antiparallel direction of both chains permit a three-dimensional hydrogen-bonding network. Blackwell and Kolpak proposed that adjacent parallel corner chains along the *a* axis are linked by intermolecular O2H···O6 bonds, adjacent parallel center chains along the *a* axis by intermolecular O3H···O6 bonds, and adjacent antiparallel corner and center chains by intermolecular O2H···O2 bonds. [5, 6]

In 1978, Kolpak et al. determined the crystal structure of mercerized cellulose from X-ray diffraction. [34] The structure was essentially identical to that of regenerated cellulose. There was some indication of a difference in the conformation of the hydroxymethyl groups of mercerized and regenerated cellulose.

In the mid 1990s, the model of cellulose II involving two different conformations for the hydroxymethyl groups has been questioned by the determination of the crystal structure of β-cellotetraose hemihydrate independently by Gessler et al. [35, 36] and Raymond et al. [37] and that of methyl β-cellotrioside monohydrate 0.25 ethanolate. [38] Both cellodextrins adopted an antiparallel packing of the molecules within the unit cell. Surprisingly, the two molecules had the same *gt* hydroxymethyl conformation, but different sets of conformational angles. The resulting cellulose II structure, which was deduced from the work on the oligomers, was, therefore, substantially different from the Sarko/Blackwell structures. [39]

In 1999, Langan, Nishiyama, and Chanzy [40] revised the structure and hydrogen-bonding system in regenerated cellulose from neutron fiber diffraction. Two samples of cellulose II were prepared: one hydrogenated by mercerization in NaOH/H_2O and the other deuterated by mercerization in NaOD/D_2O. The reported unit cell parameters, using the X-ray data of Kolpak and Blackwell for regenerated cellulose, are given in **Table 4.3**.

Table 4.3 Unit cell parameters for regenerated cellulose (cellulose II). [40]

a (Å)	8.01
b (Å)	9.04
c (Å)	10.36
γ (deg)	117.10

Langan et al. [40] also highlighted the following characteristics of regenerated cellulose:

- The antiparallel origin and center chains have different conformations, but the hydroxymethyl groups of both chains are near the *gt* conformation.
- Whereas the sugar of the origin chain is unstrained, the sugar of the center chain is strained, resulting in a relative displacement of the two chains along the *c* axis of 2.4 Å, compared to 2.2 Å with previous models.

- Regenerated cellulose forms a three-dimensional hydrogen- bonding network. Intramolecular hydrogen bonds are O3–H···O5 in each chain with O3–H···O6 as minor component. Intermolecular hydrogen bonds are O2–H···O6 in origin sheets, O6–H···O2 in center sheets, and both O6–H···O6 (due to the nearly equivalent conformation of the hydroxymethyl groups of origin and center chains), and O2–H···O2 in sheets containing origin and center chains, thereby forming a three-dimensional network of hydrogen bonds. This network appears stronger than the two-dimensional network of cellulose I.

In 2001, the same group of workers revised the crystal structure of mercerized cellulose from synchrotron X-ray diffraction. [40] The reported $P2_1$ unit cell parameters are given in **Table 4.4**. They are very close to those reported for regenerated cellulose. [32]

Langan et al. [40] also obtained the following results on mercerized cellulose:

- The antiparallel origin and center chains in mercerized cellulose have different conformations with the hydroxymethyl groups of both chains near the *gt* position, as for regenerated cellulose.

Table 4.4 Unit cell parameters for mercerized cellulose (cellulose II). [40]

a (Å)	8.10
b (Å)	9.03
c (Å)	10.31
γ (deg)	117.10

- The conformations of each type of chain in regenerated and mercerized cellulose are also similar.
- Regenerated and mercerized cellulose differ in the conformation of the hydroxymethyl groups of the centre chains. This is presumed to result from the observed difference in the amount of hydroxymethyl group disorder: ~30% for regenerated cellulose and ~10% for mercerized cellulose.

Summary on crystal structures

To conclude this review on cellulose polymorphs, **Figure 4.10** shows a comparison of the monoclinic unit cells for cellulose Iβ and II. For the two polymorphs, the repeat distance along the chain axis, *c*, is about 10.31–10.38 Å, corresponding to the glucose dimer. This is valid for all cellulose polymorphs.

The crystal structures proposed for cellulose Iα, cellulose Iβ, and cellulose II are compared in **Table 4.5**.

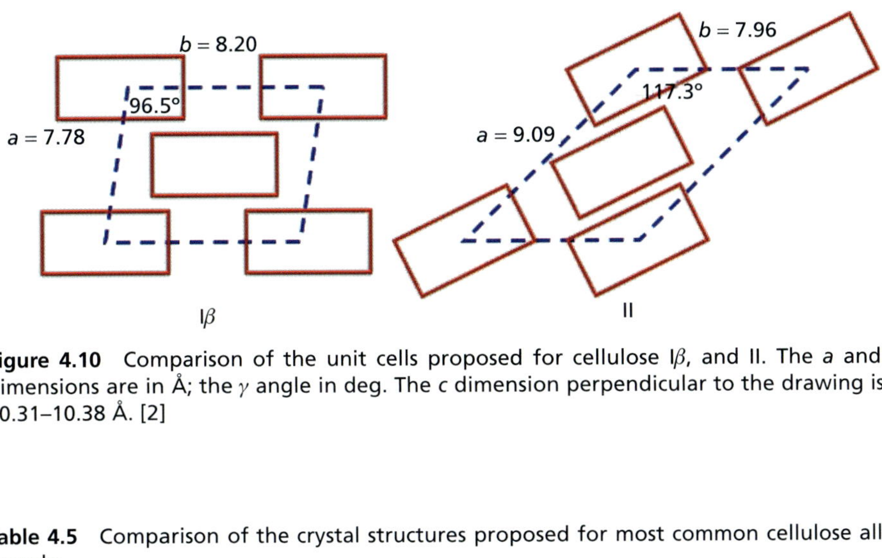

Figure 4.10 Comparison of the unit cells proposed for cellulose Iβ, and II. The *a* and *b* dimensions are in Å; the γ angle in deg. The c dimension perpendicular to the drawing is ~ 10.31–10.38 Å. [2]

Table 4.5 Comparison of the crystal structures proposed for most common cellulose allomorphs.

Type	Unit cell	Chains	Repeat distance (Å)	Space group	Asymmetric unit
Iα	One-chain triclinic	Parallel	~10.35	$P1$	Two glucosyl
Iβ	Two-chain monoclinic	Parallel	=10.35	$P2_1$	Two glucosyl
II	Two-chain monoclinic	Antiparallel	~10.35	$P2_1$	Two glucosyl

In most crystal structures, glucose repeats as a dimer, with one set of conformational angles corresponding to a twofold helical symmetry (cellulose Iβ and cellulose II) or two sets corresponding to a onefold symmetry (cellulose Iα).

4.2.3 Morphological structure of native cellulose

Introduction

The morphological structure of native cellulose is intended to describe the organization of crystals into microfibrils, layers (or lamellae), cell walls, fibers, tissues, or other cellulose morphologies (**Figure 4.11**).

A two-phase model called the *fringed micelle model* was proposed for cellulose in plant cell walls towards 1940. [6, 42, 43, 44] This model assumed that the polymer chains included segments aligned over distances corresponding to the lengths of the crystallites, and more disordered segments belonging to amorphous regions. Since the chains are very long, they were visualized as contributing to several crystallites and amorphous regions. It is now accepted that the fringed micelle model does not apply to native cellulose. A related model called *fringed fibril model* concentrated more on the fibril character of cellulose. [45, 46] It assumed that fibrils were strings of crystallites formed by well aligned segments of hydrogen-bonded cellulose chains and linked together in the string by more disordered segments of the

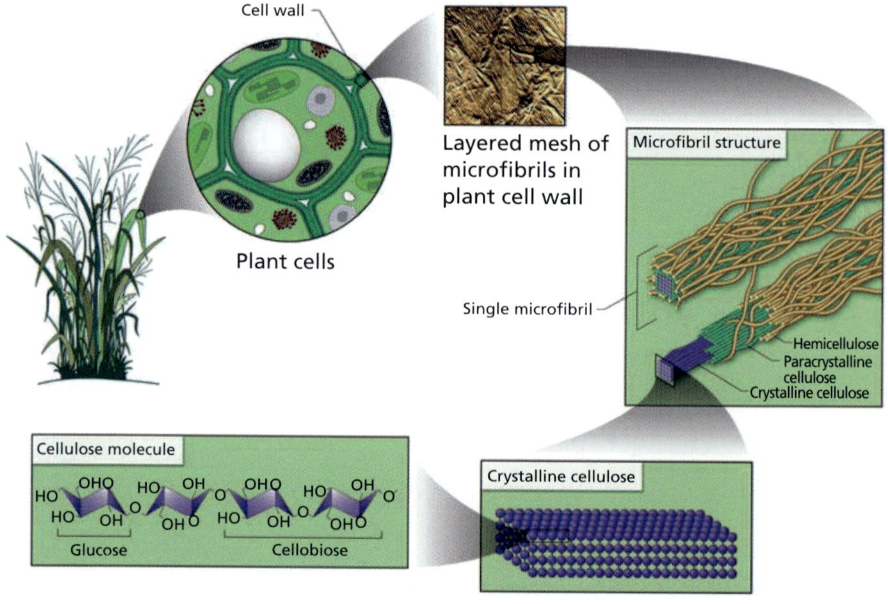

Figure 4.11 Cellulose: from macromolecules to plants. [41]

same chains. [45] The fibrils themselves were interlinked by fringe-shaped elements of fibrils. [47] The fringed fibril model has been used extensively in discussing experimental data.

The difficulty with these two-phase models is that in the vast majority of plants, the dimensions of the fibrillar domains are so small that the concept of phase boundary is no longer valid. [48]

Native cellulose from higher plants, algae, fungi, bacteria, amoebae, and tunicates is synthesized by the coordinated action of enzymatic polymerization associated with crystallization into nascent cellulose microfibrils. [49] This mechanism allows the generation of highly extended chains that can crystallize into microfibrils including two crystalline phases, cellulose Iα and Iβ. The microfibrils are then assembled to form higher order structures such as layers, cell walls, and fibers. In some plants, such as cotton, the microfibrils organize into macrofibrils 60–300 nm wide, which are then organized into fibers. [2] Cotton and wood pulp originate from single plant cells, whereas bast fibers obtained from the stems of certain plants (e.g. ramie, hemp, flax, and jute) come from bundles of individual fiber cells. [50]

Microfibrils

Electron microscopy, and especially transmission electron microscopy, has revealed the existence of very thin, crystalline fibrils, called microfibrils, in native cellulose. These microfibrils, or microcrystals, are biosynthesized by enzymatic terminal complexes located at the cell plasma membranes. It is believed that each native microfibril is in fact a single crystal and that all the earlier stories of elementary fibril have not been proven.

The microfibrils are ~2 to 50 nm across, depending on the synthesizing organism (**Table 4.6**), [29, 51, 52] and their lengths can reach several μm. [5] Therefore, they are examples of bionanofibers. Algae such as *Micrasterias* and *Valonia*, and tunicates, such as *Halocynthia*, synthesize relatively large microfibrils. In Micrasterias, the microfibrils occur in crisscrossed bands consisting of a number of parallel ribbon-like microfibrils. [53] The individual microfibrils with rectangular cross-section had thicknesses of 5 nm for a width of ~20 nm, but in some instances, two or three microfibrils merged to yield larger monocrystalline domains reaching up to 60 nm in lateral size. In *Valonia*, the microfibrils are 20 nm wide, with a squarish cross-section. In *Halocynthia*, the microfibrils are 10 nm wide, with parallelogram cross-section having substantially acute angles. [54, 55] If the microfibrils are hydrolyzed with acid, the acute angles are cut, and a 6-side section is observed. Ramie, wood, and primary wall synthesize relatively small microfibrils 5 nm, 3–4 nm, and 2–3 nm wide, respectively. The cross-sections of the small microfibrils are unknown. [51] Microfibrils are usually straight with extended cellulose molecules running parallel to the long axis of the

Table 4.6 Width and cross-section of microfibrils of various natural sources and calculated number of glucan chains per microfibril. [29, 39, 53, 54, 56, 57, 58]

	Width (nm)	Cross-section	Glucan chains per microfibril
Alga *Micrasterias*	Up to 60	Rectangular	600–700
Alga *Valonia*	20	Squarish	~1000
Tunicate *Halocynthia*	10	Parallelogram or truncated parallelogram	600
Ramie, cotton, flax	5	Not known	~80
Wood	3-4	Not known	30–40
Plant primary cell wall	2-3	Not known	~30–36

microfibril. [29] An approximate number of glucan chains in the various microfibrils can be deduced from their lateral dimensions.

High resolution observations of cellulose microfibrils were first obtained in Japan in 1984 by Sugiyama et al. [59] using high resolution transmission electron microscopy (TEM). Lattice images from ultrathin (~28 nm) sections of cellulose microfibrils in the cell wall of *Valonia* were obtained one year later by the same authors. [27] Cellulose microfibrils in transverse view are visible as squarish fragments. The diffraction pattern shows spots corresponding to ***equatorial*** (i.e. parallel to the chain axis) crystallographic planes with *d* spacings of 0.60, 0.53, and 0.39 nm. It also shows an apparent preferred orientation of the fragments, so that the 0.60-nm planes lie parallel to the cell wall surface (plasma membrane), as well as two orthogonal orientations of the 0.39-nm planes. Such a ***uniplanar orientation***, which has been previously investigated by X-ray diffraction in whole cells, [60, 61] exists in the cell wall of *Valonia* even within a single lamella.

The study of Sugiyama et al. [27] also revealed that high-resolution lattice images of *Valonia* microfibrils do not show any smaller crystalline substructures. It appears that one single microfibril is an individual crystal. Comparing the optical diffraction patterns obtained from different areas of one single lamella, it is observed that one microfibril has only one single direction of the 0.39-nm planes, while within one lamella, there are two directions for the 0.39-nm planes. Therefore, the axial directions of adjacent microfibrils in each single lamella of the cell wall are opposite to each other.

In 1994, Sugiyama et al. [62] investigated the orientation of the triclinic phase of cellulose in the cell wall of *Valonia*. They showed that the *Valonia* microfibrils are not twisted along their axis. In 1996, Kim et al. [53] studied the orientation of the monoclinic cellulose Iβ (planes indexed as $(hkl)_m$) of the alga *Micrasterias*. They found that that the 0.53 nm lattice planes $(110)_m$ lay parallel to the cell wall surface. In 1997, Koyama et al. [63] analyzed algal celluloses from different origins with reference to crystalline features. Three types of algal cellulose were identified: Iα-rich/large microfibril/0.6 nm-oriented type (e.g. *Valonia*); Iβ-dominant/flat ribbon microfibril/0.53 nm-oriented type (e.g. *Closterium*); and Iβ-dominant/small microfibril/random-oriented type (e.g. *Gelidium*). In 1998, Helbert et al. [54] analyzed the molecular organization of cellulose Iβ microfibrils from *Halocynthia* (tunicate cellulose). One edge of the parallelogram-shaped cross-section was parallel to the 0.6 nm lattice planes, while the other edge did not correspond to a crystallographic plane. The 0.6 nm lattice planes in tunicates were found to be oriented parallel to the cell wall surface. [63, 64] In 1999, Imai et al. [65] studied crystalline feature of cellulose microfibrils of the alga *Glaucocystis* and identified a new type of algal cellulose: Iα-rich/large microfibril/0.53 nm-oriented.

As a comprehensive scheme, **Figure 4.12** shows how the chains of the monoclinic and triclinic cells are arranged in a square cross-section of a cellulose microfibril. [2]

Atomic force microscopy (AFM) has confirmed the conclusions from TEM. [22, 23, 24] AFM images have illustrated the perfect alignment of the cellulose chains along the long axis of the *Valonia* microfibril without chain-folding. [24] AFM have also revealed the surface chain arrangements and differences between the triclinic (Iα) and monoclinic (Iβ) domains. [22, 23, 49]

The ***polarity*** of cellulose chains in a microfibril was discussed for a long time before two groups independently demonstrated the parallel packing by

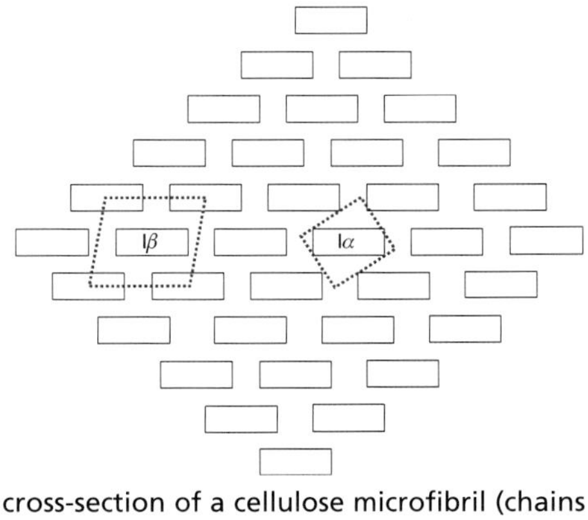

Figure 4.12 A square cross-section of a cellulose microfibril (chains depicted as rectangles) showing the monoclinic and triclinic unit cells of Iβ and Iα. The *c* axis is perpendicular to the plane of the drawing. The arrangement of the chains is such that the edges of the microfibril correspond to the diagonals $(1\bar{1}0)_m$ and $(110)_m$ of the Iβ monoclinic unit cell, or to the sides $(100)_t$ and $(010)_t$ of the Iα triclinic unit cell. [2]

electron microscopy in the mid 1980s. Hieta et al. [66] used the silver-labeling of the reducing ends of microfibrils, while Chanzy and Henrissat [67] used the unidirectional enzymatic degradation of cellulose microfibrils by a cellobiohydrolase. In both cases, the microscopic work has been achieved on *Valonia* microfibrils because of their high crystallinity and large width. These microscopic studies supported earlier crystallographic models that proposed a parallel packing of the chains in the unit cell. In 1997, the parallel up packing in cellulose Iα and Iβ demonstrated by Koyama et al. [9] allowed them to identify the directionality of cellulose chains in microfibrils. It was proved that the reducing ends of the growing chains point away from the synthesizing organism and, therefore, that polymerization by the enzymatic terminal complex takes place at the nonreducing ends of the growing chains (**Figure 4.13**).

A large percentage of non-crystalline cellulose corresponds to the crystal-surface chains: from ~11% in *Valonia*, one-third to two-thirds in higher plants, to more than 70% in plant primary wall. [29, 56] In addition, crystalline defects such as dislocations, chain ends, and twists are expected to occur along the microfibrils. [29] In 2002, the conformational features of crystal-surface cellulose from higher plants were studied by Viëtor et al. [56] Especially from [13] C NMR, it has appeared that surface chains from higher plants differ in some way from interior crystal chains, in contrast to surface chains from algae and animals. Viëtor et al. [56] showed that the surface chains from higher plants have a C6 conformation different from that in the crystal so that O6 is not in the correct position for the hydrogen bond from O2. They also proposed a model in which alternate glucose residues are tran-

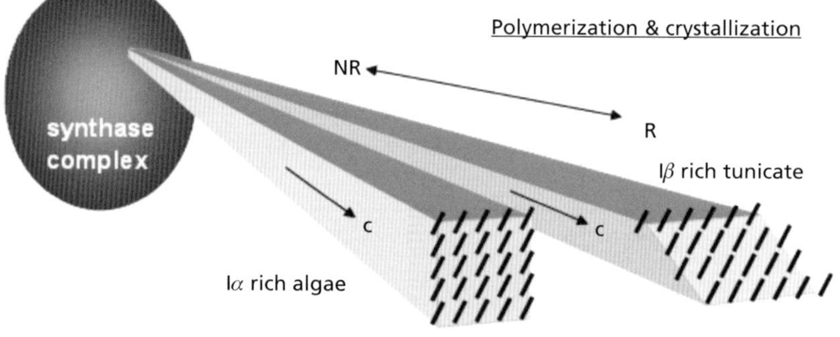

Figure 4.13 Schematic representation of cellulose Iα and Iβ-rich microfibrils spun from synthase complexes with their non-reducing (NR) ends towards the synthases. Short lines in the cross-sections of microfibrils show cellulose molecules. c= crystallographic c axis; cross-section Iα bounded by (100) and (010); cross-section Iβ bounded by (1$\bar{1}$0). Reproduced by courtesy of Junji SUGIYAMA. [49]

siently or permanently twisted away from the flat-ribbon conformation of the chain, weakening the O3–H···O5 hydrogen bond. Reduced intramolecular hydrogen bonding presumably allows the surface chains to form more hydrogen bonds to external molecules.

Plant cell walls

Extracellular matrix

The plant cell wall (**Figure 4.14**) is an extracellular matrix that encloses each cell in a plant. [68] The formation and differentiation of the cell wall plays a crucial role in plant morphogenesis. [69] The walls of plant cells are generally thicker, stronger, and more rigid than the extracellular matrix produced by animal cells. In evolving relatively rigid walls, which vary from 0.1 to many micrometers in thickness, early plant cells adopted a sedentary lifestyle that has persisted in all current plants.

The composition of the cell wall depends on the cell type. [68] The cell walls of growing cells are called ***primary cell walls***. They are relatively thin and only semi-rigid to accommodate subsequent cell growth. Once growth stops, the cell wall no longer needs to be extensible: sometimes the primary wall is retained without major modifications, but more often, a thicker, rigid, ***secondary cell wall*** is constructed by depositing new layers between the primary wall and the plasma membrane. Secondary wall formation is restricted to specialized cells and provides mechanical strength and rigidity to support aerial structures and hydrophobicity for transport functions. [69] Secondary walls may have a composition similar to that of the primary wall or be notably different. The most common additional polymer in secondary wall is ***lignin***, a complex network of phenolic compounds found in the walls of

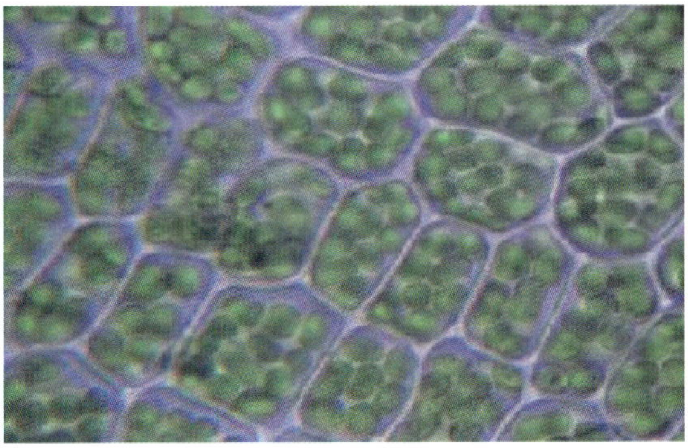

Figure 4.14 Wall of a green plant cell (size of plant cells varies from 10 to 100 μm). Reproduced by permission of Thomas DREPS under the license described in http://creativecommons.org/licenses/by-sa/2.0/de/deed.en. [70]

the xylem vessels and fiber cells of woody tissues. When the plant cell dies, the wall remains.

Even if the plant cell walls differ in composition and organization, their underlying structure is remarkably consistent: tough microfibrils of cellulose are embedded in a highly crosslinked amorphous matrix. The microfibrils provide tensile strength to the wall while the matrix provides resistance to compression, as in all fiber-composites. [68]

The tensile strength of cell walls prevents the cells from taking up water by osmosis and bursting and allows them to generate an internal hydrostatic pressure called ***turgor pressure***. [68] Turgor pressure is the driving force for cell expansion during growth.

Primary cell wall

When viewed with electron microscopy, cell walls appear to be a network of extended polysaccharides. [57] In higher plants, the dominant structural features are cellulose microfibrils with diameters of ~3 nm, which are crosslinked by single-chain polysaccharides.

Structural analysis of primary cell wall polysaccharides has revealed the presence of three major classes of polysaccharides: cellulose, hemicelluloses, and ***pectins***. [57] Hemicelluloses are branched polysaccharides containing backbones of neutral sugars that can form hydrogen bonds to the surface of cellulose microfibrils. The predominant hemicelluloses in many primary walls are ***xyloglucans*** (XG), which have a glucan backbone substituted with mono-, di- and triglycosyl side chains. [71] Other hemicelluloses found in primary and secondary cell walls include ***glucoronoarabinoxylans*** (GAX), which have a xylan backbone with glucuronic acid and arabinose side chains. Pectins are branched or unbranched polysaccharides containing D-galacturonic acid residues. The simplest pectin is ***homogalacturonan*** (HG), an unbranched polymer of α-1,4-linked-D-galacturonic acid. ***Rhamnogalacturonan I*** (RGI) is a complex pectic polysaccharide with a backbone composed of alternating α-1,2-linked-L-rhamnose-α-1,4-linked-D-galacturonic acid. ***Rhamnogalacturonan II*** (RGII) is a highly branched pectic polysaccharide with a backbone made of 1,4-linked galacturonic acid residues. RGI has been suggested to function as a scaffold to which other pectins, such as RGII and HG, are covalently attached.

Sets of microfibrils are arranged in layers or lamellae with each microfibril about 20-40 nm from its neighbors. [72] The primary cell wall consists of several such lamellae arranged in a plywood-like network (**Figure 4.15**).

In addition to polysaccharides, walls contain lesser amounts of structural glycoproteins (hydroxyproline-rich extensins), phenolic esters, minerals (e.g. calcium and boron), enzymes, and proteins (expansins) that are believed to have a role in regulating wall expansion. [71]

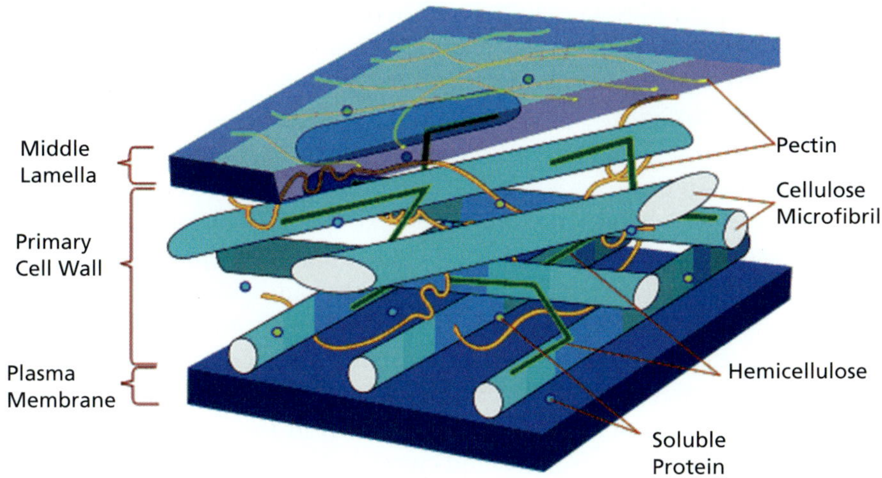

Figure 4.15 Primary cell wall diagram. [73]

One of the driving forces for assembly of the overall structure is believed to be the hydrogen bonding of hemicellulose to the surface of cellulose microfibrils. [57] However, the organization and interactions of wall components is not known with certainty, and there is still debate about how wall organization is modified to allow cells to expand and grow. [71]

The final shape of a growing plant cell is determined by controlled cell expansion. [68] Expansion occurs in response to turgor pressure in a direction that depends in part on the arrangement of the cellulose microfibrils in the wall. Cells anticipate their future morphology by controlling the orientation of microfibrils that they deposit in their walls. During their synthesis, nascent cellulose chains assemble spontaneously into microfibrils that form a lamella, in which all the microfibrils are approximately aligned (see **Figure 4.15**). Each new lamella forms internally to the previous one, resulting in a wall composed of concentrically arranged lamellae, with the oldest on the outside. The most recently deposited microfibrils in elongating cells commonly lie perpendicular to the axis of cell elongation (***transverse microfibrils***). Although the orientation in the outer lamellae may be different, it is the orientation in these inner lamellae that has presumably a dominant influence on the direction of cell expansion.

Microtubules

The mechanism that dictates this orientation is connected with the ***microtubules*** in plant cells. The microtubules are arranged in the cortical cytoplasm with the same orientation as the cellulose microfibrils which are being deposited in the cell wall in that region. They form a cortical array close to the cytosolic face of the plasma membrane. This array of microtubules can determine the orientation of newly formed microfibrils, which, in turn, determines directional cell expansion and, therefore, the final shape of the cell, and, ultimately, of the plant. [68]

Wall proteins

The organization of cellulose microfibrils and their complex interaction with other cell wall components results in a extensive and dynamic network that can be modified by the action of several cell wall proteins, such as endoglucanases and expansins. [69, 74] Even if the microfibrils may be in the proper position to permit cell expansion, these proteins contribute to the continuous breaking and remaking of bonds necessary to maintain the integrity of the cell wall during cell expansion. Many genes involved in the synthesis of cell wall polysaccharides have been isolated and characterized.

Pores

The cell wall morphology involves the existence of pores and interfaces. These pores vary widely in size and shape, and play a key role in the accessibility of native cellulose fibers to swelling agents and chemical reactants. Void volumes from 0.7 to more than 4% and pore widths from 0.5 to 15 nm have been reported. [5, 6, 8]

Biosynthesis

Pectins and hemicelluloses are assembled in the **Golgi apparatus** by glycosyltransferases that catalyze the transfer of a glycosyl residue from a nucleotide sugar (XDP-sugar) to an acceptor molecule. [75] These polysaccharides are transported in vesicles that fuse with the plasma membrane and then are released and integrated into the pre-existing wall (**Figure 4.16**). Cellulose is synthesized by rosette terminal complexes located in the plasma membrane. Cell wall polysaccharides may be modified as the cell grows and develops.

The formation of the lignified secondary wall is illustrated in **Figure 4.17**.

The secondary wall is typically composed of an outer layer called S1, a central layer called S2, and an inner layer called S3.

Representative example: Wood cell wall

In a typical wood cell, such as a tracheid (xylem cell), the cell wall is composed of the lignified middle lamella (ML), which is shared by adjacent cells and ensures the adhesion between cells, the primary wall (P), the three layers of the secondary wall (S1, S2, and S3) and a lumen. [8, 77, 78, 79]

The primary wall, which is 0.1–0.2 μm thick, is thinner than that of cotton, and constitutes only 1–4 % by weight of the wood fiber. [8] Besides cellulose, it contains hemicelluloses, pectins, and proteins. It is built up by several lamellae.

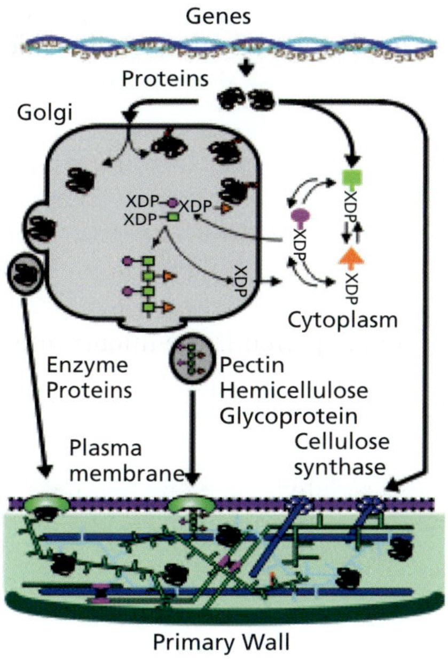

Figure 4.16 Plant cell wall model. Reproduced by permission of Malcolm O'Neil of Complex Carbohydrate Research Center. [76]

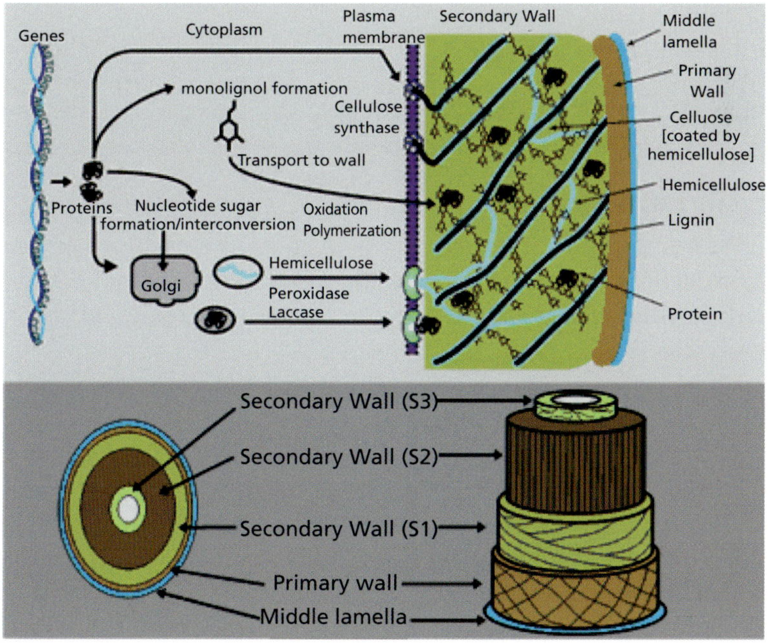

Figure 4.17 Secondary cell wall model. Reproduced by permission of Malcolm O'Neil of Complex Carbohydrate Research Center. [71]

The cellulose microfibrils in the inner lamellae have a nearly transverse orientation, whereas those in the outer lamellae form an irregular network.

The three layers of the secondary wall are built up by lamellae formed by ordered, parallel microfibrils embedded in lignin and hemicelluloses. [5, 6, 8, 77] The S2 layer, which is 1–5 μm thick, contains ~80–85% by weight of the total cellulose content of the wood fiber. The S1 and S3 layers are thin, 0.2–0.3 μm and 0.1 μm thick, respectively. The ordered parallel arrangement of microfibrils differs between the layers. In the S1 layer the microfibrils are tightly wound in helices at a nearly transverse orientation (60–80° to the cell axis). In the S2 layer the microfibrils are arranged in parallel, extended helices at a relatively small angle (5–30°) with respect to the cell axis. In the S3 layer, they are arranged in tightly wound helices at a nearly transverse orientation (60–90° to the cell axis).

4.3 Biosynthesis

The biological pathway of cellulose synthesis has been difficult to elucidate especially in plants.

Biosynthesis of cellulose is a multi-step process involving ***terminal complexes*** (TCs) containing ***cellulose synthase*** enzymes organized in spinnerets at the cell membrane. Cellulose synthases use α-linked ***uridine diphosphoglucose*** (UDPGlc) as substrate to polymerize glucose. Each terminal complex spins a crystalline microfibril consisting of parallel, hydrogen-bonded glucan chains.

The section covers at first the terminal complexes responsible for cellulose microfibrils, then the glucose polymerization by cellulose synthases, and, finally, the assembly of nascent glucan chains into microfibrils.

4.3.1. Terminal complexes

In 1964, Preston proposed the existence in the plasma membrane of a cellulose synthase complex composed of a number of subunits, each of which would catalyze the synthesis of one glucan chain that associates with adjacent chains to form a microfibril. [80, 81]

In 1976, Brown and Montezinos first imaged a linear particulate complex at the tip of a growing microfibril in freeze-fractured ***plasma membranes*** of the alga, *Oocystis apiculata*. [82] This terminal complex consists of a rectangular array of three linear rows of ~30 particles each. [83]

Since 1976, only two types of structures responsible for cellulose synthesis have been identified in freeze-fracture plasma membranes of many organisms: [81, 84, 85]

- Linearly arranged terminal complexes in single or multiple rows that are present in bacteria, the cellular slime mold *Dictyostelium discoideum* and some algae, such as *Glaucocystis* and *Valonia*. [83]

- Hexagonal structure terminal complexes with six-fold symmetry, called ***rosettes***, that are present in mosses, ferns, algae, such as *Micrasterias*, and vascular plants. [86]

The rosettes are assembled in the Golgi apparatus and then transported to the plasma membrane. [87] Furthermore, it is believed that the rosettes move through the plasma membrane synthesizing multiple cellulose chains. [88]

The TC geometry correlates with microfibril size and shape: [81] in some cellulosic algae, the microfibrils are the largest and most highly crystalline. *Valonia*, for example, synthesizes a microfibril of more than 1000 glucan chains, and would have more than 1000 catalytic units within each linear terminal complex. [89] In this instance, the length of the linear complex controls the microfibril width. [81] In plants, most microfibrils are much smaller than in algae, containing mostly about 36 glucan chains. [89] Each rosette (~25 nm-diameter complex comprising six subunits) is believed to contain a number of catalytic units, possibly six per subunit. [84] In some algae, such as *Micrasterias*, which also possess rosettes, the large microfibril size is created by an aggregation of rosettes into higher structures. [89] In the bacterium *Acetobacter xylinum*, the cellulose-synthesizing sites (subunits of the terminal complex) are observed as a linear row of 3.5 nm pores that cover, each, a 10 nm particle. [90] Each particle synthesizes a 1.5 nm-wide (sub-elementary) microfibril predicted to consist of 16 glucan chains. The microfibrils from different particles associate to form a ribbon. [90]

4.3.2 Glucose polymerization by cellulose synthases

Cellulose synthase substrate

Cellulose synthases are ***glycosyltransferases*** that catalyze the transfer of sugar moieties from activated donor molecules (glycones) to specific acceptor molecules (aglycones), forming glycosidic bonds. [81, 89-92] These activated molecules are referred to as ***nucleotide sugars*** or ***sugar nucleotides***.

The only cellulose synthase substrate is uridine 5′-diphospho-glucose (UDPGlc), located in the cytoplasm of many plant and bacterial cells (**Figure 4.18**).

UDPGlc is thought to bind to an active site of the enzyme on the cytoplasmic face of the plasma membrane. [84]

UDPGlc can be formed along a pathway in which α-D-glucose is first phosphorylated to α-D-glucose-6-phosphate by the enzyme glucokinase, then α-D-glucose-6-phosphate is isomerised to α-D-glucose-1-phosphate by the enzyme phosphoglucomutase, and, finally, α-D-glucose-1-phosphate is converted to α-linked UDPGlc by the enzyme UDPGlc pyrophosphorylase, as shown in **Equations 4.1–4.3** (ATP: adenosine triphosphate; UTP:

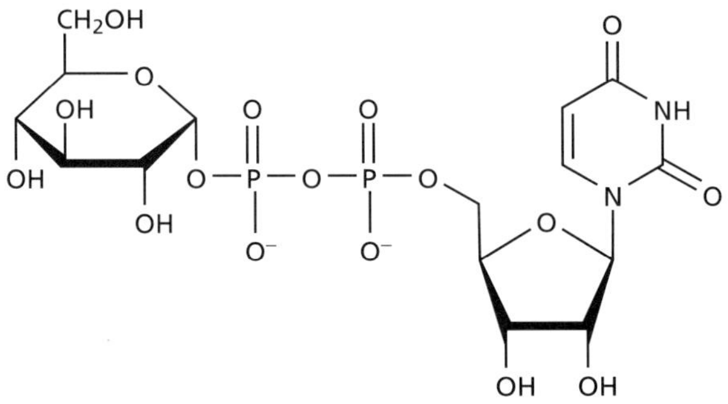

Figure 4.18 Uridine 5′-diphospho-glucose (UDPGlc).

uridine triphosphate; P_i: inorganic phosphate (a pyrophosphate in the last reaction). [93, 94]

$$\alpha\text{-D-glucose} + \text{ATP} \xrightarrow{1} \alpha\text{-D-glucose-6-P} + \text{ADP (1: Glucokinase)} \quad (4.1)$$

$$\alpha\text{-D-glucose-6-P} \xrightarrow{2} \alpha\text{-D-glucose-1-P} \quad \text{(2: Phosphoglucomutase)} \quad (4.2)$$

$$\alpha\text{-D-glucose-1-P} + \text{UTP} \xrightarrow{3} \text{UDPGlc} + P_i \quad \text{(3: UDPGlc pyrophosphorylase)} \quad (4.3)$$

The pathway shown in **Equations 4.1–4.3** is the only known source for UDPGlc in the bacterium *Acetobacter xylinum.*

An alternative pathway for synthesis of UDPGlc is catalyzed by sucrose synthase (SuSy), which converts UDP and sucrose into UDPGlc and fructose.

In plants such as *Arabidopsis thaliana*, UDPGlc channeled from sucrose degradation is the substrate for cellulose synthesis. [95] Sucrose, the main carbon source for cellulose biosynthesis, is translocated to the cellulose-producing cells or drawn away from photosynthetic cells. [96]

Thus, a key player in regulating the flux of incoming carbon into cellulose appears to be SuSy. Following the detection of a plasma membrane associated form of SuSy, pm-SuSy, from developing cotton fibers, it was proposed that the membrane association of SuSy facilitates the direct channeling of substrate to cellulose synthase proteins. [84, 96, 97] Both citrus and cotton SuSy have been shown to play an essential role in cellulose synthesis in secondary cell walls. [96-98] SuSy gene appears to be universally required and available for cellulose synthesis in all developing and growing cells in poplars. [96]

Plants possess many enzymes involved in nucleotide sugar interconversion. [91] By these enzymatic reactions, plants synthesize activated sugars that are required for cell wall biosynthesis. Enzyme activity appears to be regulated at the level of gene transcription in a cell-type and differentiation-dependent manner. It has been suggested that the control of these enzymes regulates glycosylation patterns in response to developmental, metabolic, and stress-related stimuli. [99] Regulated nucleotide sugar interconversion supports the concept that substrate channeling may be an important control point in polysaccharide biosynthesis. [57]

Cellulose synthases

We refer to cellulose synthases as the proteins responsible, each, for the polymerization of one β-1,4-glucan chain. Cellulose synthases are also called catalytic subunits of the multimeric enzyme complex (TC).

Processive β-glycosyltransferases

Cellulose synthases are the active proteins in cellulose biosynthesis. They are processive β-glycosyltransferases. They have specificities for both donor and acceptor molecules, and for the type of linkage they form (α- or β-linkage). They can be ***processive*** (catalyzing the addition of multiple sugar residues) or ***non-processive*** (catalyzing the addition of only one sugar residue).

Cellulose synthases use α-linked UDPGlc as substrate. They catalyze the formation of cellulose chains with β-1,4-linkages, involving a conformational inversion of the anomeric carbon. Therefore, they are ***inverting*** glycosyltransferases classified as family 2 glycosyltransferases (GT2). [100] UDP glycosyltransferases mediate the transfer of glycosyl residues from activated nucleotide sugars to acceptor molecules. [100]

Processive β-glycosyltransferases represent a large family of enzymes present in organisms from all domains of life. [101]

Conserved motifs

Although the gene sequences coding for processive glycosyltransferases are quite divergent, these enzymes all have at least three aspartic acid residues (D) and a QxxRW motif (Q: glutamine; R: arginine; W: tryptophan; x: any amino acid). [101] In chitin synthase, the motif is conserved as QRRRW, while in all the NodC (root nodulation factor) proteins, the sequence is conserved as QQLRW. On the other hand, proteins that polymerize a long-chain product (polysaccharide), such as cellulose, carry the QRxRW sequence, while proteins that synthesize short-chain product (oligosaccharide), such as NodC, carry the QQxRW sequence. It appears that the second residue of the motif may be involved in determination of the glucan chain length. [102] In non-processive glycosyltransferases, only two conserved aspartic acid residues have been found. [90]

Processive glycosyltransferases are characterized by two ***conserved domains*** (sequence and structure motifs), termed domains *A* and *B*, whereas non-processive enzymes are characterized by a single conserved domain, domain

A. Two aspartic acid residues in domain A, and a single aspartic acid residue and the QxxRW motif in domain B were found to be strictly conserved in the sequences analysed. [102] The aspartic acid residues are believed to be involved in the catalytic reaction, while the QxxRW motif is thought to be responsible for the processive mechanism. [102]

Cellulose synthase family

The cellulose synthase family (not including the cellulose synthase from tunicate animals) can be divided into the following subfamilies as a function of their organism source: [101, 103]

- ***Euryarchaeota*** putative cellulose synthases.
- ***Cyanobacteria*** (blue-green algae) cellulose synthases CcsA2 and CcsA1.
- ***Firmicutes/fibrobacteres*** putative cellulose synthases.
- ***Amoeba*** cellulose synthases, ***dinoflagellate*** putative cellulose synthases, and putative ***fungi*** cellulose synthases.
- Plant cellulose synthases CesA.
- ***Bacterial*** cellulose synthases BcsA.
- Cellulose synthase-like proteins (Csl) from different organisms.

To the above list should be added the cellulose synthases from animal tunicates (**Figure 4.19**). Indeed, *Ciona savignyi*, a common tunicate, shows conserved features found in all cellulose synthases, including plants, but is most similar to cellulose synthases from bacteria, fungi, and amoebas. [104]

Polymerization models

Cellulose synthases are membrane proteins with a number of trans-membrane regions. The globular fragment of the protein is probably involved in the catalysis. UDPGlc is thought to bind to an active site of the enzyme on the cytoplasmic face (PF) of the plasma membrane with the glucan chain being extruded through the membrane. [84]

Models of glucose polymerization with multiple substrate binding sites have been proposed in order to account for the 180° rotation of adjacent glucose residues in cellulose chains. [106, 107, 108] The two-domain architecture of processive β-glycosyltransferases, coupled to at least three conserved aspartates spread over the two domains, led to the development of a two-catalytic center model in which the A and B domains were each proposed to function as glycosyl transfer centers.

Two substrate molecules were predicted to bind to the enzyme such that the glucose moiety of each molecule was positioned 180° with respect to the other. During chain elongation, two β-1,4-glycosidic bonds were formed

Figure 4.19 Tunicates. [105] The genes that *Ciona* uses to produce the fibrous tunic are similar to the genes involved in metabolizing cellulose. Reproduced by permission of Nick HOBGOOD under the Creative Commons Attribution ShareAlike 3.0 License.

either sequentially or simultaneously, and the growing chain was elongated two glucose units at a time. The effective addition of a cellobiose occurred via the double addition of glucoses.

The initial model involving two centers for glycosyl transfer featured chain elongation at the reducing end with three substrate binding sites, two in domain A and one in domain B, attached to the growing chain. [106, 109] However, electron micro-diffraction and tilting analysis combined with silver labeling of the microfibril reducing ends provided evidence that cellulose biosynthesis takes place at the non-reducing end of the growing chain. [84, 110, 111] When non-reducing end elongation was demonstrated, the two-center model evolved to a putative model featuring two substrate binding sites. [102, 107, 110] The C4 hydroxyl of the sugar to be added was predicted by Koyama et al. to be activated by aspartic acid (**Figure 4.20**). [110]

The two-center model has been recently questioned by the three-dimensional structures of UDP-sugar glycosyltransferases that suggest a single substrate binding site. [84, 111] In the model with single substrate binding site,

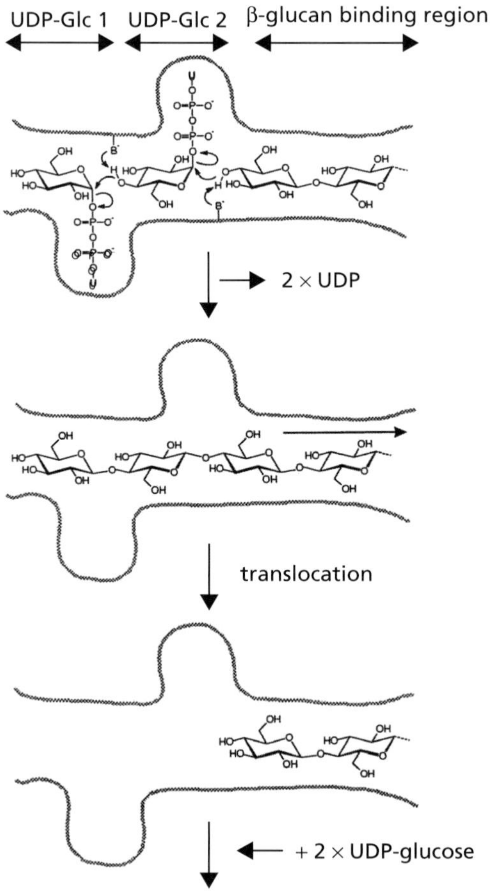

Figure 4.20 Two-binding site model for cellulose chain polymerization by addition of sugar residues at the nonreducing end. The C4 hydroxyl group is activated by general base catalysis provided by aspartate residues (B^-). Reproduced by permission. Copyright (1997) National Academy of Sciences, USA. [110]

the A domain of cellulose synthases and related processive family 2 glycosyltransferases binds the UDP-sugar, and the B domain binds the acceptor molecule, the aspartates (D) in the A and B domains forming a single center for glycosyl transfer. [111] Inverting transferases, including cellulose synthases, are assumed to use a single displacement mechanism with nucleophilic attack by the acceptor molecule at the C1 (anomeric) carbon of the donor sugar. [106, 111] Such a mechanism is presumed to require a base (aspartate residue) to activate the sugar acceptor for nucleophilic attack by deprotonation due to the poor nucleophilic character of sugar hydroxyls. For most enzymes the reaction also involves additional carboxylate(s) (aspartate residue(s)) to coordinate a divalent metal ion on the phosphate group(s) of the nucleotide (**Figure 4.21**).

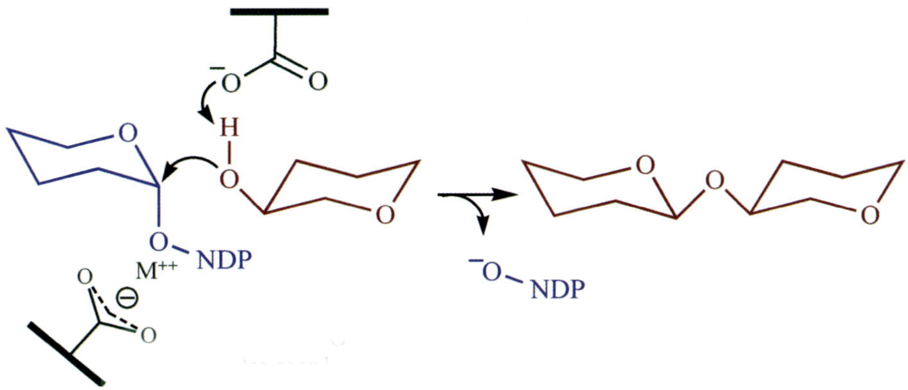

Figure 4.21 Putative mechanism for an inverting nucleotide-sugar glycosyltransferase, such as cellulose synthases. The mechanism is believed to require at least two catalytic carboxylates, one to activate the sugar acceptor and at least one to coordinate a divalent metal ion associated with the NDP-sugar (nucleoside diphosphate sugar). For cellulose synthases, it is presumed that at least two aspartates (D) in the A domain bind UDPGlc and one aspartate (D) in the B domain binds the sugar acceptor. [111]

Polymerization of cellulose in bacteria

The gram-negative bacterium *Acetobacter xylinum* produces pure cellulose, and has served as a model organism. It is now included in the new genus *gluconacetobacter* as *G. xylinus.*

The biosynthetic pathway for *Acetobacter xylinum* cellulose synthesis involves the formation of UDPGlc from glucose via the glucose-6-phosphate and glucose-1-phosphate intermediates. [90] In 1958 already, Glaser [112] presented evidence that the glycosyl donor in the synthesis of cellulose by a cell-free preparation of *Acetobacter xylinum* was UDPGlc. It was in this bacterium that UDPGlc was first implicated in cellulose biosynthesis. [81, 85]

In 1982, Aloni et al. [113] achieved high rates of in vitro synthesis of cellulose by preparing membranes from *Acetobacter xylinum* in the presence of PEG (polyethylene glycol) and GTP (guanosine triphosphate). [80] Benziman group showed that PEG precipitated a soluble diguanylate cyclase enzyme that converted GTP to an activator of the cellulose synthase: cyclic diguanylic acid (c-di-GMP). [89] The discovery of c-di-GMP, a cyclic nucleotide, allowed an acceleration of in vitro research. In 1989, Lin et al. synthesized cellulose II in vitro from *Acetobacter* extracts and purified the enzymes using trypsin and product entrapment. [81, 89] Such purification led the Brown group to identify two types of subunits of molecular mass 83- and 93-kD (kilo Dalton) in the purified fraction. [90, 94] By means of a radioactively labelled UDPGlc substrate, the 83-kD polypeptide was shown to be the ***catalytic subunit***. From amino acid sequences obtained from the 83-kD polypeptide, Saxena, Lin, and Brown [114] identified in 1990 the gene called *acsA* that encodes the catalytic subunit. Also in 1990, by analysis of cellulose-deficient mutants, Wong et al. [115] isolated an ***operon*** (group of coordinately controlled genes) of four genes called *bcsA-D*. The *bcsA* gene isolated by Wong shows high homology to the *acsA* gene cloned by Brown group.

In 1991, Saxena et al. [116] identified a second gene encoding the 93-kD polypeptide. Later in 1994, Brown group [117] isolated a cellulose-synthesizing operon (acs operon) and characterized the genes in the operon. The acsA and acsB genes, which were initially thought to be two separate single ***open reading frames*** (ORF), [118] were believed to fuse into a single ORF coding for a 168-kD polypeptide (AcsAB). [117] However, the protein is presumably cleaved into the A and B proteins after translation. [89]

It is now accepted that the acsB/bcsB gene encodes a regulatory subunit that binds c-di-GMP. The concentration of cellulose synthase activator c-di-GMP is regulated by phosphodiesterases. [90] The genes for the c-di-GMP-synthesizing diguanylate cyclase and phosphodiesterases have been identified. [90]

The catalytic subunit of the Acetobacter xylinum cellulose synthase contains a large, catalytic, globular fragment facing the cytoplasma. The dimensions of this catalytic fragment are presumably roughly 60 x 60 x 70 Å. [90] The predicted globular structure has a mixed topology of β strands and α helices (10 strands and 8 helices) and contains a central, elongated cavity. The QxxRW motif is present in a region close to the central cavity. [90] Besides, it is known that the catalytic subunit is glycosylated. [94]

The *acsC* and *acsD* genes code for associated polypeptides of molecular mass 138- and 17-kD, respectively. [90] The *acsAB* and *acsC* genes are likely to be essential for cellulose production in vivo. [117] The *acsD* gene appears to play a major role in the crystallization of cellulose, for when this gene is disrupted, only cellulose II is assembled in agitated cultures. [81, 117] The AcsC and AcsD polypeptides are believed to have a structural rather than enzymatic function. According to a model proposed by the Brown group, they are organized with the AcsAB cellulose synthase in a membrane cellulose-synthesizing complex. [90, 119] Mutant analysis also led to the identification of *acsAII,* a second gene involved in cellulose synthesis. [120]

As mentioned before, the reducing end of the growing chains points away from the bacterium, so that polymerization by cellulose synthase takes place at the nonreducing end of the growing chains. [110] This mechanism is likely to be also valid for a number of processive glycosyltransferases that belong to the GT2 family. In 2003, Imai et al. confirmed the mechanism in β-chitin. [121]

The *Acetobacter xylinum* genes involved in cellulose synthesis have been renamed *AxCes,* according to a new nomenclature proposed by Delmer [89] (first letters for Genus and species, followed by *Ces* for cellulose synthase, followed by *A* for genes encoding a presumed catalytic subunit, a final number indicating the timing of the first report of the gene). In agreement with this nomenclature, the *acsA* and *acsB* genes have been termed *AxCesA* and *AxCesB*, respectively, and the *acs* operon, *AxCesA-D*.

In 2013, Morgan et al. [122] at the University of Virginia School of Medicine reported a crystallographic snapshot of bacterial cellulose synthesis and membrane translocation. The researchers presented the crystal structure of a cellulose synthase complex from the bacterium *Rhodobacter sphaeroides* containing a translocating polysaccharide. The crystal structure revealed how the catalytic subunit of the complex formed a cellulose-conducting channel, and suggested a model for the coupling of cellulose synthesis and translocation in which the nascent polysaccharide is extended by one glucose molecule at a time. The enzyme complex is unique in that it both produces cellulose polymers (by attaching glucose molecules) and pushes them outside the cell simultaneously; usually the division of labor is different, with production and movement either handled separately or handled by different enzymes. [123] The discovery could have major ramifications for both biofuel production and the battle against bacterial infections.

Polymerization of cellulose in plants

Genes encoding cellulose synthases

Although the *Acetobacter xylinum* gene for cellulose synthase was cloned in 1990, it is not before 1996 that the first cellulose synthase genes were identified from higher plants. [124] A major difficulty in all searches for in vitro cellulose synthase activity in plasma membranes derived from higher plants was the very high levels of ***callose*** (β-1,3-glucan) synthase activity in these membranes. [80] Furthermore, rosettes disassembly occurs concomitantly with the loss of cellulose synthase activity upon cell rupture. [84] Thus, early attempts to purify the enzyme were severely hampered.

The identification of the *AxCesA-D* operon led different groups to screen cDNA libraries from plants to search for homologous genes, using *AxCesA* as a probe. These searches were first unsuccessful because the genes were strongly different or showed homology in limited, conserved regions, as yet unidentified. This homology was revealed true further to the identification in 1995 of the conserved motifs in β-glycosyltransferases, using hydrophobic cluster analysis. [81, 89, 125] In particular, the D, D, D, QxxRW motif was found characteristic for β-glycosyltransferases.

In 1996, in a search for gene sequences with the D, D, D, QxxRW motif, Pear et al. [124] identified two genes (cDNA clones) in cotton fibres (*CelA1* and *CelA2*, renamed *GhCesA1* and *GhCesA2* [89]) that are homologous to the *CesA* bacterial genes encoding the catalytic subunit of cellulose synthase. The reported cotton genes are both highly expressed during active secondary wall cellulose synthesis. [124] The full-length ORF of *GhCesA1* encodes a 974 amino acid polypeptide of ~110 kDa that, like the bacterial CesA proteins, is presumed to be a membrane-bound protein with eight trans-membrane helices at the N-terminus and at the C-terminus that border a central, cytoplasmic domain. However, GhCesA1 and GhCesA2 differ from the bacterial CesAs in that they contain two large plant-specific insertions within the central domain: a ***conserved plant-specific region*** (CR-P) and a ***hypervariable region*** (HVR). Furthermore, the plant CesA proteins have an extended N-terminal region and a shorter C-terminal region compared to bacterial CesAs. [84, 89]

Since 1996, a large number of processive β-glycosyltransferase genes have been identified in plants. They all code for proteins that contain the D, D, D, QxxRW motif. Based on predicted protein sequences, these genes have been grouped into seven families:

- the *CesA* family that encode catalytic subunits of cellulose synthase;
- six families of structurally related genes of unknown function designated as the cellulose synthase-like genes (*CslA*, *CslB*, *CslC*, *CslD*, *CslE* and *CslG*). [90, 126]

The Csl proteins are likely to catalyze the biosynthesis of non-cellulosic polysaccharides. The seven families form the *CesA* superfamily, which includes over forty genes. [127]

Ten *CesA* genes (called ***isoforms***) encode cellulose synthases in *Arabidopsis thaliana* (At) for which full genome sequence is available. There are also a number of *CesA* genes in cotton (Gh), maize (Zm), rice (Os), quaking aspen (Pt), gray poplar (Pt/Pa), and other plants. [126]

The ten *Arabidopsis CesA* genes and rice (Os) *CesA7* gene range in size from 3.5 to 5.5 kb, with approximately 10 small introns (non-coding regions) separating the exons (coding regions). [128]

Plant CesA proteins contain at their N-terminus an amino acid domain that is similar to a ***zinc finger*** (a DNA-binding protein motif that contains residues that coordinate a zinc ion which creates a finger-like loop) or LIM transcription factor. It is presumed that this domain might mediate protein-protein interaction in the CesA complex. [128]

According to a hypothetical model for the three-dimensional structure of CesA proteins, the eight trans-membrane helices of the proteins form a pore in the plasma membrane, through which the growing chain passes. [128] The N-terminus, with the putative protein-protein domain, would be localized in the cytoplasm.

In 2000, identification of nine members of the *CesA* gene family of maize (*Zea mays*), designated *ZmCesA1-9*, has allowed Holland et al. [129] to initiate a comparative study with ***homologs*** from *Arabidopsis* and other plant species. The study has shown that closely related *CesA* genes, i.e. those found to pair in sequence trees (***cladograms***), are not clustered but are found at different chromosomal locations in both *Arabidopsis* and maize.

Furthermore, sequence comparisons between the CesA proteins indicate that these cluster in groups in which ***orthologs*** (homologs in different species) are often more similar than ***paralogs*** (homologs in the same species). Genes were found to be expressed in unique cell types engaged in either primary or secondary wall synthesis.

The *AtCesA* genes have been shown by mutant analyses to play distinct roles in the cellulose synthesis process. [57, 84, 130, 131] Functional specialization within this family includes differences in gene expression (e.g. primary or secondary wall), regulation, and, possibly, catalytic function (e.g. chain initiation or elongation). [84] In 2003, Taylor et al. [132] provided evidence that AtCesA4, AtCesA7 and AtCesA8 proteins interact with each other, presumably forming part of the same cellulose-synthesizing complex in secondary cell walls, and favouring its correct assembly. A similar situation appears to occur in primary walls, where AtCesA1, AtCesA3, and AtCesA6 proteins all are required. Three not functionally redundant *CesA* genes from rice, *OsCesa4*, *OsCesA7*, *OsCesA9* have also been reported to be required for cellulose synthesis in the secondary wall. [133] All this suggests that three CesA proteins are a general requirement of cellulose synthesis in most, if not all, plant cell walls. [57, 88, 134]

In 2003, Gardiner, Taylor, and Turner [88] showed that the correct localization of CesA proteins to sites of secondary wall deposition requires the presence of cortical microtubules, but that ***actin*** filaments are unlikely to play a direct role. [134] It should be noted that microtubules and actin filaments are the two most important types of filaments in the cytoskeleton; both are made of globular protein basic units that can assemble and disassemble rapidly in the cell. [135] Gardiner et al. [88] observed that in young xylem vessels of *Arabidopsis*, the three CesA proteins localize within the cell, whereas in older vessels, all three proteins colocalize with bands of microtubules. Microtubules arrays are required continually to maintain correct CesA protein localization. [88]

In 2004, Burton et al. [136] provided evidence that the *CesA* gene family from barley (*Hordeum vulgare*) has at least eight members which are distributed across the genome. In most tissues, *HvCesA1*, *HvCesA2*, and *HvCesA6* predominate, and their relative abundance is very similar; these genes appear to be coordinately transcribed. A second group, comprising *HvCesA4*, *HvCesA7*, and *HvCesA8*, also appears to be coordinately transcribed, but in specific maturing tissues.

As reported by Joshi et al. [96] in 2004, the *CesA* gene family from aspen (*Populus tremuloides*) has at least seven members (*PtrCesA1–7*). Among these, *PtrCesA1*, *PtrCesA2*, and *PtrCesA3* appear to be secondary cell wall *CesAs* because of their high degree of amino acid sequence similarity with AtCesA8, AtCesA7, and AtCesA4 proteins, respectively, whereas *PtrCesA4*, *PtrCesA5*, *PtrCesA6*, and *PtrCesA7* appear to be primary cell wall *CesAs* because of their high degree of similarity with *AtCesA1*, *AtCesA3*, and *AtCesA6*.

In 2006, Paredez et al. [137] demonstrated that cellulose synthase complexes containing AtCesA6 are organized in the cell membrane by a functional association with cortical microtubules. Inhibition of microtubule polymerization changed the distribution and pattern of moving CesA complexes in the membrane, indicating a relatively direct mechanism of guidance of cellulose deposition by the cytoskeleton.

In 2007, Persson et al. [138] presented genetic evidence for three unique components in primary cell-wall cellulose synthase complexes in *Arabidopsis*. Indirect evidence suggests that the complex associated with primary wall cellulose deposition consists of CesA1, CesA3, and CesA6. However, phenotypes associated with mutations in *CesA1* and *CesA6* suggest unequal contribution by the different CesAs to overall enzymatic activity of the complex. Persson et al. showed that the primary complex requires three unique types of components, CesA1-, CesA3-, and CesA6-related, for activity.

Non-catalytic molecules have been recently involved in plant cellulose biosynthesis and will be reviewed in the next sections.

Korrigan

In 1998, Nicol et al. [139] discovered a plasma membrane endo-1,4-β-***glucanase*** from *Arabidopsis* that is required for correct cell wall assembly and cell elongation. They identified it as Korrigan cellulase. In 2000, Zuo et al. [140] confirmed the role of Korrigan (KOR) protein during cell plate formation. In 2001, Lane et al., [141] Sato et al., [142] and His et al. [143] independently found that the Korrigan endo-1,4-β-glucanase was required for cellulose synthesis in at least primary cell walls, suggesting that there are parallels with *Agrobacterium tumefaciens* and other bacteria which also require endo-1,4-β-glucanases for cellulose synthesis. [144] A model of plant cellulose biosynthesis involving Korrigan includes a three-step process (**Figure 4.22**) [96]:

- Plasma-membrane-associated SuSy (sucrose synthase) channels UDPG to CesA proteins.
- Coordinately expressed CesAs polymerize glucose into glucan chains while liberated UDP is recycled to SuSy.
- Korrigan acts as an editor or monitor of glucan chain conversion to cellulose microfibril, cleaving defective glucan chains. [88, 119, 145]

In another scenario (see next paragraph), Korrigan would be involved in cellulose synthesis by removing putative sterol-cellodextrin primers from growing glucan chains. [130, 146]

Sitosterol-β-glucoside primer

In 2002, Peng et al. [146] proposed a model in which the lipid sitosterol-β-glucoside (SG) serves as a primer for plant cellulose synthesis catalyzed by CesA

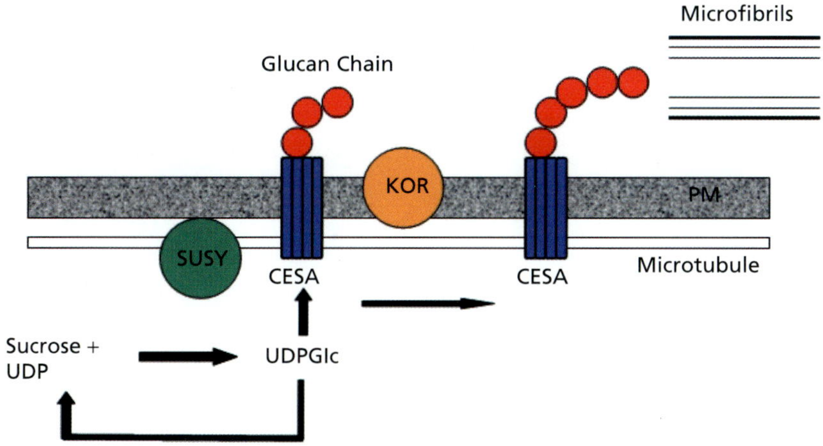

Figure 4.22 Recent model of plant cellulose biosynthesis. Plasma membrane (PM)-associated SuSy (sucrose synthase) channels UDPGlc to CesA proteins that form glucan chains; glucan chains self assemble into microfibrils with Korrigan acting as an editor/monitor of this process; the multimeric CesA complexes may be moving across the plasma membrane with the help of microtubules. [96]

proteins. In their in vitro experiments using cotton-fiber membranes and yeast membranes expressing GhCesA1, they were able to identify sitosterol-cellodextrins (SCDs), i.e. sitosterol-linked oligosaccharides, as lipid intermediates in the initiation of cellulose synthesis. They also suggested that the Korrigan cellulase required for plant cellulose synthesis may cleave SG from the growing chain.

The primer-related polymerization model proposed by Read and Bacic [147] is shown in **Figure 4.23**.

The Read and Bacic model involves the following steps.

- UDP-glucosyltransferase (UGT) transfers a glucose residue from UDPGlc onto a β-sitosterol molecule in the cytoplasmic side of the plasma membrane, forming SG.
- An initiating CesAi subunit transfers additional glucose residues onto SG, forming SCD.
- SCd "flips" to the outer side of the plasma membrane.
- SCD is cleaved by Korrigan cellulase.
- The resulting oligosaccharide binds to an elongating CesAe subunit, and is extended into a cellulose chain by addition of UDPGlc provided by sucrose synthase.

However, no exogenous primer was required to initiate in vitro synthesis of cellulose, raising doubts on the proposed involvement of SG as a primer. [57]

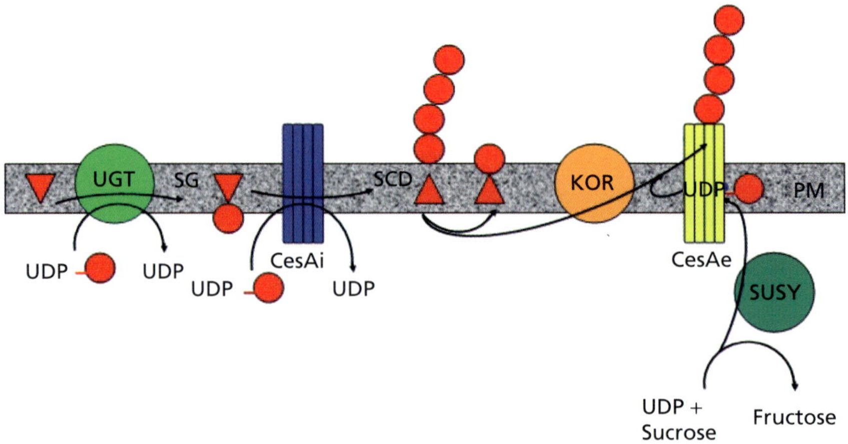

Figure 4.23 Primer-involving model for plant cellulose polymerization. PM, plasma membrane; red triangle, sitosterol; red circle, glucose; green circle, UDP-glucosyltransferase; SG, sitosterol-β-glucoside; CesAi, initiating cellulose synthase; CesAe, elongating cellulose synthase; SCD, sitosterol cyclodextrin; KOR, Korrigan; SUSY, sucrose synthase. [147]

Another problem with this proposal is that the catalytic domain of Korrigan cellulase is predicted to be outside the plasma membranes, whereas catalytic domains of CesA proteins are located towards cytoplasmic side. [96]

In 2004, Schrick et al. [148] showed a link between sterol biosynthesis, the cell wall, and cellulose in *Arabidopsis*. In their work, three sterol biosynthesis mutants exhibit a deficiency in cellulose, but are not reduced in pectin and sugars of the cell wall and cytosol, suggesting that sterols are crucial for cellulose synthesis. Possibly sterol production correlates with the metabolic states of cells, which, in turn, are exploited to regulate developmental processes, such as cell differentiation. [149]

4.3.3 Chain assembly into microfibrils

The unique structure and properties of cellulose result not only from its linear glucan chain but also from the assembly of the chains into a microfibril. The terminal complex organization appears most important to determine the size and crystal structure of the microfibril formed.

Chain assembly in Acetobacter xylinum

Under standard culture conditions, *Acetobacter xylinum* produces a 40–60 nm wide ribbon, which is extruded parallel to the long axis of the cell through about one hundred pores in a linear row, and twisted in a right-handed manner. [150] The cell rotates about its longitudinal axis and is propelled forward as the ribbon elongates at a rate of about 2 μm min^{-1}. [85, 94] The ribbons contain both cellulose Iα and Iβ in a parallel up packing. [110]

In contrast, *Acetobacter xylinum* produces band-like cellulose assemblies under some conditions. [150] The band-like assemblies consist of strand-like entities, each of which is extruded perpendicularly to the longitudinal axis of the cell. Hirai et al. [150] reported that two kinds of band-like assemblies, coarse and dense structures, are produced at 4 °C, whereas ribbons are produced by the same colonies at 28 °C. The dense band-like assembly, produced in an early stage of incubation at 4 °C, is composed of cellulose II, while the coarse band-like assembly, produced after a longer incubation time at 4 °C, does not give any crystalline reflections. It can be noted that the in vitro product from a cytoplasmic membrane fraction of *Acetobacter* was cellulose II. [151]

This raises the question of the nature and localization of the proteins in the cellulose-synthesizing particles within the linearly arranged terminal complexes (**Figure 4.24**).

It has been proposed that each particle is made up of the AcsAB, AcsC, and AcsD proteins based on the *acs* operon. [119] According to this model, the cellulose synthase catalytic subunit and the c-di-GMP-binding protein (AcsAB) are present in the bacterial cytoplasmic membrane. These proteins possibly interact with the AcsC and AcsD proteins in forming a transenvelope complex, which constitutes the cellulose-synthesizing particle. The AcsC protein may form a pore in the bacterial outer membrane through which cellulose must be secreted. [117] Based on cellulose II production by the *acsD* mutant, the AcsD protein is suggested to play a role in crystallization of the microfibrils by assisting in aligning the glucan chains so that they interact and crystallize under standard conditions into cellulose I. [117]

The mechanism of production of the cellulose II band-like assemblies is not clear, even if a putative folded-chain model were proposed. [150]

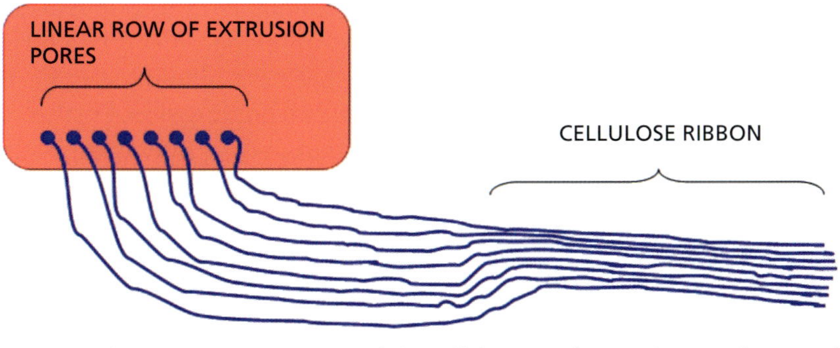

Figure 4.24 Schematic representation of the cellulose production in *Acetobacter xylinum*. Glucan chain aggregates are spun from a linear row of pores. [152]

Chain assembly in plants

Rosette terminal complexes were first reported in higher plants in 1980. [83] Single, hexagonal rosette structures were later found to be exclusive to all higher plant cellulose assembly. The size and hexameric structure of the rosette fit with a model in which each subunit of the rosette synthesizes six glucan chains, leading to a 36-chain microfibril. [80] This is in agreement with the hypothesis that truly crystalline cellulose I requires an aggregate of at least 35 chains. [153]

In 1995, a hypothetical model for plant cellulose biosynthesis involving a multisubunit cellulose synthase complex, SuSy for UDPGlc synthesis, and a microtubule was proposed by Delmer and Amor. [80] The model showed a cellulose synthase complex at the plasma membrane. Each complex, responsible for the synthesis of one microfibril, had one synthetic unit including a catalytic subunit per glucan chain. The catalytic subunit substrate, UDPGlc, was synthesized by a membrane-associated form of SuSy. In 1999, Kimura et al., [154] using immuno-cytochemical labelling of proteins attached to freeze-fractured replicas, confirmed the long-held hypothesis that cellulose synthase is a component of the rosette terminal complex in vascular plants. [155]

Several explanations have been suggested for the requirement for expression of at least two different *CesA* genes in the same cells. [156] One hypothesis is that two non-identical proteins are required to catalyze β-1,4-glucan polymerization. [157] However, another hypothesis results from consideration of the geometry of the planar multisubunit rosette complex (**Figure 4.25**).

In 2002 Doblin et al. proposed a model for the structure of the rosette. [84] Each rosette (~25 nm in diameter comprising six subunits) is believed to contain possibly six synthetic units per subunit, each of which polymerizes a glucan chain that associates with chains of the same rosette to form one microfibril. The simultaneous presence of three different CesA proteins to produce a functional cellulose synthase complex was possibly explained by the geometrical constraints associated with assembling 36 proteins into the rosette. [84, 158] From theoretical consideration of the rosette assembly, at least two different types of interaction could be envisaged, one between CesA proteins within a rosette subunit and one between rosette subunits. [84] If spontaneous rosette self-assembly occurred via interaction between CesA proteins alone, at least two different CesA isoforms were required: the α subunits that have binding sites for three CesA subunits, and the β subunits that have only two binding sites. Furthermore, two different types of α isoforms could be distinguished: α_1 interacting with two β isoforms only, and α_2 interacting with two β isoforms and one α_2 isoform. This type of model explained the need for three different types of CesA isoforms within a cell-type.

The association of CesA proteins has been suggested to be mediated by their zinc finger domains. [84] All members of the CesA family contain two putative zinc fingers at their N-terminus that are thought to mediate protein-protein

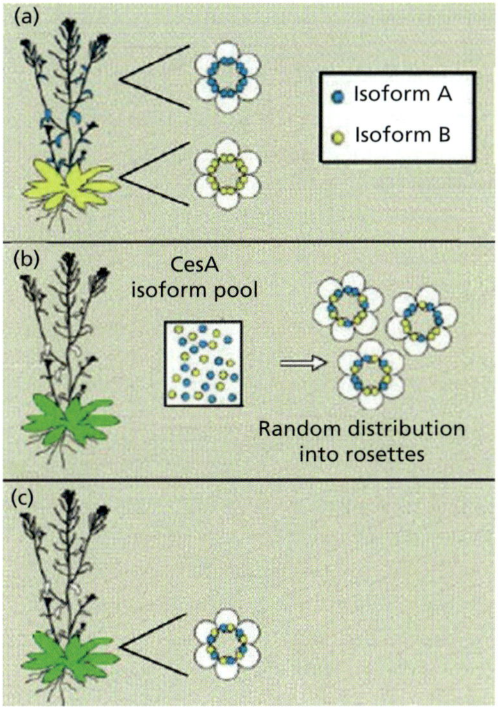

Figure 4.25 Models for distribution of CesA isoforms within a plant and consequences for the cellulose synthase rosette. (a) Different isoforms are expressed in different tissues. Rosettes contain only one type of isoform. (b) Different isoforms are expressed in the same cells. The isoforms are functionally redundant and contribute to a pool of CesA proteins. Rosettes contain randomly distributed isoforms. (c) Different isoforms are expressed in the same cells, but the isoforms are not redundant. One isoform of each type is required to synthesize each glucan chain. Reproduced by permission of Elsevier. [157]

interaction. [84] Kurek et al. [158] have shown that the zinc finger domain of GhCesA1 can interact with itself to form homodimers or heterodimers with the zinc finger domain of GhCesA2 in a redox-dependent manner. [84] The zinc finger domain can exist in either a reduced or oxidized form. Under reduced conditions, it coordinates two zinc ions. Under oxidized conditions, the zinc finger domain and also the full-length protein can dimerize. These results have led to a model in which the process of plant multimeric complex (TC) assembly and function may involve oxidative dimerization of CesA proteins via their zinc finger domains. [84, 158] In support of this hypothesis, it has been proposed that the OsCesA4 zinc finger domain can interact with OsCesA7 or OsCesA9 for the cellulose-synthesizing complex assembly. [133]

In 2005 Saxena and Brown [159] proposed a revised model for the rosette structure. The model takes into consideration two levels of assembly of the cellulose synthases. In the first level, assembly and processing of three different homodimers, composed each of two identical cellulose synthases, occurs to form a linear array of six particles, presumably within the cytoplasmic portion of the TC. In the next level, the linear arrays are arranged in a rosette with a six-fold symmetry. In the model, each linear row produces glucan chains that can associate by van der Waals forces to form a glucan sheet. Formation of glucan sheets is the first step in cellulose crystallization. In the second step, six separate glucan sheets are directed into the exit channel of the TC, where they pass through the rosette aperture and are then hydrogen-bonded into the crystalline cellulose I microfibril.

In 2002, the bimolecular fluorescence complementation (BiFC) assay was implemented to confirm the interaction in plant living cells. [160] Using this technique, Desprez et al. [161] demonstrated in 2007 that the primary AtCesA proteins (CesA1, CesA3, and CesA6) can interact in vivo, and, therefore, are present in the same complex. Using the same technique, Timmers et al. [162] indicated in 2009 that although all secondary AtCesA proteins (CesA4, CesA7, and CesA8) can interact with each other, only CesA4 is able to form homodimers. They proposed a model for the secondary rosette structure (**Figure 4.26**). They also indicated that the zinc-fingers are not essential for the interaction between the CesA proteins.

Until recently, no biochemical data had shown any non-CesA proteins to be associated with the complex: however, TED6 (Tracheary Element Differentiation-Related6), a plasma membrane-localized protein of unknown function, has been demonstrated to interact with the CesA7 subunit of the secondary wall complex. [163, 164] TED appears to have a general role in secondary cell-wall formation in vessels, but whether it has a specific function in cellulose deposition is unclear.

In 2010, Gu et al. [165, 166] identified a cellulose synthase-associated protein required for cellulose biosynthesis. Cellulose synthase-interactive protein

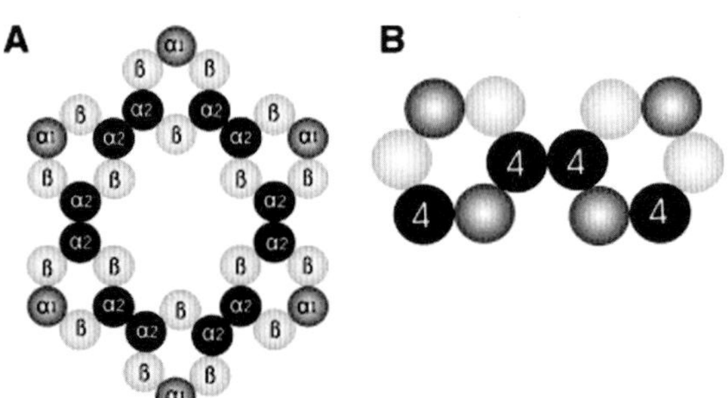

Figure 4.26 Proposed models for the structure of the rosette. **(A)** Six subunits, containing six CesA polypeptides, interact to form a rosette as suggested by Doblin et al. **(B)** The modified model as suggested by Timmers et al., CesA4 (4) interacts with all isoforms, the homodimerization links the subunits together, the two other positions are filled by CesA7 and CesA8. [162]

1 (CSI1) was identified in a two-hybrid screen for proteins that interact with cellulose synthase (CesA) isoforms involved in primary plant cell wall synthesis. Mutations in CSI1 cause defective cell elongation in hypocotyls and roots and reduce cellulose content. CSI1 is associated with CesA complexes, and *csi1* mutants affect the distribution and movement of CesA complexes in the plasma membrane. As the first non-CesA proteins associated with the cellulose synthase complexes (CSCs; also called terminal complexes), CSI1 opens up many opportunities to unravel the remaining mysteries of cellulose biosynthesis in plants. Research towards (1) the identification of novel components in CSCs, (2) the investigation of the interactions between minimal components in CSCs, and (3) a better understanding in assembly, delivery, and regulation of CSCs will increase our knowledge of how plants synthesize cellulose and will provide perspective on more efficient lignocellulosic biorefineries. [166]

4.4 Properties and applications

Cellulose is a biodegradable, non-toxic, solid polymer. Its melting point is far above its decomposition temperature. However, it has been shown that if the heating was faster than the thermal degradation, molten cellulose could be obtained. [167] Cellulose is not thermoplastic. It is insoluble in water and most organic solvents. The polymer does swell, however, in many polar liquids, including water. Cellulose is a hygroscopic substrate which absorbs ~8–14% water at 20 °C and 60% relative humidity (RH). [50]

Mechanical, physical and environmental properties of cellulose will be reviewed subsequently in relation to some of its applications.

4.4.1. Properties

Mechanical properties

Theoretical values

Several modes of deformation can occur within cellulose fibers. [8] The ordered regions will show the typical elastic deformations of a crystalline solid, with a high modulus. In 1960, Treloar [168] reported theoretical calculations leading to an elastic modulus of 56 GPa in the axial direction and lower in the other directions. In 1962, Sakurada et al. [169] deduced a value of 137 GPa from an experimental study. In 1985, Northolt et al. [170] demonstrated that not only the fiber structure but also the deformation mechanism was similar to that found in high performance aramid fibers. In 1995, Nishino et al. [171] estimated the elastic modulus of the crystalline regions of cellulose allomorphs by x-ray diffraction. The values for cellulose I and II were 138 and 88 GPa, respectively. In 2005, by modeling crystal and molecular deformation, Eichhorn et al. [172] reported the theoretical crystal modulus for cellulose II as 98 GPa. This value is in agreement with other studies and with an experimental measurement using synchrotron

X-ray diffraction. In 2006, Tanaka et al. [173] used the molecular simulation technique to derive values between 124 and 155 GPa for the modulus of a cellulose Iβ crystal.

It is now well accepted that estimates of the crystal modulus of cellulose I are in the range of 130–145 GPa. [174]

Experimental values

Experimental values of the tensile properties of various cellulose structures are shown in **Table 4.7**.

The following should be noted in the table: 1) The measured modulus of 143 GPa for tunicate whiskers is close to the theoretical value. [175] 2) The value of 114 GPa obtained for a single bacterial cellulose filament is higher than previous reported values, but lower than estimates of the crystal modulus of cellulose I. [174] These discrepancies have been explained in terms of the crystallinity and structural composition of the samples. [174] 3) The value of 80 GPa for flax was obtained using elementary flax fibers. [176]

Native cellulose fibers at the macroscopic level show a wide diversity in mechanical properties (**Table 4.8**).

Important structural parameters in determining the mechanical properties of native fibers are the degree of polymerization, crystallinity, degree of orientation of the microfibrils to the fiber axis, size and shape of the microfibrils, and the arrangement of the microfibrils in the fiber. The tensile properties of cellulose fibers increase with degree of polymerization, crystallinity, and degree of orientation. [6, 8] The orientation of the microfibrils with respect to the fiber axis causes a lowering of the tensile modulus compared to that of axial microfibril. In flax, jute, and sisal, the microfibrils form helices at an angle of ~10° with respect to the fiber axis, whereas in cotton the angle is ~21°, resulting in a further lowering of the modulus.

Cotton's strength is about 25% greater wet than dry, in contrast to most animal and synthetic fibers. [8] This has been explained in the following manner. In the dry state, there will be strong hydrogen bonding between the cellu-

Table 4.7 Experimental values of the tensile properties of various cellulose structures.

	Elastic modulus (GPa)	Method
Tunicate whiskers [175]	143	Raman spectroscopic technique
Bacterial cellulose filament [174]	114	
Flax [176]	80[1)]	Single fiber tension

1) 1-2 GPa for tensile strength

Table 4.8 Tensile properties of native cellulose fibers. [177]

	Elastic modulus (GPa)	Tensile strength (GPa)
Flax	27–100	0.34–1.03
Jute	26–65	0.39–0.77
Ramie	60–128	0.40–0.94
Cotton	5–13	0.28–0.6
Wood	10–40	1

lose molecules on the surfaces of neighboring microfibrils, inhibiting slippage and creating a shear resistance. As the moisture content increases, with the absorption of water between the microfibrils, slippage will become easier, until in the wet state the system can be regarded as fully lubricated with a relieve of stress concentrations and a more uniform sharing of the load.

Viscose fibers are produced via the dissolution of a cellulose derivative called xanthate in aqueous NaOH solutions, while ***lyocell fibers*** are produced via the direct dissolution of cellulose in organic solvents (NMMO). Textile (standard) viscose has lower strength (tenacity) than cotton. Its wet strength is only ~60% of the dry value. [50] Stretched (high-tenacity) viscose is much stiffer than textile viscose and generally stronger than cotton. Although lyocell fibers also consist of regenerated cellulose, they are considerably stronger, either dry or wet, than viscose fibers. [50] In regenerated cellulose fibers, the degree of orientation, the structure across the cross-section, and the form of the crystals would be primarily responsible for the mechanical diversity. [8] The mechanism of deformation in all these materials has been tentatively related to the amorphous cellulose linking the crystalline elements. In the wet state, the amorphous chain segments would be plasticized by water and free to deform as a low-modulus rubber. In the dry state, hydrogen-bonding between the amorphous segments would make the fiber stiffer with a higher modulus.

Cell walls synthesized by plants and animals are high-performance biocomposites consisting of a matrix reinforced by fibrous polymers. [178] Cellulose and chitin are typical examples of these reinforcing elements, which occur as long microfibrils ~2–50 nm wide (see § 4.2.3.2). Along the cellulose microfibrils, the main imperfections are the twists, which have been recognized as the main defects. [39] These crystal imperfections have been exploited to produce by acidic treatment monocrystals called ***whiskers*** having the same diameter as the starting microfibril but much shorter length. [4] Depending on their origin, their properties can differ, but their mechanical strength remains of great interest (**Table 4.9**).

Table 4.9 Strength and stiffness of cellulose nanocrystal compared to some other synthetic fibers. [179]

Material	Tensile strength (GPa)	Modulus (GPa)
Cellulose nanocrystal	7.5	145
Glass fiber	4.8	86
Steel wire	4.1	207
Graphite whisker	21	410
Carbon nanotubes	11–73	270–970

Physical properties and water absorption

Physical properties

Cellulose is a semi-crystalline polymer. It is non-melting and not thermoplastic. Thermal degradation starts at ~180 °C. Up to this temperature, the amorphous regions of cellulose are in the glassy state, without large-scale molecular motion, and mechanical properties are maintained. Amorphous cellulose is plasticized by water, as illustrated by cellulose sponges, which are rigid when dry and supple when wet. Sponge suppleness in the wet state means that water is capable to lower the glass transition temperature of the cellulose-water system below room temperature. Actually, water can be considered as the most efficient plasticizer of cellulose. [180] **Table 4.10** gives reported thermal data on cellulose. [5, 50]

Pure dry cellulose is a good insulator with a resistivity of 10 [18] Ωcm. Conductivity increases with increasing water and ionic impurities content (**Table 4.11**). [5]

Table 4.10 Thermal data on cellulose. [5, 50]

Thermal decomposition	>180 °C
Glass transition temperature (extrapolated)	230 – 245 °C
Ignition point	>290 °C
Heat of combustion	17.46 J g^{-1}
Heat of crystallization	18.7–21.8 kJ mol^{-1} glucose
Specific heat	1.00–1.21 J g^{-1} K^{-1}
Heat of transition Cell I → Cell II	38.1 J g^{-1}
Coefficient of thermal conductivity	0.255–0.920 kJ m^{-1} h^{-1} K^{-1}

Table 4.11 Resistivity of cellulose. [5]

Substrate	Relative humidity (%)	Resistivity (Ωcm)
Pure dry cellulose	0	10^{18}
Cotton	52	2.10^{7}
	62	10^{7}
	75	7.10^{5}
	85	3.10^{5}
Viscose fiber	75	3.10^{6}
	85	10^{2}

Table 4.12 Dielectric constant of cellulose. [5]

Substrate	Relative humidity (%)	Dielectric constant	
		1 kHz	100 kHz
Cotton	0	3.2	3.0
	45	7.1	4.4
	65	10.0	6.0
Viscose fiber	0	3.6	3.5
	45	5.4	4.7
	65	8.4	5.3

Dielectric constants of cellulose at two frequencies are given in **Table 4.58.** Dielectric constants of cellulose at two frequencies are given in **Table 4.12.** They are dependent on humidity conditions.

Cellulose fibers are usually white and opaque. They exhibit strong birefringence due to the preferential orientation of the molecules along the fiber axis. **Table 4.13** gives, with respect to the fiber axis, the parallel (n_1) and perpendicular (n_2) refractive indices for cellulose I and II. [5]

Many cellulose derivatives form liquid crystalline phases, both in solution (***lyotropic mesophases***) and in melt (***thermotropic mesophases***). [2, 7] In 1976, Werbowyj and Gray [181] first reported that aqueous solutions of 30% hydroxypropylcellulose form lyotropic mesophases that display iridescent colors characteristic of the ***chiral*** (having different left-handed and right-handed forms) ***nematic*** (having long-range orientational order so that extended molecules can move parallel to each other) state, also called ***cholesteric*** state. [2] Typically, cellulosic liquid crystals are of interest because they have the potential of opening a route to high strength, high modulus fibers. [2, 7]

Table 4.13 Refractive indices of cellulose. [5]

	n_1 1)	n_2 2)
Completely oriented cellulose I	1.618	1.544
Cotton	1.576–1.595	1.527–1.534
Completely oriented cellulose II	1.578	1.523
Viscose fiber	1.529–1.547	1.512–1.520

1) Parallel to the fiber axis; 2) Perpendicular to the fiber axis.

Water absorption

Cellulose-water interaction plays a central role in cellulose isolation and processing, as well as in papermaking. [5] Cellulose swelling in water is the most representative example of intercrystalline swelling.

Cellulose exhibits a high hygroscopic resulting from the interaction between its OH groups and water molecules, but is not dissolved in water because the cellulose molecules prefer their equals to water. Cellulose-water interaction can be described as a competition in hydrogen bonding between cellulose OH groups, on one side, and cellulose OH groups and water molecules or clusters, on the other side. [5]

Cellulose-water interaction strongly depends on the supramolecular structure of the substrate. For example, on swelling in water, native fibers show an increase in cross-sectional area of 20–35%, whereas regenerated fibers show one of 55–70%. [6]

Water sorption isotherms of cellulose have been studied extensively. [5, 6, 8] The amount of water vapor sorbed at definite temperature and relative humidity (water vapor regain) is one of the important criteria for describing the textile properties of cellulose fibres. [5] It is commonly used to characterize cellulose accessibility. [6]

A typical water sorption isotherm is shown in **Figure 4.27**. The sigmoidal shape of the curve results from the superposition of two processes: (1) a mono- and multilayer adsorption of water molecules interacting with accessible OH groups in amorphous regions and crystal surfaces, and (2) a capillarity condensation (physical sorption) in the pore system enlarged by the sorption and swelling process itself. [5, 6] The sum of bound water originating from process (1) and free water originating from process (2) is total water sorption.

As illustrated in **Figure 4.28** for cotton, sorption and desorption isotherms display hysteresis. This hysteresis possibly arises from positional changes of structural entities due to the previous sorption and/or from a pertinacious binding of the last traces of water on full desorption. [5]

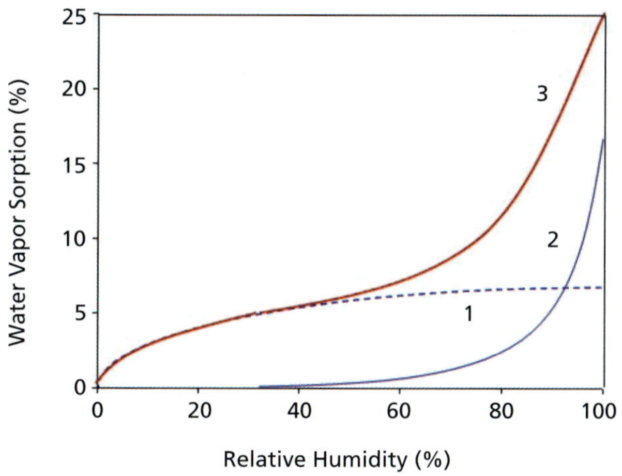

Figure 4.27 Water sorption isotherm of cellulose (3) showing total sorbed water (3) as the sum of bound water (1) and free water (2). [5, 182]

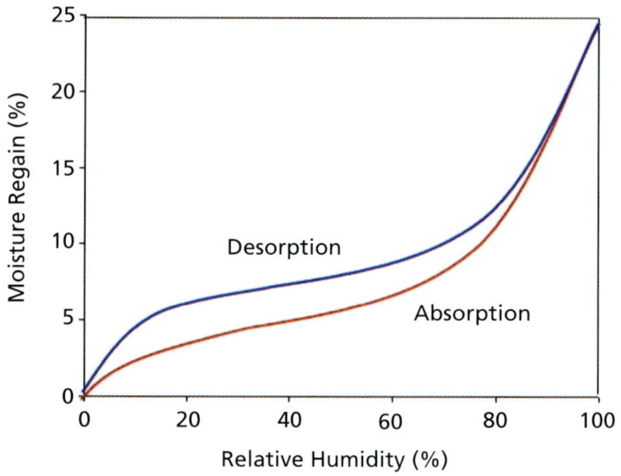

Figure 4.28 Hysteresis displayed by water sorption and desorption isotherms of cotton at 25 °C. [183]

Specific surface areas can be calculated from water sorption isotherms. Examples of such calculations using the Brunauer-Emmett-Teller (BET) method are presented in **Table 4.14**.

As shown in **Table 4.14**, the specific surface areas accessible to water vapor increase with decreasing crystallinity. Values range from 135 m^2/g for microcrystalline cotton cellulose to more than 400 m^2/g for amorphous cellulose. In this range, the value for cotton is ~150 m^2/g and that for cellulose II rayon ~250 m^2/g.

Table 4.14 Monomolecular adsorption of water vapor and specific surface areas of various cellulosic materials calculated by the BET method. [6, 8]

Substrate	Temperature (°C)	Wm[1)] (g/g)	RVP[2)] at Wm	Specific surface area[3)] (m^2/g)
Microcrystalline cotton cellulose[4)]	21	0.0273	0.21	135 [184]
Cotton linters	20	0.0318	0.22	157 [185]
Cotton	21	0.0327	0.21	162 [184]
Cotton	25	0.0282	0.19	139 [185]
Microcrystalline mercerized cotton cellulose[4)]	21	0.0345	0.20	171 [184]
Mercerized cotton	21	0.0456	0.21	226 [184]
Cellophane	25	0.0587	0.22	288 [185]
Viscose rayon	25	0.0515	0.20	254 [185]
Amorphous cellulose	21	0.0875	0.19	434 [184]

1) Weight of adsorbate per unit weight of adsorbent when a monomolecular layer has been just formed. 2) Relative vapor pressure. 3) The cross-sectional area of the water molecule taken as 0.148 nm^2. 4) Highly crystalline regions of the original cellulose after acid hydrolysis of the more accessible material.

Environmental properties

Cellulose is biodegradable and non-toxic to living organisms. Its biodegradation is an essential step in the carbon cycle, which ensures the carbon balance in the biosphere.

The enzymes that hydrolyze the β-1,4 linkages in cellulose are called ***cellulases*** (see Chapter 5). [8, 52] They are produced by many microorganisms, such as bacteria and fungi, but also by plants and some invertebrate animals.

Traditionally, cellulases have been divided into two groups, ***endoglucanases*** and ***exoglucanases*** (or ***cellobiohydrolases***), according to their respective capacity to cleave the β-1,4-glucosidic bond either internally or at one of the ends of the cellulose chain. In addition, the hydrolytic enzymes β-1,4-***glucosidases*** are capable to cleave cellobiose into glucose.

Cellulases are modular enzymes. They are members of the ***glycoside hydrolase*** (GH) families of enzymes that hydrolyze the polysaccharides in plant cell walls. The efficient degradation of cellulosic substrates generally requires multienzyme systems which include either a collection of free cellulases or/and multicomponent complexes called ***cellulosomes***.

The rapid enzymatic degradation of cellulose, if it can be achieved, should offer an enormous commercial potential. In particular, a great interest lies in the use of cellulosic biomass as a renewable source of energy via breakdown to sugars that can then be converted to liquid fuel. [186]

4.4.2 Applications of cellulose fibers

Cellulose can be found in nature in many different organisms including the whole vegetable kingdom, and the corresponding cellulosic fibers differ in nature and properties (molecular environment, dimensions, resistance, etc.). This influences their main uses and applications.

Wood pulp and manmade fibers

As a starting point, it is interesting to know that most of the cellulosic biomass exploited in the world (essentially wood) is used for production of energy (burning) and furniture. This represents ~1700 million tons per year. Compared to that, production of wood pulp is only 150 million tons per year, and about 98% of it is used for paper production. What remains, ~3.8 million tons, gives specialty and dissolving wood pulps for mainly production of regenerated cellulosic fibers like viscose (**Figure 4.29**).

Viscose fibers production is booming in China and India, due to increased demand of textile and nonwovens markets. For the first time, in 2009 China accounted for more than 50% of global demand, and more than 50% of global production. [188] Lyocell/Tencel fibers are now exclusively from Lenzing, an Austrian company. Production of regenerated cellulose fibers by the dissolving of wood pulps, spinning, and regeneration enable different shapes and properties (**Figure 4.30**). New properties can also be obtained through the addition of active molecules directly in the cellulose solution, such as biocides and anti-UV additives. Cellulose dissolution/regeneration processes are more detailed in the book, *Cellulose Science and Technology*. [189]

Among interesting and growing areas, wood fiber composites (WFC) are developing fast. Those products are mainly composed of HDPE (high density polyethylene), or PP (polypropylene) or PVC (polyvinyl chloride) mixed with ~40–60% wood. Their main application is the manufacture of decking due to their increased durability and comfort compared to wood. The WFC market is expected to grow from ~1.5–2 million tons currently up to ~10 million tons in the following 5 years mainly due to the increase of the Chinese market.

Wood pulps, eventually chemically modified, are also used as additives in concrete and other building materials.

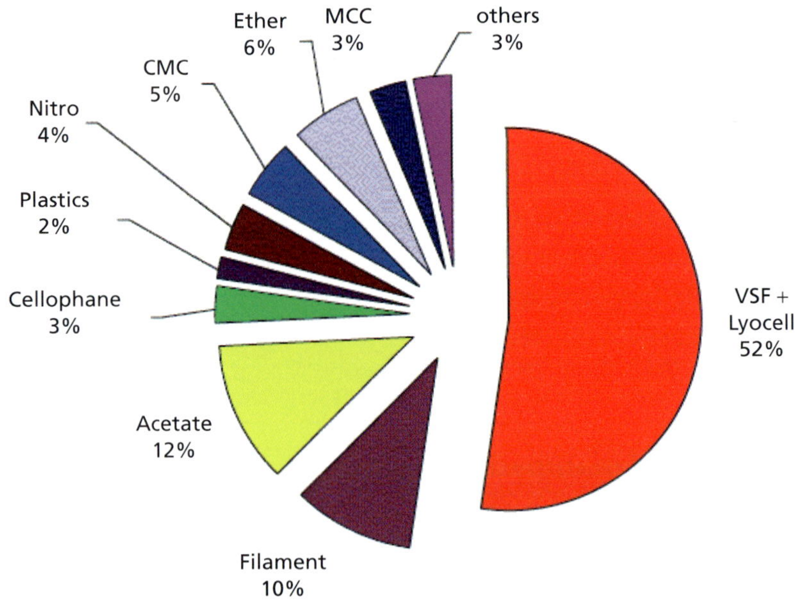

Figure 4.29 Main specialty and dissolving pulp market shares (Nitro = nitrocellulose; Acetate=cellulose acetate; Plastics = other esters than acetate; MCC=microcrystalline cellulose; Ether = various cellulose ethers; VSF = viscose staple fibers; CMC = carboxy methyl cellulose). [187]

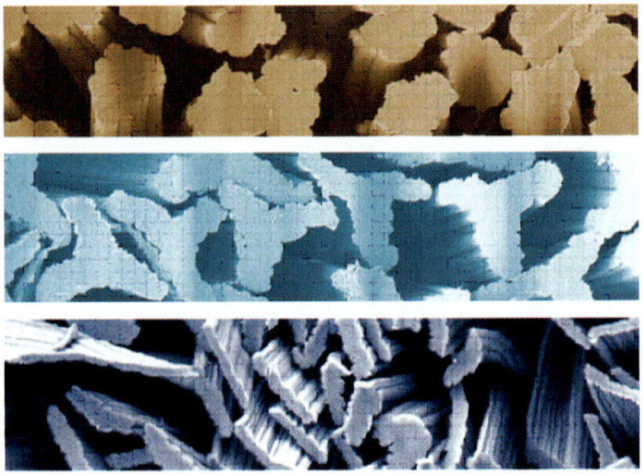

Figure 4.30 Different sections of viscose fibers for different applications: standard for textile (top), trilobal for tampons (middle), flat for nonwovens (bottom) (courtesy of KELHEIM FIBERS). [190]

Staple cotton and linters

Compared to other non-woody natural fibers, staple cotton (long fibers of length up to 5cm) is by far the most important in use with a production of ~26 million tons in 2005 (more than 40% of it from China and India). Application for these long fibers is almost 100% for textile production, and cotton is the most used fiber in textile industry, even compared to polyesters or polyamides. As cotton culture is known to have a very negative impact on the environment, the tendency is to switch to organic cotton (non-transgenic and grown without pesticides); however, this still represents only 0.03% of the total production of cotton.

Cotton shrubs also provide the market with another grade of fibers called cotton linters (~18 million tons a year). These are very short fibers (~1.5 mm) covering cotton seeds and mechanically separated from the seeds after ginning the long staple fibers.

Linters are almost pure cellulose with high DP (~1000–3000) and narrow distribution. Therefore, linters can be used like a dissolving pulp for viscose production, especially in China where most cotton plants are integrated within viscose plants. They also can be used like fluff pulps for diapers, and also for specialty papers for their touch and strength.

Other non-woody fibers

Vegetal fibers (wood and cotton excluded) include abaca, bamboo, banana, coir, flax, hemp, jute, kenaf, ramie, roselle, and sisal.

At ~2.5 million tons per year, jute is widely used for production of bags, rugs, and carpets. [191] Fibers like hemp, kenaf, ramie, sisal, and flax have garnered interest over the last 10 years for the production of technical textiles and composites. Due to new regulations concerning greenhouse gas emissions, life-cycle analysis, and the recycling of materials, natural fibers are interesting in the fields of construction and car industry: they are biobased, light weight with good mechanical resistance, offer good thermal and acoustic insulation, and can advantageously replace glass or carbon fibers in composites. [192]

As with wood fiber composites (WFC), the key parameters are the dispersion of the fibers in the matrix and the compatibility at the fiber-matrix interface. Generally, coupling agents are used in the form of maleated polyolefins, which can be added to the composition at a ratio of a few percent. Several known chemical companies are supplying different grades depending on polymer type, chain length, and number of maleic residues. [193]

It is noteworthy that lignin prevents adhesion with the plastic matrix in composites. However, after extraction and isolation, lignin is added in formulations as an adhesion promoter. [194, 195] Lignin can also be used as filler in composites, or reversibly as a matrix for panel boards.

Whiskers from *Valonia* or diverse biobased fibers have been widely studied in the laboratories for production of composites and especially bionanocomposites in which both the nanosized reinforcement and the matrix are biobased. [196, 197] But until now their use has been hindered by yields and cost of production and great difficulty in processing, due to their tendency to aggregate and form hydrogen bonds. [198]

Since 1995 the world market of technical textiles has been growing rapidly (3–6% per year). It is now estimated at 21 million tons and has been greatly diversified with end uses in various industries, such as construction, packaging, medical, leisure, and filtration. [199, 200] The market shares of vegetal fiber materials are expected to grow even more in the future. Good examples of the expanding market for nonwovens are mats made of hemp, or bast and flax being used in building insulation, sometimes thermally bonded with e.g. thermofusible polyester fibers, sometimes needlepunched.

4.6 References

[1] P.M. Collins and R.J. Ferrier, *Monosaccharides-Their Chemistry and Their Roles in Natural Products*, John Wiley & Sons, Chichester, 1995

[2] A.D. French, N.R. Bertoniere, R.M. Brown, H. Chanzy, D. Gray, K. Hattori and W. Glasser, *Encyclopedia of Polymer Science and Technology–Cellulose*, John Wiley & Sons, Inc., 2003

[3] S. Nishikawa and S. Ono, *Proc. Math. Phys. Soc.* Tokyo 7, 131, 1913

[4] S. Perez and B. Mackie, 2001 in http://glyco3d.cermav.cnrs.fr/glyco3d/lessons/cellulose/index.html

[5] D. Klemm, B. Philipp, T. Heinze, U. Heinze and W. Wagenknecht, *Comprehensive Cellulose Chemistry*, Volume 1: *Fundamentals and Analytical Methods*, Wiley-VCH, Weinheim, 1998

[6] H. Krässig, *Cellulose – Structure, Accessibility and Reactivity*, Polymer Monographs, Volume 11, Gordon and Breach Science Publishers, Amsterdam, 1993

[7] R.D. Gilbert, ed., *Cellulosic Polymers, Blends and Composites*, Hanser/Gardner Publications, Inc., Cincinnati, 1994

[8] T.P. Nevell and S.H. Zeronian, eds., *Cellulose Chemistry and Its Applications*, Ellis Horwood Ltd, Chichester, 1985

[9] M. Koyama, W. Helbert, T. Imai, J. Sugiyama and B. Henrissat, *Proc. Natl. Acad. Sci. USA* 94, 9091, 1997

[10] K.H. Meyer and H.F. Mark, Z. Phys. Chem. B2, 115, 1929; K. H. Meyer and L. Misch, *Helv. Chim. Acta 20*, 232, 1937; K. H. Meyer and L. Misch, *Ber. 70B*, 266, 1937

[11] C.A. Jentzen, Doctor's Dissertation, Appleton, 1964, in http://etd.gatech.edu/theses/submitted/ipstetd-217/unrestricted/jentzen_ca.pdf

[12] K.H. Gardner and J. Blackwell, *J. Biopolymers 13* 1975, 1974

[13] A. SARKO and R. MUGGLI, *Macromolecules 7*, 486, 1974

[14] R.H. ATTALA and D.L. VANDERHART, *Science 223*, 283, 1984

[15] D.L. VANDERHART and R.H.ATTALA, *Macromolecules 17*, 1465, 1984

[16] F. HORII, H. YAMAMOTO, R. KITAMARU, M. TANAHASHI and T. HIGUCHI, *Macromolecules 20*, 2946, 1987

[17] H. YAMAMOTO, F. HORII and H. ODANI, *Macromolecules 22*, 4130, 1989

[18] E.M. DEBZI, H. CHANZY, J. SUGIYAMA, P. TEKELY and G. EXCOFFIER, *Macromolecules 24*, 6816, 1991

[19] J. SUGIYAMA, R. VUONG and H. CHANZY, *Macromolecules 24*, 4168, 1991

[20] Y. NISHIYAMA, P. LANGAN and H. CHANZY, *J. Am. Chem. Soc. 124*, 9074, 2002

[21] Y. NISHIYAMA, J. SUGIYAMA, H. CHANZY and P. LANGAN, *J. Am. Chem. Soc. 125*, 14300, 2003

[22] A.A. BAKER, W. HELBERT, J. SUGIYAMA and M.J. MILES, *J. Struct. Biol. 119(2)*, 129, 1997

[23] A.A. BAKER, W. HELBERT, J. SUGIYAMA and M.J. MILES, *Appl. Phys. A. 66*, S559, 1998

[24] A.A. BAKER, W. HELBERT, J. SUGIYAMA and M.J. MILES, *Biophys. J. 79*, 1139, 2000

[25] K.R. ANDRESS, *Z. Phys. Chem. 34*, 190, 1929

[26] J.F. REVOL and D.A.I. GORING, *Polymer 24*, 1547, 1983

[27] J. SUGIYAMA, H. HARADA, Y. FUJIYOSHI and N. UYEDA, *Planta 166*, 161, 1985

[28] H. CHANZY, B. HENRISSAT, R. VUONG and J. F. REVOL, *Holzforschung 40* (Suppl.), 129, 1986

[29] H. CHANZY, in J.F. Kennedy, G.O. Phillips, and P.A. Williams, eds., *Cellulose Sources and Exploitation*, Ellis Horwood Ltd, Chichester, 1990

[30] N.H. KIM, T. IMAI, M. WADA and J. SUGIYAMA, *Biomacromol. 7*, 274, 2006

[31] T. OKANO and A. SARKO, *J. Appl. Polym. Sci. 30*, 325, 1985

[32] F.J. KOLPAK and J. BLACKWELL, *Macromolecules 9*, 273, 1976

[33] A. STIPANOVIC and A. SARKO, *Macromolecules 9*, 851, 1976

[34] F.J. KOLPAK, M. WEIH and J. BLACKWELL, *J. Polymer 19*, 123, 1978

[35] K. GESSLER, N. KRAUSS, T. STEINER,C. SANDMAN and W. SAENGER, *Science 266*, 1027, 1994

[36] K. GESSLER, N. KRAUSS, T. STEINER, C. BETZL, A. SARKO and W. SAENGER, *J. Am. Chem. Soc. 117*, 11397, 1995

[37] S. RAYMOND, A. HEYRAUD, D. TRAN QUI, Å KVICK and H. CHANZY, *Macromolecules 28*, 2096, 1995

[38] S. RAYMOND, B. HENRISSAT, D. TRAN QUI, Å KVICK and H. CHANZY, *Carbohydr. Res. 277*, 209, 1995

[39] H. CHANZY, Personal communication, 2008

[40] P. LANGAN, Y. NISHIYAMA and H. CHANZY, *J. Am. Chem. Soc. 121*, 9940, 1999., P. LANGAN, Y. NISHIYAMA and H. CHANZY, *Biomacromolecules 2*, 410, 2001

[41] Cellulose Structure and Hydrolysis Challenges in https://public.ornl.gov/site/gallery/originals/Fig2_Cellulose_Structure_a.jpg

[42] A. FREY-WYSSLING, *Protoplasma 25*, 261, 1936

[43] K.H. MEYER and A.J.A. VAN DER WYK, *Z. Elektrochem. 47*, 353, 1941

[44] H.F. MARK, *J. Phys. Chem. 44*, 764, 1940

[45] K. HESS, H. MAHL and E. GÜTTER, *Kolloid Z. 155*, 1, 1957

[46] J.W.S. HEARLE, *J. Polym. Sci. 28*, 432, 1958

[47] J.W.S. HEARLE, *Fiber Structure*, J.W.S. Hearle and R.H. Peters, eds., Butterworth, London, 1963

[48] R.H. ATALLA, Proceedings of the second Tricel symposium on *Trichoderma reesei cellulases and other hydrolases*, Espoo 1993, ed. by P. Suominen and T. Reinikainen. Foundation for Biotechnical and Industrial Fermentation Research 8, 25, 1993 in www.fpl.fs.fed.us/documnts/pdf1993/atall93b.pdf

[49] J. SUGIYAMA and T. IMAI, *Trends Glycosci. Glycotechnol. 11*, 23, 1999

[50] H. KRÅSSIG, J. SCHURZ, R.G. STEADMAN, K. SCHLIEFER, W. ALBRECHT, M. MOHRING and H. SCHLOSSER, in *Ullmann's Encyclopedia of Industrial Chemistry–Cellulose*, Wiley-VCH Verlag Gmbh & Co. KGaA, 2004

[51] H. CHANZY, Personal communication, 2005

[52] E.A. BAYER, H. CHANZY, R. LAMED and Y. SHOHAM, *Curr. Opin. Struct. Biol. 8*, 548, 1998

[53] N.H. KIM, W. HERTH, R. VUONG and H. CHANZY, *J. Struct. Biol. 117*, 195, 1996

[54] W. HELBERT, Y. NISHIYAMA, T. OKANO and J. SUGIYAMA, *J. Struct. Biol. 124*, 42, 1998

[55] Y.V. DAELE, J.F. REVOL, F. GAILL and G. GOFFINET, *Biol. Cell 76*, 87, 1992

[56] R.J. VIETOR, R.H. NEWMAN, M.A. HA, D.C. APPERLEY and M.C. JARVIS, *Plant J. 30*, 721, 2002

[57] C. SOMERVILLE, S. BAUER, G. BRININSTOOL, M. FACETTE, T. HAMANN, J. MILNE, E. OSBORNE, A. PAREDEZ, S. PERSSON, T. RAAB, S. VORWERK and H. YOUNGS, *Science 306*, 2206, 2004

[58] D.P. DELMER, *Annu. Rev. Plant Physiol. Plant Mol. Biol. 50*, 245, 1999

[59] J. SUGIYAMA, H. HARADA, Y. FUJIYOSHI and N. UYEDA, *Mokuzai Gakkaishi 30*, 98, 1984

[60] R.D. PRESTON, *The Physical Biology of Plant Cell Walls,* Chapman and Hall, London, 1974

[61] F. TANAKA and K. Okamura, *J. Polym. Sci. Polym. Phys. Ed. 15*, 897, 1977

[62] J. SUGIYAMA, H. CHANZY and J.F. REVOL, *Planta 193*, 260, 1994

[63] M. KOYAMA, J. SUGIYAMA and T. ITOH, *Cellulose 4*, 147, 1997

[64] Y. HORIKAWA, T. ITOH and J. SUGIYAMA, *Cellulose 13*, 309, 2006

[65] T. IMAI, J. SUGIYAMA, T. ITOH and F. HORII, *J. Struct. Biol. 127*, 248, 1999

[66] K. HIETA, S. KUGA and M. USADA, *Biopolymers 23*, 1807, 1984

[67] H. CHANZY and B. HENRISSAT, *FEBS Lett. 184*, 285, 1985

[68] B. ALBERTS, A. JOHNSON, J. LEWIS, M. RAFF, K. Roberts and P. WALTER, *Molecular Biology of the Cell*, 4th Edition, Garland Publishing, New York, 2002, in www.ncbi.nlm.nih.gov/books/bv.fcgi?call=bv.View..ShowTOC&rid=mboc4.TOC&depth=2

[69] A.I. CANO-DELGADO, K. METZLAFF and M.W. BEVAN, Development 127, 3395, 2000 in http://dev.biologists.org/content/127/15/3395.full.pdf

[70] Chloroplasts in http://commons.wikimedia.org/wiki/File:Chloroplasten.jpg

[71] Complex Carbohydrate Research Center, The University of Georgia, *Plant Cell Walls*, 2007 in www.ccrc.uga.edu/~mao/intro/ouline.htm

[72] B. ALBERTS, A. JOHNSON, J. LEWIS, M. RAFF, K. ROBERTS and P. WALTER, *Molecular Biology of the Cell*, 4th Edition, Garland Publishing, New York, 2002 in www.ncbi.nlm.nih.gov/books/bv.fcgi?rid=mboc4.section.3603

[73] http://en.wikipedia.org/wiki/Cell_wall

[74] S. G. SAUPE, College of St. Benedict/St. John University, *Plant Physiology (Biology 327)*, 2009 in http://employees.csbsju.edu/SSAUPE/biol327/Lecture/cell-wall.htm

[75] Complex Carbohydrate Research Center, The University of Georgia, *Plant Cell Walls*, 2007 in www.ccrc.uga.edu/~mao/biosyn/text.htm

[76] Complex Carbohydrate Research Center, The University of Georgia, *Plant Cell Walls*, 2007 in www.ccrc.uga.edu/~mao/cellwall/main.htm

[77] U.P. AGARWAL, Planta 224, 1141, 2006, in www.fpl.fs.fed.us/documnts/pdf2006/fpl_2006_agarwal003.pdf

[78] E. SJOSTROM, *Wood Chemistry. Fundamentals and Applications*, 2nd Edition, Academic Press, San Diego, 1993

[79] C. PLOMION, G. LEPROVOST and A. STOKES, *Plant Physiol 127*, 1513, 2001

[80] D.P. DELMER and Y. AMOR, *Plant Cell 7*, 987, 1995

[81] R.M. BROWN *Jr.*, *Pure Appl. Chem. 71*, 767, 1999

[82] R.M. BROWN Jr. and D.L. MONTEZINOS, *Proc. Natl. Acad. Sci. USA 73*, 143, 1976

[83] R.M. BROWN Jr., in *Experimental Phycology. Cell Walls and Surfaces, Reproduction, Photosynthesis*, W. Wiessner, D. G. Robinson, and R. C. Starr, eds., Springer-Verlag, Berlin, 1990

[84] M.S. DOBLIN, I. KUREK, D. JACOB-WILK and D.P. DELMER, *Plant Cell Physiol. 43*, 1407, 2002

[85] T.P. NEVELL and S.H. ZERONIAN, eds., *Cellulose Chemistry and Its Applications,* Ellis Horwood Ltd, Chichester, 1985

[86] T.H. GIDDINGS, D.L. BROWER and L.A. STAEHELIN, *J. Cell Biol. 84*, 327, 1980

[87] C.H. HAIGLER and R.M. BROWN Jr., *Protoplasma 134*, 111, 1986

[88] J.C. GARDINER, N.G. TAYLOR and S.R. TURNER, *Plant Cell, 15*, 1740, 2003

[89] D.P. DELMER, Ann. Rev. Plant Physiol. *Plant Mol. Biol. 50*, 245, 1999

[90] I.M. SAXENA, T.DANDEKAR and R. M. BROWN, Jr., *Mechanisms in Cellulose Biosynthesis*, in www.esf.edu/ce/conferences/Cell_pdf/Saxena.PDF

[91] http://www.cazy.org/GlycosylTransferases.html

[92] W.D. REITER and G.F. VANZIN, Plant Mol. Biol. 47, 95, 2001

[93] D. KLEMM, H.P. Schmauder and T. HEINZE, in *Biopolymers–Cellulose*, Wiley-VCH, 2001

[94] S. BIELECKI, A. KRYSTYNOWICZ, M. TURKIEWICZ and H. KALINOWSKA, in *Biopolymers–Bacterial Cellulose*, Wiley-VCH, 2001

[95] SRI International, *MetaCyc Pathway: Cellulose Biosynthesis* in http://biocyc.org/META/NEW-IMAGE?type=PATHWAY&object=PWY-1001

[96] C.P. JOSHI, S. BANDHARI, P. RANJAN, U.C. KALLURI, X. LIANG, T. FUJINO and A. SAMUGA, *New Phytologist, 164*, 53, 2004

[97] Y. AMOR, C.H. HAIGLER, S. JOHNSON, M. WAINSCOTT and D.P. DELMER, Proc. *Natl. Acad. Sci. USA 92*, 9353, 1995

[98] A. KOMATSU, T. MORIGUCHI, K. KOYAMA, M. OMURA and T. AKIHAMA, *J. Exp. Bot. 53*, 61, 2002

[99] G. J. SEIFERT, Curr. Opin. *Plant Biol. 7*, 277, 2004

[100] J. ROSS, Y. LI, E.K. LIM and D.J. BOWLES, *Genome Biol. 2*, 3004, 2001 in http://genomebiology.com/2001/2/2/reviews/3004

[101] D. NOBLES, *The Processive β-glycosyltransferase Family Tree*

[102] I.M. SAXENA and R.M. BROWN Jr., *Cellulose 4*, 35, 1997

[103] Tree of Life Web Project in http://tolweb.org/tree

[104] A.G. MATTHYSSE, K. DESCHET, M. WILLIAMS, M. MARRY, A.R. WHITE and W.C. SMITH, *Proc. Natl. Acad. Sci. USA 101*, 986, 2004

[105] http://commons.wikimedia.org/wiki/File:Tunicate_medley_komodo.jpg

[106] I.M. SAXENA, R.M. BROWN Jr., M. FEVRE, R.A. GEREMIA and B. HENRISSAT, *J. Bacteriol. 177*, 1419, 1995

[107] N. CARPITA and C. VERGARA, *Science 279*, 672, 1998

[108] M.S. BUCKERIDGE, C.E. Vergara and N.C. CARPITA, *Plant Physiol. 120*, 1105, 1999

[109] R.M. BROWN Jr., I.M. SAXENA and K. KUDLICKA, *Trends Plant Sci. 1*, 149, 1996

[110] M. KOYAMA, W. HELBERT, T. IMAI, J. SUGIYAMA and B. HENRISSAT, *Proc. Natl. Acad. Sci. U.S.A. 94*, 9091, 1997

[111] S.J. CHARNOCK, B. HENRISSAT and G.J. DAVIES, *Plant Physiol. 125*, 527, 2001

[112] L. GLASER, *J. Biol. Chem. 232*, 627, 1958

[113] Y. ALONI, D.P. DELMER and M. BENZIMAN, *Proc. Natl. Acad. Sci. USA 79*, 6448, 1982

[114] I.M. SAXENA, F.C. LIN and R.M. BROWN Jr., *Plant Mol. Biol. 15*, 673, 1990

[115] H.C. WONG, A.L. FEAR, R.D. CALHOON, G.H. EICHINGER, R. MAYER, D. AMIKAM, M. BENZIMAN, D.H. GELFAND, J.H. MEADE, A.W. EMERICK, R. BRUNER, B.A. BEN-BASSAT and R. TAL, *Proc. Natl. Acad. Sci. U.S.A. 87*, 8130, 1990

[116] I.M. SAXENA, F.C. LIN and R.M. BROWN Jr., *Plant Mol. Biol. 16*, 947, 1991

[117] I. M. SAXENA, K. KUDLICKA, K. OKUDA and R.M. BROWN, *J. Bacteriol., 176*, 5735, 1994

[118] Saccharomyces Genome Database (SGD)in www.yeastgenome.org/help/glossary.html

[119] S. KIMURA, H.P. CHEN, I.M. SAXENA, R.M. BROWN Jr. and T. ITOH, *J. Bacteriol. 183*, 5668, 2001

[120] I.M. SAXENA and R.M. BROWN Jr., *J. Bacteriol. 177*, 5276, 1995

[121] T. IMAI, T. WATANABE, T. YUI and J. SUGIYAMA, *Biochem. J. 374*, 755, 2003

[122] J.L.W. MORGAN, J. STRUMILLO and J. ZIMMER, *Nature 493*, 181, 2013 in http://www.ncbi.nlm.nih.gov/pubmed/23222542

[123] University of Virginia, 2012 in http://news.virginia.edu/content/uva-research-may-offer-big-benefits-biofuels-and-battling-infections

[124] J.R. PEAR, Y. KAWAGOE, W.E. SCHRECKENGOST, D.P. DELMER and D.M. STALKER, *Proc. Natl. Acad. Sci. USA 93*, 12637, 1996

[125] I.M. SAXENA, R.M. BROWN Jr., M. FEVRE, R.A. GEREMIA and B. HENRISSAT, *J. Bacteriol. 177*, 1419, 1995

[126] T.A. RICHMOND and C.R. SOMERVILLE, *Plant Physiol. 124*, 495, 2000

[127] Carnegie Institution for Sciences, Plant Biology in http://www-ciwdpb.stanford.edu or http://carnegiedpb.stanford.edu

[128] T. RICHMOND, Genome Biology 1: reviews3001, 2000 in http://genomebiology.com/2000/1/4/reviews/3001

[129] N. HOLLAND, D. HOLLAND, T. HELENTJARIS, K.S. DHUGGA, B. XOCONOSTLE-CAZARES and D. P. DELMER, *Plant Physiol. 123*, 1313, 2000

[130] BioCyc Database Collection in http://biocyc.org; MetaCyc (a member of BioCyc database collection) in http://metacyc.org

[131] T. ARIOLI, L. PENG, A. S. BETZNER, J. BURN, W. Wittke, W. HERTH, C. CAMILLERI, H. HÖFTE, J. PLAZINSKI, R. BIRCH, A. CORK, J. GLOVER, J. REDMOND and R.E. WILLIAMSON, *Science 279*, 717, 1998

[132] N.G. TAYLOR, R.M. Howells, A.K. HUTTLY, K. Vickers and S.R. TURNER, *Proc. Natl. Acad. Sci. 100*, 1450, 2003

[133] K. TANAKA, K. MURATA, M. YAMAZAKI, K. ONOSATO, A. MIYAO and H. HIROCHIKA, *Plant Physiol. 133*, 73, 2003

[134] N.A. ECKARDT, *Plant Cell 15*, 1685, 2003

[135] H.W. WANG and E. NOGALES, *Nature 435*, 911, 2005

[136] R.A. BURTON, N.J. SHIRLEY, B.J. KING, A.J. HARVEY, G.B. FINCHER, *Plant Physiol. 134*, 224, 2004

[137] A.R. PAREDEZ, C.R. SOMERVILLE and D.W. EHRHARDT, *Science 312*, 1491, 2006

[138] S. PERSSON, A. PAREDEZ, A. CARROLL, H. PALSDOTTIR, M. DOBLIN, P. POINDEXTER, N. KHITROV, M. AUER and C.R. SOMERVILLE, *PNAS 104* 15566, 2007

[139] F. NICOL, I. HIS, A. JAUNEAU, S. VERNHETTES, H. CANUT and H. HÖFTE, *Embo Journal 17*, 5563, 1998

[140] J. ZUO, Q.W. NIU, N. NISHIZAWA, Y. WU, B. KOST and N.H. CHUA, *Plant Cell 12*, 1137, 2000

[141] D.R. LANE, A. WIEDEMEIER, L. PENG, H. HÖFTE, S. VERNHETTES, T. DESPREZ, C.H. HOCART, R.J. BIRCH, T.I. BASKIN, J.E. BURN, T. ARIOLI, A.S. BETZNER and R. E. WILLIAMSON, *Plant Physiol. 126*, 278, 2001

[142] S. SATO, T. KATO, K. KAKEGAWA, T. ISHII, Y.G. LIU, T. AWANO, K. TAKEBE, Y. NISHIYAMA, S. Kuga, S. SATO, Y. NAKAMURA, S. Tabata and D. SHIBATA, *Plant Cell Physiol. 42*, 251, 2001

[143] I. HIS, A. DRIOUICH, F. NICOL, A. JAUNEAU and H. HÖFTE, *Planta 212*, 348, 2001

[144] M. MOLHOJ, S. PAGANT and H. HÖFTE, *Plant Cell Physiol. 43*, 1399, 2002

[145] D. P. DELMER and C.H. HAIGLER, *Metab. Eng. 4*, 22, 2002

[146] L. PENG, Y. KAWAGOE, P. HOGAN and D. DELMER, *Science 295*, 147, 2002

[147] S. M. READ and T. BACIC, *Science 295*, 59, 2002

[148] K. SCHRICK, S. FUJIOKA, S. TAKATSUTO, Y.D. STIERHOF, H. STRANSKY, S. YOSHIDA and G. JURGENS, *Plant J. 38*, 227, 2004

[149] K. SCHRICK, Keck Graduate Institute in http://faculty.kgi.edu/kschrick

[150] A. HIRAI, M. TSUJI and F. HORII, *Cellulose 9*, 105, 2002

[151] T.E. BUREAU and R.M. BROWN Jr., *Proc. Natl. Acad. Sci. USA 84*, 6985, 1987

[152] The R. Malcolm BROWN Jr. Laboratory for Cellulose and Biofuels Research, The University of Texas at Austin in www.botany.utexas.edu/facstaff/facpages/mbrown/aceto.htm

[153] H. CHANZY, K. IMADA and R. VUONG, *Protoplasma 94*, 299, 1978

[154] S. KIMURA, W. LAOSINCHAI, T. ITOH, X. CUI, C.R. LINDER and R.M. BROWN Jr., *Plant Cell 11*, 2075, 1999

[155] T. ITOH, S. KIMURA and R.M. BROWN Jr., *Cellulose 11*, 385, 2004

[156] W.R. SCHEIBLE, R. Eshed, T. RICHMOND, D. DELMER and C. SOMERVILLE, *Proc. Natl. Acad. Sci. USA 98*, 10079, 2001

[157] R.M. PERRIN, *Curr. Biol. 11*, R213, 2001

[158] I. KUREK, Y. KAWAGOE, D. JACOB-WILK, M. DOBLIN, and D.P. DELMER, *Proc. Natl. Acad. Sci. USA 99*, 11109, 2002

[159] I.M. SAXENA and R. M. BROWN Jr., *Ann. Bot. 96*, 9, 2005

[160] C. D. HU, Y. CHINENOV and T.K. KERPPOLA, MOL. *Cell 9 789*, 2002

[161] T. DESPREZ, M. JURANIEC, E.F. CROWELL, H. JOUY, Z. POCHYLOVA, F. PARCY, H. HOFTE, M. GONNEAU and S. VERNHETTES, *Proc. Natl. Acad. Sci. USA 104*, 15572, 2007

[162] J. TIMMERS, S. VERNHETTES, T. DESPREZ, J.P. VINCKEN, R.G.F. VISSER and L.M. TRINDADE, *FEBS Lett. 583*, 978, 2009 in http://www.sciencedirect.com/science/article/pii/S0014579309001495

[163] S. ENDO, E. PESQUET, M. YAMAGUCHI, G. TASHIRO, M. SATO, K. TOYOOKA, N. NISHIKUBO, M. UDAGAWA-MOTOSE, M. KUBO, H. FUKUDA and T. DEMURA, Plant Cell 21, 1155, 2009 in http://www.plantcell.org/cgi/content/short/21/4/1155

[164] A.H. LIEPMAN, R. WIGHTMAN, N. GESHI, S.R. TURNER and H.V. SCHELLER, Plant J. 61, 1107, 2010 in http://onlinelibrary.wiley.com/doi/10.1111/j.1365-313X.2010.04161.x/full

[165] Y. GU, N. KAPLINSKY, M. BRINGMANN, A. COBB, A. CARROLL, A. SAMPATHKUMAR, T.I. BASKIN, S. Persson and C.R. SOMERVILLE, *Proc. Natl. Acad. Sci. USA 107*, 12866, 2010 in http://www.pnas.org/content/107/29/12866.abstract

[166] The Pennsylvania State University, Department of Biochemistry & Molecular Biology, Ying GU, 2011 in http://bmb.psu.edu/directory/yug13

[167] S.B. NORDIN, J.O. NYREN and E.L. BACK, *Text. Res. J. 44*, 152, 1974

[168] L.R.G. TRELOAR, *Polymer 1*, 290, 1960

[169] L. SAKURADA, Y. NUKUSHINA, and T. ITO, *J. Polym. Sci. 57*, 651, 1962

[170] M.G. NORTHOLT and H. DE VRIES, *Angew. Makromol. Chem. 133*, 183, 1985

[171] T. NISHINO, K. TAKANO and K. NAKAMAE, *J. Polym. Sci., Polym. Phys. 33*, 1647, 1995

[172] S.J. EICHHORN, R.J. YOUNG and G.R. DAVIES, *Biomacromolecules 6*, 507, 2005

[173] F. TANAKA and T. IWATA, *Cellulose 13*, 509, 2006

[174] H.C. HSIEH, H. YANO, M. NOGI and S.J. EICHHORN, *Cellulose 15*, 507, 2008, in www.springerlink.com/content/p26m3343m0478750/

[175] A. STURCOVA, G.R. DAVIES and S.J. EICHHORN, *Biomacromolecules 6*, 1055, 2005

[176] J. ANDERSONS, E. SPARNINS, R. JOFFE and L. WALLSTROM, *Compos. Sci. Technol. 65*, 693, 2005

[177] T. PEIJS and F. VILASECA, *Cellulose-Based Nanocomposites, A review*, Queen Mary, University of London, 2005 in www.forestprod.org/woodfiber05vilaseca.pdf

[178] A. DUFRESNE, *Production of carbohydrate nanofillers, Their potential interest*, CERIG/EFPG, 2006 in http://cerig.efpg.inpg.fr/EFPG-days/2005/page15.htm

[179] J. SIMONSEN, College of Forestry, Oregon State University in http://wood-science.oregonstate.edu/people/faculty/simonsen-john

[180] SPONTEX internal reports, 1993, 1994

[181] R.S. WERBOWYJ and D.G. GRAY, *Mol. Cryst. Liq. Cryst. (Lett.) 34*, 97, 1976

[182] R. MEREDITH and J.W. PRESTON, *Fibre Science, Manchester 23*, T163, 1932

[183] A.R. URQUHART, *Moisture in Textiles*, Textile Book Publishers Inc., New York, 1960

[184] S.H. ZERONIAN, M.L. COOLE, K.W. ALGER and J.M. CHANDLER, *J. Appl. Polym. Sci., Appl. Polym. Symp. 37*, 1053, 1983

[185] A.J. STAMM, *Wood and Cellulose Science*, Ronald Press, New York, 1964

[186] A.L. DEMAIN, M. NEWCOMB and J.H.D. WU, *Microbiol. Mol. Biol. Rev. 69*, 124, 2005

[187] P. DOHRN, Sateri, Commercial presentation, 2008

[188] M. BLAGOEV, S. BIZZARI and Y. INOGUCHI, *Rayon and Lyocell Fibers,* SRI-Consulting, 2010 in http://www.sriconsulting.com/CEH/Public/Reports/541.3000/

[189] J.L. WERTZ, O. BEDUE and J.P. MERCIER, *Cellulose Science and Technology*, EPFL Press, Lausanne, 2010

[190] Kelheim Fibres, *New Fibres,* 2010 in http://www.kelheim-fibres.de/produkte/neue_fasern_uk.php

[191] FAO Corporate Document Repository, Economic and Social Development Department, *Medium Term Prospects for Agricultural Commodities* in http://www.fao.org/docrep/006/y5143e/y5143e1g.htm

[192] K. PICKERING, *Properties and Performance of Natural-Fibre Composites*, Woodhead Publishing, Cambridge, 2008

[193] T.J. KEENER, R.K. STUART and T.K. BROWN, *Compos. A 35*, 357, 2004 in http://www.eastman.com/Literature_Center/W/WA114.pdf

[194] N. GRAUPNER, *J. Mater. Sci. 43*, 5222, 2008

[195] W. TIELLEMANS and R.P. Wool, *Polym. Compos. 26*, 695, 2005

[196] K. OKSMAN and M. SAIN, eds., *Cellulose Nanocomposites, Processing, Characterization and Properties*, ACS Symposium Series 938, American Chemical Society, Washington, DC, 2006

[197] K. OKSMAN, *Bio-nanocomposites; Processing and Relationship between the Structure and Properties*, Norwegian University of Science and Technology in www.ivt.ntnu.no/ipm/forsk/Plast%20og%20komposittt/web-page_NANOMAT.htm

[198] M.A.S. AZIZI, F. ALLOIN and A. DUFRESNE, *Biomacromolecules 6*, 612, 2005

[199] Direction Générale des Entreprises, Développement & Conseil, *Etude sur les Textiles Techniques,* 2005 in http://www.industrie.gouv.fr/biblioth/docu/dossiers/sect/pdf/textrap1.pdf

[200] Yarns and Fibers Exchange, *Innovations in Intelligent Apparel and Technical Textiles,* 2010 in http://www.yarnsandfibers.com/revamp_ir/report_fullstory.php3?id=517&p_type=45&story_type=F

Chapter 5

Enzymatic and non-Biological Degradation of Cellulose

5.1 Introduction

Cellulose is a chemically stable polymer. Its glycosidic bonds are strong and stable under various reaction conditions. However, cellulose can be degraded in nature by enzymes as an essential part of the carbon cycle, and non-biologically by acids, alkalis, oxidants, mechanical means, thermal energy, and radiations. [1, 2]

The major barrier to the development and cost reduction of biochemical cellulosic ethanol, and the implementation of related biorefineries, is the recalcitrance of the cell wall to the enzymatic attack. In particular, the compact structure of crystalline cellulose and its embedding in a matrix of lignin and hemicelluloses impede enzymatic access. Therefore, in order to reduce complexity and costs of the exploitation of cellulose feedstock, improved biochemical systems to access and hydrolyze cellulose have to be developed, yields in the fermentation steps have to be improved, and biomass pretreatment conditions have to be optimized. The objective is to bridge the current cost gap between cellulosic ethanol and corn ethanol, which is estimated to be between \$0.7 and \$1.5 per gallon.

Understanding the mechanisms of cellulose degradation is essential for the development of most lignocellulosic biorefineries.

5.2 Enzymatic hydrolysis

In a typical cellulose-degrading ecosystem, a variety of cellulolytic bacteria and fungi work in association with related microorganisms to convert insoluble cellulose to soluble sugars, primarily cellobiose and glucose, which are then assimilated by the cell. [1] For catalyzing this process, the cellulolytic microbes produce a wide diversity of enzymes, known as ***cellulases***. Cellulases catalyze the hydrolysis of the β-1,4 glucosidic linkages in cellulose. They are members of the ***glycoside hydrolase*** (***GH***, also called ***glycosidase***) families of enzymes that hydrolyze the polysaccharides in plant cell walls. [2] Due to the structural complexity and rigidity of cellulosic substrates, their efficient degradation generally requires multienzyme systems. Such systems

include either a collection of free cellulases or/and multicomponent complexes called ***cellulosomes***.

The commercial potential for the efficient enzymatic hydrolysis of cellulose is enormous because of the overwhelming interest in the use of cellulosic biomass as a renewable source of energy to produce biofuels and biobased products. [3]

5.2.1 Cellulases

Organism source

Cellulases are produced mainly by microorganisms, such as bacteria, fungi and protozoa, but also by plants and some invertebrate animals. [4, 5, 6, 7] Although it is widely accepted that most multicellular animals (Metazoa) do not have endogeneous cellulases, relying instead on intestinal symbiotic microorganisms for cellulose digestion, cellulases have been found recently in the animal genomes of termites (insects), abalone (a mollusc), and sea squirts (tunicates). [8] Most likely, cellulases are not produced by vertebrate animals. Ruminants, for example, have highly specialized digestive tracts where mixtures of bacteria and protozoa degrade cellulose under anaerobic conditions. [7]

Endo- and exo-action

Cellulases are enzymes that hydrolyze the β-1,4 linkages in cellulose: [9, 4]

$$\mathrm{H(4\text{-}O\text{-}\beta\text{-}Glc\text{-}1)_{x+y}OH + H_2O \rightarrow H(4\text{-}O\text{-}\beta\text{-}Glc\text{-}1)_xOH + H(4\text{-}O\text{-}\beta\text{-}Glc\text{-}1)_yOH}$$

Historically, they have been divided into two groups, ***endoglucanases*** and ***exoglucanases*** (or ***cellobiohydrolases***), according to their respective capacity to cleave the β-1,4-glucosidic bond either internally or at one of the ends of the cellulose chain [1] The endoglucanases are commonly characterized by a groove or a cleft into which a linear cellulose chain can fit in a random manner. [10] Usually, cellobiohydrolases (CBHs) possess tunnel-like active sites, which can only accept a substrate chain via its terminal regions. [10] These exo-acting enzymes act by threading the cellulose chain through the tunnel, where successive cellobiose units are removed in a sequential manner. [10] The sequential cleavage of a polymeric chain is termed ***processivity***. Although this division based on substrate specificity is instructive, some cellulases show both modes of action and a classification based on their structural and mechanistic properties was proposed by Henrissat [11] in 1991. This was updated and extended to enzymes that degrade, modify, or create glycosidic bonds (***carbohydrate-active enzymes***). [2, 12, 13, 14]

Cellulases, like carbohydrate-active enzymes in general, are composed of independently folding, structurally and functionally discrete units, called ***modules*** or domains. [1, 2] Two essential modules are the ***catalytic module*** and the ***carbohydrate-binding module***.

Catalytic module

In the system proposed by Henrissat [11] in 1991, glycoside hydrolases are classified into families according to the amino acid sequence similarities of their catalytic modules. Because there is a direct relationship between sequence and folding similarities, such a classification reflects the structural features of these enzymes, helps to reveal the evolutionary relationships between them, and provides a tool to derive mechanistic information. [15] The approach revealed that enzymes with the same name, EC number, and catalytic activity could be found in different sequence-based families. [16] Endoglucanases, for example, are found in 13 and more families and display a range of protein folds including (**Table 5.1**):

- ***$(\alpha/\alpha)_6$ barrels***, consisting each of six inner and six outer ***α helices*** forming a barrel-like structure, such as in GH 8, 9 and 48;
- ***$(\beta/\alpha)_8$ barrels***, consisting each of 8 repeating units of β/α module, in which the 8 ***β strands*** form an inner parallel ***β sheet*** arranged in a barrel structure, which is surrounded by the 8 α helices, such as in GH 5 and 10;
- ***β jelly rolls,*** variants of ***Greek key*** topology (in which typically three antiparallel β strands connected by hairpins are followed by a longer connection to the fourth strand, which lies adjacent to the first) with both ends of a ***β sandwich*** (two β sheets that pack together, face-to-face, in a layered arrangement) or a ***β barrel*** (β sheet that twists and coils to form a closed structure) fold being crossed by two interstrand connections, such as GH 7 and 12; and
- *β barrels*, such as in GH 45. [16-20]

Table 5.1. Fold of the catalytic module, enzyme type, organism source and mechanism for representative GH families. [1, 15]

GH family (clan)	Fold	Enzyme type	Organism source	Mechanism[1)]
5 (A)	$(\beta/\alpha)_8$ barrel	Mainly endoglucanases	Bacteria, fungi	Retaining
6	β/α barrel	Endoglucanases and CBHs	Bacteria, fungi	Inverting
7 (B)	β jelly-roll	Endoglucanases and CBHs	Fungi	Retaining
8 (M)	$(\alpha/\alpha)_6$ barrel	Mainly endoglucanases	Bacteria	Inverting
9	$(\alpha/\alpha)_6$ barrel	Mainly endoglucanases	Bacteria, fungi	Inverting
10	$(\beta/\alpha)_8$ barrel	Xylanases	Bacteria, Eukaryota	Retaining
12 (C)	β jelly-roll	Endoglucanases	Bacteria, fungi	Retaining
45	β barrel	Endoglucanases	Bacteria, fungi	Inverting
48	$(\alpha/\alpha)_6$ barrel	Processive endoglucanases and/or CBHs	Bacteria	Inverting

GH: glycoside hydrolase; CBH: cellobiohydrolase 1) retention or inversion of the configuration of the anomeric carbon (see section on mechanism)

Some families include enzymes with different substrate specificities. [13] Family GH-7 contains only fungal enzymes, whereas family GH-8 contains only bacterial enzymes. Because the fold of proteins is better conserved than their sequences, some of the sequence-based families can be grouped in clans. [15] The largest clan (GH-A) includes families which share a $(\beta/\alpha)_8$ barrel fold.

One further benefit of the sequence-based families of carbohydrate-active enzymes is that these can be easily searched for in genomes. [16] The availability of a number of completely sequenced genomes like that of *Arabidopsis* has enabled the census of all the carbohydrate-active enzymes in each organism. [21]

Carbohydrate-binding module

In addition to their catalytic module, polysaccharide-degrading enzymes frequently bear ancillary, non-catalytic modules. [13]

The commonest ancillary module is the carbohydrate-binding module (CBM), sometimes called cellulose-binding domain (CBD). A CBM is defined as contiguous amino acid sequence within a carbohydrate-active enzyme with a discreet fold having carbohydrate-binding activity. [22] Exceptions are CBMs of cellulosomal ***scaffoldin*** proteins and rare instances of independent putative CBMs. Most cellulases have a catalytic module and a CBM joined by a highly glycosylated and presumably flexible linker peptide. [23] Removal of the CBM results in a significantly reduced enzymatic activity on crystalline cellulose probably due to a decreased binding capacity, but the activity on soluble cellulose oligomers is retained. [23] Like the catalytic modules, the CBMs also form distinct families of related amino acid sequences. Representative structures for each CBM family have been elucidated either by crystallography or by NMR spectroscopy (**Table 5.2**). [1]

The family 1 CBMs of ~40 residues are found almost exclusively in fungi, whereas the family 2 CBMs of ~100 residues, as well as the family 3 CBMs of ~150 residues, are found in bacteria. CMB1 is found either at the N-terminal or at the C-terminal extremity of the enzymes known to contain such a domain. [24] The CBMs of family 2 can be classified in two subfamilies according to substrate specificities: CMB2, which binds cellulose, and CMB2b, which inter-

Table 5.2 Representative carbohydrate-binding modules (CBMs) of families 1, 2 and 3. [1, 22]

Family	1	2a[1)]	3a	3c
Organism	*Trichoderma reesei*	*Cellulomonas fimi*	*Clostridium thermocellum*	*Thermomonospora fusca*
Protein	Cel7A (GH7)[2)]	Exoglucanase Cex (GH10)[2)]	Cellulosomal scaffoldin CipA	Endoglucanase E4 (GH9)[2)]
Size[3)]	36	110	155	143
Dominant fold	Three-stranded antiparallel β sheet	Nine-stranded β barrel	Nine-stranded β sandwich	Nine-stranded β sandwich
Preferred substrate	Crystalline cellulose	Crystalline cellulose	Crystalline cellulose	Single cellulose chain
Proposed binding	Three coplanar aromatic residues [24]	Three coplanar aromatic residues [25, 26]	Coplanar aromatic and polar residues [27, 28]	Coplanar aromatic and polar residues [28, 29, 30]

1) CBM2a binds cellulose, whereas CBM2b binds xylan [25, 26] ,2) glycoside hydrolase family 3) Number of amino acid residues

acts specifically with xylans. [25] CMB2 is found either at N-terminal or at the C-terminal extremity of cellulases and ***xylanases***. The CBMs of family 3 are separated into two functionally different types. One type (family 3a and 3b) binds strongly to crystalline cellulose, whereas another (family 3c) fails to bind crystalline cellulose but serves in a helper role in the hydrolysis of a single cellulose chain by feeding it into the active site of the neighboring catalytic module. [28, 29, 30, 31] The families 3a (cellulosomal scaffoldins) and 3b (mainly free enzymes) are closely similar in their primary structures. The CMB3 domain is mainly found C-terminal to the catalytic domain, which corresponds to a wide range of bacterial GH like families 9, 5, and 10. [29]

The major function of the CBMs is to deliver the catalytic domain to the crystalline substrate. The binding appears extremely stable, although the enzyme may undergo a lateral diffusion on the substrate. Some CBMs also appear to disrupt the noncovalent interactions between the chains of the crystalline substrate, while others bind preferentially to noncrystalline substrates. [1]

In 2006, Ding et al. [32] demonstrated the use of CBMs as molecular probes for mapping the chemistry and structure of complex carbohydrate-containing materials, and then for imaging these macromolecules using transmission and scanning transmission electron microscopy at the molecular level of resolution. Three types of CBMs from families 3, 6, and 20 that exhibit different carbohydrate specificities (*Ct*CBM3 from *Clostridium thermocellum* CipA with high affinity for crystalline cellulose, *Ct*CBM6 from *Clostridium thermocellum* xylanase with high affinity for xylan but also binding to xyloglucan and noncrystalline cellulose, and *Sl*CBM20 from *Streptomyces lividans* α-amylase with high affinity for starch) were each fused with either ***green fluorescent protein*** (GFP) and ***red fluorescent protein*** (RFP) and used for double-labeling fluorescence microscopy studies of primary cell walls and other complex carbohydrate target molecules.

In February 2013, the CAZY database collated 66 CBM families. [22]

Mechanism

At the catalytic level, cellulose hydrolysis, as well as in general glycosidic bond hydrolysis, proceeds either with net ***inversion*** or ***retention*** of the configuration of the anomeric carbon (**Figure 5.1**). In both cases, it is catalyzed primarily by two carboxyl groups located in the active site of the enzyme. [1] One catalytic residue of the enzyme is a general acid (proton donor) and the other is a nucleophile (base). [15]

Mechanism is dictated by the location of these residues within the 3-D structure and, hence, by the sequence. [13] This involves the conservation of the stereochemical mechanism within each sequence-based family. Inverting enzymes use a single-displacement mechanism, in which the catalytic acid gives protonic assistance to leaving group departure while the catalytic base deproto-

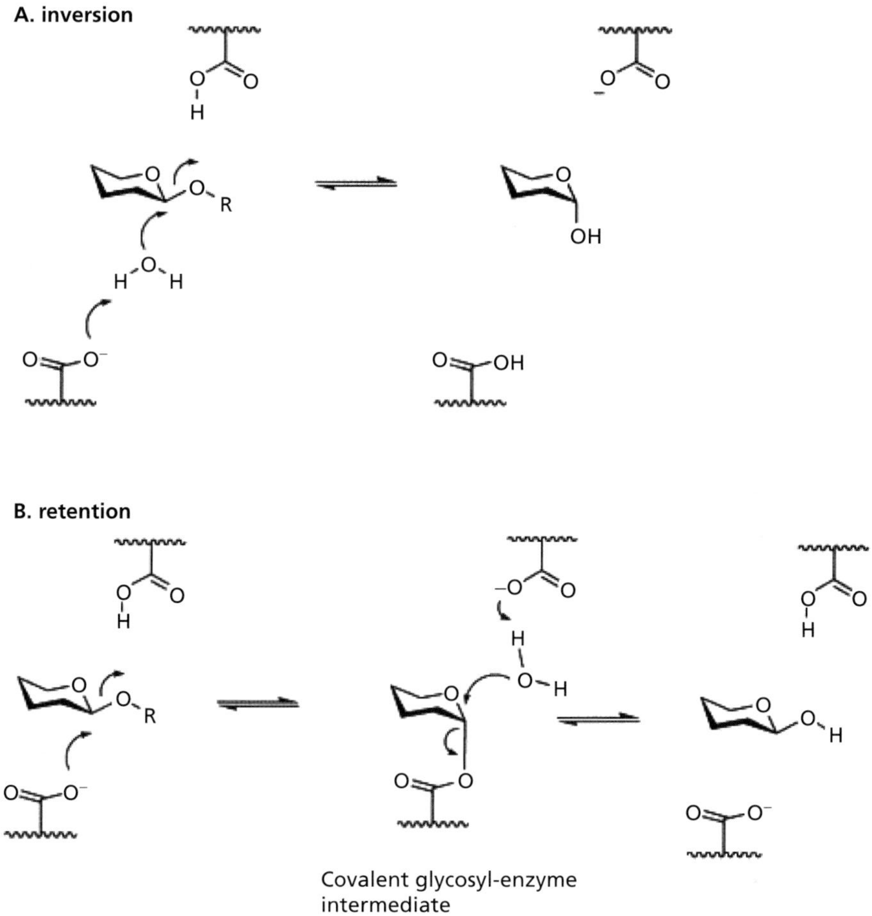

Figure 5.1 The two mechanisms for glycoside hydrolases. Reproduced by permission of Elsevier. [9]

nates a water molecule for nucleophilic substitution at the anomeric center. [9] The acid and base residues are typically separated by a distance of ~9.5 Å in order to accommodate the water molecule in addition to the substrate. [33] Retaining enzymes function via a double-displacement mechanism, in which a covalent enzyme-substrate intermediate is formed, and then hydrolysed. [1, 9] This requires a catalytic nucleophile for the formation of the intermediate and a catalytic acid which first protonates the leaving group to assist departure and then functions as a base, deprotonating the incoming water nucleophile. In this case, the substrate undergoes ring distortion within the active site of the enzyme in order to reach a twisted boat conformation. [1] In retaining enzymes, the two catalytic residues are located on the two opposite sides of the glycosidic bond to hydrolyze and are separated by a distance of ~5.5 Å. [33]

In April 2013, the CAZy (Carbohydrate-Active Enzymes) database collated 132 GH families. [15]

Nomenclature

The nomenclature scheme has been developed for designating enzymes that hydrolyze the polysaccharides in plant cell walls. [2] These enzymes are predominantly β-1,4 glycanases. The designation for an enzyme starts with the three letters (identical for gene and protein) designating its preferred substrate (e.g. *cel* and Cel for cellulase gene and protein, respectively). These three letters are followed by the family number of the enzyme (e.g. Cel5). If an organism produces multiple enzymes from the same family, these will be designated by e.g. Cel5A, Cel5B, with the final capital letters indicating the order in which the enzymes were first reported.

As the scheme does not distinguish between endo- and exo-action, particular enzymes will be referred to as endoglucanase CelA, cellobiohydrolase Cel6A, and so on. Furthermore, two similar enzymes from different organisms will be differentiated by indicating the organism of origin (e.g. CfCel9B from *Cellulomonas fimi*).

The active site of a cellulase consists of multiple binding subsites for glucose units. These subsites are designated by the –3, –2, –1, +1, +2, +3, system, where –1 is the subsite for the ***glycone*** (sugar component of a glycoside) and +1 is the subsite for the ***aglycone*** (nonsugar component of a glycoside) of the bond being hydrolyzed. [2]

5.2.2 Noncomplexed cellulase systems

In order to degrade crystalline cellulose substrates, microorganisms produce multienzyme systems. Generally two types of systems occur. One type consisting of a collection of independent extracellular cellulases that act synergistically is produced by ***aerobic*** bacteria and fungi. [34, 35] The second type, consisting of an enzyme complex and produced by ***anaerobic*** bacteria and fungi, will be reviewed in the next section. [36-38]

Cellulases from the soft-rot fungus *Trichoderma reesei* and the soft-rot fungus *Humicola insolens* have been intensively studied. Cellulases from *T. reesei* are commercial especially under the trade name *Celluclast* from Novozymes. Also, *H. insolens* cellulases are commercial especially under the trade name *Celluzyme* from Novozymes.

Solubilization of native cellulose

As long ago as 1950, Reese et al. [39] postulated that the degradation of native cellulose required a system of enzymes working together. According to this early concept, biological solubilization of cellulose presumably took place in at least two steps: (1) conversion of the native cellulose molecule into linear anhydroglucose chains, i.e. physical disruption of insoluble cellulose; [40] and (2) hydrolysis of the β-1,4 glucosidic linkage to form soluble sugars. [39] The term ***amorphogenesis*** has been used to describe physical changes, such as swelling, segmentation, and destratification of cellulose, that enhance enzymatic hydrolysis and render crystalline cellulose more accessible to cellulase. [40]

Reese's original hypothesis of a multienzyme system required for digesting crystalline cellulose has been confirmed by a number of studies. [39] From these studies, it has appeared that three major classes of enzymes need to cooperate to digest crystalline cellulose into glucose: endoglucanases, which cleave the cellulose chains randomly, cellobiohydrolases, which recurrently cleave cellobiose from the ends of the cellulose chain, and ***β-glucosidases***, which hydrolyze cellobiose and other soluble cellodextrins into glucose. [41]

It has been observed that the heterogeneous structure of cellulose causes a rapid decrease in rate as hydrolysis proceeds. [40] The declining reactivity of residual cellulose during enzymatic hydrolysis is presumably a result of factors such as: less surface area, fewer accessible chain ends, and/or adsorption of inactive cellulase on cellulose.

When cellulase systems act in vitro on insoluble cellulose, three processes occur simultaneously: (1) physical and chemical changes in the solid-phase cellulose; (2) primary hydrolysis, involving the release of soluble intermediates from reacting cellulose; and (3) secondary hydrolysis, involving hydrolysis of soluble intermediates. [40] Endoglucanase activity is presumably primarily responsible for chemical changes (changes in the DP and chain end concentration) in solid-phase cellulose, while exoglucanase (or CBH) activity is presumably primarily responsible for solubilization.

Trichoderma reesei (Hypocrea jecorina)

Fungal cellulases of the genus *Trichoderma* have been extensively studied due especially to the high levels of cellulase produced by the species in this genus. All *T reesei* are developed from a single isolate, named in recognition of Reese's work. Most commercial cellulases are produced from *Trichoderma* spp., with a few also produced by *Aspergillus niger.* [40] The cellulases of *Trichoderma* are known to work at acid pH. [42]

The *T. reesei* cellulolytic enzyme system consists of at least two cellobiohydrolases (Cel7A and Cel6A, formerly CBHI and CBHII, respectively), five endoglucanases, and two β-glucosidases (**Table 5.3**). [40, 41] Cel7A and Cel6A are the main components of the enzymatic system, representing ~50% and ~20% of total cellulolytic enzyme. Cel7B and Cel5A are the most abundant of the endoglucanases. [43]

The structure of Cel7A, Cel6A, and endoglucanase Cel7B (formerly EGI), like most fungal cellulases, displays a catalytic domain and a CBM connected by a glycosylated peptide linker (**Figure 5.2**).

Table 5.3. Trichoderma reesei cellulases. [41]

Enzyme	Former name	Molecular weight (kDa)	Position of the CBM
Cel7A	CBHI	57	C
Cel6A	CBHII	53	N
Cel7B	EGI	55	C
Cel5A	EGII	50	N
Cel12A	EGIII	25	-
Cel61A	EGIV	55	C
Cel45A	EGV	36	C

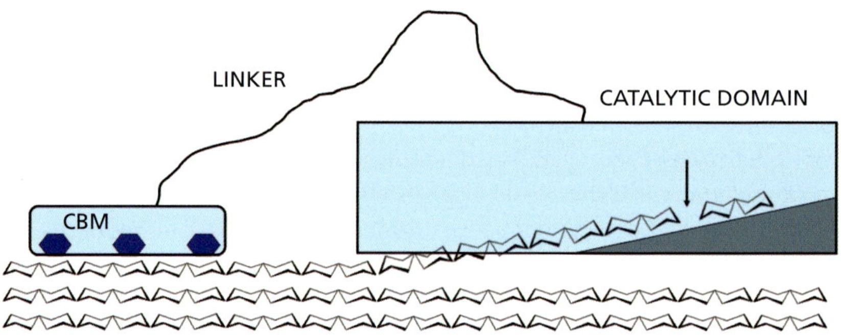

Figure 5.2 Schematic representation of a fungal cellulase consisting of a catalytic domain and a CBM (carbohydrate binding module) linked by a glycosylated peptide. The hexagons in the CBM are the three aromatic residues that interact with the glucose rings of the cellulose chain. The gray area represents the loops covering the active site. [41]

The catalytic domain structures of Cel7A and Cel6A are very different, but both display tunnel-shaped topologies formed by disulfide bridges. [36]

Cel7A

The crystal structure of Cel7A catalytic domain was solved in 1994 by Divne et al. [44]. The structure revealed a β sandwich of 12-14 β strands in two sheets with a 50 Å-long tunnel formed by the inner β strands and four long surface loops covering the active site. [41, 44, 45, 46] The catalytic residues are glutamate 212 and 217, which are located on opposite sides of the active site. [10] Ten well-defined binding subsites for glucose units have been identified along the 50Å-long tunnel. [45] Four tryptophan residues form a glucose-binding platform in the tunnel.

In 1998, T. IMAI et al. [47] studied the mode of action of Cel7A on *Valonia* cellulose microcrystals. After incubation with Cel7A, most of the initially smooth and well defined microcrystals showed fibrillation. However, some microcrystals remained superficially intact. Such crystals displayed polar morphology: one end was extremely narrowed or pointed. The narrowing of the microcrystals occurred at their reducing side. This was confirmed by the visualization of selective reducing end labeling (silver labeling) at the pointed ends of microcrystals. Consequently, the processivity of Cel7A action is toward non-reducing ends from reducing ends, contrary to the traditional view, according to which exoglucanases released cellobiose from the non-reducing ends.

The probable steps involved in Cel7A action on cellulose include binding to cellulose via the carbohydrate-binding module (CBM), recognition of a free cellulose chain end via aromatic groups at the entrance of the catalytic domain tunnel, initial threading of the chain into the active site tunnel to form a catalytically active complex (CAC), and decrystallization from the substrate, hydrolysis reaction, cellobiose expulsion, and reformation of the CAC. [48]

Cel6A

In 1985, Chanzy and Henrissat [49] demonstrated the unidirectional degradation of *Valonia* cellulose microcrystals subjected to Cel6A action. Electron microscopy revealed the formation of a pointed tip at the non-reducing end of a microcrystal (**Figure 5.3**). The synergism between Cel6A and endoglucanase such as Cel5A was also clearly demonstrated.

The crystal structure of Cel6A catalytic domain was solved in 1990 by Rouvinen et al. [50] Cel6A catalytic domain consists of a β/α barrel structure, similar to triose phosphate isomerase (TIM), but with seven instead of eight β strands. [41] A 20 Å-long substrate-binding tunnel adjacent to the α/β barrel is formed by two well-ordered loops covering the active site. [36] It contains four glucose unit-binding subsites. [41] Two additional subsites close to the tunnel entrance have been identified. A tryptosan residue (Trp 272) at the +4 subsite is critical in the degradation of crystalline cellulose.

The tunnel-shaped topology of these enzymes allows for a structural explanation of the processivity of exoglucanases. The tunnel retains a single cellulose chain and prevents it from re-adhering to the cellulose crystal. [9, 51] The catalytic sites of both CBHs are within this cellulose-binding tunnel near the outlet so that β-1,4 glucosidic bonds are cut with retaining (Cel7A) or inverting (Cel6A) mechanisms. [40] Both enzymes can cleave several bonds following a single adsorption event before dissociation of the enzyme substrate complex.

T. reesei Cel7A and Cel6A have a partial endo character. [10, 40, 51] It has been suggested that some exoglucanases are capable of both endo-and exo-actions, due to temporary conformational changes of loops on the tunnel

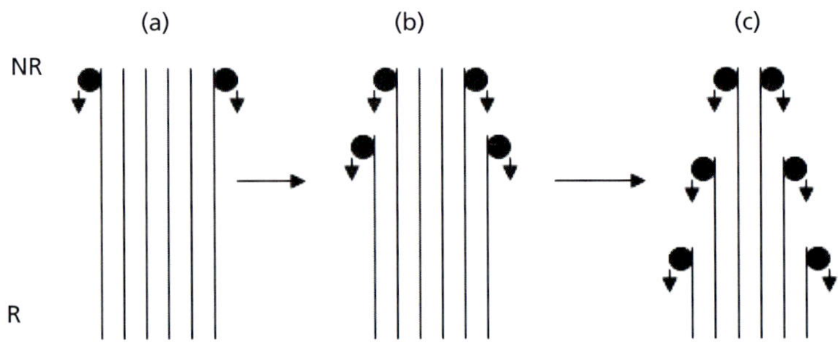

Figure 5.3 Schematic representation of the exo-attack from Cel6A giving rise to pointed tips at the non-reducing (NR) ends of cellulose microcrystals. (a) The enzyme attacks accessible NR ends at the crystal surface. In (b) degradation of the outer layer exposes a new surface of NR ends to the enzymes. As this proceeds (c), a sharpened tip appears at the NR end of the microcrystal. [49]

structure that encloses the active sites. This proposal has been supported by the observation that one of the loops of *T. reesei* Cel6A has substantial mobility and that the resulting tunnel could be either more tightly closed or more open. [52] The open conformation is likely to correspond to the endo-action. In addition, it should be noted that Cel6A contains fewer loops along the catalytic tunnel and has greater endoglucanase activity than Cel7A.

It is worth noting here that Koyama et al. [53] demonstrated the parallel up packing in cellulose Iα and Iβ by a combination of staining the reducing ends of cellulose chains and microdiffraction-tilting electron analysis. **Figure 5.4** shows electron micrographs of microcrystals with one end labeled with silver and the other end eroded to a pointed tip due to selective degradation with a cellobiohydrolase attacking from the non-reducing end.

CBMs

Removal of the ~40-residue *T. reesei* CBM causes a significant reduction in the rate of hydrolysis of insoluble cellulose while having little effect on that of soluble cellulose. [40] The *T. reesei* CBMs, like all fungal CBMs, belong to family 1 (CBM1), characterized by a small wedge-shaped fold featuring a cellulose binding surface with three exposed coplanar aromatic residues. [54, 55, 56, 57] These aromatic residues are critical for the binding of a CBM1 onto crystalline cellulose and the presence of a tryptophan (Trp or W) instead of a tyrosine (Tyr or Y) residue leads to an increased affinity on crystalline cellulose. [56, 58] The spacing of the three aromatic residues coincides with the spacing of every second glucose ring on a glucan chain. [54] Thus it has been presumed that the aromatic amino acids of the CBMs form van der Waals interactions and aromatic ring polarization interactions with the glucose rings exposed on the hydrophobic (110) crystalline face of cellulose. [55]

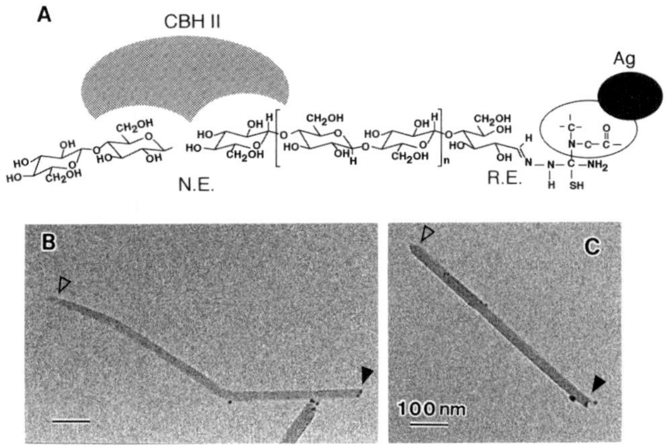

Figure 5.4 Evaluation of the chain directionality by microscopic visualization. (A) Schematic representation of the silver staining of reducing ends and selective degradation by cellobiohydrolase Cel6A (CBHII). N.E., nonreducing end, R.E. reducing end. (B and C) Electron micrographs of microcrystals of cellulose from the green alga *Cladophora* (reducing end labeled with silver: solid arrow, non-reducing end eroded with Cel6A (CBHII): open arrow). Reproduced by permission. Copyright (1997) National Academy of Sciences, USA. [53]

In 2003, Lehtiö et al. [56] were able to visualize the binding sites of CBMs on *Valonia* cellulose crystals by transmission electron microscopy using direct immuno-gold labeling of CBM fusion proteins. The electron microscopy images provided evidence that the fungal family I CBMs and one family 3 CBM from *Clostridium thermocellum* have defined binding sites on two opposite corners of Valonia crystals. In most samples these corners are worn to display significant area of the hydrophobic (110) plane in the cellulose Iα phase, which thus constitutes the binding site for these CBMs (**Figure 5.5**).

Humicola insolens

Cellulases from the fungus *Humicola insolens*, developed by Novo, work at neutral or alkaline pHs, and are therefore adapted to laundry detergents. [42] *H. insolens* produces two cellobiohydrolases, Cel7A (CBHI) and Cel6A (CBHII), and six endoglucanases, Cel7B (EGI), Cel5A (EGII), Cel12A (EGIII), Cel61A (EGIV), Cel45A (EGV), and Cel7B (EGVI). [59-61]

The structure of Cel7A has not yet been solved, but the strong sequence similarity with of this enzyme with *T. reesei* Cel7A predicts an identical folding geometry. [51] The roof of the tunnel where the active sites are located is made of two pairs of loops.

The three-dimensional structure of the catalytic core of cellobiohydrolase Cel6A was solved in 1999 by Varrot et al. [62] The enzyme displays a β/α barrel

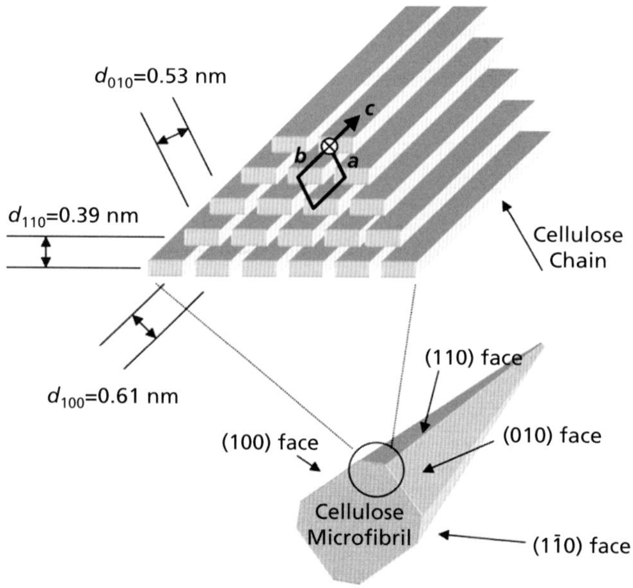

Figure 5.5 Schematic representation of the organization of the cellulose chains in the cellulose Iα phase, which is predominantly present in *Valonia* crystals (microfibrils), and the shape of the complete crystal. The obtuse corner (circle), which exposes the (110) face in worn crystals, is the proposed binding site for the CBMs. Reproduced by permission. Copyright (2003) National Academy of Sciences, USA. [56]

fold with a high degree of similarity with *T. reesei* Cel6A. The active-site residues are located within a tunnel whose roof consists of a pair of large loops. [51]

As demonstrated with oligosaccharides as substrates, the Cel6A structure undergoes several conformational changes upon substrate binding, the most significant of which is a closing of the two active site loops (residues 174–196 and 397–435). [63, 64] This complex has provided the first three-dimensional demonstration of conformational changes in this class of enzymes.

In 2000, Boisset et al. [51] investigated the digestion of dispersed cellulose ribbons from bacterial cellulose with cloned cellobiohydrolases Cel6A and Cel7A from *H. insolens*, either alone or in combination and in the presence of an excess of β-glucosidase. Both Cel6A and Cel7A were effective in partially converting the ribbons into soluble sugars, Cel7A being more active than Cel6A. In combination, these enzymes showed substantial synergy. TEM observations indicated that Cel7A induced a thinning of the cellulose ribbons, whereas Cel6A cut the ribbons into shorter elements, indicating an endo mode of attack. Thus, Cel6A can be described as an endo-processive cellobiohydrolase, whereas Cel7A is essentially a processive cellobiohydrolase. The synergy from the mixture of Cel7A and Cel6A can be explained by the partial endo character of Cel6A. These results, combined with those of the degradation of

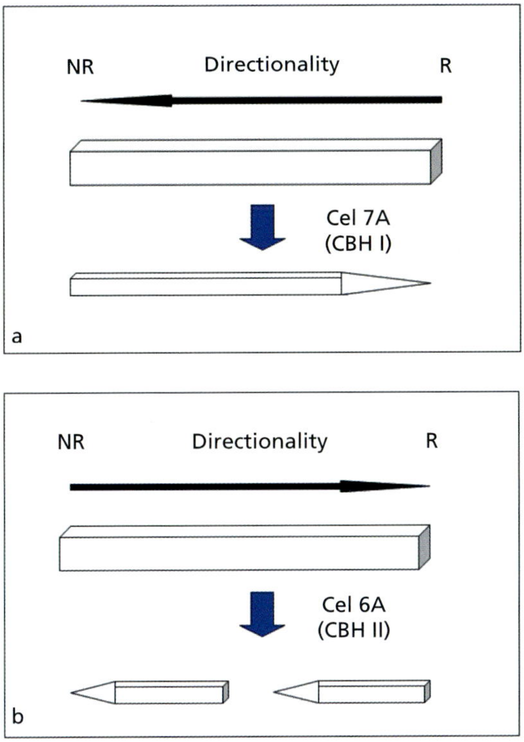

Figure 5.6 Schematic representation of the action of *Humicola insolens* Cel7A and Cel6A toward an isolated crystal of cellulose. Endo attack corresponds to crystal cutting, whereas processivity corresponds to crystal thinning and pointed tips oriented toward the end from which erosion occurs. NR and R, nonreducing and reducing ends of the crystal. [51]

Valonia cellulose microcrystals, [47, 51] suggest the schematic representation shown in **Figure 5.6**.

Kinetics and activity

The enzymatic degradation of solid cellulose is a complex process which occurs at a solid-liquid phase boundary where the enzymes are the mobile components. [41] Substrate factors influencing the kinetics of enzymatic hydrolysis of cellulose include crystallinity, degree of polymerization, and accessibility. [40]

Cellulose hydrolysis rates mediated by fungal cellulases are typically 3–30 times faster for amorphous cellulose as compared to highly crystalline cellulose. [40] Thus, it would be expected that crystallinity should increase over the course of reaction due to a more rapid removal of the amorphous fraction. [65] Nevertheless, several studies have found that crystallinity does not increase during enzymatic hydrolysis. [40] Consequently, it is difficult at this stage to conclude that crystallinity is a critical factor in determining the hydrolysis rate. It is obvious that in a substrate made of a very high number of microfibrils, the degradation of one microfibril on the surface will not change the substrate crystallinity. [42] The loss in crystallinity can only be seen if there is only one microfibril left or if the cellulose microfibrils are beforehand dispersed. In this case, the change in the diffraction pattern is spectacular. [66]

The surface area of cellulose accessible to cellulases has been shown to be the most important factor. [41] This is supported by a study of cellulose degradation by chimeric cellulosomes, indicating that substrate recalcitrance is not necessarily a function of its crystallinity but reflects the overall accessibility of reactive sites. [67]

It has been observed that the heterogeneous structure of cellulose gives rise to a rapid decrease in rate as hydrolysis proceeds. [40] It would seem logical to expect that the declining reactivity of residual cellulose during hydrolysis is a result of factors such as less surface area and fewer accessible chain ends or adsorption of inactive cellulases. At a macroscopic level, both the accessible area of cellulose and cellulase adsorptive capacity have been reported to decrease with increasing hydrolysis. At a microscopic level, the *T. reesei* Cel7A disrupts fibers, resulting in more surface area, while Cel5A appears to smooth fibers, resulting in less surface area.

Measurements of specific activities of *T. reesei* Cel7A, Cel6A, and Cel7B on various insoluble celluloses vary substantially even for similar enzyme preparations and substrates. [40]

Measurements of specific activities of *T. reesei* Cel7A, Cel6A, Cel7B, and β-glucosidase on soluble glucans also vary. [40] However, the following trends may be noted from the reported measurements: 1) the rate of hydrolysis catalyzed by Cel7B, Cel7A, and Cel6A increases with increasing soluble substrate chain length, whereas activity of β-glucosidase decreases with increasing chain length; 2) higher rates are reported for Cel7B in comparison with Cel7A and Cel6A. It should be also noted that specific activities of Cel7B, Cel7A, and Cel6A on soluble substrates are higher than activities on insoluble substrates. Consequently, the hydrolysis from cellulose to soluble glucans is slower than the hydrolysis from soluble glucans to cellobiose and glucose.

Synergism

The enzymes involved in cellulose degradation appear to act synergistically, i.e. in such a way that the activity of a mixture of enzymes is higher than the sum of the activities of the individual components. The synergy factor is defined as the ratio of the activity of the mixture of components to the sum of the activities of individual components. Several types of synergism in the degradation of cellulose by free enzymes have been described. [51, 40]

- Endo- and exo-acting enzymes. This type of synergy is particularly important for degradation of crystalline cellulose. In such cooperation, the action of endoglucanases is to increase the number of chain ends which then become substrates for processive exoglucanases. In addition, β-glucosidases can work in synergy with cellulases by removing cellobiose. Recently, Nutt [41] proposed a new mechanism for this type of synergism whereby endoglucanases, in addition to creating cuts in amorphous cellulose, also "polish" the cellulose surface by removing shorter chains, thereby supporting the processive action of CBHs.

- Exoglucanase and exoglucanase. This synergy has been confirmed with fungal and bacterial cellobiohydrolases. Based on the observation that Cel6A and Cel7A from *T. reesei* and *H. insolens* act, respectively, from the nonreducing and reducing ends of the substrate, it was proposed that the differences in both the chain end preference and in the directionality of action of the two cellobiohydrolases were responsible for the exo-exo synergy. [51, 68] This synergy could also be explained by the partial endo activity of one or more cellobiohydrolases from a microorganism as shown by the study by Boisset et al. [51] on *H. insolens* Cel6A and Cel7A.

5.2.3 Multienzyme complexes

The cellulosome concept

Many cellulolytic microorganisms produce multienzyme complexes called cellulosomes that efficiently degrade cellulose and related plant cell wall polysaccharides. [69, 70] The cellulosome concept was established in 1983 by Lamed and Bayer et al. [71] in the anaerobic thermophilic bacterium *Clostridium thermocellum*, wherein the cellulases were found to be organized into a high molecular weight cellulolytic complex. The cellulosomes are composed of subunits, each of which comprises a set of interacting functional modules. A multifunctional integrating subunit called scaffoldin is responsible for organizing the catalytic subunits into the multienzyme complex. This is accomplished by the interaction of the ***dockerin*** domains present in the enzymatic subunits with the reiterated ***cohesin*** domains of the scaffoldin subunit. The high-affinity cohesin-dockerin interaction is responsible for the cellulosome architecture. The scaffoldin also bears a CBM that anchors the cellulosome to its substrate. In any given species, the interaction among the cohesins and dockerins usually proceeds in a non-specific manner. [72, 73]

Types of cellulosome

The simplest cellulosomes, such as those produced by the mesophilic *Clostridium cellulolyticum*, include eight enzymes tightly bound to the scaffoldin. [72] The most elaborate cellulosomes, such as those produced by rumen bacteria, contain several interacting scaffoldins that can integrate dozens of enzymes.

Two major types of cellulosome have so far been identified in relation with the cellulosomal gene cluster arrangement in the microorganisms. [74, 70]

In this first type of cellulosome system, many of the cellulosomal genes are arranged in a cluster on the genome, consisting of a **primary** (enzyme-incorporating) **scaffoldin** gene and various dockerin-containing enzyme genes. [74] This enzyme-linked gene cluster arrangement is found in many mesophilic clostridia, such as *C. cellulolyticum*, [75] *C. cellulovorans*, [76] *C. josui*, [77] and *C. acetobutylicum*. [78] Hydrophilic domains present in the scaffoldin proteins from these bacteria have been proposed to fulfill a cell-surface-anchoring role. [79, 80]

In the other type of cellulosome system, a more elaborate arrangement allowing amplification of the number of enzymes in the complex comprise two or more genes that encode scaffoldins, at least one of which anchors the cellulosome onto the cell surface. [74, 81] The genes for the enzymes are distributed elsewhere on the genome. [74] In addition to the cohesins, each anchoring protein usually bears an S-layer homology (SLH) module which is known to bind to the cell surface. [69] This multiple-scaffoldin gene cluster arrangement was first observed for *C. thermocellum*. [82] Similar multiple-scaffoldin gene clusters have been described for *Acetivibrio cellulolyticus*, [81] *Ruminococcus flavefaciens*, [79] and *Bacteroides cellulosolvens*. [83]

Scaffoldin carbohydrate-binding module

The CBM of scaffoldins binds the cellulosome to the carbohydrate substrate.

In 1996, the crystal structure of a family 3a CBM from the primary scaffoldin (CipA, cellulosome-integrating protein) of *C. thermocellum* was solved by Tormo et al. [84] The CBM forms a nine-stranded β sandwich with a jelly-roll topology and binds a calcium ion. [85]

Two defined surfaces, located on opposite sides of the molecule contain conserved, surface-exposed residues. One of these faces forms a planar linear strip of aromatic and polar residues which are presumed to interact with crystalline cellulose. The other conserved residues are contained in a shallow groove of unknown function. The major interaction with cellulose likely involves the planar strip of aromatic residues, which align along one cellulose chain. In addition, polar amino acid residues are proposed to anchor the CBM to two adjacent cellulose chains.

In 2000, Shimon et al. [86] determined the crystal structure of a family 3a CBM from the scaffoldin (CipC) of *C. cellulolyticum*. The structure revealed a nine-stranded β sandwich with a calcium-binding site, a putative cellulose-binding surface and a shallow groove. [87] The *C. cellulolyticum* CipC CBM structure is very similar to the *C. thermocellum* CipA CBM.

The CBMs on most other known enzyme-incorporating scaffoldins belong to family 3a. However, *R. flavefaciens* primary scaffoldin (ScaA) lacks an identifiable CBM. [88]

Cohesin-dockerin interaction

The incorporation of the enzymes into the cellulosome is a function of the reiterated cohesins borne by the scaffoldin. For most species of scaffoldin, these cohesins have been classified as ***type I cohesins*** on the basis of sequence homology. [89, 90] The type I cohesin module is composed of ~150 amino acid residues. The crystal structures of one type I cohesin from *C. thermocellum* and one from *C. cellulolyticum* have been determined. [91, 92, 93] The *C. thermocellum* domains form a nine-stranded β sandwich with a jelly-roll topology, similar to the fold of the family 3a scaffoldin CBM. [91, 92] Despite a low sequence similarity of 32%, the *C. cellulolyticum* domain has a fold similar to those of *C. thermocellum* cohesins suggesting that all type I cohesins have a similar fold. [92, 94] However, observed differences in polarity between the cohesin surfaces of the two species may explain the cross-species specificity of cohesin in recognizing the cognate dockerin. [3, 93]

The corresponding dockerin domain of the enzymes, classified as ***type I dockerin***, contains ~70 amino acids and is characterized by a 22-residue duplicated docking sequence, which is similar to the **EF-hand motif** (calcium-binding motifs composed of two helices, E and F, joined by a loop) of various calcium-binding proteins. [89, 95] This repeated sequence contains a 12-residue calcium-binding loop. The solution structure of the dockerin from the *C. thermocellum* enzyme CelS revealed a flexible unique-fold protein that consists of two calcium-binding loop-helix motifs connected by a linker; the E helices entering each loop of the classical EF-motif are absent from the dockerin. [96, 97] Both calcium-binding segments of type I dockerin are required for cohesin recognition.

In 2003, Carvalho et al. [95] reported the crystal structure of the cohesin-dockerin complex (association constant $K_a = 6.2\text{x}10^6\ M^{-1}$) from *C. thermocellum* (**Figure 5.7**). The structure reveals that the β-sheet cohesin (chain A) interacts predominantly with one helix of the dockerin (chain B). Whereas the structure of the cohesin remains basically unchanged, the loop-helix-helix-loop-helix motif of the dockerin undergoes conformational change and ordering compared to its solution structure to display a near-perfect twofold symmetry. However, the classical 12-residue EF-hand coordination to two calcium ions is maintained.

The structure of the type I cohesin-dockerin complex reveals that protein-protein recognition is mainly mediated by hydrophobic interactions between one of the faces of the cohesin and helices 1 (Ser-11B–Leu-22B) and 3 (Ser-45B–Leu-56B) of the dockerin; there are relatively few direct hydrogen bonds between the two proteins. [95] In the dockerin, the residues that make direct hydrogen bonds with the cohesin are Leu-22B, Arg-23B, Ser-45B, Thr-46B, and Arg-53B. Conservation of the residues involved in hydrophobic interactions and hydrogen bonding in both cohesin and dockerin is consistent with

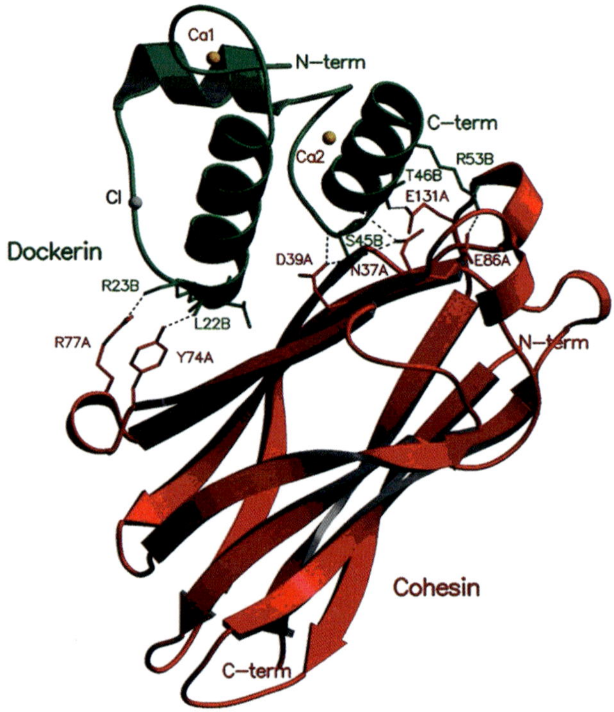

Figure 5.7 Structure of the type I cohesin–dockerin complex from *C. thermocellum*. [95] Cohesin (chain A) is in red, dockerin (chain B) in green. The residues involved in domain contacts are shown as stick models. The cohesin of the complex consists of a nine-stranded β-barrel, defined by two β-sheets. The dockerin forms three α-helices (Ser-11B–Leu-22B, Thr-28B–Asp-36B, and Ser-45B–Leu-56B), in a conformation defined by a loop-helix motif followed by a helix-loop-helix motif. The two Ca^{2+} ions, Ca1 and Ca2, are in orange, and Cl^- ion, Cl in gray. Reproduced by permission. Copyright (2003) National Academy of Sciences, USA.

the inability of the cohesins of *C. thermocellum* to discriminate between the different dockerins. In addition, replacements of hydrogen bonding residues in *C. thermocellum* dockerins with other residues in *C. cellulolyticum* dockerins provide a partial explanation for the cross-species specificity of the dockerin in recognizing the cognate cohesin.

The scaffoldin of *C. thermocellum* also contains at its C-terminal end a special type of dockerin domain, characterized by a duplicated sequence of 22 residues, and termed ***type II dockerin***. [89, 90] This domain resembles the type I dockerins, although its amino acid sequence is more divergent from the consensus. [88] The type II dockerin fails to bind to the type I cohesins of the same scaffoldin, but interacts with a different type of cohesin, termed ***type II cohesin***, identified on the basis of sequence homology. These cohesins are somewhat different from type I cohesins, having an additional segment and diversity in the latter half of the sequence. The type II cohesins were identified as component parts of anchoring proteins in *C. thermocellum*. [82, 90, 98]

Crystal structures of type II cohesins from *C. thermocellum*, [99, 100] *A. cellulolyticus*, [101] and *B. cellulosolvens* [102] display the β jelly-roll topology of type I cohesins. A unique α helix and two regions disrupting two strands were also observed and have been proposed to play roles in the type II interaction and specificity. [103]

In 2006, Adams et al. [103] solved the structure of an ultra-high-affinity ($K_a = 1.44x10^{10}\ M^{-1}$) complex between type II dockerin, together with its neighboring X module from the primary scaffoldin of *C. thermocellum*, and a type II cohesin associated with the bacterial cell surface (**Figure 5.8**). The structure reveals roles for the X module in dockerin stability and enhanced cohesin recognition. The type II interaction involves one face of the cohesin and both helices of the dockerin, and comprises significant hydrophobic character and a complementary hydrogen-bond network. An explanation for type I-type II specificity is revealed by the observed differences in dockerin orientation, the physiochemical surface properties of the type I and II interfaces, and the role of the X module in the type II interaction.

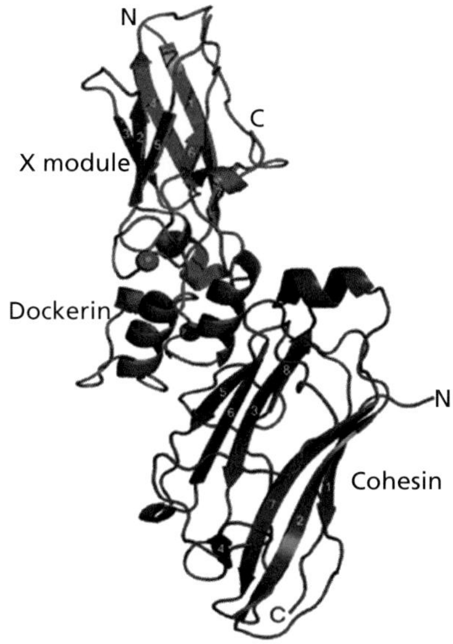

Figure 5.8 Structure of the cellulosomal complex between type II cohesin and type II dockerin module X modular pair. [103] The type II cohesin is in blue, type II dockerin in green, and X module in magenta. The β strands of the X module and type II cohesin are numbered in yellow. The N and C termini are shown, and the calcium ions are depicted as orange spheres. Reproduced by permission. Copyright (2006) National Academy of Sciences, USA.

Clostridium cellulolyticum

C. cellulolyticum is an anaerobic, mesophilic, soil bacterium that is able to grow on cellulose as its sole carbon source. [104]

Twelve genes encoding key components of *C. cellulolyticum* cellulosomes are clustered. [105, 106] Among them, the first upstream encodes the scaffoldin CipC, the second encodes the major processive endocellulase Cel48F, and the fifth encodes the major cellobiohydrolase Cel9E which has retained some capacity for random attack mode. [107, 108] The other cellulosomal enzymes encoded by the clustered genes include Cel8C, Cel9G, Cel9H, Cel9J, Man5K, Cel9M, and Cel9N. [108] CipC contains eight reiterated cohesins, a family 3a CBM, and two domains of unknown function. [27, 75, 109] Three isolated genes encoding the cellulosomal cellulases CelA and CelD and noncellulosomal cellulase CelI were found elsewhere on the chromosome. [106]

The enzymatic composition of the *C. cellulolyticum* cellulosomes is probably regulated by the amounts of the different dockerin-containing enzymes, since the cohesin-dockerin interaction is not enzyme specific. [110] Among the >13 dockerin-containing proteins identified in cellulosome preparations, [75] Cel9E is comprised of a family 4 CBM, an immunoglobulin (Ig)-like domain and a GH9 catalytic module, Cel9G, [104] Cel9H, and Cel9J are comprised of a GH9 catalytic module and a family 3c CBM attached to the catalytic module, [75] while Cel48F possesses a GH48 catalytic module but no CBM. [107, 108]

The crystal structures of the endoglucanase Cel9G [104, 111] and the processive endocellulase Cel48F [112, 113, 114] have been determined. The catalytic module and CBM of Cel9G form a flat platform onto which crystalline cellulose would bind and be fed into the active-site cleft. [104] The active site of Cel48F is composed of a 25-Å long tunnel, followed by an open cleft. [112, 113] The cellulose chain would slide through the tunnel during the processive mechanism in order to supply the leaving group site with sugar residues after the catalytic cleavage.

In 2005, Hammel et al. [72] proposed a model of a *C. cellulolyticum* cellulosome and its interaction with the cellulose substrate (**Figure 5.9**). Following the binding of the cellulosome to cellulose through the powerful family 3a CBM, the scaffoldin linkers between the cohesins would undergo large scale rearrangement to adjust the positions of the enzyme subunits to the topography of the substrate. In this context, the relatively weak CBM borne by some cellulosomal enzymes would play a role in maintaining an extended conformation of the whole complex.

Clostridium thermocellum

C. thermocellum is a cellulolytic, thermophilic, anaerobic bacterium. [93]
In 1993, Fujino et al. [82] reported the organization of a *C. thermocellum* gene cluster encoding the primary scaffoldin CipA and a putative anchoring protein. Sequencing of the DNA downstream of *cipA* revealed three open read-

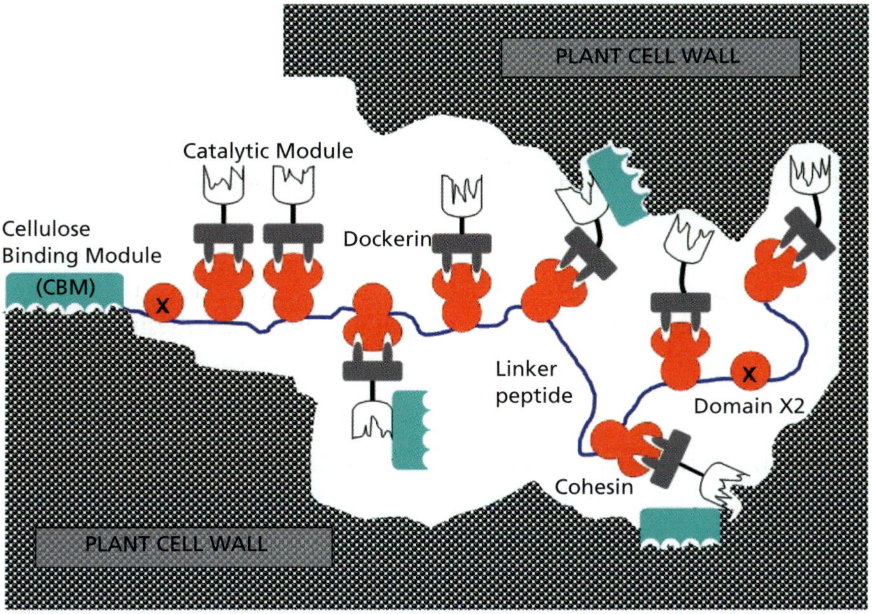

Figure 5.9. Proposed schematic representation depicting a functional model of a cellulosome from *Clostridium cellulolyticum* and the interaction of its components with the cellulose substrate. [72] The scaffoldin is bound to the substrate through the family 3a CBM. Some cellulosomal enzymes bear a CBM that mediates a relatively weak interaction with the substrate.

ing frames (ORFs), termed ORF1, ORF2, and ORF3. [82] The polypeptides ORF1p, ORF2p, ORF3p encoded by these ORFs contained a SLH module. In addition, a cohesin domain was detected on ORF3p, suggesting that the protein might be involved in attachment of the cellulosome to the cell surface by binding the dockerin domain on CipA. ORF3p was renamed OlpA (for outer layer protein).

In 1995, Lemaire et al. [98] identified ORF1p (renamed OlpB) as a new outer layer protein of *C. thermocellum*. One year later, Leibovitz et al. [90] reported a new type of cohesin, termed type II, that specifically binds the type II dockerin of CipA. This type II cohesin domain is present in one, two, and four copies in the anchoring proteins SdbA (for scaffoldin dockerin binding), ORF2p, and OlpB, respectively. In contrast, the cohesin domain of OlpA does not bind the type II dockerin of CipA.

The arrangement of *C. thermocellum* scaffoldin-related genes, as recently described, is shown in **Figure 5.10**. [115] The four genes *cipA*, *olpB*, *orf2*, and *olpA* are clustered, whereas *sdbA* is located elsewhere on the genome. The proteins OlpB, Orf2p, and SdbA are anchoring scaffodins, whereas OlpA binds individual cellulases to the cell surface.

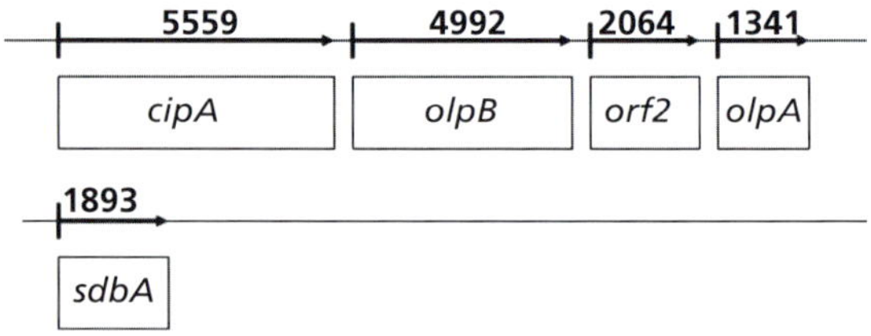

Figure 5.10. Organization of *Clostridium thermocellum* scaffoldin-related genes. [115] *cipA* and the three genes *olpB*, *orf2*, and *olpA* are located in tandem within a gene cluster. *sdbA* is located elsewhere on the chromosome. The proteins OlpB, Orf2p, and SdbA serve as anchoring proteins. In contrast, OlpA binds individual cellulases to the cell surface. The numbers above the genes refer to its size (in base pairs).

A schematic representation of the cellulosome from C. thermocellum is shown in **Figure 5.11**. [116]

The C. thermocellum cellulosome is characterized by a primary scaffoldin that contains nine copies of type I cohesins, each of which binds to a complementary type I dockerin, contained on each enzymatic subunit. [83, 115, 117] The primary scaffoldin also contains a type II dockerin that interacts selectively with complementary type II cohesins, contained on at least three different anchoring scaffoldins. The three known anchoring scaffoldins carry

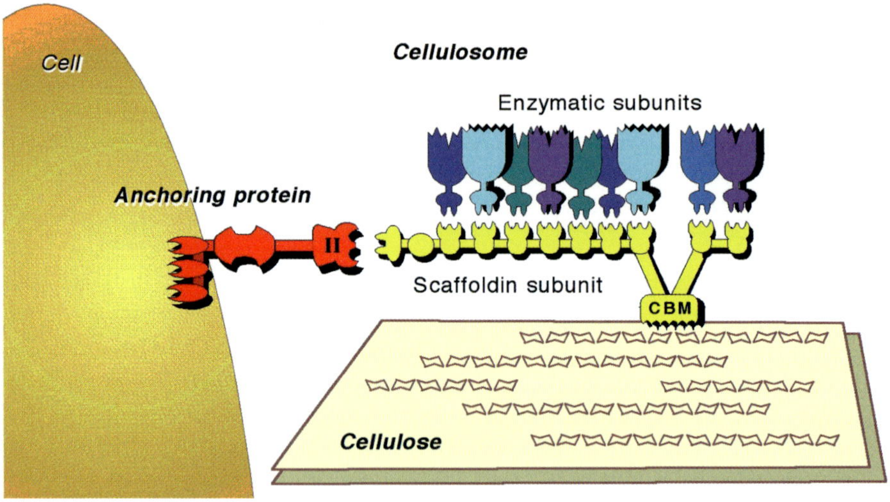

Figure 5.11. Schematic diagram of the *Clostridium thermocellum* cellulosome. [116] The type I dockerins mediate attachment of the enzymatic subunits to the primary scaffoldin subunit, which is comprised of nine cohesins, a CBM, a hydrophilic domain of unknown function and a type II dockerin. The scaffoldin binds through its type II dockerin to a type II cohesin-containing protein that anchors the cellulosome through an SLH module to the cell surface. Courtesy of Edward A. BAYER and Raphael LAMED.

different numbers (one, two, and four) of type II cohesins, but each bears an SLH module. [115] In *C. thermocellum*, the type I cohesin-dockerin interaction is apparently responsible for the incorporation of the enzymes into the cellulosome, whereas the type II cohesin-dockerin interaction would mediate the anchoring of the cellulosome onto the cell surface. Individual cellulosomes consist of up to nine enzymes attached to the primary scaffoldin.

In *C. thermocellum*, the nine cohesins of CipA are unable to discriminate between the individual dockerins present in the various enzymes so that any cellulosome may comprise a different set of enzymatic subunits. [95] To date, at least 20 cellulosomal enzymes are known to be produced by *C. thermocellum*. [118] These catalytic subunits include mostly endoglucanases, exoglucanases, xylanases, and other hemicellulases.

In 1999, Boisset et al. [119] investigated the digestion of bacterial cellulose ribbons and *Valonia* cellulose microcrystals by the *C. thermocellum* cellulosome. The cellulosome was very effective in catalyzing the complete digestion of bacterial cellulose, but the complete degradation of *Valonia* microcrystals was achieved more slowly (**Figures 5.12** and **5.13**). It was suggested that the rapid degradation of bacterial cellulose was due to a very efficient synergy between the various enzyme components. The degraded *Valonia* substrate assumed various shapes, ranging from thinned-down microcrystals to microcrystals where one end is pointed, and the other is intact. In the case of bacterial

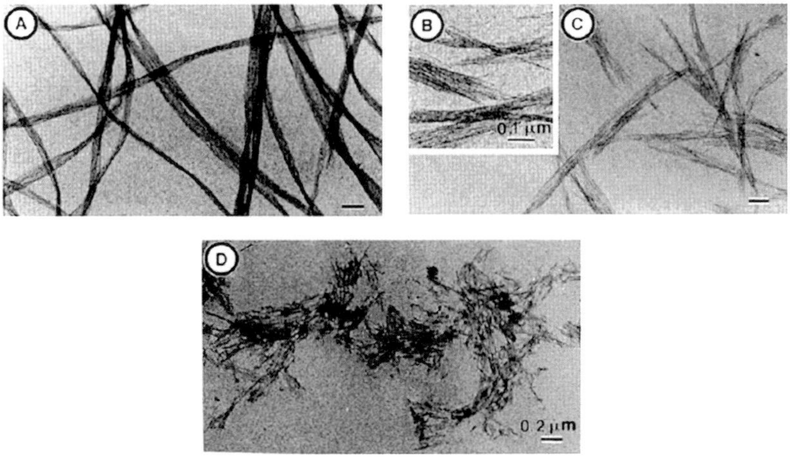

Figure 5.12. TEM micrographs of *Clostridium thermocellum* cellulosome-induced degradation of bacterial cellulose ribbons. (A) Untreated cellulose ribbons. ((B) and (C) Cellulose ribbons after 3h of digestion (D) After 6.5 h of digestion. (A), (C) and (D) are at the same magnification; the magnification of (B) is twice that of the others. Reproduced with permission, from C. BOISSET, H. CHANZY, B. HENRISSAT, R. LAMED, Y. SHOHAM, and E. A. BAYER, 1999, *Biochemical Journal*, **340**, 829-835. © The Biochemical Society (http://www.biochemj.org). [119]

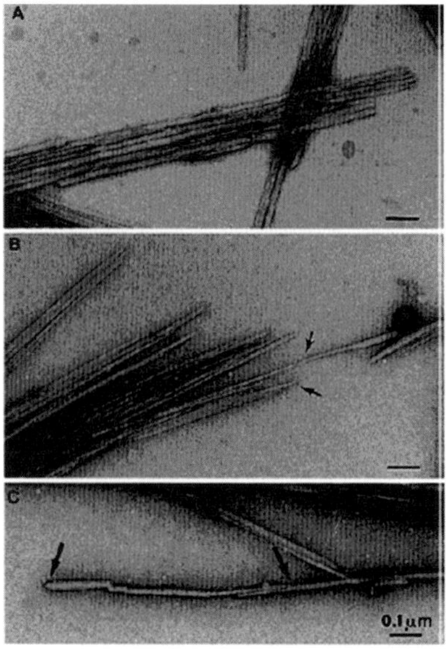

Figure 5.13. TEM micrographs of *Clostridium thermocellum* cellulosome-induced degradation of *Valonia* cellulose microcrystals. (A) Untreated microcrystals. (B) and (C) Microcrystals after 16 days of digestion. The arrows in (B) indicate pointed microcrystals, typical of the unidirectional action of cellulases. The arrows in (C) show the two tips of a unidirectionally eroded microcrystal. Reproduced with permission, from C. BOISSET, H. CHANZY, B. HENRISSAT, R. LAMED, Y. SHOHAM, and E. A. BAYER, 1999, *Biochemical Journal*, **340**, 829–835. © The Biochemical Society (http://www.biochemj.org). [119]

cellulose degradation, the resulting coordinated multi-enzyme attack seems to be so efficient and rapid that the features of cellulose crystal degradation (spicing, thinning, and tip sharpening) appear to occur simultaneously. Thus, the substrate would be degraded in a near-uniform manner, along both its length and width. In the case of *Valonia* microcrystals, only moderate crystal thinning and tip sharpening were observed, in contrast with the observations reported for fungal cellulases. Thus, cellulosome action on cellulose crystals was presumed to exhibit a relatively limited processive character. The results were consistent with a mechanism whereby initial attack by a cellulosome results in its "commitment" towards complete degradation.

Designer cellulosomes

The complexity and diversity of native cellulosomes prevents detailed analysis of the structural features responsible for their enhanced activity. [72] In addition, the lack of specificity of the cohesin-dockerin interaction within a species is an obstacle for the construction of homogeneous simplified cellulosomes containing more than one enzyme. There has been interest in construct-

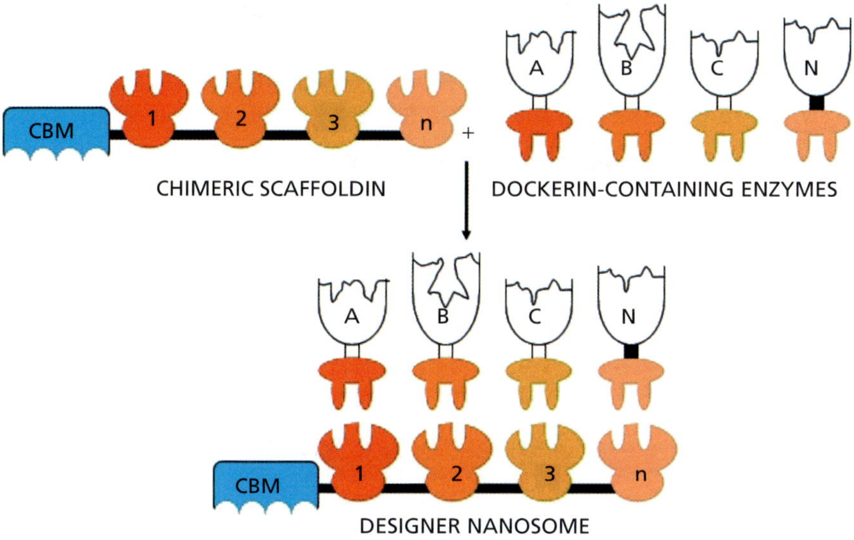

Figure 5.14 Schematic representation of a designer cellulosome (nanosome). [121] A chimeric scaffoldin is prepared that contains an optional CBM and multiple (n) cohesins of different dockerin specificities. The dockerin counterparts include distinct modules as part of the polypeptide chains of the desired enzymatic component (e.g. enzymes A, B, C, and N).

ing designer cellulosomes (or designer nanosomes), in which enzymes can be incorporated into defined positions, for both basic and applied purposes.

Such artificial cellulosomes are designed to comprise recombinant chimeric scaffoldin constructs and selected dockerin-containing enzyme hybrids, as a platform for promoting synergism among enzyme components (**Figure 5.14**). [69, 120]

In practice, a chimeric scaffoldin is produced that bears cohesins of different dockerin specificities. The designer cellulosome is constructed by mixing in solution the chimeric scaffoldin and complementary dockerin-containing components. Thus, in these cellulosomes, the composition and architecture of the complex can be controlled. The concept has potential applications as a molecular Lego in biotechnology and nanotechnology. [69]

5.2.4 Conclusion

The recalcitrance of cellulosic substrates, linked with their structural complexity and rigidity, have given rise to a phenomenal diversity of degradative enzymes, the cellulases. [1] Thus, multiple enzyme systems are required to efficiently degrade cellulose. Such systems comprise either a collection of free enzymes and/or multicomponent complexes thereof, known as cellulosomes.

Microbial breakdown of cellulose and related by-products is a highly complex process, requiring multiple types of synergistic catalytic activities. [122] Industrial biomass utilization processes will also require a combination of activities in order to degrade lignocellulosic feedstocks, but with the additional complication that thermal and chemical pretreatment processes, which improve physical substrate access to the enzymes, alter the substrate chemically and form or solubilize inhibitory compounds that can negatively impact enzyme performance.

5.3 Non-biological degradation

Despite its chemical stability, cellulose can be degraded non-enzymatically by acids, alkalis, oxidants, thermal energy, mechanical means, and radiations.

Like cellulases, acids catalyze the hydrolysis of the glycosidic linkages in cellulose. Acid hydrolysis of cellulose leads to high yields of glucose under certain conditions. [4] There are two acid hydrolysis processes commonly used for converting cellulosic biomass to ethanol: dilute acid hydrolysis and concentrated acid hydrolysis. [123, 124] Both technologies have long industrial histories. [125]

Cellulose loses weight upon reaction with alkalis, which is a phenomenon that is well known from pulping and textile industries. [126] Alkaline degradation of cellulose has also been studied in relation to transformation of biomass to various low-molecular-weight products. Alkalis cause degradation of cellulose with a completely different mode of action than acids. Alkaline degradation occurs by two primary ways: ***endwise degradation*** (peeling) and alkaline hydrolysis of glycosidic bonds.

The complete oxidation of cellulose yields carbon dioxide and water. [4] Much less drastic oxidation, as discussed in this chapter, leads to solid, and often fibrous, products that are insoluble in water. These materials were called ***oxycelluloses***. The partial oxidation of cellulose has stimulated immense interest, much of which stems of a need to prevent damage to cellulose during industrial processes such as bleaching of textile fibers and wood pulp. Occasionally, oxidation is used to reduce the degree of polymerization of cellulose, notably in the ageing of the alkali cellulose used in the manufacture of viscose rayon. With partial oxidation, products containing new functional groups, such as carboxy, aldehyde, and keto groups, can be formed while the glycosidic bonds remain essentially intact. [127] Cellulose is particularly susceptible to oxidation under alkaline conditions.

Thermal degradation of cellulose and cellulosic materials has attracted considerable attention at different times and for different reasons. [4] Early interest in the thermal degradation of cellulose was stimulated by lively academic discussions about the origin of bituminous coal and whether it is derived from the lignin or polysaccharides of wood. This subject was also investigated in connection with earlier processes for production of chemicals and fuel by destructive distillation or gasification of wood. The textile and pulp and paper industries have been concerned about ageing, deterioration, and degradation of cellulosic materials at elevated temperatures for practical reasons, such as the performance of tire yarn and electrical insulation paper, and the problems involved in manufacturing or drying of cellulosic materials. For several decades, biomass gasification technology has been under intensive development. [128] Cellulosic biomass can be gasified into synthesis gas (consisting primarily of carbon monoxide, carbon dioxide, and hydrogen), which can then be converted to fuels and chemicals through Fischer-Tropsch synthesis or fermentation.

The mechanical degradation of cellulose has not been studied as much as other types, mainly because it does not cause serious problems in industrial processes. [4] However, input of mechanical energy into cellulose may involve not only macroscopic disintegration and loss of crystallinity but also, if energy is sufficiently high, chain cleavage. [127]

Studies on the interaction of visible and ultraviolet light and high energy, ionizing radiations, with cellulose have gained in momentum mainly because of their industrial significance. [4] Light promotes the deterioration of cellulosic products, and certain dyes and other additives greatly accelerate this process.

Acid hydrolysis

Mechanism and kinetics

Acid hydrolysis of cellulose was first reported by C. Calvert [129] in 1855, at the same time as the discovery of mercerization by J. Mercer. [127]

The mechanism of the reaction is the same as that of simple glycosides (**Figure 5.15**). [4, 130]

Acid hydrolysis comprises three steps:

- rapid protonation of the glycosidic oxygen, forming a conjugated acid;
- transfer of the positive charge to C1 with consequent formation of a carbonium ion, and cleavage of the glycosidic bond;
- rapid attack on the carbonium ion by water to give the free sugar residue and to regenerate the proton.

The acid hydrolysis of glycosidic bonds follows a first-order reaction. [4, 127, 130] The rate constant depends on acid concentration, temperature, and chemical environment of the glycosidic bond. [127] Thus, hydrolysis of cellobiose is characterized by one rate constant, whereas that of cellotriose needs two constants to be described due to the different environment of its two bonds.

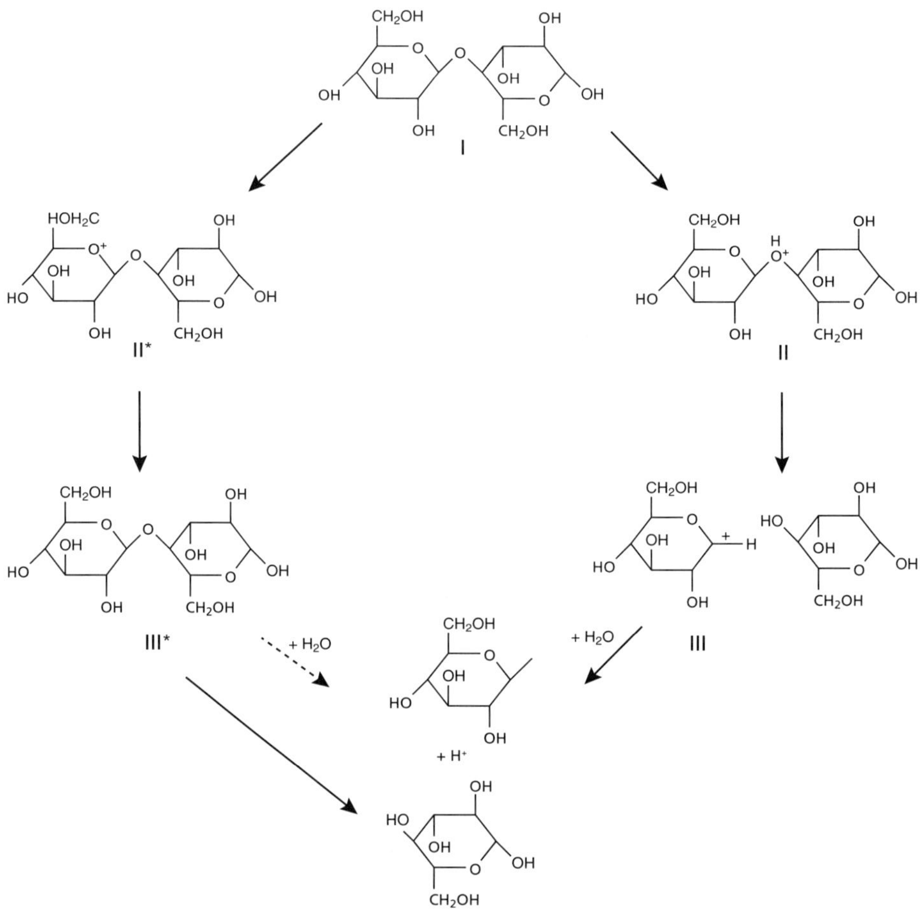

Figure 5.15 Mechanism of acid-catalyzed hydrolysis of β-glycosidic linkages. [130] The dominant pathway involves the intermediates II and III.

Action on cellulose

Acid-catalyzed hydrolysis of cellulose is controlled not only by the hydrolytic chemical reaction (temperature and acid concentration) but also by the physical state of the cellulose. [130] The first-order kinetic equation generally applies to reactions in homogeneous phase. Therefore, first-order reaction is justified for hydrolysis of soluble oligosaccharides. In actual hydrolysis of cellulose with dilute acids, a heterogeneous reaction occurs yielding ***hydrocellulose***, a product with reduced degree of polymerization (DP) and higher crystallinity. The rate of hydrolysis of crystalline cellulose is one to two orders of magnitude lower than that of hydrolysis of soluble model compounds. When the DP of a cellulose is plotted against time of acid hydrolysis, the DP falls sharply initially and then becomes roughly constant at a value characteristic of the cellulose, which is called its ***leveling-off degree of polymerization*** (***LODP***) and can be used as an indication of the length of the crystallites present in cellulose. [4] Furthermore, cellulose, once it is dissolved into concentrated sulfuric acid and re-precipitated, showed a reaction rate two orders of magnitude higher than that of untreated cellulose, i.e. about the same magnitude as cornstarch. The hydrolysis of cellulose is also greatly influenced by its degree of crystallinity and swelling.

A kinetic model for dilute-acid hydrolysis of cellulose has been proposed by Q. Xiang et al. [131] Under low temperature and acid concentration, the cellulose structure stays in stable crystalline form. Therefore, the prevailing reaction mode is endwise hydrolysis. When temperature and/or acid concentration is increased to a certain level, the structure becomes unstable by breakage of hydrogen bonding. Once the crystalline structure is disrupted, acid molecules can penetrate into additional regions of the material. Finding effective means of disrupting hydrogen bonding before or during hydrolysis may lead to a novel biomass saccharification process. [130]

5.3.2 Alkaline degradation

Before reviewing the degradation of cellulose by alkalis, their action on monosaccharides will be described. [4] Monosaccharides dissolved in dilute alkali will isomerize by a series of keto-enol ***tautomerizations***. The tautomerism of aldoses and ketoses with enediols is the basic mechanism of the reaction of monosaccharides with alkali (**Figure 5.16**).

As an example, a solution of D-glucose containing alkali forms several products, including D-mannose and D-fructose.

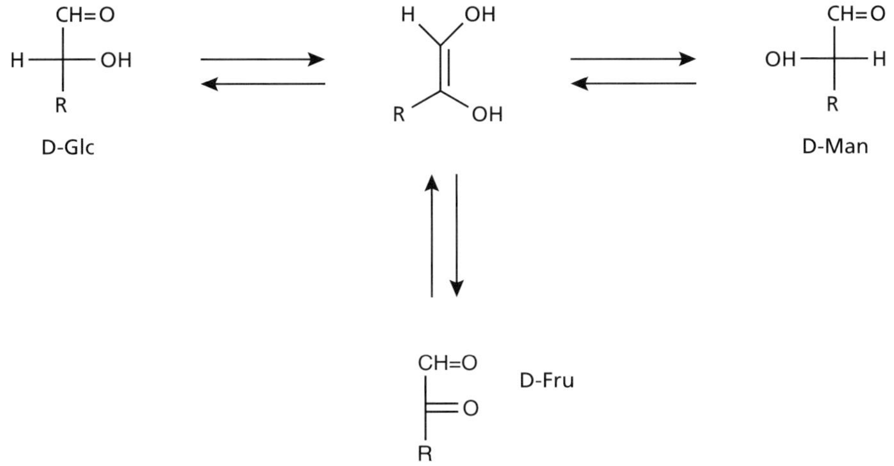

Figure 5.16 Scheme of reaction of monosaccharides with dilute alkali involving aldose-ketose isomerization by a series of keto-enol tautomerizations. [4]

Products resulting from such keto-enol tautomerizations participate in the slow formation of deoxyaldonic (saccharinic) acids. [4] The formation of these acids involves particularly a benzilic acid rearrangement, by which 1,2-diketones are transformed into α-hydroxycarboxylic acids in the presence of a strong base. This arrangement derives its name from the conversion of benzil to benzilic acid in the presence of alkali.

The alkaline degradation of cellulose is generally accepted as occurring in two primary ways, namely endwise degradation (peeling) and alkaline hydrolysis of glycosidic bonds. In this section only alkaline degradation under non-oxidative conditions will be considered because cellulose is very susceptible to oxidative degradation in the presence of alkali (see § 5.3.3).

Endwise degradation

At temperatures below 140–170°C the glycosidic bonds in cellulose are stable towards alkalis. [4, 132] However, at temperatures already below 100 °C, a stepwise peeling of monomer units starting from the reducing ends leads to a reduction in chain length (**Figure 5.17**). [127, 133]

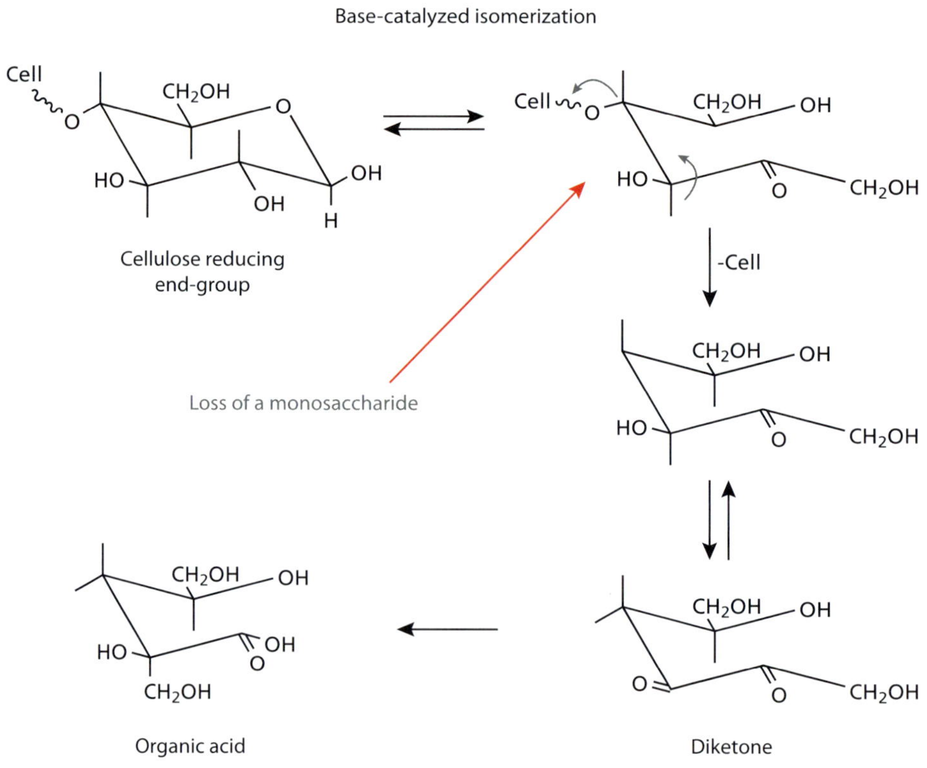

Figure 5.17 Mechanism of peeling. Reproduced by permission of Michel MUGUET, Air Liquide. [133]

If the erosion of cellulose molecules from their reducing ends were to continue, the whole material would eventually dissolve. [4, 134] It is well known that this does not happen. If it did happen, the scouring of cotton textiles and the kraft process for wood pulping would be impossible. Stabilization of the cellulose (stopping reaction) is brought about by a competing reaction which begins with the elimination of the hydroxyl group at C3 in the reducing terminal chain unit. The terminal unit is then converted into an alkali-stable 3-deoxyhexonic acid unit.

The extent of peeling can be minimized by the oxidation or the reduction of the hemiacetal end groups of cellulose. [132] Polysulfide and anthraquinone can oxidize these reducing ends into stable carboxyl end groups.

Alkaline hydrolysis

When cellulose is heated in alkaline conditions at temperatures above 140 °C, a considerable fall in DP is observed due to the random cleavage of glycosidic bonds. [4, 132] New reducing ends are formed by this alkaline hydrolysis and peeling is reactivated (secondary peeling), thus causing much greater weight loss than at lower temperatures.

Most of the mechanistic studies of alkaline hydrolysis of cellulose have involved model compounds. [4, 135] A glycosidic linkage may be broken either between the oxygen atom and the anomeric carbon atom (C1) of the glycosyl group or between the oxygen atom and the aglycone. Both reactions exist, but cleavage of the oxygen-glycosyl bond greatly predominates. The major proposed mechanism for glycosidic bond cleavage for phenyl glycosides and alkyl glycosides is an $S_N$icB(2) mechanism, which refers to a nucleophilic substitution by an internal nucleophile: the conjugate base of the C2 hydroxyl group (**Figure 5.18**). [135]

The $S_N$icB(2) mechanism requires that the pyranose ring flips from its more stable $^4C_1$ conformation to the $^1C_4$ conformation; this permits backside attack of the ionized hydroxyl group at C2 on C1, displacing the aglycone. [135] Such a ring flip would be impossible for crystalline cellulose and probably difficult for amorphous cellulose. Studies with disaccharide models have shown that heating with alkali leads to a predominance of glycosyl-oxygen bond cleavage by way of a number of mechanisms, including S_N1, $S_N$icB(2), and $S_N$icB(2)-ro. The latter involves ring opening of the glycone as an intermediate step (**Figure 5.18**). The ring opening mechanism could be the dominant glycosyl-oxygen bond cleavage mechanism for 1,5-anhydrocellobiitol models, including a conformationally rigid model. [135] Since pyranose ring flips appear difficult for carbohydrate polymers, random chain cleavages of cellulose may also occur largely by an $S_N$icB(2)-ro mechanism.

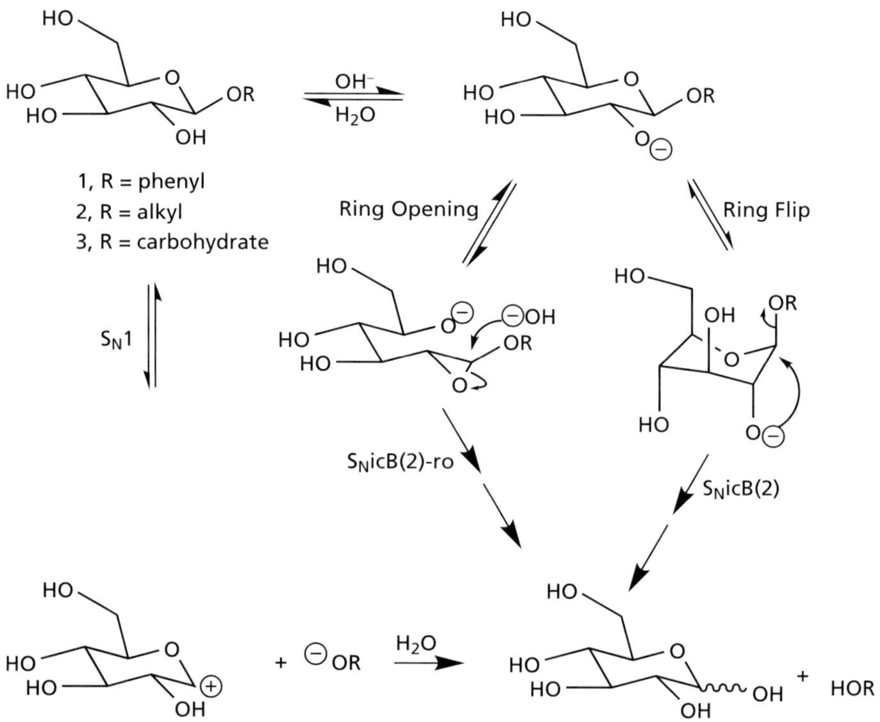

Figure 5.18 Putative mechanisms for alkaline glycosyl-oxygen bond cleavage. Reproduced by permission of Institute Director Dr. Ron ROUSSEAU, and The Institute of Paper Science and Technology at Georgia Tech. [135]

5.3.3 Oxidative degradation

Oxidative degradation of cellulose proceeds via various routes comprising a number of parallel and consecutive reactions. [127] It usually starts with the conversion of glucose units to a relatively unstable derivative followed by ring opening and/or chain cleavage. The oxidation of cellulose is complex for three reasons: (1) the three hydroxyl groups in the monomer have different reactivities, (2) different regions the cellulosic material have different accessibilities, and (3) different oxidants behave in different ways. An important characteristic of the various types of oxycellulose is the extent of oxidation, usually expressed as the oxygen consumption that is the number of atoms of oxygen consumed per anhydroglucose unit.

There are numerous ways in which the glucose units of cellulose can be oxidized while the glycosidic bonds remain significantly intact. [4] For example, periodate ions and dinitrogen tetroxide share the unusual ability to react with the whole bulk of the cellulose material without destroying its fibrous structure. More generally, only the accessible regions are attacked at a reasonable rate, and the fibers fall to powder when only a small proportion of the glucose units have been oxidized. Considerable dissolution of material often takes place simultaneously. The disintegration of cellulose fibers is due to chain scission, which occurs for a variety of reasons but always occurs when the oxidizing solution is alkaline.

The following discussion is centered on oxidative chain degradation by sodium hypochlorite (NaOCl), and by atmospheric oxygen and by hydrogen peroxide.

Oxidation by sodium hypochlorite

The oxidation of cellulose by sodium hypochlorite was extensively studied due to the need for preventing the overbleaching of cotton textiles. [4] It was established as early as in 1925 that both the rate of oxidation and the nature of the products depended on the pH of the solution. The rate of oxidation passed through a maximum near pH 7, suggesting that both undissociated hypochlorous acid and hypochlorite ions participated in the reaction.

In 1962, Lewin and Epstein [136, 137] made a major contribution in understanding the nature of the reaction. [4] The main attack occurs randomly in the accessible regions of the fibers. The carboxyl group content of the oxidized cottons increases, while the aldehyde and ketone group contents decrease with increasing pH from 5 to 10. There is considerable chain scission, and only 40% of the oxygen consumed is accounted for by the functional groups in the oxycellulose at all pH values. The rest is used in the further oxidation of soluble, short-chain material. One glycosidic bond was broken for every 26 atoms of oxygen consumed. The degradation and oxidation reactions were suggested to occur simultaneously. [137]

The oxidation of cellulose by hypochlorite is catalyzed by various substances such as the reduced forms of vat dyes and transition-metal ions in their lower valence states. [4] A common characteristic of these substances is that they are readily oxidized. Catalysis by transition-metal ions suggests that the hypochlorite oxidation of cellulose occurs by a free radical mechanism.

The hypothesis of a free radical mechanism was tested by Epstein and Lewin [4, 136, 137] They found that the reaction rate could be expressed by the equation (**Equation 5.1**):

$$-\mathrm{d}c/\mathrm{d}t = k[\mathrm{HClO}]^2[\mathrm{ClO}^-]^{1/2} \tag{5.1}$$

where c = the total concentration of hypochlorite. [4] The value of the function $[\mathrm{HClO}]^2[\mathrm{ClO}^-]^{1/2}$ can be calculated at any pH from the dissociation constant of hypochlorous acid. When the function is plotted against pH, it passed through a maximum at pH 6.9, which was also the pH at which the oxidation rate appeared maximal. Therefore, Epstein and Lewin suggested a free radical mechanism that fit **Equation 5.1**. Since this original suggestion, other equations have been proposed for which free radical mechanisms can be written, but differing from each other and from that derived from **Equation 5.1**. [138, 139, 140] A free radical mechanism at neutral pH was confirmed by electron paramagnetic resonance (EPR). [140] It involves essentially Cl˙ radicals.

Regenerated cellulose based materials, in particular sponges, with improved resistance to bleach (sodium hypochlorite) have been developed. [141] The materials contain protective agents having nitrogen atoms α to carbon atoms and being fixed to the cellulose lattice. [141]

Oxidation by atmospheric oxygen and by hydrogen peroxide

Environmental concerns have enhanced interest in chlorine-free bleaching systems. Oxygen-alkali systems are of particular interest since the by-products are environmentally benign. [142]

The oxidation of alkali cellulose with atmospheric oxygen has long been used for the reduction of the DP of wood pulp to a suitable level during the viscose process. [4] Alkaline peroxide is now an established bleaching agent for cellulosic textiles, and the oxygen bleaching of wood pulp is now widely used for removing residual lignin. [4, 127, 143, 144] While in the first process degradation of cellulose is the purpose of the operation, in the latter two every effort is made to preserve carbohydrates. Unfortunately, the use of oxygen as a bleaching agent degrades carbohydrates as well as lignin, resulting in a lower yield, and possibly lesser pulp strength. [142]

Gaseous oxygen and hydrogen peroxide appear to be very similar in their reaction with cellulose under alkaline conditions. [4] The reaction is non-specific and is accompanied by chain scission. The main functional groups formed are ketone groups, but aldehyde and carboxyl groups are also present. Strong catalytic effects of transition-metal ions have been observed. It seems to be established that Fe(II), Co(II), and Mn(II) are very effective in low concentrations, but actually retard oxidation in high concentrations. Catalysis is inhibited by magnesium ions. This has been useful in preventing excessive degradation of carbohydrates during the oxygen bleaching of wood pulp and the peroxide bleaching of textiles.

The oxidation of alkali cellulose by oxygen (autoxidation) involves the formation of peroxide radicals by the carbonyl-containing units in the original cellulose. [4, 143] The initial rate of oxidation is directly proportional to the number of carbonyl groups present. The function of the alkali is to form a dicarbonyl intermediate which is then converted into a hydroperoxide (ROOH) with oxygen. The hydroperoxide can decompose into an alkoxy radical (RO˙) or a peroxy radical (ROO˙). The latter has been detected by EPR and is generally accepted as the main species responsible for cellulose degradation by some form of radical chain reaction.

Several oxygen species are present during basic oxygen delignification, including dioxygen, hydroxyl radical, and superoxide anion radical ($^{\cdot}O_2^{-}$). [142, 145] The literature suggests that hydroxyl radicals are the main species responsible for carbohydrate degradation. In particular, hydroxyl radicals, generated photochemically from hydrogen peroxide in aqueous base, have been shown to cleave glycosidic bonds in methyl β-D-glucoside and methyl β-cellobioside as model compounds. [142] Hydroxyl radicals, generated from the same system, are also responsible for the cleavage of glycosidic bonds in 1,4-anhydrocellobitol and 2-methoxytetrahydropyran by substitution reactions between hydroxyl radicals and the model compounds at the anomeric carbon. [142]

Once the glycosidic bonds are broken, reducing carbohydrates undergo reactions, forming aldonic acids and aldoses. The results on model compounds suggest that hydroxyl radicals are responsible for random cleavage of carbohydrate chains by attack at anomeric linkages. Therefore, limiting hydroxyl radical formation is essential to hinder carbohydrate degradation during oxygen delignification.

5.3.4 Thermal degradation

It is well known that the application temperature range for cellulose is limited not by melting but by a rapid chemical decomposition. The thermal degradation (or pyrolysis) of cellulose occurs through dehydration, hydrolysis, oxidation, decarboxylation, and transglycosylation. [146] It can be represented by two main pathways (**Figure 5.19**), one occurring at lower temperatures (~200 °C), the other at higher temperatures (~300 °C). [147]

The low-temperature pathway results in decomposition of the glycosyl units of cellulose by evolution of water, carbon dioxide, and carbon monoxide and formation of char. [4] The concept of "active" cellulose was introduced as intermediate between cellulose and reaction products by several authors. [148-150] The thermal degradation of cellulose can be accelerated in the presence

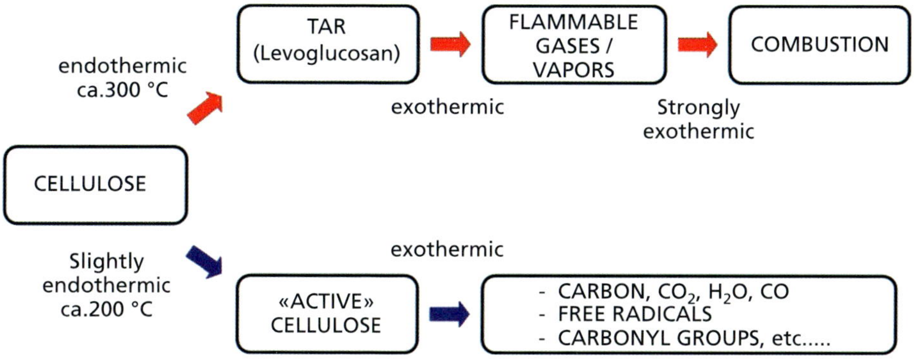

Figure 5.19 The low- and high- temperature pathways of thermal degradation of cellulose. [147]

of water, acids, and oxygen. [146, 151] As the temperature increases, the degree of polymerization decreases, free radicals appear and carbonyl, carboxyl, and hydroperoxide groups are formed. Thermal degradation rates increase as heating continues. Overall pyrolysis reactions are endothermic. During this low-temperature pathway (or char-forming pathway), exothermic reactions of char and volatiles with atmospheric oxygen are manifested as ***glowing combustion*** (a type of flameless combustion). [151]

By the high-temperature pathway (or tar forming pathway), a completely different type of reaction occurs that results in depolymerization of the molecule by cleavage of the glycosidic linkages to produce mainly the volatile levoglucosan (1,6-anhydro-β-D-glucopyranose) in the form of a tar fraction. [4, 146, 151]

This reaction, however, is accompanied by some decomposition of the original molecule and the anhydro sugar products. The flammable volatiles vigorously produced during the high-temperature pathway are available for ***flaming combustion***. [151] As temperature increases to around 450 °C, the production of volatile compounds is complete. [146, 151] The continuing weight loss above 450 °C is due to degradation of the remaining char, which is oxidized to CO_2, CO, and H_2O. This is referred to as ***afterglow***.

5.3.5 Mechanical degradation

Pioneering work was performed in the 1940s by Hess et al., [152, 153] who disintegrated native cellulose in a ball mill or a vibrating ball mill and obtained an X-ray amorphous material of low *DP*. [127] It was discovered later that a prolonged ball milling of dissolving pulp for one week resulted in a complete loss of the crystalline X-ray peaks and a reduction of the specific viscosity in Cuam.

As shown by ESR (electron spin resonance) spectroscopy, chain scission by mechanical treatment of cellulose proceeds by homolytic cleavage of bonds with formation of free radicals. [127, 154, 155] These mechano-induced free radicals were found to be capable of initiating graft copolymerization reactions. [156] Of these radicals, alkoxy radicals appeared to be the most stable at ambient temperature. [154]

Schleicher and Kunze [157] investigated the effect of dry grinding (e.g. in cutting mills) on structure and reactivity of pulps. [127] They observed that the DP of both cotton linters and dissolving pulps was reduced by dry grinding. Furthermore, grinding was found to result in a higher carbonyl content, a lower degree of order, a lower water retention value, and a different pore structure.

The effect of mechanical treatment and hydration on the order of cellulose was investigated by Wormald et al. [158] using ball milling followed by cyclic wetting and drying. [127] The results, monitored by ^{13}C-CP/MAS NMR spectroscopy, showed that a large part of the disorder induced by the mechanical treatment is reversible, and reordering upon hydration leads to the cellulose I initially present. The C4 signals corresponding to the reversibly disordered cellulose chains were observed in the amorphous region, together with signals from cellulose chains on the surface of ordered regions. Wormald et al. concluded that the behavior of cellulose upon milling was in accordance with the concept of ordered cellulose microfibrils containing amorphous cellulose mainly as surface layers and reversible lattice distortions.

Cellulose fibers suspended in liquid media are susceptible to mechanical degradation by shear forces. [127] The drop in *DP*, however, is much less pronounced than on dry grinding or milling of cellulose. The shearing of an aqueous suspension of cellulose fibers by beating results mainly in fibrillation. In relation with the effect on *DP*, Marx-Figini et al. [159] studied the degradation by high-speed stirring of cellulose in n-butanol suspensions using samples with a *DP* ranging from 300 to 7,500. Only the samples with a *DP* above 3,000 were degraded. In case of degradation, a limiting *DP* close to 3,000 was observed. The authors assumed that the bonds cleaved by mechanical forces were the weak links present in high *DP* cellulose.

5.3.6 Radiation degradation

Sunlight promotes the deterioration of cellulosic products, particularly cotton fabrics, and certain dyes or other additives such as pigments accelerate this process. [4] This phenomenon is generally referred to as ***phototendering***, i.e. the process by which organic fibers and textiles lose strength and flexibility as a result of exposure to sunlight. [160] It is the ultraviolet (UV) component of the sun's spectrum which affects fibers, causing chain degradation and, hence, loss of strength.

High-energy radiation (electron beam and γ-rays) are employed to sterilize a variety of cellulosic products used especially in food packaging and health care. [4] High-energy radiations have the ability to remove electrons from saturated molecules since the energy of the quanta ($h\nu$) is greater that the binding energy of an electron. Labile ionic and radical states can be thus produced. Such free radicals have been employed to initiate graft polymerization onto cellulose.

Interestingly, high-energy radiation can be also applied for the pretreatment of dissolving pulp in the viscose process. [161, 162] This pretreatment can significantly reduce the amounts of chemicals used and the production of hazardous pollutants.

Early systematic investigations revealed that the quantum energy of the radiation is the crucial factor in determining the mechanism and rate of degradation. [127]

IR, visible, and UV radiation

It is well known that only light that is absorbed can produce a chemical change. With highly purified cellulose there is no absorption in the IR and visible region and, hence, no detectable degradation by light of these wavelengths. [4, 127]

Pure cellulose is not directly ***photolyzed*** (or photodissociated) by UV wavelengths greater than 300 nm. [4, 127] By contrast, the 253.7 nm wavelength can induce photodegradation and produce free radicals in cellulose. [163] The absorption of this wavelength radiation has been ascribed both to carbonyl groups present as impurities and to the so-called acetal chromophore. [4, 164, 165] According to Parra et al., [166] molecular orbital calculations support assignment of the circular dichroism of unsubstituted saccharides and polysaccharides to oxygen-centered transitions of the acetal chromophore.

Direct degradation by 253.7 nm radiation appears to be independent of the presence of oxygen. [4] This type of degradation also appears to be greatly diminished by the presence of vat dyes. This effect of vat dyes contrasts with the way in which they accelerate degradation of cellulose by radiations of wavelengths greater than 300 nm. UV light initiates significant modifications to the cellulosic material, as shown by increases in solubility, reducing power, formation of carboxyl groups, and loss in *DP*. It was also noted that the main gaseous products evolved in the initial stages of irradiation are carbon dioxide, carbon monoxide, and hydrogen. [167] The mechanism of the direct photolytic degradation of cellulose remains uncertain, as well as the initial light absorption process. [4, 168, 169] If carbonyl groups are responsible for the absorption, the primary process would be C-C bond cleavage α to the carbonyl group.

Although light of wavelengths greater than 300 nm cannot cause direct degradation of cellulose, certain dyes and related compounds, lignin or metal ions, are able to absorb light of near-UV or visible wavelengths and induce degradation of cellulose. [4] These processes can be classified as photosensitized degradation of cellulose. The mechanism of phototendering presumably involves a dye-substrate initiation process. [4] The presence of lignin also promotes the formation of free radicals and chemical oxidation when wood cellulose is exposed to UV light. Lignin is suggested to take up the role of sensitizer. Metal ion sensitizers such as Fe^{2+}, Fe^{3+}, and Ce^{4+} have been used to produce free radicals which can initiate graft polymerization onto cellulose.

High-energy radiation

Degradation of cellulose by high-energy radiation is characterized by rapid decrease in *DP* due to random chain cleavage, formation of carbonyl groups and carboxyl groups, and relatively unchanged crystallinity and morphology up to high radiation doses. [4, 127] The effect of high-energy radiation on organic polymers is predominantly to produce ionization and electronic excitation. In cellulose, the primary process of bond cleavage due to ionization and excitation radiation is followed by energy dissipation/transfer over relatively long distances, and secondary radical reactions, often proceeding as a radical chain reaction. The key underlying processes are hydrogen abstraction leading to radical formation preferentially at C1 and C4 with subsequent cleavage of the glycosidic bond, and dehydration leading frequently to carbonyl group formation. In the presence of water, hydrogen atoms and hydroxyl radicals are formed as water radiolysis products, which participate in subsequent chain reactions. The intramolecular transfer of high energy in cellulose has been demonstrated to occur over distances equivalent to several cellobiose units. [170, 171, 172]

The use of high-energy radiation for pretreating dissolving wood pulp prior to its conversion to viscose has generated a tremendous interest amongst viscose producers. Typically, dissolving pulp is electron treated off-line before going to the viscose process. [161, 162, 173] This treatment can give rise to a significantly lower consumption of chemicals (in particular carbon disulfide, sodium hydroxide, and sulfuric acid) used for producing viscose, and a reduced environmental pollution. The key phase of the electron treatment is a controlled degradation of the dissolving pulp down to a *DP* conventionally obtained after the alkali-cellulose preripening step, thereby making this step unnecessary. [127]

5.3.7 Conclusion

Degradation of the cellulose macromolecules can be brought about with various kinds of energy input, i.e. chemical, thermal, mechanical, or radiation energy, and can proceed via many reaction routes. [174] In nearly all modes of cellulose degradation, except perhaps high-energy radiations, cellulose supramolecular and morphological structure plays a decisive role in determining the efficiency of a degradation process.

All the modes of non-biological cellulose degradation have the potential to be used as lignocellulosic biomass pretreatment technologies (see Chapter 7).

5.4 References

[1] E. A. Bayer, H. Chanzy, R. Lamed, and Y. Shoham, *Curr. Opin. Struct. Biol. 8*, 548, 1998

[2] B. Henrissat, T. T. Teeri, and R. A. J. Warren, *FEBS Lett. 425*, 352, 1998

[3] A. L. Demain, M. Newcomb, and J. H. D. Wu, *Microbiol. Mol. Biol. Rev. 69*, 124, 2005

[4] T.P. Nevell and S.H. Zeronian, eds., *Cellulose Chemistry and Its Applications*, Ellis Horwood Ltd, Chichester, 1985

[5] D. Klemm, B. Philipp, T. Heinze, U. Heinze, and W. Wagenknecht, *Comprehensive Cellulose Chemistry*, Vol. 1&2, Wiley-VCH, Weinheim, 1998

[6] N. Lo, H. Watanabe, and M. Sugimura, *Proc. R. Soc. Lond. B (Suppl.) 270*, S69, 2003

[7] G. Smant, J. P. W. G. Stokkermans, Y. Yan, J. M. De Boer, T. J. Baum, X. Wang, R. S. Hussey, F. J. Gommers, B. Henrissat, E. L. Davies, J. Helder, A. Schots, and J. Bakker, *Proc. Natl. Acad. Sci. USA. 95*, 4906, 1998

[8] A. Davison, and M. Blaxter, *Mol. Biol. Evol. 22*, 1273, 2005

[9] M. Schülein, Biochim. *Biophys. Acta 1543*, 239, 2000

[10] A. Grassick, P. G. Murray, R. Thompson, C. M. Collins, L. Byrnes, G. Birrane, T. M. Higgins, and M. G. Tuohy, *Eur. J. Biochem. 271*, 4495, 2004

[11] B. Henrissat, *Biochem. J. 280*, 309, 1991

[12] B Henrissat and A. Bairoch, *Biochem. J. 293*, 781, 1993

[13] B. Henrissat and G. J. Davies, *Plant Physiol. 124*, 1515, 2000

[14] The Carbohydrate-Active EnZymes database (CAZy): an expert resource for Glycogenomics in www.cazy.org

[15] CAZy Glycoside Hydrolase Family Server in www.cazy.org/fam/acc_GH.html

[16] G. J. Davies and B. Henrissat, Biochem. Soc. Trans. 30, 291, 2002 in http://www.biochemsoctrans.org/bst/030/bst0300291.htm

[17] J. Walshaw and A. Mills, Birkbeck College, University of London, *Protein folds*, 1995 in www.cryst.bbk.ac.uk/PPS95/course/8_folds/

[18] C. Jurgens, A. Strom, D. Wegener, S. Hettwer, M. Wilmanns, and R. Sterner, *Proc. Natl. Acad. Sci. USA 97*, 9925, 2000

[19] P. M. Alzari, H. Souchon, and R. Dominguez, *Structure, 4*, 265, 1996

[20] The_scop_authors/scop@mrc-lmb.cam.ac.uk, Glossary of terms used in the brief descriptions of protein folds, 2009 in http://scop.berkeley.edu/gloss.html

[21] B. Henrissat, P. M. Coutinho, and G. J. Davies, *Plant Mol. Biol. 47*, 55, 2001

[22] CAZy Carbohydrate-Binding Module Family Server in www.cazy.org/fam/acc_CBM.html

[23] V. Receveur, M. Czjzek, M. Schülein, P. Patine, and B. Henrissat, *J. Biol. Chem. 277*, 40887, 2002

[24] ExPASy (Expert Protein Analysis System) server of the Swiss Institute of Bioinformatics, Prosite Documentation PDOC00486, CBM1 Domain Signature and Profile in http://au.expasy.org/cgi-bin/prosite-search-ac?PDOC00486

[25] P. J. Simpson, H. Xie, D. N. Bolam, H. J. Gilbert, and M. P. Williamson, *J. Biol. Chem. 275*, 41137, 2000

[26] ExPASy, Prosite Documentation PDOC00485, CBM2 Domain Signature and Profile in http://kr.expasy.org/cgi-bin/nicedoc.pl?PS51173

[27] J. TORMO, R. LAMED, A. J. CHIRINO, E. MORAG, E. A. BAYER, Y. SHOHAM, and T. A. STEITZ, *Embo J. 15*, 5739, 1996

[28] ExPASy, Prosite Documentation PDOC51172, CBM3 Domain Profile in http://au.expasy.org/cgi-bin/prosite-search-ac?PDOC51172

[29] J. SAKON, D. IRWIN, D. B. WILSON, and P. A. KARPLUS, *Nat. Struct. Biol. 4*, 810, 1997

[30] D. IRWIN, D. H. SHIN, S. ZHANG, B. K. BARR, J. SAKON, P. A. KARPLUS, and D. B. WILSON, *J. Bacteriol. 180*, 1709, 1998

[31] S. Y. DING, E. A. BAYER, D. STEINER, Y. SHOHAM, and R. LAMED, *J. Bacteriol. 181*, 6720, 1999

[32] S. Y. DING, Q. XU, M. K. ALI, J. O. BAKER, E. A. BAYER, Y. BARAK, R. LAMED, J. SUGIYAMA, G. RUMBLES, and M. E. HIMMEL, *BioTechn. 41*, 435, 2006

[33] B. HENRISSAT, I. CALLEBAUT, S. FABREGA, P. LEHN, J. P. MORNON, and G. DAVIES, *Proc. Natl. Acad. Sci. USA. 92*, 7090, 1995

[34] R. H. DOI, A. KOSUGI, K. MURASHIMA, Y. TAMURA, and S. O. HAN, *J. Bacteriol. 185*, 5907, 2003

[35] C. BOISSET, H. CHANZY, B. HENRISSAT, R. LAMED, Y. SHOHAM, and E.A. BAYER, *Biochem. J. 340*, 829, 1999

[36] E. A. BAYER, E. SETTER, and R. LAMED, *J. Bacteriol. 163*, 552, 1985

[37] W. H. SCHWARZ, *Appl. Microbiol. Biotechnol. 56*, 634, ,2001

[38] D.L. BLUM, I.A. KATAEVA, X.L. LI, and L.G. LJUNGDAHL, *J. Bacteriol. 182*, 1346, 2000 in http://www.ncbi.nlm.nih.gov/pmc/articles/PMC94422/

[39] E. T. REESE, R. G. H. SIU, and H. S. LEVINSON, *J. Bacteriol. 59*, 485, 1950

[40] Y. H. P. ZHANG, and L. R. LYND, *Biotech. Bioeng. 88*, 797, 2004

[41] A. NUTT, Thesis, *Hydrolytic and oxidative mechanisms involved in cellulose degradation*, Acta Universitatis Upsaliensis Uppsala, 2006

[42] H. CHANZY, Personal communication ,2008

[43] J. KARLSSON, M. SIIKA-AHO, M. TENKANEN, and F. TJERNELD, *J. Biotechnol. 99*, 63, 2002

[44] C. DIVNE, J. STAHLBERG, T. REINIKAINEN, L. RUOHONEN, G. PETTERSSON, J. K. KNOWLES, T. T. TEERI, and T. A. JONES, *Science 265*, 524, 1994

[45] C. DIVNE, J. STAHLBERG, T. T. TEERI, and T. A. JONES, *J. Mol. Biol. 275*, 309, 1998

[46] M. Ike, Y. Ko, K. Yokoyama, J. Sumitani, T. Kawaguchi, W. Ogasawara, H. Okada, and Y. Morikawa, *J. Mol. Cat. B: Enzymatic 47*, 159, 2007

[47] T. Imai, C. Boisset, M. Samejima, K. Igarashi, and J. Sugiyama, *FEBS Lett. 432*, 113, 1998

[48] S.P.S. Chundawat, G.T. Beckham, M.E. Himmel, and B.E. Dale, *Ann. Rev. Chem. Biomol. Eng. 2*, 212, 2011 in http://www.annualreviews.org/doi/abs/10.1146/annurev-chembioeng-061010-114205

[49] H. Chanzy, and H. Henrissat, *FEBS Lett. 184*, 285, 1985

[50] J. Rouvinen, T. Bergfors, T. Teeri, J. K. Knowles, and T. A. Jones, *Science 249*, 380, 1990

[51] C. Boisset, C. Fraschini, M. Schülein, B. Henrissat, and H. Chanzy, Appl. *Environ. Microbiol. 66*, 1444, 2000

[52] J. Zou, G. J. Kleywegt, J. Stahlberg, H. Driguez, W. Nerinckx, M. Claeyssens, A. Koivula, T. T. Teeri, and T. A. Jones, *Structure, 7*, 1035, 1999

[53] M. Koyama, W. Helbert, T. Imai, J. Sugiyama, and B. Henrissat, *Proc. Natl. Acad. Sci. USA 94*, 9091, 1997

[54] J. Kraulis, G. M. Clore, M. Nilges, T. A. Jones, G. Petterson, J. Knowles, and A. M. Gronenborn, *Biochemistry 28*, 7241, 1989

[55] M. L. Mattinen, M. Linder, T. Drakenberg, and A. Annila, *Eur. J. Biochem. 256*, 279, 1998

[56] J. Lehtiö, J. Sugiyama, M. Gustavsson, L. Fransson, M. Linder, and T. T. Teeri, *Proc. Natl. Acad. Sci. USA 100*, 484, 2003

[57] J. Lehtiö, Thesis, *Functional studies and engineering of family 1 carbohydrate-binding modules,* Royal Institute of Technology, Stockholm 2001 in http://kth.diva-portal.org/smash/get/diva2:8989/FULLTEXT01

[58] T. Reinikainen, L. Ruohonen, T. Nevanen, L. Laaksonen, P. Kraulis, T. A. Jones, J. K. Knowles, and T. T. Teeri, *Proteins 14*, 475, 1992

[59] M. Schulein, J. *Biotechnol. 57*, 71, 1997

[60] CAZy Glycoside Hydrolase Family 6 in www.cazy.org/fam/GH6.html

[61] K. Igarashi, T. Ishida, C. Hori, and M. Samejima, *Appl. Environ. Microbiol. 74*, 5628, 2008

[62] A. Varrot, S. Hastrup, M. Schulein, and G. J. Davies, *Biochem. J. 337*, 297, 1999

[63] A. Varrot, M. Schulein, and G. J. Davies, *Biochemistry 38*, 8884, 1999

[64] RCSB PDB, Cellobiohydrolase II (Cel6A) from Humicola insolens in Complex with Glucose ans Cellotetraose in www.rcsb.org/pdb/explore.do?structureId=2bvw

[65] L. R. Lynd, P. J. Weimer, W. H. Van Zyl, and I. S. Pretorius, *Microbiol. Mol. Biol. Rev. 66*, 506, 2002

[66] H. Chanzy, and B. Henrissat, *Carbohydr. Polym. 3*, 161, 1983

[67] H. P. Fierobe, E. A. Bayer, C. Tardif, M. Czjzek, A. Mechaly, A. Belaich, R. Lamed, Y. Shoham, and J. P. Belaich, *J. Biol. Chem. 277*, 49621, 2002

[68] B. K. Barr, Y. L. Hsieh, B. Ganem, and D.B. Wilson, *Biochemistry 35*, 586, 1996

[69] E. A. Bayer, Weizmann Institute of Science, *The Cellulosome Complex* in www.weizmann.ac.il/Biological_Chemistry/scientist/Bayer

[70] E. A. Bayer, J. P. Belaich, Y. Shoham, and R. Lamed, *Annu. Rev. Microbiol. 58*, 521, 2004

[71] R. Lamed, E. Setter, R. Kenig, and E. A. Bayer, *Biotechnol. Bioeng. Symp. 13*, 163, 1983

[72] M. Hammel, H. P. Fierobe, M. Czjzek, V. Kurkal, J. C. Smith, E. A. Bayer, S. Finet, and V. Receveur-Bréchot, *J. Biol. Chem. 280*, 38562, 2005

[73] A. Mechaly, H. P. Fierobe, A. Belaich, J. P. Belaich, R. Lamed, Y. Shoham, and E. A. Bayer, *J. Biol. Chem. 276*, 9883, 2001

[74] Q. Xu, E. A. Bayer, M. Goldman, R. Kenig, Y. Shoham, and R. Lamed, *J. Bacteriol. 186*, 968, 2004

[75] A. Belaich, G. Parsiegla, L. Gal, C. Villard, R. Haser, and J. P. Belaich, *J. Bacteriol. 184*, 1378, 2002

[76] Y. Tamaru, S. Karita, A. Ibrahim, H. Chan, and R. H. Doi, *J. Bacteriol. 182*, 5906, 2000

[77] M. Kakiuchi, A. Isui, K. Suzuki, T. Fujino, E. Fujino, T. Kimura, S. Karita, K. Sakka, and K. Ohmiya, *J. Bacteriol. 180*, 4303, 1998

[78] F. Sabathe, A. Belaich, and P. Soucaille, *FEMS Microbiol. Lett. 217*, 15, 2002

[79] M. T. Rincon, T. Cepeljnik, J. C. Martin, R. Lamed, Y. Barak, E. A. Bayer, and H. J. Flint, *J. Bacteriol. 187*, 7569, 2005

[80] R. H. Doi, A. Kosugi, K. Murashima, Y. Tamaru, and S. O. Han, *J. Bacteriol. 185*, 5907, 2003

[81] Q. Xu, Y. Barak, R. Kenig, Y. Shoham, E. A. Bayer, and R. Lamed, *J. Bacteriol. 186*, 5782, 2004

[82] T. Fujino, P. Beguin, and J. P. AUBERT, *J. Bacteriol. 175*, 1891, 1993

[83] S. Y. Ding, E. A. Bayer, D. Steiner, Y. Shoham, and R. Lamed, *J. Bacteriol. 182*, 4915, 2000

[84] J. Tormo, R. Lamed, A. J. Chirino, E. Morag, E. A. Bayer, Y. Shoham, and T. A. Steitz, *Embo J. 15*, 5739, 1996

[85] RCSB PDB, Bacterial Type 3A Cellulose-Binding Domain in www.rcsb.org/pdb/explore.do?structureId=1NBC

[86] L. J. SHIMON, S. PAGES, A. BELAICH, J. P. BELAICH, E. A. BAYER, R. LAMED, Y. SHOHAM, and F. FROLOW, Acta Crystallogr., Sect.D 56, 1560, 2000

[87] RCSB PDB, Crystal Structure of a Family IIIA CBD from Clostridium cellulolyticum in www.rcsb.org/pdb/explore.do?structureId=1g43

[88] M. T. RINCON, J. C. MARTIN, V. AURILIA, S. I. MCCRAE, G. J. RUCKLIDGE, M. D. REID, E. A. BAYER, R. LAMED, and H. J. FLINT, J. Bacteriol. 186, 2576, 2004

[89] E. A. BAYER, Y. SHOHAM, and R. LAMED in http://141.150.157.177:8080/prokPUB/chaphtm/297/11_01.htm (no longer available site): Ref. 83; E. A. BAYER, Y. SHOHAM, and R. LAMED in *The Prokaryotes*, M. Dworkin, S. Falkow, E. Rosenberg, K. H. Schleifer, and E. Stackebrandt (eds.) Springer New York, 2006; E. A. B BAYER, Y. SHOHAM, and R. LAMED, *Cellulose-decomposing Bacteria and Their Enzyme Systems. The Prokaryotes: An Evolving Electronic Resource for the Microbiological Community*, 3rd ed., M. Dworkin et al. (eds.) Springer-Verlag, New York, 2001

[90] E. LEIBOVITZ, and P. BEGUIN, *J. Bacteriol. 178*, 3077, 1996

[91] L. J. W. SHIMON, E. A. BAYER, E. MORAG, R. LAMED, S. YARON, Y. SHOHAM, and F. FROLOW, *Structure 5, 381*, 1997

[92] G. A. TAVARES, P. BEGUIN, and P. M. ALZARI, *J. Mol. Biol. 273*, 701, 1997

[93] S. SPINELLI, H. P. FIEROBE, J. P. BELAICH, B. HENRISSAT, ans C. CAMBILLAU, *J. Mol. Biol. 304*, 189, 2000

[94] RCSB PDB, Cohesin Module from the Cellulosome of Clostridium cellulolyticum in www.rcsb.org/pdb/explore.do?structureId=1G1K

[95] A. L. CARVALHO, F. M. V. DIAS, J. A. M. PRATES, T. NAGY, H. J. GILBERT, G. J. DAVIES, L. M. A. FERREIRA, M. J. ROMÃO, and C. M. G. A. FONTES, *Proc. Natl. Acad. Sci. USA 100*, 13809, 2003

[96] B. L. LYTLE, B. F. VOLKMAN, W. M. WESTLER, M. P. HECKMAN, and J. H. HU, *J. Mol. Biol. 307*, 745, 2001

[97] RCSB PDB, Solution Structure of the Type I Dockerin Domain from the Clostridium thermocellum Cellulosome in www.rcsb.org/pdb/cgi/explore.cgi?pdbId=1daq

[98] M. LEMAIRE, H. OHAYON, P. GOUNON, T. FUJINO, and P. BEGUIN, *J. Bacteriol. 177*, 2451, 1995

[99] A. L. CARVALHO, V. M. PIRES, T. M. GLOSTER, T. P. TURKENBURG, J. A. PRATES, L. M. FERREIRA, M. J. ROMAO, G. J. DAVIES, C. M. FONTES, and H.J. GILBERT, *J. Mol. Biol. 349*, 909, 2005

[100] RCSB PDB, Structure of the Type II Cohesin from Clostridium thermocellum SdbA in www.rcsb.org/pdb/explore.do?structureId=2BM3

[101] I. Noach, R. Lamed, Q. Xu, S. Sosenheck, L. J. Shimon, E. A. Bayer, and F. Frolow, *Acta Crystallogr. D Biol. Crystallogr. 59*, 1670, 2003

[102] I. Noach, F. Frollow, H. Jacoby, S. Rosenheck, L. W. Shimon, R. Lamed, and E. A. Bayer, *J. Mol. Biol. 348*, 1, 2005

[103] J. J. Adams, G. Pal, Z. Jia, and S. P. Smith, *Proc. Natl. Acad. Sci. USA 103*, 305, 2006

[104] D. Mandelman, A. Belaich, J. P. Belaich, N. Aghajari, H. Driguez, and R. Haser, *J. Bacteriol. 185*, 4127, 2003

[105] H. Maamar, L. Abdou, C. Boileau, O. Valette, and C. Tardif, *J. Bacteriol. 188*, 2614, 2006

[106] S. Pages, O. Valette, L. Abdou, A. Belaich, and J. P Belaich, *J. Bacteriol. 185*, 4727, 2003

[107] C. Reverbel-Leroy, S. Pages, A. Belaich, J. P. Belaich, and C. Tardif, *J. Bacteriol. 179*, 46, 1997

[108] C. Gaudin, A. Belaich, S. Champ, and J. P. Belaich, *J. Bacteriol. 182*, 1910, 2000

[109] S. Pages, A. Belaich, H. P. Fierobe, C. Tardif, C. Gaudin, and J. P. Belaich, J. *Bacteriol. 181*, 1801, 1999

[110] S. Perret, A. Belaich, H. P. Fierobe, J. P. Belaich, and C. Tardif, *J. Bacteriol. 186*, 6544, 2004

[111] RCSB PDB, The *Crystal Structure of Endoglucanase 9G from Clostridium cellulolyticum* in http://pdbbeta.rcsb.org/pdb/explore.do?structureId=1g87

[112] G. Parsiegla, M. Juy, C. Reverbel-Leroy, C. Tardif, J. P. Belaich, H. Driguez, and R. Haser, *Embo J. 17*, 5551, 1998

[113] G. Parsiegla, C. Reverbel-Leroy, C. Tardif, J. P. Belaich, H. Driguez, and R. Haser, *Biochemistry 39*, 11238, 2000

[114] RCSB PDB, *Processive endocellulase CelF of Clostridium cellulolyticum* in www.rcsb.org/pdb/explore.do?structureId=1FCE

[115] T. W. Dror, A. Rolider, E. A. Bayer, R. Lamed, and Y. Shoham, *J. Bacteriol. 185*, 5109, 2003

[116] R. Lamed, The George S. Wise Faculty of Life Science, Tel Aviv University, Molecular Microbiology and Biotechnology, in www.tau.ac.il/lifesci/departments/biotech/members/lamed/lamed.html

[117] U. T. Gerngross, M. P. Romaniec, T. Kobayashi, N. S. Huskisson, and A. L. Demain, *Mol. Microbiol. 8*, 325, 1993

[118] T. W. Dror, A. Rolider, E. A. Bayer, R. Lamed, and Y. Shoham, *J. Bacteriol. 187*, 2261, 2005

[119] C. Boisset, H. Chanzy, B. Henrissat, R. Lamed, Y. Shoham, and E. A. Bayer, *Biochem. J. 340*, 829, 1999

[120] E.A. Bayer, E. Morag, and R. Lamed, *Trends Biotechnol. 12*, 378, 1994

[121] H. P. Fierobe, A. Mechaly, C. Tardif, A. Belaich, R. Lamed, Y. Shoham, J. P. Belaich, and E. A. Bayer, *J. Biol. Chem. 276*, 21257, 2001

[122] M.D. Sweeney and F. Xu, *Catalysts 2*, 244, 2012 in http://www.mdpi.com/2073-4344/2/2/244

[123] http://www.ravenbiofuels.com/technology/cellusosic.asp

[124] http://bluefireethanol.com/technology/

[125] B. Kamm, P.R. Gruber, and M. Kamm, eds., *Biorefineries - Industrial Processes and Products, Status Quo and Future Directions*, Vol. 1, Wiley-VCH, Weinheim, 2006 in www.wiley-vch.de/templates/pdf/3527310274_c01.pdf

[126] I. Pavasars, J. Hagberg, H. Boren, and B. Allard, *J. Polym. Environ. 11*, 39, 2003 in http://www.springerlink.com/content/x8343r5766g2632n/fulltext.pdf

[127] D. Klemm, B. Philipp, T. Heinze, U. Heinze, and W. Wagenknecht, *Comprehensive Cellulose Chemistry, Vol. 1: Fundamentals and Analytical Methods*, Wiley-VCH, Weinheim, 1998

[128] http://www1.eere.energy.gov/biomass/fy04/gasification_wet_biomass.pdf

[129] C. Calvert, *Pharm. J. Trans.14*, 364, 1850

[130] Q. Xiang, Y. Y. Lee, P. O. Petterson, and R. W. Torget in *Applied Biochemistry and Biotechnology*, Humana Press, Vol. 105-108, 505, 2003 in www.nrel.gov/docs/gen/fy03/34484.pdf

[131] Q. Xiang, J. S. Kim, and Y. Y. Lee in *Applied Biochemistry and Biotechnology*, Humana Press, Vol. 106, 337, 2003 in www.ingentaconnect.com/content/hum/abab/2003/00000106/F0030001/art00009

[132] http://cerig.efpg.inpg.fr/EFPG-days/2003/page05.htm

[133] http://cerig.efpg.inpg.fr/EFPG-days/2003/page62.htm

[134] http://cerig.efpg.inpg.fr/EFPG-days/2003/page39.htm

[135] http://smartech.gatech.edu/bitstream/1853/1846/1/tps-552.pdf

[136] J. A. Epstein and M. Lewin, *J. Polym. Sci. 58*, 991, 1962

[137] M. Lewin and J. A. Epstein, *J. Polym. Sci. 58*, 1023, 1962

[138] O. P. Singh, *Text. Dyer Printer 15*, 35, 1982

[139] T. P. Nevell and O. P.Singh, *Text. Research J. 56*, 270, 1986

[140] J. DUCOMMUN, SPONTEX internal report, 1997

[141] C. CHALVIN, J. DUCOMMUN, and J. L. WERTZ, US Patent 5852066, 1998

[142] D. F. GUAY, B. J.W. COLE, R. C. FORT, M. C. HAUSMAN, and J. M. GENCO, 2002 TAPPI Fall Conference & Trade Fair, in http://www.umche.maine.edu/Pilot/docs/2002tpc_reactions.pdf

[143] O. SAMUELSON, Pure & Appl. Chem. 55, 695, 1983 in www.iupac.org/publications/pac/1983/pdf/5504x0695.pdf

[144] A. J. RAGAUSKAS, L. A. LUCIA, and H. JAMEEL, *High selectivity oxygen delignification, Final technical report*, Georgia Institute of Technology, 2005 in www.osti.gov/energycitations/product.biblio.jsp?osti_id=859934 or in www.osti.gov/bridge/servlets/purl/859934-In9vDe/859934.PDF

[145] D. ZHANG, *Characterization and enhancement of carboxylic groups in softwood kraft pulps during oxygen delignification*, Ph.D. thesis, Georgia Institute of Technology, 2006 in http://etd.gatech.edu/theses/available/etd-08092006-142806/unrestricted/dongcheng_zhang-200612_phd.pdf.pdf

[146] www.fpl.fs.fed.us/documnts/pdf1989/levan89a.pdf

[147] T. HAKKARAINEN, E. MIKKOLA, B. ÖSTMAN, L. TSANTARIDIS, and H. BRUMER, P. PIISPANEN, InnoFireWood, State-of-the-art, 2005 in www.vtt.fi/virtual/innofirewood/stateoftheart/ifw-stateoftheart.pdf

[148] A. G. W. BRADBURY, Y. SAKAI, and F. SHAFIZADEH, *J. Appl. Polym. Sci. 23*, 3271, 1979

[149] R. CAPART, L. KHEZAMI, and A. K. BURNHAM, Assessment of Various Kinetic Models for the Pyrolysis of a Microgranular Cellulose, *Thermochimica Acta*, 2004 in www.llnl.gov/tid/lof/documents/pdf/304202.pdf

[150] S. KAYAMA, M. SUZUKI, and A. TSUTSUMI, *Reaction Mechanism of Tar Evolution in Biomass Steam Gasification for Hydrogen production*, WHEC 16, 2006 in www.cder.dz/a2h2/Medias/Download/Proc%20PDF/PARALLEL%20SESSIONS/%5BS08%5D%20Production%20-%20Biomass/15-06-06/355.pdf

[151] www.fpl.fs.fed.us/documnts/pdf2001/white01a.pdf

[152] K. HESS, H. KIESSIG, and J. GUNDERMANN, *Z. Physik. Chem. B49*, 64, 1941

[153] K. HESS, E. STEURER, and E. FROMM, *Kolloid-Z. 98, 148 and 290*, 1942

[154] D. N. S. HON, J. Appl. *Polym. Sci. 23*, 1487, 1979

[155] D. N. S. HON, and K. S. V. SRINIVASAN, *J. Appl. Polym. Sci. 28*, 1, 1983

[156] D. N. S. HON, J. Polym. Sci.: *Polym. Chem. Ed. 18*, 1857, 1980

[157] H. SCLEICHER and J. KUNZE in J. F. Kennedy, G. O. Phillips, and P. O. Williams, eds., *Cellulose and Cellulose Derivatives, Physico-chemical Aspects and Industrial Applications*, Woodhead Publishing Ltd, Abington, 1995

[158] P. Wormald, K. Wickholm, P. T. Larsson, and T. Iversen, *Cellulose 3*, 141, 1996

[159] M. Marx-Figini and R. V. Figini, *Angew. Makrom. Chem. 224*, 179, 1995

[160] http://en.wikipedia.org/wiki/Phototendering

[161] K. Fischer, W. Goldberg, and M. Wilke, *Lenzinger Ber. 59*, 32, 1985

[162] T. M. Stepanik, D. E. Ewing, and R. Whitehouse, *Rad. Phys. Chem. 57*, 377, 2000

[163] N. S. Hon, *J. Polym. Sci., Polym. Chem. Ed. 14*, 2497, 1976

[164] http://smartech.gatech.edu/dspace/bitstream/1853/1711/1/3134_002_1975.pdf

[165] A. Bos, *J. Appl. Polym. Sci. 16*, 2567, 1972

[166] A. Parra, and E. S. STEVENS, *Carbohydr. Polym. 41*, 111, 2000

[167] A. Bos, and A. S. Buchanan, *J. Polym. Sci., Polym. Chem. Ed. 11*, 833, 1973

[168] G. Buschle-Diller, and S. H. Zeronian, *J. Appl. Polym. Sci. 47*, 1319, 1993

[169] B. George, E. Suttie, A. Merlin, and X. Deglise, *Polym. Degrad. Stab. 88*, 268, 2005

[170] J. C. Arthur Jr., and T. Mares, *J. Appl. Polym. Sci. 9*, 2581, 1965

[171] J. C. Arthur Jr., D. J. Stanosis, T. Mares, and O. Hinojosa, *J. Appl. Polym Sci. 11*, 1129, 1967

[172] S. Singh, O. Hinojosa, and J. C. Arthur Jr., *J. Appl. Polym. Sci. 15*, 1129, 1971

[173] I. Graveson, H. Mobius, and D. A. Weightman, European Patent EP1390411, 2004

[174] M.C. Area and H. Cheradame, *BioResources* 6, 5307, 2011 in http://www.ncsu.edu/bioresources/BioRes_06/BioRes_06_4_5307_Area_Cheradame_Paper_Aging_Degrad_Review_2177.pdf

Chapter 6

Hemicelluloses and Lignin, Other Key Constituents of Biomass

6.1 Introduction

Knowing that the overall objective of a lignocellulosic biorefinery is to maximize the added value generated from the entire biomass, its economic viability is highly dependent on how ***hemicelluloses*** and ***lignin*** present in biomass are valorized.

Since hemicelluloses are present under various forms in all green plants and since lignin is present under various forms only in ***vascular plants***, [1] it may be useful to show first a general classification of green plants based on the Tree of Life Web Project [2] (**Figure 6.1**), and then a more detailed classification of some selected ***angiosperms*** which we shall meet in this book and particularly in this chapter (**Figure 6.2**). Angiosperms, or flowering plants, are the largest group of plants.

Green plants include all organisms known as green algae and land plants. Land plants, often called higher plants, have been divided into seed plants

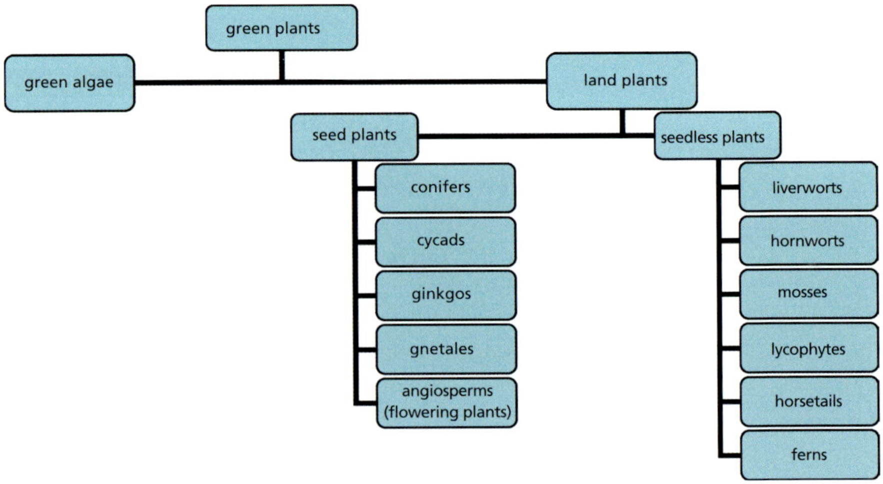

Figure 6.1 Classification of green plants. [3]

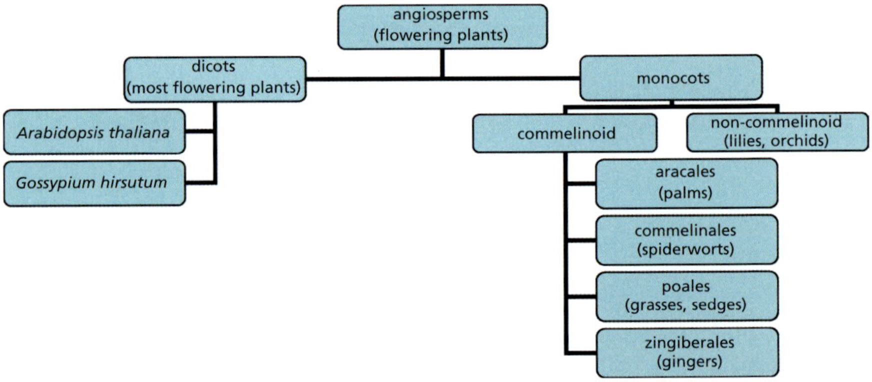

Figure 6.2 Classification of some selected angiosperms (flowering plants). [3]

and seedless plants. Seed plants include angiosperms and a group known as ***gymnosperms*** including conifers, cycads, ginkgos and gnetales. Seedless plants include liverworts, hornworts, mosses, lycophytes, horsetails and ferns. Vascular plants include seed plants as well as ferns, lycophytes, and horsetails amongst seedless plants.

Angiosperms include ***dicotyledons***, commonly called dicots and ***monocotyledons*** commonly called monocots. The dicots are represented by (1) the majority of the woody angiosperms such as oak, apple and cherry trees; (2) shrubs such as lilac; and (3) herbaceous plants such as potato, daisy, and *Arabidopsis thaliana* (mouse-ear cress also called thale-cress). Cotton herbs or shrubs (*Gossypium hirsutum*) also fall into this category. Monocots are divided into ***commelinoid*** (or commelinid) ***monocots*** and ***non-commelinoid*** (or non-commelinid) ***monocots***. Commelinoid monocots include the orders Arecales (palms), Commelinales (spiderwort), Poales (grasses and sedges) and Zingiberales (gingers). The grass family, or *Poaceae* (formerly Gramineae), which has been intensively studied by molecular biologists, includes cereal grasses as well as hay and pasture plants. Non-commelinoid monocots include lilies and orchids.

6.2 Hemicelluloses

6.2.1 Structure and occurrence

Overview

Hemicelluloses are one of the three major components of biomass, representing ~20–40% of biomass by weight. Hemicelluloses have been originally defined as soluble in aqueous alkali but insoluble in water. [4]

Hemicelluloses are defined structurally as plant cell wall polysaccharides that have a backbone of β-1,4-linked-D-***pyranosyl*** residues with an equatorial configuration at C1 and C4. [4, 5] From the backbone, short side chains of other sugars protrude. [4]

Hemicelluloses are a heterogeneous group of branched matrix polymers of relatively low molecular weight which are associated with cellulose and other polymers in plant cell walls. They bind tightly to the surface of the cellulose microfibrils and to each other by hydrogen bonding, and can be referred to as cross-linking ***glycans***. [7] In contrast with cellulose, hemicelluloses contain a variety of sugars including 5-carbon sugars such as xylose and arabinose, 6-carbon sugars such as glucose, mannose, galactose, galacturonic acid and glucuronic acid, and the 7-carbon sugar 4-*O*-methyl glucuronic acid [8] (**Figure 6.3**). Xylose is the second most abundant sugar in the biosphere.

Hemicelluloses can be divided into four major groups (**Figure 6.4**):

- ***xyloglucans*** (glucose backbone and xylose-containing branches); which are the predominant hemicelluloses in many primary cell walls;
- ***xylans***; including (1) ***glucuronoxylans*** (xylose backbone and glucuronic acid branches) found in the secondary cell walls of dicots and (2) ***arabinoxylans*** (xylose backbone and arabinose branches) and ***glucuronoarabinoxylans*** (xylose backbone and branches of arabinose and glucuronic acid) found in primary walls of commelinid monocots; xylans are considered to be the second most abundant biopolymer in the plant kingdom; [9]
- ***mannans*** (mannose backbone) and ***glucomannans*** (backbone of mannose and glucose), also including ***galactomannans*** (mannose backbone and galactose branches) found as major seed storage compounds in some

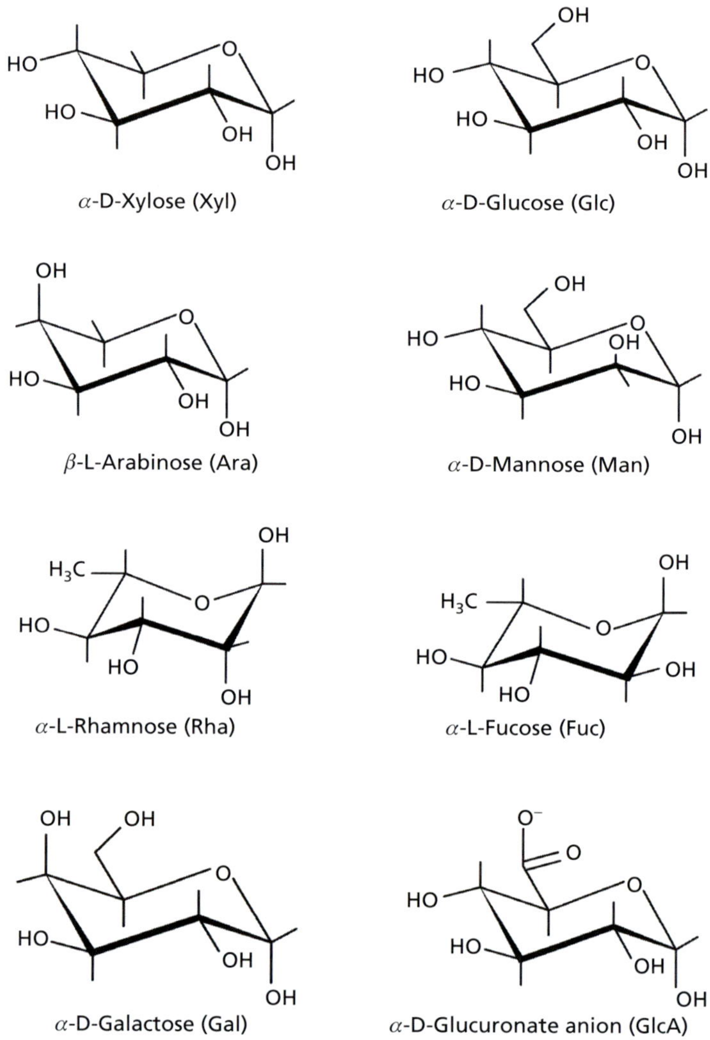

Figure 6.3 Common component sugars of hemicelluloses.

dicots such as legume species, and ***galactoglucomannans*** (backbone of mannose and glucose, and galactose branches) found in the secondary cell walls of gymnosperms; [10, 11]

- ***β-1,3;1,4-glucans*** (called also mixed-linkage glucans), which are highly prevalent in grasses.

The most important biological role of hemicelluloses is their contribution to strengthening the cell wall by interaction with cellulose and, in some walls with lignin. [5, 12]

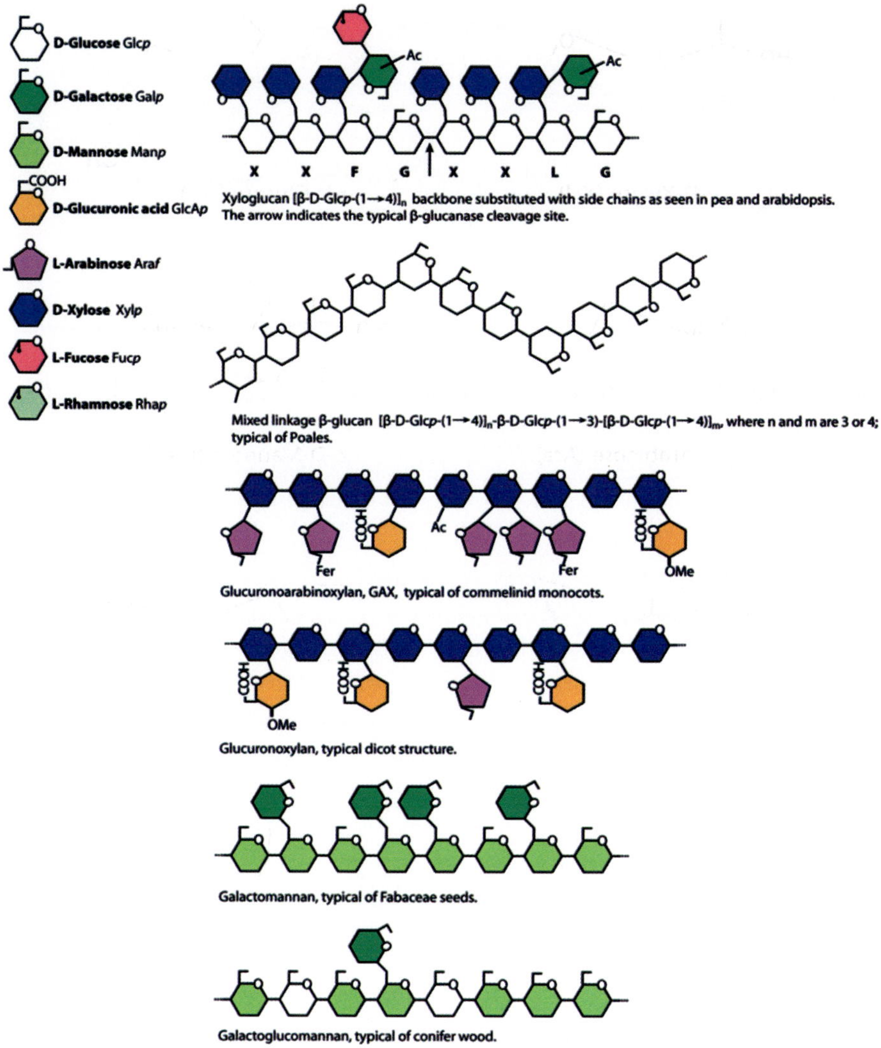

Figure 6.4 Schematic structures of hemicelluloses including xyloglucans, β-1,3;1,4-glucans, glucuronoarabinoxylans, glucuronoxylans, galactomannans and galactoglucomannans. The letters under the xyloglucan structure represents the symbols used for the most common side chains. "Fer" represents esterification with ferulic acid, which is characteristic of xylans in commelinid monocots. Fabaceae, or Leguminosae, are dicot plants, commonly known as the legume family, bean family or pea family. Reproduced by permission of H.V. SCHELLER and P. ULVSKOV, *Annu. Rev. Plant Biol.* 61, 263–89, 2010. [5]

Xyloglucans

Xyloglucans (XyGs) have been found in every land plant species, including mosses, but have not been found in Charophytes, green algae closely related to land plants (**Figure 6.5** for an overview of plant phylogeny). [5, 13]

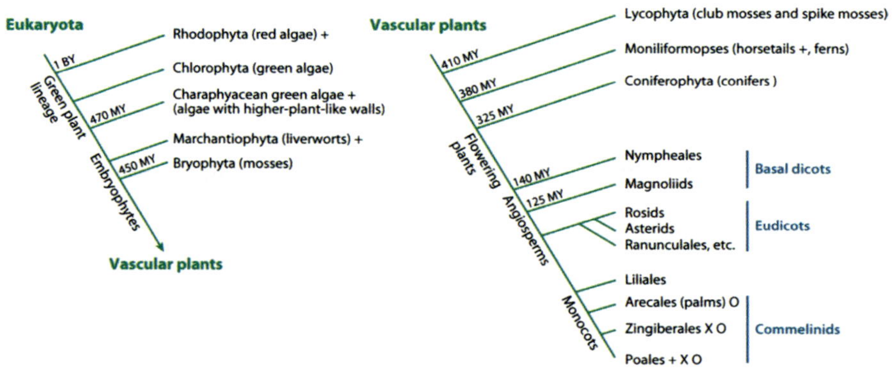

Figure 6.5 Simplified phylogenetic tree illustrating the main taxa mentioned in the chapter. Most taxa are not mentioned in the figure. Estimates of the age of some diversifications are indicated. Symbols designate hemicellulose patterns: +, β-1,3;1,4-glucans have been found in members of the group; X, taxa where glucuronoarabinoxylans are the major hemicelluloses in primary walls (as opposed to XyGs in other groups); O, taxa where xylans are esterified with ferulic acid. Reproduced by permission of H.V. SCHELLER and P. ULVSKOV, Annu. Rev. Plant Biol. 61, 263–89, 2010. [5]

XyGs are the predominant hemicellulosic polysaccharides in the primary walls of dicots and non-graminaceous monocots where they may account for about 20–25% of the dry weight of the primary wall (**Table 6.1**). [14, 15, 16]

They are polymers consisting of repetitive segments of four residues of a β-1,4-glucan backbone substituted on the first three positions with α-1,6 xylose. [17] The basic structure of XyGs is shown in **Figure 6.4**.

A special one-letter code is used to denote the different XyG side chains. [18, 19] According to this nomenclature, an unbranched glucosyl residue is designated G, while an α-xylose-1,6-β-glucose segment is designated X. The β-galactose-1,2-α-xylose-1,6-β-glucose segment and the

Table 6.1 Occurrence of hemicelluloses in plant cell walls.[5]

Amount of hemicelluloses (% w/w)	Primary cell wall			Secondary cell wall		
	Grasses	Conifers	Dicots	Grass	Conifers	Dicots
XyGs	1-5	10	20-25	Minor	~0[(1)]	Minor
Glucuronoxylans	~0	~0	~0	~0	~0	20-30
Glucuronoarabinoxylans	20-40	2	5	40-50	5-15	~0
(Gluco)mannans	2	~0	3-5	0-5	~0	2-5
Galactoglucomannans	~0	present	~0	~0	10-30	0-3
β- 1,3;1,4-glucans	2-15	0	0	Minor	0	0

(1) ~0, absent or minor

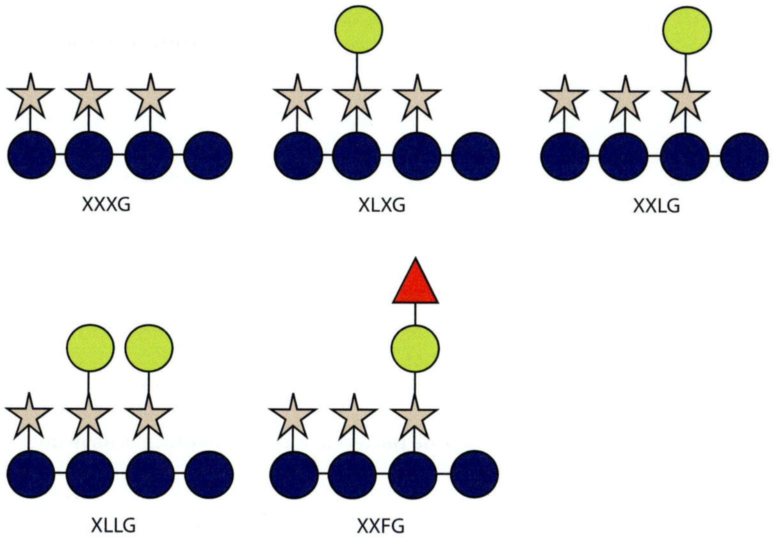

Figure 6.6 Structures of segments that build up XyGs in many land plants. The building blocks are composed of glucosyl units (blue circle) joined together by β-1,4- linkages, to which α-1,6-linked xylosyl units (star) may be added. Galactose (yellow circle) and fucose (red triangle) are incorporated into these structures via β-1,2- and α-1,2- linkages, respectively. © 2009 VON SCHANTZ et al.; licensee BioMed Central Ltd. [20]

α-arabinose- 1,2- α-xylose-1,6-β-glucose segment are designated L and S respectively. The letter F refers to the α-fucose-1,2-β-galactose-1,2-α-xylose-1,6-β-glucose segment of the fucosylated nonasaccharide (XXFG) commonly found in XyGs (**Figure 6.6**).

XyGs are classified as XXXG- or XXGG type based on the number of backbone residues that carry side chains with the XXXG type having three consecutive glucosyl residues with xylose attached and a fourth unbranched residue. [18]

The glucosidic bonds of the unbranched glucosyl residues in XXXG-type and XXGG-type XyGs are cleaved by many endo-1,4-β-glucanases. [4] Endoglucanase treatment of XXXG-type XyGs typically generates a series of well-defined oligosaccharides that have a tetrasaccharide backbone. In contrast, the types of oligosaccharide generated from XXGG-type XyGs depend on the substrate specificity of the enzyme and on the presence or absence of *O*-acetyl groups on the unbranched glucosyl residues.

The primary cell walls of a wide range of dicots, non-graminaceous monocots, and gymnosperms contain fucosylated XyGs with a XXXG-type structure. [4] The typical structures of the subunits in these XyGs are: XXFG, XXXG, and XLFG.

The branching pattern of XyGs is of both functional and taxonomic significance. [5] Less branched XyGs are less soluble and this may correlate with functional aspects in the families with the lowly substituted XyGs. The most profound difference in XyG structure is charge versus uncharged side chains, with the former found in mosses and liverworts whereas vascular plant XyGs are neutral.

Vascular plants can be categorized into whether XXGG or XXXG is the predominant, repeating XyG oligosaccharide. [5] The less substituted XXGG structure predominates in cell walls of commelinid monocots, which are also distinguished by the low content of XyGs in the primary walls, typically 1–5% compared with 20% in dicots. These figures refer to expanding cells. Oligosaccharides of the XXGG type are also found in ***solanaceous*** XyGs, which are further characterized by lacking fucosyl residues and featuring the arabinose-containing S chain. S is also found in XXXG-type units in olive. Furthermore, the F and S chains coexist in an XXXG-type XyG produced by *Nerium oleander*. [21] It was concluded that while the XyG subunit structures correlate with taxonomical groupings, the XyG structures in vascular plants do not mirror evolution in any simple way. [21, 5]

The structure of XyG has been shown to differ in a tissue specific manner. [4] Fucosyl residues are typically absent from seed XyGs but present on the XyGs in the vegetative portions of the same plant. XLFG subunits are more abundant in pea leaf XyG than in pea stem XyG. Changes have also been observed in the relative abundance of different subunits and the extent of their *O*-acetylation when suspension-cultured plant cells are grown in the presence or absence of light. It was also shown that different cells and different parts of the same wall are differentially labeled by an antibody that recognizes fucosylated XyGs. It has been concluded from these observations that the extent to which XyG is fucosylated in a tissue or cell is, at least in part, metabolically controlled. The role of fucosyl residues in XyGs is a subject of debate. XyGs from solanaceous plants are not fucosylated, showing that fucose is not essential for function.

Xylans

Xylans are a complex group of hemicellulosic polysaccharides with the common feature of a backbone of β-1,4-linked xylose residues (**Figure 6.4**). [5, 22] A common modification of xylans is substitution with glucuronosyl and 4-*O*-methyl glucuronosyl residues α-1,2-linked to the xylosyl residues of the backbone. [5] Such xylans are generally known as glucuronoxylans (GX) and are the main hemicellulosic component in the secondary walls of dicots. In commelinid monocots such as grasses, xylans are the major hemicellulosic component of the primary walls, constituting about 20% of the wall. These xylans usually contain many arabinose residues attached to the backbone and are known as arabinoxylans (AX) and glucuronoarabinoxylans (GAX). Branching patterns in xylans, like those in xyloglucans, depend on species and tissues. [22] Arabinofuranose substitutions in grass walls are mostly from O3 of the backbone residues, and xylose residues doubly substituted with arabinofuranose at both O2 and O3 are

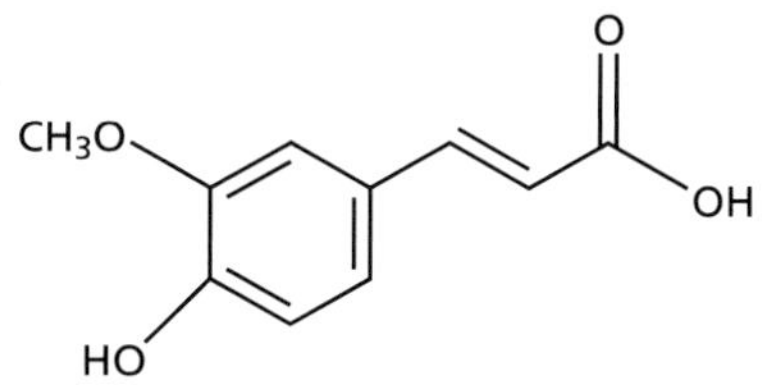

Figure 6.7 Structure of ferulic acid, a hydroxycinnamic acid.

common in grass ***endosperm***. Gymnosperm walls also contain arabinoxylans in relatively high amounts.

Unlike xyloglucans, xylans do not have a repeated structure. [5] Most xylans are acetylated to various degrees. [23] Acetyl groups are attached to O3 of xylose residues and to a lesser extent to O2.

It has been found that the reducing end of xylans from dicots and conifers contains the conserved oligosaccharide sequence xylose-1,4-xylose-1,3-rhamnose-1,2-galacturonic acid-1,4-xylose. [5, 24]

An important feature of grass xylans is the presence of ***ferulic acid*** esters attached to O5 of some of the arabinofuranosyl residues (**Figure 6.7**). [5]

Feruloylated glucuronoarabinoxylans are found throughout the commelinid monocots and not in other groups of seed plants. While glucuronoarabinoxylans are generally the most abundant noncellulosic polysaccharides in commelinid species, the basal commelinid order Arecales (palms) is an exception. The palm cell walls resemble the walls of the non-commelinid seed plants, i.e. with pectins as the major matrix polysaccharides, most likely with fucosylated xyloglucans, but in addition, small amounts of xylans bearing feruloyl esters. These walls, thus, appear to represent a transition form. Ferulate esters bound to cell walls have been also described in gymnosperms, but it is not clear to which polysaccharide they are bound.

Ferulate esters are important because they can be oxidatively crosslinked in a variety of ways. [5] Ferulate dimers are easily detected in grass walls and likely represent intra- and intermolecular linkages in and between glucoronoarabinoxylan molecules. Ferulate can also be crosslinked with lignin, and it is likely that glucuronoarabinoxylans and lignin become covalently crosslinked through these linkages. Crosslinking through ferulate esters is assumed to render the cell wall recalcitrant to digestion by microorganisms and herbivores. However, the ferulate esters make grass cell walls also recalcitrant to enzymatic saccharification prior to fermentation into biofuels.

Mannans and glucomannans

β-1,4-Polysaccharides containing mannose are widely distributed and are the main hemicelluloses in Charophytes (**Figure 6.4**). [5] The backbones may consist entirely of mannose, as in mannans and galactomannans, or with mannose and glucose in a non-repeating pattern as in glucomannans and galactoglucomannans. Mannans and glucomannans are often acetylated. Mannans have been much studied in their role as seed storage compounds, but they are found in variable amounts in all cell walls. In gymnosperms, galactoglucomannans are major components of secondary cell walls. Mannans appear to have been very abundant in early land plants and are still abundant in mosses and lycophytes. In seed plants, mannans and glucomannans are generally much less abundant and other hemicelluloses have likely replaced them.

Xylans and mannans compose most of hemicellulosic species found in trees with various proportions: Hardwoods (deciduous woods) contain 10–30% glucuronoxylan and 2–5% glucomannans, whereas softwoods (coniferous woods) contain 5–10% glucuronoxylan and 20% galactoglucomannans.

Mixed-linkage glucans

β-1,4-Glucans with interspaced single β-1,3-linkages are typical in grasses (**Figure 6.4**). [5] These mixed linkage glucans are dominated by cellotriosyl and cellotetrasyl units linked by β-1,3-linkages, but longer β-1,4-linked segments also occur. Mixed-linkage glucans in primary cell walls play a role in cell expansion and the amount is very growth-stage dependent.

The β-1,3;1,4-glucans have not been found in dicots but are found throughout Poales, including the most basal family. [5]

Related polysaccharides

Other complex polysaccharides that are present in most plant primary cell walls are ***pectins***. Pectins, which form a gel matrix around the cellulose and hemicelluloses scaffold, are heterogeneous and highly hydrated polysaccharides that contain α-1,4-linked-D-galacturonic acid residues (**Figure 6.8**). [25]

Three major pectic polysaccharides have been characterized. They are:

- ***Homogalacturonan*** (***HG***), which is a linear chain of α-1,4-linked-D-galacturonic acid residues where some of the carboxyl groups are methyl esterified or partially *O*-acetylated. It accounts for ~65% of plant pectins.

- Substituted galacturonans: ***apiogalacturonan*** (***AGA***), which has β-apiose and β-apiobiose residues attached to C2 or C3 of the galacturonan backbone, and ***xylogalacturonan*** (***XGA***), which has β-xylose residues attached to C3 of the galacturonan backbone.

- ***Rhamnogalacturonan I*** (***RG-I***) and ***rhamnogalacturonan II*** (***RG-II***): RG-I has a backbone of the repeating (1,4)-α-D-galacturonic acid-(1,2)-

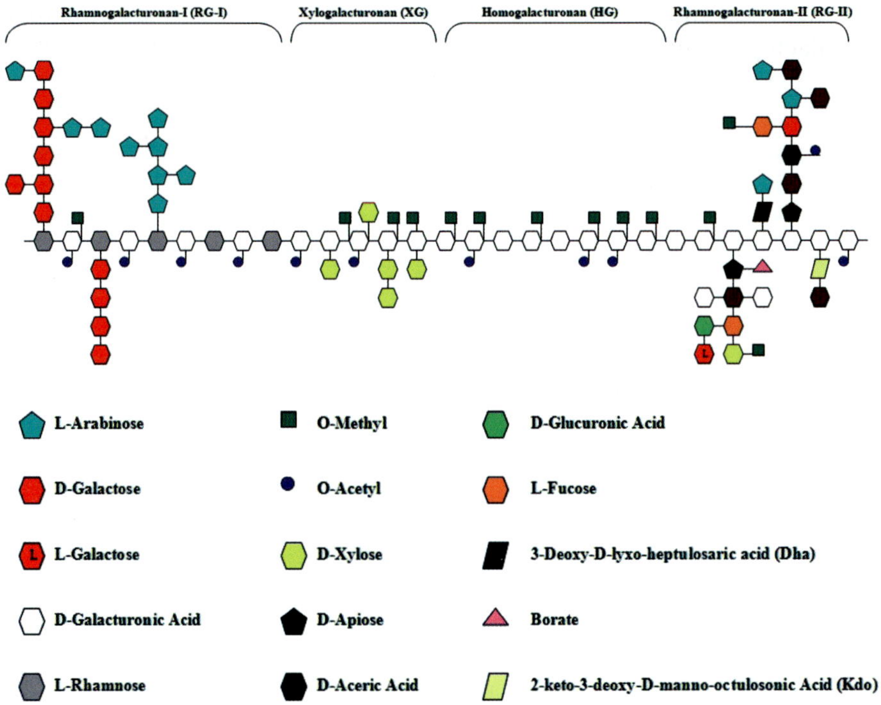

Figure 6.8 Pectic polysaccharides: Rhamnogalacturonan I and II (RG-I and RG-II), Xylogalacturonan (XG), and Homogalacturonan (HG). [26]

α-L-rhamnose disaccharide, in which 20–80% of the rhamnosyl residues can be substituted with (1,5)-α-L-arabinan, (1,4)-β-D-galactan and/or arabinogalactans. [27] From rhamnose, side chains of various neutral sugars branch off. The neutral sugars are mainly D-galactose and L-arabinose. RG-II has an α-1,4-linked homogalacturonan backbone that is substituted with four structurally different oligosaccharide side chains. [28, 29, 30] At least twelve different glycosyl residues are present in RG-II. RG-II is present in the primary cell wall of all higher plants predominantly in the form of a dimer that is cross-linked by a borate diester between two apiosyl residues. [30]

The formation of pectin networks by covalent and ionic linkages between the different pectins has been often assumed. [27, 31]

6.2.2 Biosynthesis

General

In contrast to cellulose that is synthesized at the cell surface in the plasma membrane, hemicelluloses, like most noncellulosic polysaccharides, are synthesized in the Golgi apparatus and delivered to the wall by secretory vesicles. [32, 33, 34, 35]

Hemicelluloses are synthesized from ***sugar nucleotides*** by ***glycosyltransferases*** (***GTs***) located in the Golgi membranes. [5, 36] Many GTs needed for biosynthesis of XyGs and mannans are known. In contrast, the biosynthesis of xylans and β-1,3;1,4-glucans remains elusive.

The majority of the enzymes involved in noncellulosic polysaccharide synthesis are integral membrane proteins. [32] Several Golgi proteins that are involved in glucomannan and XyG biosynthesis have been identified, including some ***glycan synthases*** that show sequence similarities to the ***cellulose synthases***, and several GTs that add sidechains to the polysaccharide backbones. [35] Recent progress in identifying the proteins needed for polysaccharide biosynthesis should lead to an improved understanding of these complex processes, and possibly to an ability to manipulate them in order to generate plants that have improved properties for human uses.

Xyloglucans

Considerable progress has been made in the identification of GTs involved in the biosynthesis of xyloglucans. [5, 37] The XyG fucosyltransferase (named FUT1; GT37) was one of the first cell wall biosynthetic enzymes to be identified. [38, 39] The α-1,2-FUTase directs addition of fucose residues to terminal galactose residues on XyG side chains. A XyG galactosyltransferase (named MUR3; GT47) that attaches a β-galactose residue specifically to the third xylose residue within the XXXG core structure has also been identified. [40] The galactosyltransferase has been shown to be highly specific for the third galactose in the repeating unit. Therefore, the second galactose in the XXXG must be incorporated by a different galactosyltransferase. A GT in the GT47 family has been shown to be responsible for the activity, but the final evidence has not yet been reported. [5, 41]

The xylose residues in XyGs are transferred from UDP-xylose onto β-1,4-glucan chains by α-1,6-xylosyltransferases (XT), which are retaining enzymes from family GT34. [5, 42] Two enzymes, XXT1 and XXT2, have been identified in *Arabidopsis* and shown to be involved in the synthesis of XyGs. [43]

The backbone of XyGs is apparently synthesized by members of ***cellulose synthase-like*** (***CSL***) proteins belonging to the CSLC family. [5] CSL proteins are a group of plant GTs that are predicted to synthesize β-1,4-linked polysaccharide backbones. [44]

Xylans

The biosynthesis of xylans will be presented under the headings: backbone, reducing end and side chains. [24, 45, 46, 47]

Backbone

None of the CSL proteins is apparently involved in xylan biosynthesis. [5] Instead, characterization of xylan-deficient mutants *irx9* (irregular xylem 9), *irx14*, *irx10*, *irx10-like* has indicated that the corresponding GTs belonging to families GT43 and GT47 are responsible for elongation of xylan backbone. [5] Unlike the CSL proteins with multiple transmembrane segments, the members of GT43 and GT47 are predicted to be type II membrane proteins with a single N-terminal membrane anchor.

In 2010, Lee et al. [24] studied the roles of the four members of the *Arabidopsis* family GT43 (IRX9, IRX14, I9H or IRX9-L (IRX9 ***homolog***) and I14H or IRX14-L (IRX14 homolog)) in glucuronoxylan biosynthesis. Their results have shown that all four GT43 members are involved in glucuronoxylan biosynthesis and have suggested that they form two functionally nonredundant groups essential for the normal elongation of glucuronoxylan backbone.

In 2010 also, Wu et al. [48] investigated the role of eight genes from *Arabidopsis*, *IRX9*, *IRX9-L*, *IRX10*, *IRX10-L*, *IRX14*, *IRX14-L*, *FRA8* and *F8H* (*FRA8 HOMOLOG*) in glucuronoxylan synthesis. Their results have revealed a set of four genes (*IRX9*, *IRX10*, *IRX14* and *FRA8*) that perform the main role in glucuronoxylan synthesis during vegetative development, and a second set of four genes (*IRX9-L*, *IRX10-L*, *IRX14-L* and *F8H*), homologs of the first set, that are able to partially substitute for their respective homologs and normally perform a minor function.

In 2012, Lee et al. [49] provided the first biochemical evidence that *Arabidopsis* IRX9 and IRX14, which have been shown to function non-redundantly, are xylosyltransferases (XylTs) that function cooperatively in the elongation of the xylan backbone. They further extended this finding in a work demonstrating that two poplar GT43 glycosyltransferases are xylan XylTs that act cooperatively in catalyzing the successive transfer of xylosyl residues during xylan backbone synthesis. [50]

Reducing end oligosaccharide

Treatment of xylans with xylanases allows the isolation of the reducing end oligosaccharide (see **Figure 6.7**). It has been shown that IRX7, IRX8 and PARVUS proteins are specifically involved in its biosynthesis. [5]

It is unclear how many GTs are required to make the oligosaccharide. There must be at least a xylosyltransferase specific for Rha-GalA-Xyl, a rhamnosyltransferase specific for GalA-Xyl, and a galacturonic acid transferase specific for the terminal xylose. The rhamnose-specific xylosyltransferase and the galacturonic acid-specific rhamnosyltransferase are expected to be inverting enzymes, while the xylose-specific galacturonic acid transferase would be a retaining enzyme. IRX8, a member of GT8, is the most obvious candidate for the galacturonic acid transferase. PARVUS, also a member of GT8, is presumably an α-xylosyltransferase transferring the reducing end xylose to a primer, which may be a lipid. IRX7, an inverting enzyme belonging to GT47, is a candidate for the rhamnose-specific xylosyltransferase.

The function of the reducing end oligosaccharide is not yet clearly established.

Side chains

The most important side chains of xylans should be formed by α-glucuronosyltransferases and α-arabinofuranosyltransferases. [5] Both activities have been detected in vitro, but the GTs responsible for the transfer have not been identified.

Mannans and glucomannans

A mannan synthase involved in making the backbone of galactomannans in guar was shown to be a member of the CSLA family. [5] Later, work has demonstrated that several CSLA members have mannan and glucomannan synthase activity, apparently being able to utilize both GDP-mannose and GDP-glucose as substrates.

In 2011, Verhertbruggen et al. [44] gathered evidence that three CSLD proteins are also mannan synthases. Furthermore, their work revealed a complex interaction between the three GTs and brought new evidence regarding the formation of non-cellulosic polysaccharides through multimeric complexes.

A galactosyltransferase involved in making galactomannans was identified in fenugreek (a plant in the family Fabaceae or Leguminosae) and was the first GT involved in synthesis of plant cell walls for which the activity of the pure enzyme was shown. [5] Whereas the fenugreek enzyme, which belongs to GT34, is involved in making seed galactomannans, it appears to have ***orthologs*** in dicots that have not been reported to have seed galactomannans, such as *Arabidopsis*.

Mixed-linkage glucans

The biosynthesis of β-1,3;1,4-glucans has been shown to involve CSLF and CSLH proteins. [5] The corresponding gene families are absent in *Arabidopsis* and poplar and present in rice and *Brachypodium* (a wild annual grass) consistent with a grass-specific occurrence.

6.2.3 Hemicelluloses in cell walls

The structural similarity between the hemicelluloses and cellulose facilitates a strong non-covalent association of the hemicelluloses with cellulose microfibrils. [51]

Cellulose-xyloglucan network

The cellulose-XyG network is believed to be the major load-bearing structure in the primary wall. [4] XyGs coat the surface of the cellulose microfibrils, limiting their aggregation and connecting them via tethers that directly or indirectly regulate the mechanical properties of the wall.

Xyloglucan endo-transglycosylase/hydrolase

The breaking and reformation of the glucosidic bonds in the xyloglucan backbone by ***xyloglucan endotransglycosylase/hydrolases*** (***XTHs***) has been proposed as a mechanism whereby the wall can expand and grow under turgor pressure without loosing its mechanical strength. [52-58] The *xyloglucan XTH* genes encode proteins that can potentially have two distinct catalytic activities, with radically different effects on XyG: XyG endotransglycosylase (XET) activity results in the nonhydrolytic cleavage and ligation of xyloglucan chains, whereas xyloglucan endohydrolase (XEH) activity yields irreversible chain shortening. [58] XET acts similarly to the glycoside hydrolases in the first reaction step but prefers a carbohydrate acceptor over a water molecule in the second reaction step. [59] This strong preference for transglycolysation instead of hydrolysis allows XET to perform its natural function (i.e. religation of the nascent donor end of one XyG molecule to the nonreducing, acceptor, end of another XyG molecule).

Since the last major review on the XTH family by Rose et al. [56] in 2002, the first three-dimensional structure of a plant XET [60] was published in 2004. Because of the primary importance of XET activity to plant cell wall structure and dynamics, this represent one of the most long sought after structures in plant cell wall enzymology. [61] Poplar XET (PttXET16A) exhibits a curved β-sandwich (β-jellyroll) that is typical for enzymes in the glycoside hydrolase family GH16 except for the long C-terminal linker that forms the only α-helix in the molecule and a short additional β-strand. [60] The extension of the acceptor binding site, together with variations in length and conformation of loops connecting the strands in the β-sandwich, produce an active site that is unique to the XET family of enzymes.

The three-dimensional structure of a plant XEH, TmNXG1 from nasturtium (*Tropaeolum majus*, cabbage family; NXG nasturtium xyloglucanase), was published in 2007. [62] Both the PttXET16 and TmNXG1 structures display the β-jellyroll fold common to all members of GH16, but with notable differences that reflect the specialization of these enzymes toward their highly branched substrate. [58] In addition to three loop variants, an N-terminal extension, a C-terminal truncation, three small loop truncations, and a shortening of the C-terminal α-helix were observed in the TmNXG1 enzymes relative to PttXET16.

The structures of XET and XEH provide the basis for detailed structure-function studies.

Action of XTHs

It has been suggested that the structure of the cellulose-xyloglucan network includes three XyG domains: [4]

- regions of XyGs that are bound directly to the cellulose surface;
- regions of XyGs that are not in direct contact with the cellulose and act as tethers between two neighboring microfibrils (these regions are the cross-linking tethers);
- regions of XyGs that are entrapped within the cellulose microfibrils.

It is likely that XTHs act on the tether domains rather than the bound or entrapped XyG domains. [4] Two main roles have been proposed for XTHs in growing cells: [56]

- Wall restructuring (**Figure 6.9**). XTHs may reversibly (XET) or irreversibly (XEH) loosen existing wall material, enabling cell expansion. [56]

- Integration into the cell wall. XTHs may also catalyze the integration of newly synthesized XyGs into the cell wall through XET activity. [56]

The binding of cellulose with XyGs is likely to be a complex topological process. [4] Conformational energy calculations suggest that for binding to cellulose to occur the xyloglucan backbone must adopt a "flat ribbon" conformation whose surface is complementary to that of cellulose. In solution XyGs are likely to adopt a "twisted" conformation that is not complementary to cellulose. It has been presumed that binding of xyloglucan to cellulose untwists the XyG backbone.

Cellulose-xylan network

Xylans are the main hemicelluloses that cross-link with cellulose in secondary cell walls of dicot plants. [24] The precise role of xylans in the secondary cell walls is not clear. [46] However, it has been suggested that xylans coat the cellulose microfibrils and may influence the helicoidal orientation of the microfibrils.

Xylans have been shown to be a major load-bearing structure in grasses. [5] The important role of xylans in strengthening secondary walls is very clear from analysis of xylan-deficient mutants. All these contain collapsed xylem vessels and have severely impacted growth and fertility.

Xylans contribute to the recalcitrance of secondary walls to enzymatic degradation. [63] They become a barrier to efficient cellulase function and must be removed for efficient cell wall saccharification. [64] Thus, reducing the xylan content of secondary walls and altering xylan structure, molecular weight, ease of extractability, and susceptibility to enzymatic degradation are key targets for the genetic improvement of plants.

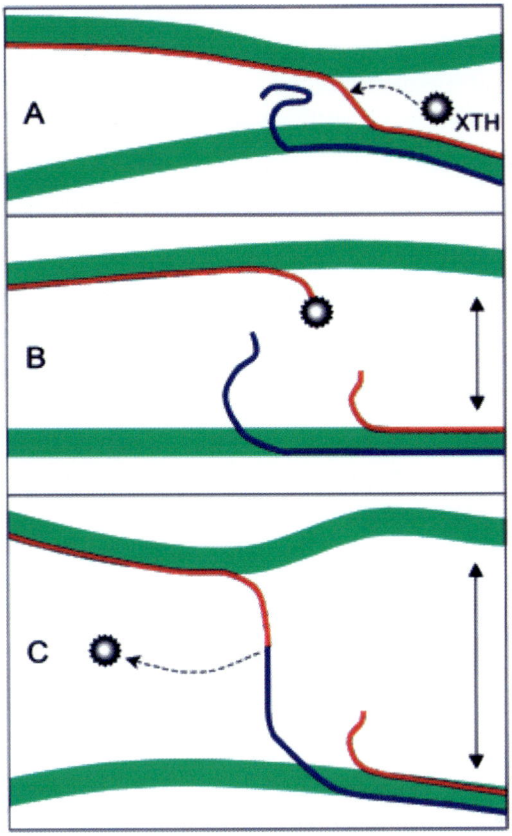

Figure 6.9 Polysaccharide-to-polysaccharide transglycosylation of xyloglucan in the cell wall showing restructuring-type transglycosylation. **(A and B)** XTH cleaves a xyloglucan chain, which cross-links two microfibrils. This xyloglucan is broken and a xyloglucan-XTH complex is formed. If the cell is turgid, the microfibrils can now move further apart, as indicated by the unbroken arrows. **(B and C)** The xyloglucan-XTH complex is now within reach of the non-reducing terminus of an adjacent xyloglucan chain. The latter acts as an acceptor substrate and a tether is thereby re-formed between the two microfibrils. Copyright © 2002, Oxford University Press. [56]

Mannans and glucomannans in cell walls

Mannans are widespread in *Arabidopsis* tissues, and may be of particular significance in both lignified and non-lignified thickened cell walls. [65] It is unclear what specific role the mannans have except for the role of galactomannans as seed storage compounds. [5] Mannans/glucomannans are highly conserved during plant evolution.

Mixed-linkage glucans

The exact role of β-1,3;1,4-glucans in the cell wall architecture of the Poaceae and other families with wall containing β-1,3;1,4-glucans remains unclear. [66]

Models of the architecture of these walls have been proposed in which there are two co-extensive, but independent, polymer neworks. Initially, β-1,3;1,4-glucans and GAX were considered to form crosslinks, but the hypothsis was abandoned. From a new study, it was concluded that much of the β-1,3;1,4-glucans are tightly associated with microfibrils, rather than forming a gel-like matrix.

6.2.4 Enzymatic hydrolysis

Hemicellulases are a diverse group of enzymes that hydrolyze hemicelluloses. [67, 68] These enzymes have many biotechnological applications and their structure/function relationships are the subject of intensive research. Recently, new structures of catalytic and non-catalytic domains of hemicellulases have been elucidated and reveal, together with biochemical studies, the principles of catalysis and specificity for these enzymes.

The degradation of cellulose and hemicelluloses is carried out by microorganisms that can be found either free in nature or as part of the digestive tract of higher animals. [67] The variable structure and organization of hemicelluloses require the concerted action of many enzymes for its complete degradation. Hemicellulases are typically modular proteins with a catalytic module and other functional modules. Essential non-catalytic modules are the ***carbohydrate-binding module*** (CBM), which facilitates the targeting of the enzymes to the insoluble polysaccharides, and the ***dockerin*** module that mediates the binding of the ***catalytic module*** via cohesin-dockerin interactions.

The catalytic modules of hemicellulases are either glycoside hydrolases (GHs) that hydrolyze glycosidic bonds, or carbohydrate esterases (CEs), which hydrolyze ester linkages of acetate or ferulic acid side groups. [67] Based on homology of their primary sequence, these catalytic modules can be classified into families. [52] The various chemical bonds hydrolyzed by hemicellulases and the classification of these enzymes are shown in **Figure 6.10** and **Table 6.2**.

The understanding of the structure/function relationship of hemicellulases has progressed considerably from the combination of high-resolution crystal structures and in-depth catalytic analysis. [67] These relationships are outlined below on the basis of the main substrate:

- β-Xylanases hydrolyze the β-1,4 linkage in the xylan backbone, yielding short xylooligomers. Most known xylanases belong to families GH 10 and 11.

- β-Xylosidases are exo-type glycosidases that hydrolyze short xylooligomers into single xylose units, and are found in families GH 3, 39, 43, 52, and 54.

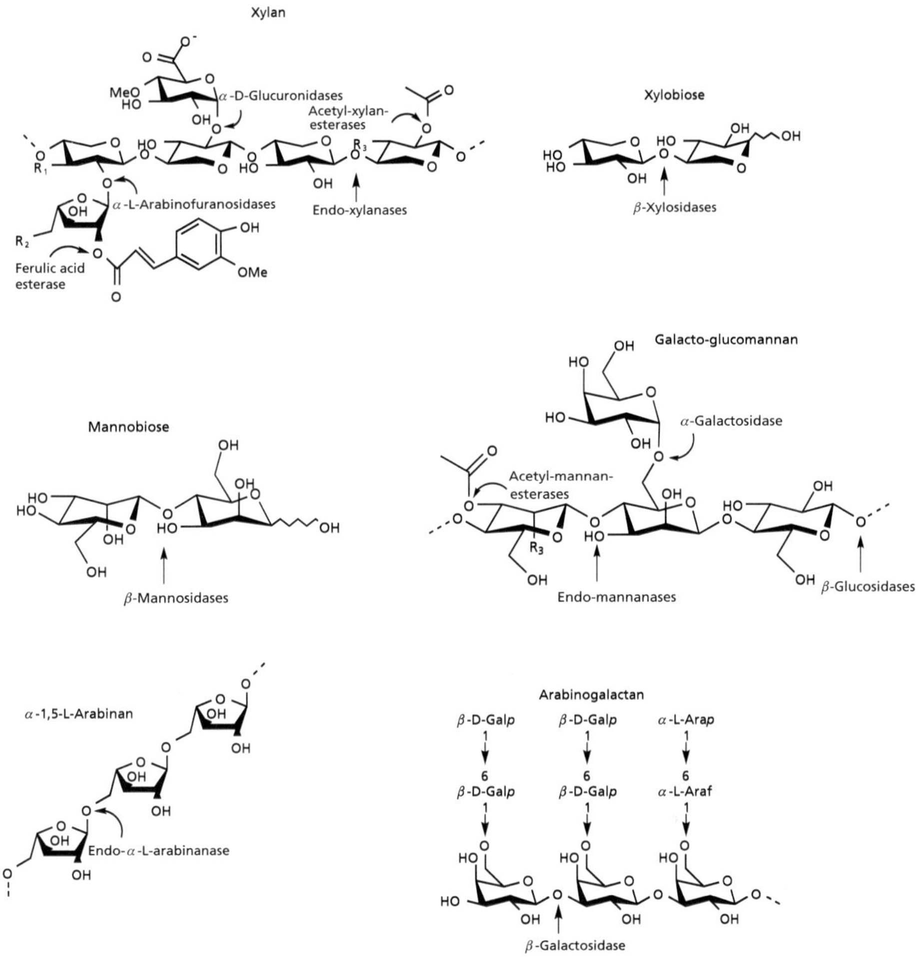

Figure 6.10 Basic structural components found in hemicelluloses and the hemicellulases responsible for their degradation. Me, methyl; R_1, OH or arabinofuranosyl group; R_2, OH or ferulic acid; R_3, OH or acetyl group. [67]

- β-Mannanases hydrolyze mannan-based hemicelluloses and liberate short β-1,4-mannooligomers, which can be further hydrolyzed to mannose by β-mannosidases. Most known mannanases belong to families GH 5 and 26.
- α-Arabinofuranosidases and α-arabinanases hydrolyze arabinofuranosyl-containing hemicelluloses and are found in families GH 3, 43, 51, 54, and 62.
- α-Glucuronidases cleave the α-1,2-glycosidic bond between 4-*O*-methyl-glucuronic acid and short xylooligomers. They are found exclusively in family GH 67.
- Acetyl xylan esterases hydrolyze the acetyl substitutions on xylose moieties.

Table 6.2 Typical hemicellulolytic enzymes. [67]

Enzyme	Substrate	Family[1)]
Endo-β-1,4-xylanase	β-1,4-Xylans	GH 5, 8, 10, 11, 43
Exo-β-1,4-xylosidase	β-1,4-Xylooligomers	GH 3, 39, 43, 52, 54
Endo-β-1,4-mannanase	β-1,4-Mannans	GH 5,26
Exo-β-1,4-mannosidase	β-1,4-Mannooligomers	GH 1, 2, 5
α-Arabinofuranosidase	β-Arabinofuranosyl-containing hemicellulose	GH 3, 43, 51, 54, 62
Endo-α-1,5-arabinanase	α-1,5-arabinan	GH 43
α-Glucuronidase	4-*O*-methyl-α-glucuronic acid-1,2-xylooligomers	GH 67
Acetyl xylan esterase	*O*-acetyl xylan	CE 1, 2, 3, 4, 5, 6, 7

1) GH, Glycoside hydrolase; CE, carbohydrate esterase

- Feruloyl esterases hydrolyze the ester bond between the arabinose substitutions and ferulic acid.

6.2.5 Properties and applications

General

As a valuable renewable resource, hemicelluloses are promised to a bright industrial future with products including food additives, plastics, cosmetics, and pharmaceuticals. [69]

Hemicelluloses in the cell wall have the primary role of interacting with other polymers to ensure the adequate properties of the wall. [5] However, they may also function as seed storage carbohydrates. Much of our knowledge of hemicelluloses comes from the study of seed polysaccharides rather than the polymers in vegetative tissues. The major storage hemicelluloses in seeds include:

- XyGs, which are important storage polymers in the seeds of nasturtium and tamarind. [70]
- Galactomannans, which are known from many economically important plants such as coconut, guar and locust bean. They are especially abundant in legumes, but also in tobacco and coffee.
- Glucomannans, which are present in the konjac plant but, in this case, the storage organ is a corm and not the seed.

- Arabinoxylans, which are present in seeds of dicots such as flax and psyllium and also in cereal endosperm.
- Mixed-linked glucans, which are present in cereal endosperm.

Seed storage hemicelluloses are used directly as products in the food industry, e.g. guar and locust bean gums (galactomannans), konjak gum (glucomannans) and tamarind gum (XyGs). [5] Furthermore, the hemicelluloses give important properties to many food and feed products. In the baking industry, the insoluble arabinoxylans affect baking quality. Mixed-linkage glucans and arabinoxylans are well-known antinutritional compounds in animal feed, and they can cause filtering and haze problems in the brewery industry due to their viscosity. To correct these problems, hemicellulases are added to feed and are used in the baking and brewery industries. Mixed-linkage glucans have a documented cholesterol-lowering effect in hypercholesterolemic humans and daily intake of mixed-linkage glucans is recommended by the U.S. Food and Drug Administration.

Hemicelluloses find also applications as sustainable films and coatings. [71] The targeted uses of hemicelluloses have primarily been packaging films for foodstuffs as well as biomedical applications. Oxygen permeability was typically comparable to values found for other biopolymer films, such as amylase and amylopectin. The modification of hemicelluloses to create more hydrophobic films reduced the water vapor permeability. However, modified hemicellulose coatings intended for food still exhibited water vapor permeabilities several magnitudes higher than those of other polymers currently used for this purpose. Research on hemicelluloses for biomedical applications has included biocompatible hydrogels and coatings with enhanced cell affinity. Several possibilities exist for chemically modifying hemicelluloses, and studies of films from modified hemicelluloses have identified other potential applications such as selective membranes.

Last but not least, hemicellulose hydrolysis results in sugars, largely pentoses, which can be chemically or biochemically converted into ethanol or a variety of chemicals.

Particular uses

Galactomannans

The largest market for guar galactomannans is in the food industry. Applications include: [72]

- baked goods: increases dough yield, gives greater resiliency, and improves texture and shelf life; in pastry fillings, it prevents "weeping" (syneresis) of the water in the filling, keeping the pastry crust crisp;
- dairy: thickens milk, yogurt, kefir, and liquid cheese products; helps maintain homogeneity and texture of ice crams and sherbets;
- meat: functions as lubricant and binder;

- dressings and sauces: improves the stability and appearance of salad dressings, barbecue sauces, relishes, ketchups, etc.;
- miscellaneous: dry soups, instant oatmeal, sweet desserts, canned fish in sauce, frozen food items, and animal feed.

Non-food industrial applications for guar galactomannans include: [72]

- textile industry: sizing, finishing and printing;
- paper industry: improved sheet formation;
- explosives industry, as waterproofing agent mixed with ammonium nitrate, nitroglycerin, etc.;
- pharmaceutical industry, as binder or as disintegrator in tablets;
- cosmetics and toiletries industries: thickener in toothpastes, conditioners in shampoos (usually in chemically modified version);

- oil and gas drilling, hydraulic fracturing;
- mining;
- hydroseeding: formation of seed bearing "guar tack."

Guar galactomannans are also known for their nutritional and medicinal effects. [72]

Locust bean galactomannans are used as a thickening agent and gelling agent in food technology. [73] They are soluble in hot water.

Glucomannans

Konjac glucomannans are water-soluble dietary fibers and can be used as a gelling agent, thickener, stabilizer, emulsifier, and film former. [74]

XyGs

Tamarind XyGs are having applications in paper, food, textile industry, etc. [75] In recent years, research has been initiated on the use of tamarind XyGs in pharmaceutical and cosmetic applications. Pharmaceutical applications include binder in tablet dosage form, ophthalmic drug delivery, sustained drug delivery, ocular drug delivery and controlled release of spheroids. A summary of hemicellulose applications is presented in **Table 6.3**.

6.3 Lignin

Lignin is a complex phenolic polymer that imparts strength, rigidity, and hydrophobicity to plant secondary cell walls. [76] Found in all vascular plants, particularly within the woody tissues, lignin makes up a substantial fraction of

Table 6.3 Applications of hemicelluloses.

Type	Application
Seed storage hemicelluloses	Food industry
Hemicelluloses in general	Films and coatings; biomedical
Hemicelluloses in general	Chemicals via hydrolysis and conversion of the sugars
Galactomannans	Food industry, textiles, paper, explosives, pharmaceuticals, cosmetics, oil and gas drilling, mining, hydroseeding
Glucomannans	Water-soluble dietary fibers
XyGs	Paper, food, textiles, pharmaceuticals, cosmetics

the total organic carbon in the biosphere. Lignin is mainly deposited in terminally differentiated cells of supportive and water-conducting tissues.

The evolution of lignin and other hydrophobic polymers some 400 million years ago was one of several critical developments in the colonization of terrestrial environments by vascular plants, allowing them to support increasingly large aerial structures, to supply these structures with water, and to protect them against desiccation. [76] Lignin is highly resistant to both mechanical disruption and enzymatic degradation. Lignin polymers make the cell wall rigid and impervious, allowing transport of water and nutrients through the vascular system and protecting plants against microbial invasion. [77] The recalcitrance of lignin to degradation has driven scientific inquiry into its structure and biosynthesis. Lignin biosynthesis and its potential manipulation have been the subject of increased interest as a result of the efforts towards developing cellulose-based bioenergy technologies. Lignin limits the yield and increases the cost of generating fermentable sugars from plant biomass. The elucidation of lignin biosynthesis, structure, regulation, and function has major implications for potential bioengineering strategies to improve feed, fiber, and bioenergy crops.

6.3.1 Structure and distribution of lignin

Lignin is the generic term for a large group of aromatic polymers resulting from the ***oxidative combinatorial coupling*** of 4-hydroxy***phenylpropanoids***. [78, 79, 80] It is the only naturally synthesized polymer with an aromatic backbone. [81] The first step in lignin polymerization involves the dehydrogenation of the lignin monomers by oxidative enzymes, such as peroxidases or laccases, with the formation of radicals. [78]

The three most abundant lignin monomers, or ***monolignols***, are *p*-coumaryl (4-hydroxycinnamyl), coniferyl (3-methoxy 4-hydroxycinnamyl) and sinapyl (3,5-dimethoxy 4-hydroxycinnamyl) alcohols (**Figure 6.11**). [76, 78]

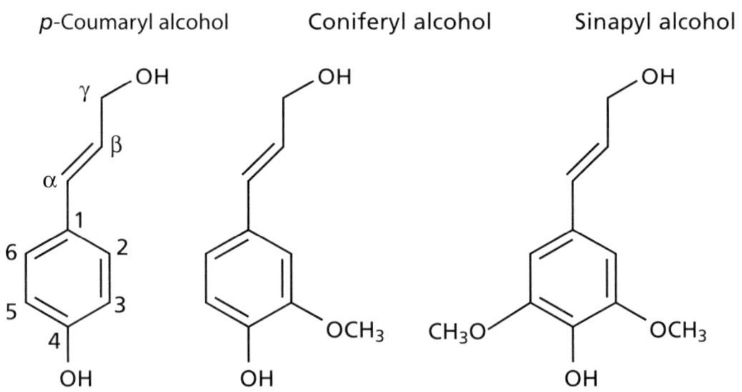

Figure 6.11 The three monolignols.

Upon incorporation into the lignin polymer, these monomers are referred to as *p*-hydroxyphenyl (H), ***guaiacyl*** (***G***) and ***syringyl*** (***S***) units, respectively.

Lignification typically occurs after the polysaccharides have been laid down in the cell wall. [77] The units in the lignin polymer are linked by a variety of chemical bonds that have different chemical properties. Lignin composition varies among plant species:

- with some exceptions, lignin from gymnosperms is composed of G-units only (with minor amounts of H-units);
- angiosperm dicot lignin is composed mainly of G- and S-units (with very little H);
- lignin from grasses (monocots) contains G- and S-units at comparable levels, and more H-units than dicots. [82]

The majority of secondary wall formation in most plants occurs in the water-conducting cells of the xylem, which tend to be enriched in G lignin, and in the structural fibers, which in angiosperms typically contain higher levels of S subunits. [76] Lignified cell walls are also found in sclereid cells, endodermal tissue in roots, and in specialized cells of anthers and some seed pods, where they are important for the dehydration-driven release of pollen and seeds, respectively. The wood of both gymnosperm and angiosperm trees is composed largely of the secondary cell walls of vascular tissue and associated fibers, and as such, of lignin. It is important to note that lignin biosynthesis is also of crucial importance in herbaceous (non-woody) species.

Less abundant units than the three main monolignols have been identified from a variety of species and may be incorporated into the polymer at varying levels. [78] Particularly common, less abundant subunits include ferulates (which form crosslinks between lignin and hemicelluloses), coniferaldehyde, and acylated monolignols containing acetate, *p*-coumarate, or *p*-hydroxybenzoate moieties. [76] Lignin composition also varies among cell types and can even be different in individual cell wall layers. Lignin composition is also influenced by environmental conditions; for example, lignin in compression wood is enriched in H-units. [77] Hence, both developmental and environmental parameters influence the composition and, thus, the structure of the lignin polymer.

Following their synthesis in the cytoplasm, lignin monomers are exported to the extracellular space (the apoplast) for incorporation into the lignin polymer. [76, 83] After reaching the apoplast, monolignols undergo single-electron oxidation by wall-bound laccases and/or peroxidases to form radical species. Then bond formation can occur between two such radicals (any combination of two monolignols, a monolignol and an elongating polymer end, or two polymer ends) in a process known as bimolecular radical coupling. Alternatively, activated monolignols can be involved in a radical transfer reaction.

The three different monolignols, and the subunits derived from them can freely form bond with one another. [76] Depending on the ring-substitution pattern of the molecules (H, G and S) and their location (free or attached to the polymer), multiple sites may be available for bond formation, resulting in a variety of potential coupling patterns, mainly including β-*O*-4, β-5 and β-β linkages as the central (β) carbon of the monolignol tail is most frequently involved (**Figure 6.12**).

6.3.2 Biosynthesis

Lignin biosynthetic pathway

During lignin deposition, monolignols are synthesized in the cytoplasm, translocated to the apoplast and polymerized into lignin (**Figure 6.13**). [83] Over the last two decades, the biosynthesis of monolignols has been a major focus of research on lignification. The monolignol biosynthetic pathway has been now relatively well elucidated, at least for angiosperms.

The monolignols are synthesized from the aromatic amino acid phenylalanine (Phe) through the general phenylpropanoid pathway supplying the precursor *p*-coumaroyl-CoA, and monolignol-specific pathways beginning at *p*-coumaroyl-CoA. [78] Phe is derived from the ***shikimate*** pathway. [84] It is synthesized within the ***plastids*** in plants. [84]

Shikimate biosynthetic pathway

Phenylalanine, like the other aromatic amino acids tyrosine (Tyr) and tryptophan (Trp), is derived from the shikimate biosynthetic pathway in the chloroplast (**Figure 6.14**). [84]

Monolignol biosynthetic pathway

The synthesis of the monolignols from Phe requires deamination, hydroxylation at one, two or three positions of the aromatic ring, methylation of one or

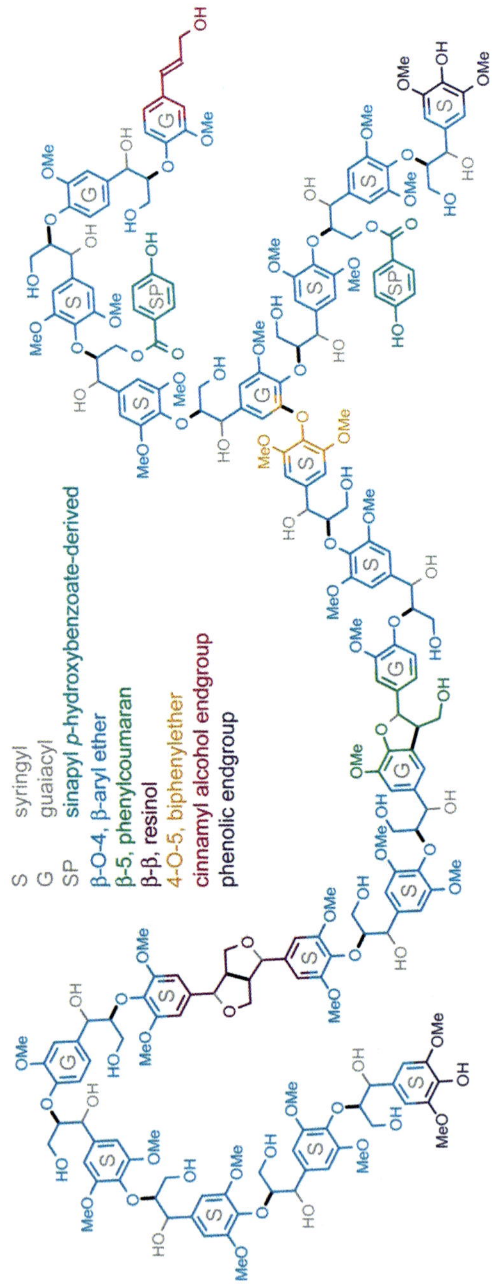

Figure 6.12 Lignin polymer from poplar, as predicted from NMR-based lignin analysis. Copyright © 2010 American Society of Plant Biologists. [78]

two of these hydroxyl groups, and two successive reductions of the monolignol side chain from a carboxylic acid first to an aldehyde and then to an alcohol (**Figure 6.15**). [76]

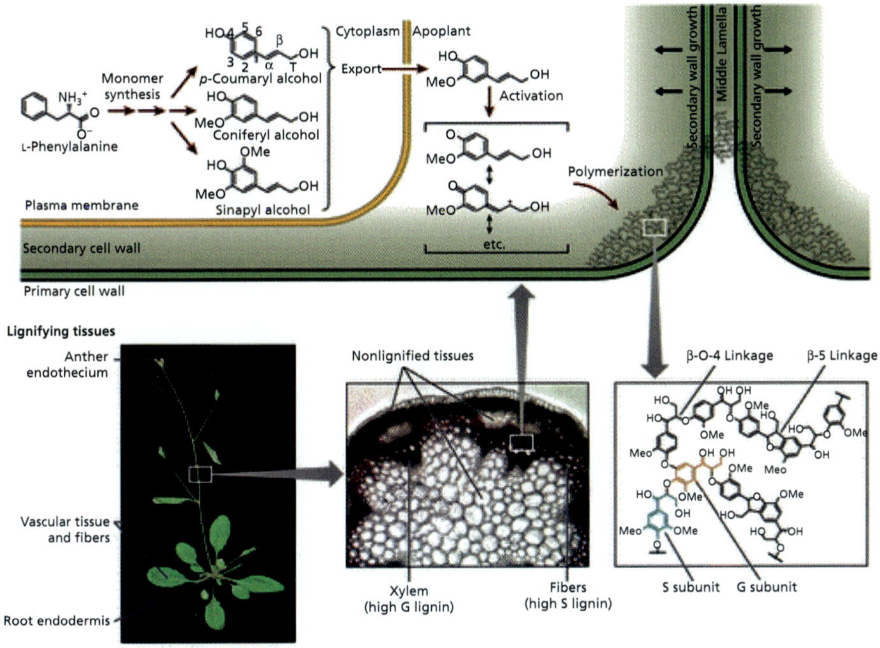

Figure 6.13 Overview of lignin biosynthesis, structure and distribution Gray arrows follow the progression from the whole organism to the tissue, cellular and molecular levels. Copyright © 2010 Annual Review, Inc. rather than (Reproduced by permission of N.D. BONAWITZ and C. CHAPPLE, *Annu. Rev. Genet.* 44, 337–63, 2000). [76]

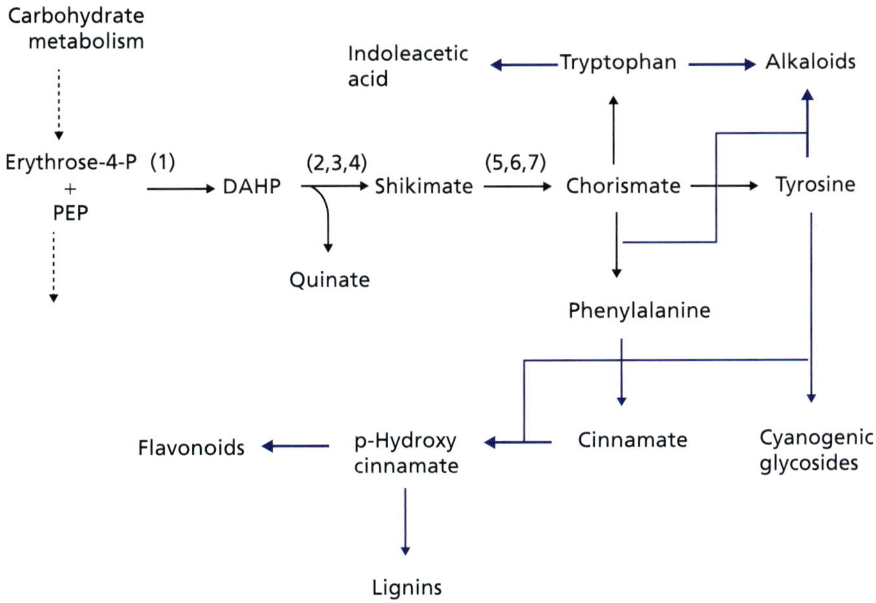

Figure 6.14 The shikimate pathway: biosynthesis of aromatic amino acids leading to lignin formation. The red numbers represent enzymes that catalyze those steps. [84]

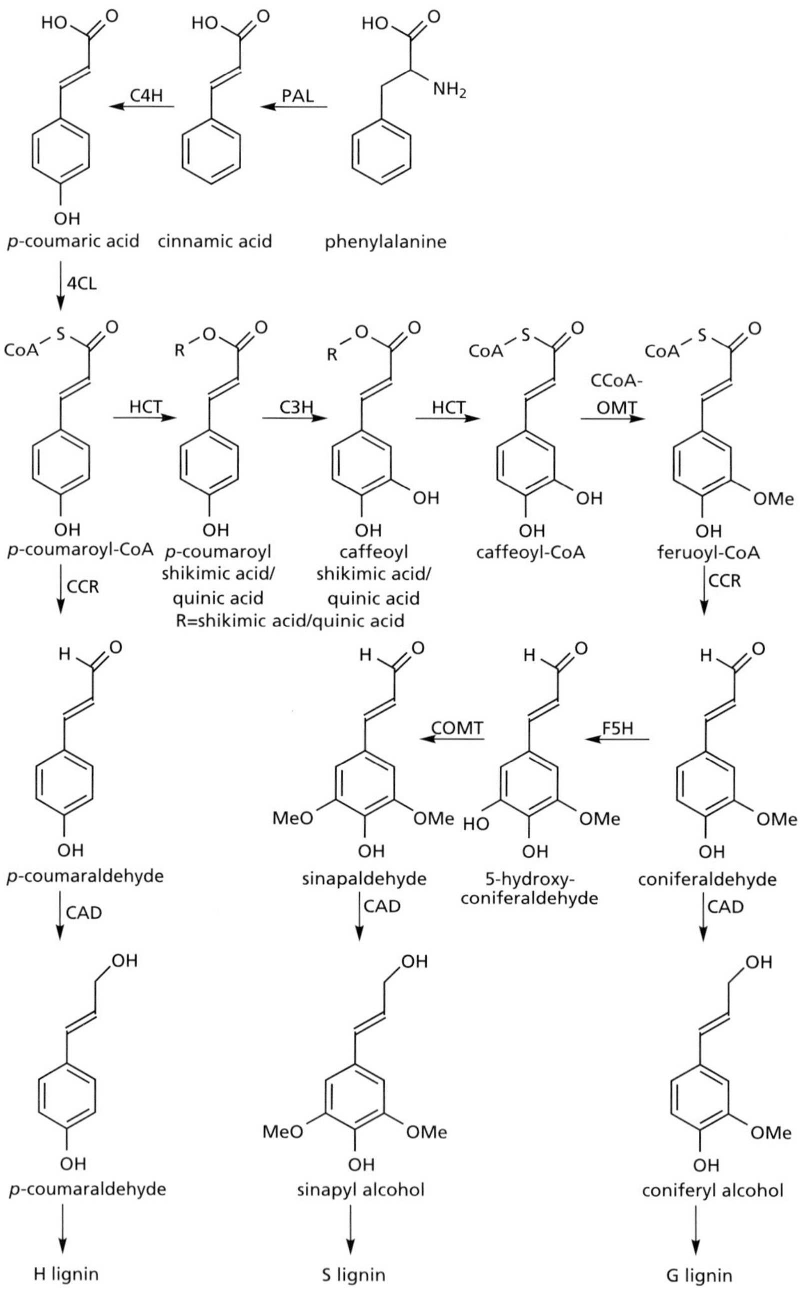

Figure 6.15 The main biosynthetic routes toward the monolignols *p*-coumaryl, coniferyl and sinapyl alcohols PAL, phenylalanine ammonia-lyase; C4H, cinnamate 4-hydroxylase, 4CL, 4-coumarate:CoA ligase, CoA, coenzyme A; C3H, *p*-coumarate 3-hydroxylase; HCT, *p*-hydroxycinnamoyl-CoA:quinate/shikimate *p*-hydroxycinnamoyltransferase; CCoAOMT, caffeoyl-CoA *O*-methyltransferase; CCR, cinnamoyl-CoA reductase; F5H, ferulate 5-hydroxylase; COMT, caffeic acid *O* methyltransferase; CAD, cinnamyl alcohol dehydrogenase. Copyright © 2010 American Society of Plant Biologists. [78]

These reactions are mediated by:

- a phenylalanine ammonia-lyase (PAL);
- three different cytochrome P450-dependent ***monooxygenases***: cinnamate 4-hydroxylase (C4H), *p*-coumarate 3-hydroxylase (C3H), and ferulate 5-hydroxylase (F5H);
- two methyltransferases: caffeoyl-CoA *O*-methyltransferase, (CCoAOMT; CoA, coenzyme A) and caffeic acid *O*-methyltransferase (COMT);
- two oxidoreductases: cinnamoyl-CoA reductase (CCR) and cinnamyl alcohol dehydrogenase (CAD). [76]

In addition to these eight enzymes, whose actions are apparent in the final pathways products, two more enzymes, 4-coumarate:CoA ligase (4CL) and *p*-hydroxycinnamoyl-CoA:quinate/shikimate *p*-hydroxycinnamoyltransferase (HCT), are required to synthesize pathway intermediates that serve as substrates for subsequent reactions. [76] 4CL is an ATP-dependent CoA ligase that catalyzes the synthesis of *p*-coumaroyl CoA. This *p*-coumaroyl CoA is then used as an acyl donor by acyltransferase HCT to synthesize *p*-coumaroyl shikimate, the substrate for 3-hydroxylation by C3H. Alternatively, *p*-coumaroyl CoA can serve as a substrate for chalcone synthase, diverting it to the flavonoid pathway, or CCR, committing it to H lignin biosynthesis. The relatively minor pathway leading to p-coumaryl alcohol (H lignin) requires only a subset of these enzymes (PAL, C4H, 4CL, CCR and CAD) whereas the synthesis of coniferyl alcohol (G lignin) requires these five plus HCT, C3H and CCoAOMT, and the synthesis of sinapyl alcohol (S lignin) requires all ten enzymes, including F5H and COMT. In many plants, some or all of these enzymes are found as multiple isoforms encoded by different genes. 267

Lignin pathway perturbations

Genes have been cloned for each of the steps of the biosynthetic pathway and the impact on lignin amount and composition has been studied in a variety of plants. [78] From these studies, it can be concluded that reducing the expression of each gene, except *F5H*, lowers lignin amount in the cell wall to varying extents, depending on the gene, the species, and the level of gene redundancy. Secondary walls with decreased lignin amounts are often characterized by collapsed vessels.

In addition to lignin content, H/G/S composition also appears to be rather flexible. [78] Silencing of *HCT* or *C3H* leads to lignin with H-units levels as high as 100 % of the total thioacidolysis (an ether-cleaving analysis method for lignins) monomers. Down-regulation of *F5H* or *COMT* strongly reduces S-unit content, whereas up-regulation of *F5H* increases the S-unit content. Furthermore, down-regulation of *CAD* increases the incorporation of cinnamaldehydes into the polymer. Shifts in lignin composition are also seen when *COMT* is down-regulated; in these lignins, 5-hydroxyconiferyl alcohol, derived from the COMT substrate, is incorporated, although it is below the detection limit in wild-type lignin. Shifts in G and S levels, as well as shifts toward more aldehydes, generally have only minor effects on plant development. Because the relative ratio of the different monomers determines the frequency of the different bonds in the polymer, all these compositional shifts have repercussions on the structure of the polymer and may thus alter the cell wall properties.

The effects of lignin pathway perturbations often go beyond alterations in lignin amount, composition, and cell wall structure. [78] Perturbing individual steps of the lignin biosynthetic pathway affects the expression level of other lignin pathway genes and also the expression of genes involved in many other biological processes. These wider effects are also reflected at the metabolic level.

Regulation of lignification

Lignification is coordinated with the production of cellulose and hemicelluloses during secondary wall formation. [76] The regulation of this process is overseen by a complex network of tissue-specific ***transcription factors*** that respond to as-yet-unidentified developmental signals triggering an expression cascade of other transcription factors and, ultimately, the enzymes and other proteins responsible for synthesizing the cell wall itself. Several different genetic approaches have contributed to the identification of this regulatory network.

The regulatory cascade explains why several of these transcription factors lead to enhanced or reduced lignification when misexpressed in plants while they do not directly regulate the lignin biosynthetic genes by binding to their promoters. [78] Engineering the expression of transcription factors has the potential to alter lignification.

Transport and polymerization

Transport

After their biosynthesis, the monolignols are translocated to the cell wall by a mechanism that is still unknown. [78] In one model, the monolignols are translocated over the plasma membrane through their 4-*O*-glucosylated form, ***coniferin*** and ***syringin***, that are deglucosylated upon their arrival by glucosidases located in the cell wall. In another model, monolignols are transported to the plasma membrane by Golgi-derived vesicles. At present, there is no convincing support for a role for monolignol glucosidases or Golgi-derived vesicles in the transport of monolignols to the cell wall.

Oxidative radicalization and dimerization

Lignin polymerization occurs via oxidative radicalization of phenols, followed by combinatorial radical coupling. [77, 78] In the first step, the monolignol phenol is oxidized, i.e. dehydrogenated. The resulting phenolic radical is relatively stable due to delocalization of the unpaired electron in the conjugated system (**Figure 6.16**).

Subsequently, two monomer radicals may couple to form a dimer, establishing a covalent bond between both subunits (**Figures 6.16** and **6.17**). Monolignol radicals favor coupling at their β positions, resulting primarily in the formation of a β-*O*-4 ether linkage (β-aryl ether), a β-5 linked structure

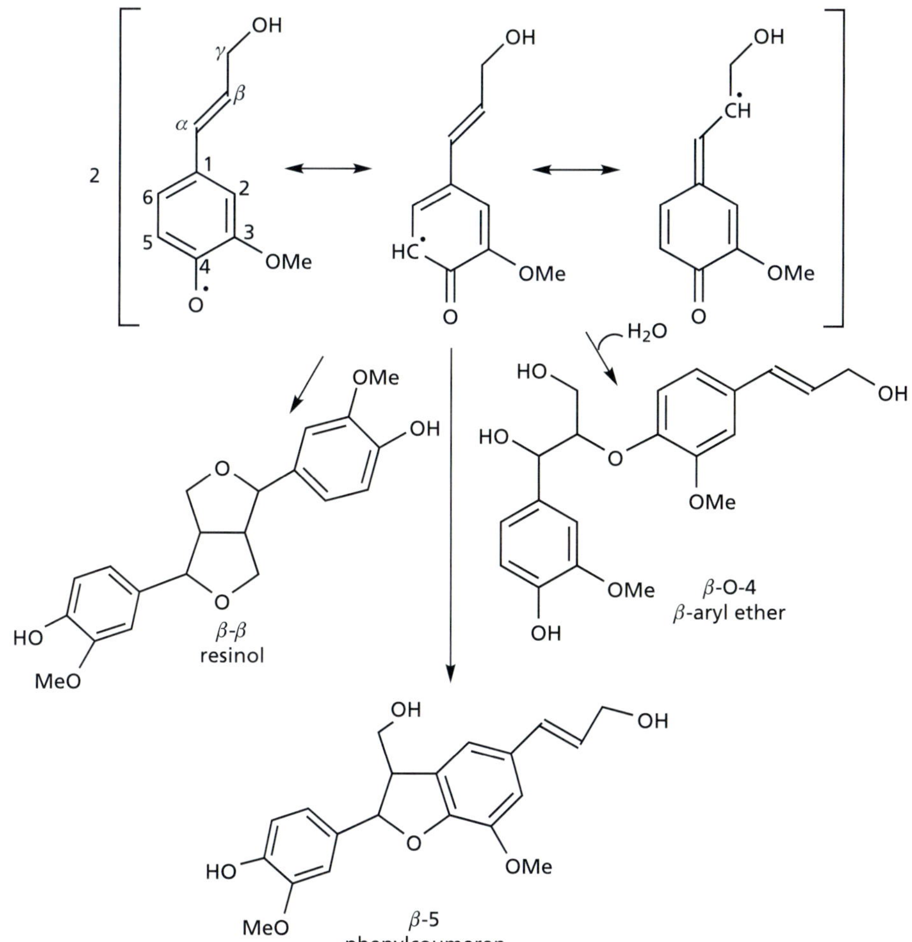

Figure 6.16 Dimerization of two dehydrogenated coniferyl monomers. Resonance forms with the unpaired electron localized at C1 or C3 are not shown because radical coupling reactions do not occur at these positions. [78]

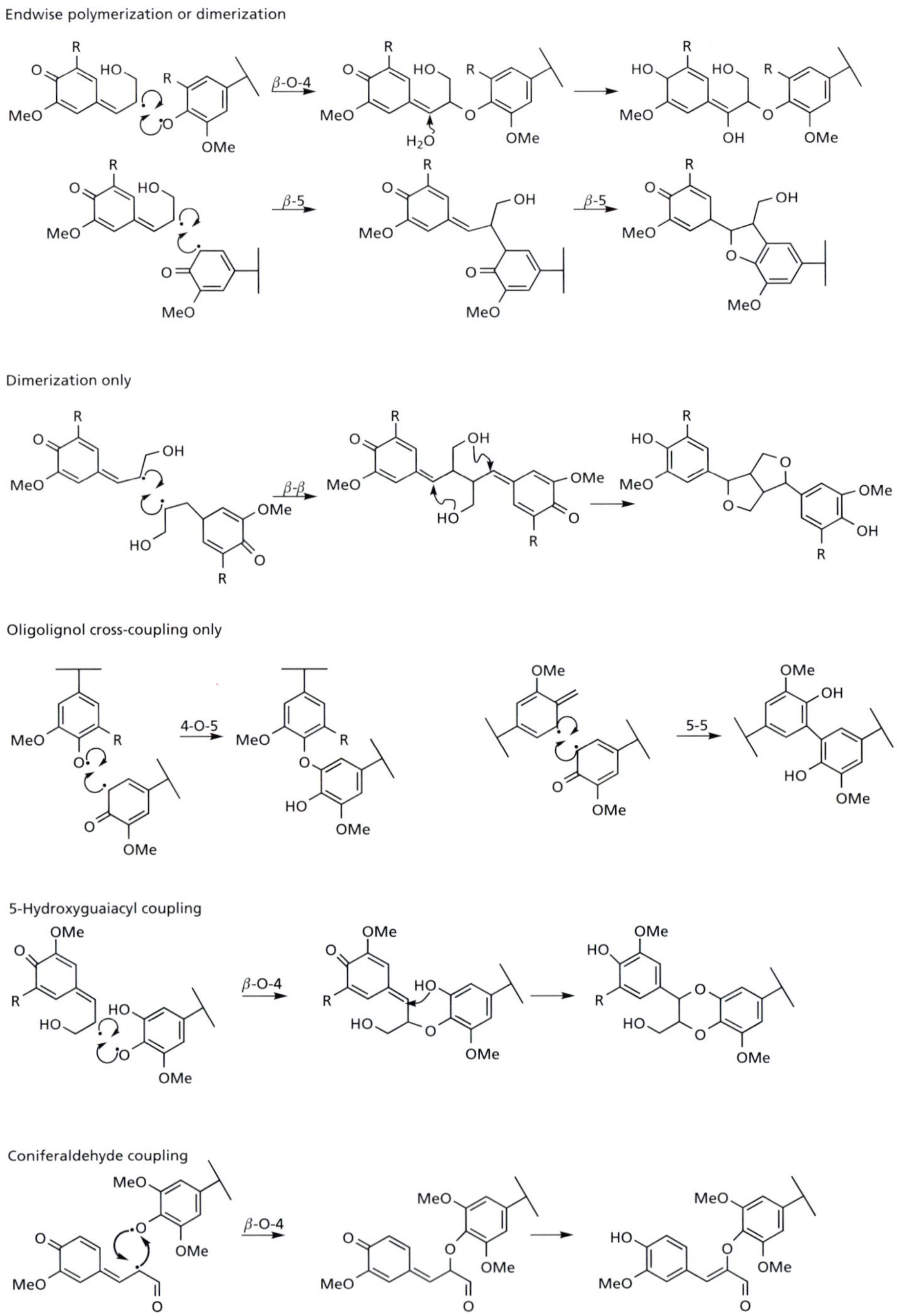

Figure 6.17 Major structural units in the lignin polymer, including less abundant monomers, and the bimolecular coupling reactions that give rise to them. Half-headed arrows show single-electron mechanisms, and full arrows show two-electron processes, such as the position of the nucleophile attack in re-aromatization reactions. Reproduced with permission of N.D. BONAWITZ and C. CHAPPLE, *Ann. Rev. Genet.* 44, 337-363, 2010. [76]

known as phenylcoumaran, and a β-β resinol structure. [76] The β-5 linked phenylcoumaran is a more rigid and more hydrophobic structure than that produced by β-*O*-4 coupling. This radical-radical coupling occurs in a chemical-combinatorial fashion; thus, the ratio of each of the possible coupling products depends largely on the chemical nature of each of the monomers and the conditions in the cell wall.

Endwise polymerization

Once formed, the dimer needs to be dehydrogenated, again to a phenolic radical, before it can couple with another monomer radical. [78] This mode of action, in which a monomer (radical) adds to the growing polymer, is called endwise coupling or polymerization: the polymer grows one unit at a time (**Figure 6.17**). During each coupling reaction, two radicals are consumed as each single electron contributes to the newly formed bond, making this type of radical polymerization different from the radical chain reactions that occur in the polymerization of several fossil-based polymers.

The simplest case of endwise polymerization is the coupling of a monolignol radical to a syringyl radical at the end of a growing polymer. [76] By far, the most commonly formed bond is between the 4-O position of the syringyl terminal unit and the β carbon of the monolignol radical. Addition of a monolignol to a terminal guaiacyl unit can also proceed via formation of a β-*O*-4 ether. Alternatively, the β carbon of the monolignol radical can react with the carbon at the 5-position of the guaiacyl subunit, which is not available in syringyl units, thereby forming a β-5 linked structure. As explained in **Figure 6.17**, endwise coupling results in the formation of a ***quinone methide*** intermediate (see **Figure 6.19**), followed by a nucleophilic addition of usually water to the quinone methide. [86]

In native lignin there are some forms of reduced structures that cannot be explained only by oxidative coupling. [86] It has been shown via model experiments that nicotinamide adenine dinucleotide (NADH), in an uncatalyzed process, reduces quinone methides to their benzyl derivates.

Oligolignol cross-coupling

Because endwise polymerization always involves the β carbon of the incoming monolignol, the common bonds remaining to be described only occur during the coupling of oligolignol end units, which do not have their β carbon available for coupling (see **Figure 6.17**). [76] Remarkably, if lignin polymerization were constrained to only β-*O*-4, β-5, and β-β bonds, the lignin polymer would be entirely linear and devoid of any branches. In fact, *branched* may not be the best term to describe the cross-linked structures seen in the lignin polymer, given that these structures are not the result of one elongating polymer end giving rise to two new elongating ends, but rather arise from the collision of two oligolignol ends to form one new end.

When such collisions do occur, there are two possible coupling modes available. If at least one oligolignol end is a guaiacyl subunit, bond formation can proceed between the phenoxy radical at one end unit and the 5-carbon of the guaiacyl unit (4-O-5 coupling). Alternatively, if both oligolignol ends are guaiacyl units, a bond can form between the carbons at the 5-positions of both subunits (5-5 coupling). Syringyl subunits do not have a 5-carbon available for bonding, and thus two terminal syringyl units cannot couple with each other. Although 5-5 linked structures possess two phenolic hydroxyl groups, addition of a monolignol to one of them results in internal trapping of the other and formation of an eight-membered ring. Thus, because, at most, one monolignol can be added to a 4-O-5 or 5-5 structure, the apparent three-way branching structures observed in the polymer at the eight-membered rings can be viewed as the collapse of two elongating ends into one. Nevertheless, lignin containing a high proportion of G subunits is more highly crosslinked than lignin rich in S subunits, presumably rendering it more rigid and more hydrophobic. Coupling of two oligomers is rare in S/G-lignins but relatively frequent in G-lignins, where 5-5 coupling accounts for ~4% of the linkages. [78] The average length of a linear lignin chain in poplar (angiosperm) is estimated to be between 13 and 20 units.

Less abundant monomers

Coniferaldehyde (in CAD-deficient plants) and 5-hydroxyconiferyl alcohol (in COMT-deficient plants) couple via similar mechanisms as G and S subunits, respectively. [76] However, the β-*O*-4 structures that each forms during polymerization are more rigid than β-*O*-4 structures formed by coniferyl and sinapyl alcohols (see **Figure 6.17**). The coupling of a coniferaldehyde subunit results in only three rotatable bonds between the rings of the two lignin subunits instead of the four usually present. The benzodioaxane structure formed when a monolignol couples with a 5-hydroxyguaiacyl end unit is much more rigid, containing only one rotatable bond between the two aromatic rings. Much less is known about the coupling and cross-coupling of p-coumaryl alcohol, which forms a minor component of the lignin polymer in most plants. [76]

Monolignol dehydrogenation

Monolignol dehydrogenation involves ***peroxidases*** and/or ***laccases***. [78] Whereas peroxidases use hydrogen peroxide as electron acceptor, laccases use molecular oxygen to oxidize their metal centers to enable catalytic oxidation. Both types of enzymes belong to large gene families of which the members have overlapping activities, making the process difficult to study. Peroxidases may differ in their substrate affinity. Because the structure of lignin depends on the availability of monolignol radicals, peroxidases specificity may determine partly the structure of the final lignin polymers. Although monolignols can be dehydrogenated via direct interaction with an electron-removing enzyme, the radicals might also be generated by radical transfer.

Conclusion

As a conclusion, the present model of lignification involves peroxidases and/or laccases to provide the oxidative capacity in the cell wall. [78] All phenolic compounds that enter the cell wall will eventually have the potential to radicalize and incorporate into the lignin polymer, subject to chemical oxidation and coupling propensity. This lignification model also explains why many other phenolic molecules can be integrated into the growing lignin polymer and opens up the possibility of tailoring lignins for industrial applications.

Evolution of lignification

It is commonly accepted that lignin evolved together with the adaptation of plants to a terrestrial life to provide them with the structural support needed for an erect growth habit. [78] The recent discovery of lignin in a bryophyte, the liverwort *Marchiantia polymorpha*, expands the distribution of lignification to nonvascular plants (**Figure 6.18**). [83] Even more strikingly, lignin was found in the cell wall of the red alga, *Calliathron cheilosporioides*, which shares a common ancestor with vascular plants over 1 billion years ago. These observations raise new questions on the evolutionary origin and history of lignification.

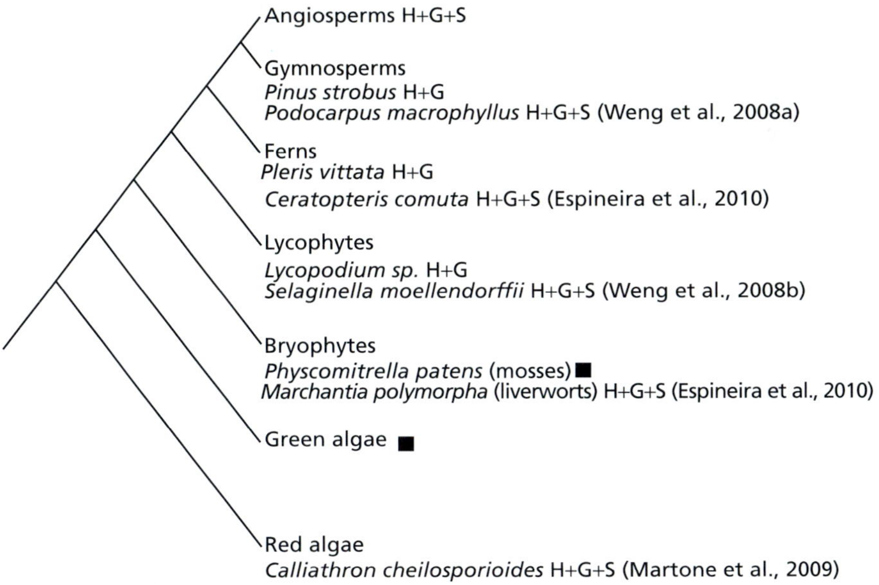

Figure 6.18 Phylogenetic tree showing the distribution of lignification and lignin monomer composition through major plant lineages. It is evident that S lignin is not restricted to angiosperms. The distribution of S lignin within lycophytes, fern and gymnosperms is not uniform. Black squares indicate no lignification. Copyright © 2010 by the American Society of Plant Biologists. [83]

Similarly, recent research on the distribution of lignin monomers in the plant kingdom challenges our perception about the evolution of S lignin monomer biosynthesis (**Figure 6.18**). [83] S lignin has been generally considered to be characteristic of the angiosperms; however, recent studies confirmed that S lignin is also present in gymnosperms and some basal vascular plants such as ***lycophytes*** and ferns. Furthermore, S lignin is also detected in liverworts and red algae, which are more distantly related to angiosperms.

The occurrence of lignification in liverworts and red algae suggests that the genes required for lignin deposition evolved in their common ancestor before the divergence of specific lineages (e.g. red algae and liverworts) and were subsequently lost in certain lineages (e.g. green algae and mosses). [83] Alternatively, red algae and liverworts lineages could have independently evolved the biochemical pathways for lignification. Similar evolutionary scenarios can also be envisaged to explain the phylogenetic pattern of S lignin. ***Convergent evolution*** of S lignin biosynthesis between the lycophyte *Selaginella moellendorffii* and angiosperms has been demonstrated.

Research trends

One new trend in lignin research is to steer biosynthetic pathways toward the biosynthesis of molecules that, after incorporation into the lignin polymer, will facilitate lignin degradation. [78] This concept of copolymerizing alternative lignin monomers may result in less-energy-demanding biomass pretreatments to make cellulose accessible.

Additional insight into the factors that determine lignin structure and degradability can be obtained by mathematical modeling. [78] Coupling of monolignols is likely to be dependent on their speed of delivery to the cell wall, their chemical characteristics, and the oxidative environment in the wall. A mathematical model has been developed that predicts the frequency of the different bond types, the number of different polymers, their lengths, and the relative abundance of the different length polymers under varying parameters. [77] This tool can be used to determine the parameters that influence lignin degradation.

Finally, field trials to examine the effects of reduced lignin on the conversion of biomass to bioethanol have recently been initiated. With the emergence of the biobased economy, the easy release of sugars from the cell wall and their conversion into biobased products is becoming ever more important.

6.3.3 Lignin in plant cell walls

The secondary cell walls of woody tissues and grasses are composed predominantly of cellulose, lignin and hemicelluloses (see **Figure 4.17**). [16] Lignin is closely mixed with the other wall components. The covalent linkages between lignin and polysaccharides form lignin-carbohydrate complexes (LCC). [87] Important LCCs involve quinone methide (MQ) and phenolic acids.

Three major classes of LCCs via QM have been reported: benzyl ether linkages, benzyl ester linkages and, to a lower extent, glycosidic linkages. [87, 88] The two first are formed via the nucleophilic addition of polysaccharide hydroxyl or carboxylic groups to the α-carbon of the methylene quinone formed during monolignol polymerization (**Figure 6.19**). This mechanism competes with the addition of other nucleophilic agents such as water.

In grasses, another type of covalent linkage occurs via phenolic acids (**Figure 6.20**). Phenolic acids, such as ferulic acid, are bifunctional molecules because of their carboxylic and hydroxyl functions. They are potential agents for wall polymer crosslinking.

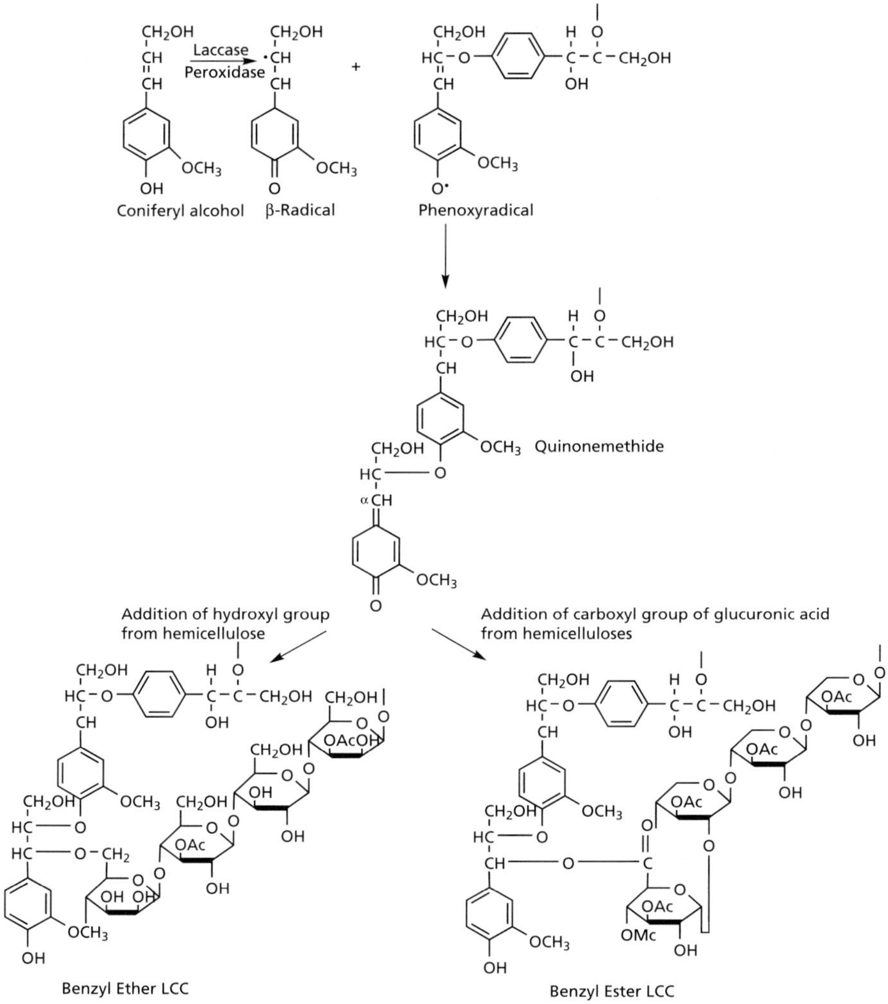

Figure 6.19 Formation of benzyl ether and benzyl ester bonds between lignin and hemicelluloses via quinone methide intermediates. [88, 89]

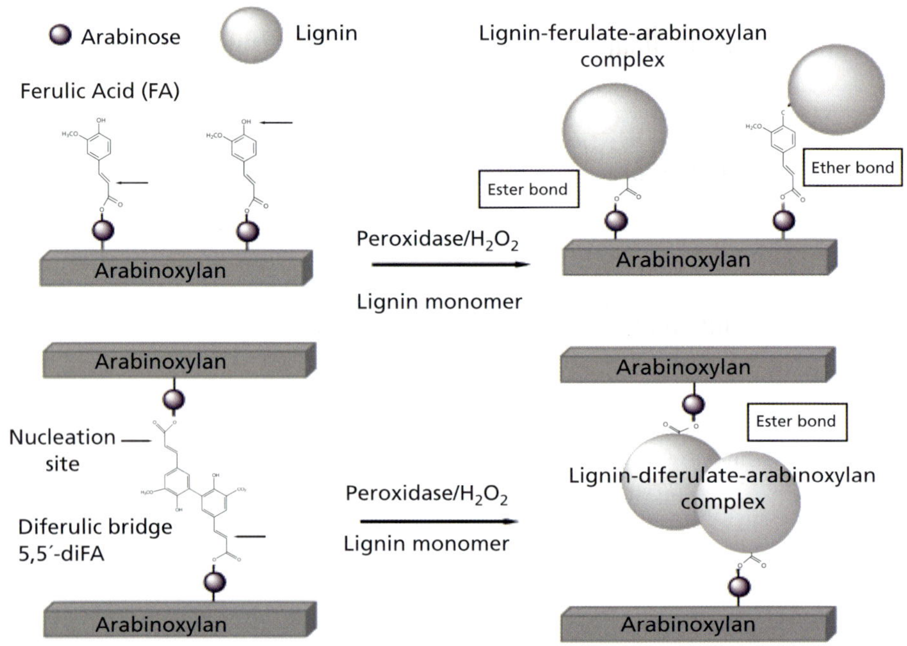

Figure 6.20 Formation of ether and ester bonds between arabinoxylans and lignin via ferulic acid in grasses. [87]

Crosslinking of the network of hemicelluloses and lignin presumably is thought to result in the elimination of water from the wall and the formation of a hydrophobic composite that limits accessibility of hydrolytic enzymes and is a major contributor to the structural characteristics of secondary walls. [16]

6.3.4 Degradation

Lignin is extremely recalcitrant to degradation. [81] By linking to both cellulose and hemicelluloses, it creates a barrier to any solutions or enzymes and prevents the penetration of lignocellulolytic enzymes into the lignocellulosic structure. However, some basidiomycetes white-rot fungi are able to degrade lignin efficiently using a combination of extracellular ligninolytic enzymes, organic acids, mediators, and accessory enzymes.

Lignocellulolytic enzyme-producing fungi

Lignocellulolytic enzyme-producing fungi are widespread, and include species from the Ascomycetes (e.g. *Trichoderma reesei*), and basidiomycetes phyla such as white-rot (e.g. *Phanerochaete chrysosporium* (**Figure 6.21**)) and brown-rot fungi (e.g. *Fomitopsis palustris*). [81] In addition, a few anaerobic species are

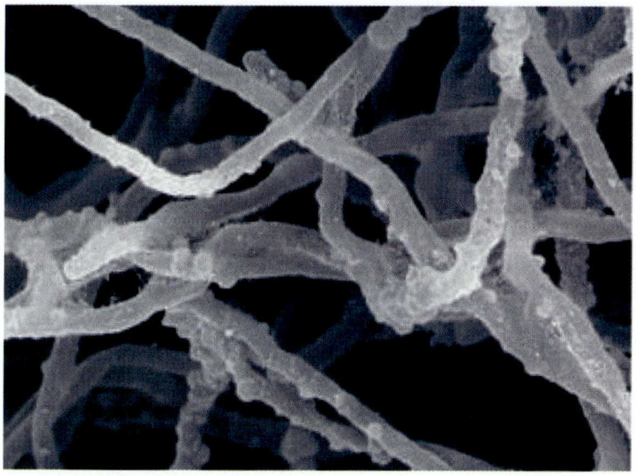

Figure 6.21 Scanning electron micrograph of *Phanerochaete chrysosporium* fungi; Mag. 5x. Photograph courtesy of UC Regents. [90]

found to be able to degrade cellulose in the gastrointestinal tracts of ruminant animals. Biomass degradation by these fungi is performed by complex mixtures of cellulases, hemicellulases, and ligninases.

In nature, efficient lignin degradation during the process of wood decay became possible mainly by basidiomycetes white-rot fungi. [81] Many white-rot fungi simultaneously attack lignin, hemicelluloses, and cellulose, whereas some other white-rot fungi preferentially work on lignin in a selective manner. Selective lignin degraders may have potential biotechnological applications when the removal of lignin is required to obtain pure cellulose such as in biopulping and in processes where the objective is to provide an unprotected carbohydrate.

In contrast to white-rot fungi, brown-rot fungi can degrade wood carbohydrates, but not oxidized lignin. Ascomycetes are mostly able to degrade cellulose and hemicelluloses, while their ability to degrade lignin is limited. [91] Various plant pathogenic fungi are able to degrade lignin by production of laccases and lignin peroxidases.

Fungal extracellular ligninases

Fungi degrade lignin by secreting enzymes collectively termed ***ligninases***. [81, 92] Ligninases can be classified as either phenol oxidases (laccases) or ***heme*** peroxidases, which are further subdivided into ***lignin peroxidases*** (***LiP***), ***manganese peroxidases*** (***MnP***), and ***versatile peroxidases*** (***VP***). White-rot fungi variously secrete one or more of the lignin-modifying enzymes in addition to other compounds necessary for effective lignin degradation. It has been shown that *P. chrysosporium* produces several LiP and MnP isoenzymes but no laccases. Although LiP is able to oxidize the non-phenolic part of lignin (80–90% of lignin composition), it is

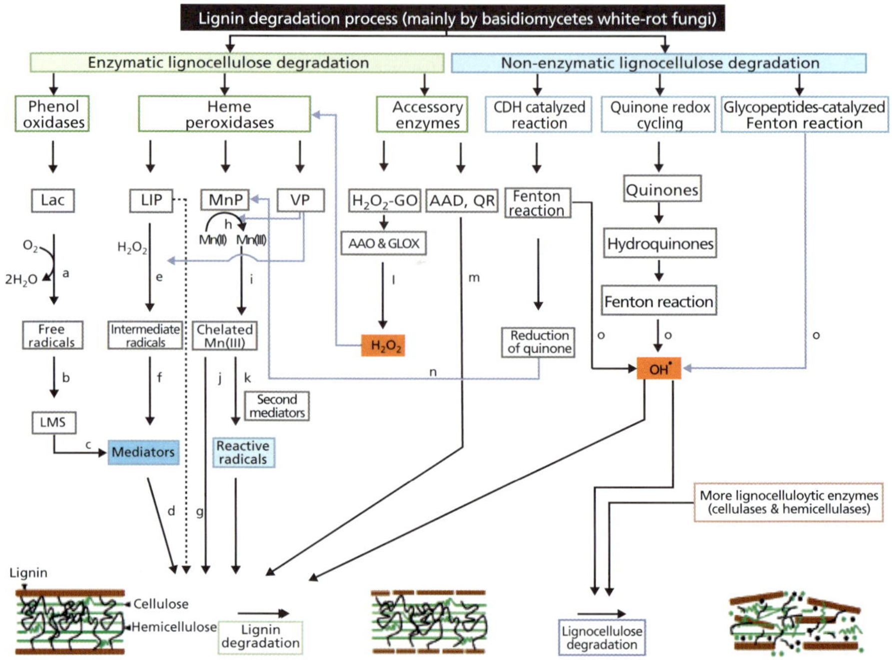

Figure 6.22 Schematic diagram of lignin degradation by basidiomycete white-rot fungi: the major steps and enzymes involved (refer to text). Lac, laccase; LMS, laccase-mediator system; LiP, lignin peroxidase; MnP, manganese peroxidase; VP, versatile peroxidase; H_2O_2-GO, generating oxidases; AAO, aryl-alcohol oxidase; GLOX, glyoxal oxidase; AAD, aryl-alcohol dehydrogenases; QR, quinone reductases; CDH, cellulose dehydrogenase. IJBMB Copyright © 2010–present. [81]

absent from many lignin degrading fungi. As oxidative ligninolytic enzymes are too large to penetrate into the wood cell wall micropores, it has been suggested that low molecular weight oxidative compounds must initiate changes to the lignin structure. **Figure 6.22** shows the major steps and enzymes involved in lignin degradation by basidiomycetes white-rot fungi.

Phenol oxidases (laccases)

Laccases (phenol oxidases, EC 1.10.3.2) are glycosylated multi-copper oxidoreductases that use molecular oxygen to oxidize various aromatic and non-aromatic compounds through a radical-catalyzed reaction mechanism. [81, 93] The enzymes catalyze the overall reaction (**Equation 6.1**): [92]

$$4 \text{ benzenediol} + O_2 \rightarrow 4 \text{ benzosemiquinone} + 2\ H_2O \quad (6.1)$$

Laccases couple the reduction of dioxygen into two molecules of water with the oxidation of a variety of substrates, such as phenols, arylamines, anilines, thiols, and lignins (**Figure 6.22, a**).

Laccases represent one of the oldest enzymes ever described. [92] The enzyme seems ubiquitous in all white-rot fungi, and its presence is less frequent in plants. The better-characterized enzymes are from the fungus *Trametes versicolor* and the Japanese lacquer tree *Rhus vernicifera* (hence the name laccase). The plant enzyme participates together with peroxidases in the biosynthesis of lignin. Fungal laccases occur often as isozomes with monomeric or dimeric protein structures. Both intracellular and extracellular isozymes may be produced from a single microorganism. The monomeric proteins have a molecular mass ranging from 50 to 110 kDa. Laccases are ~10–45% glycosylated, with the fungal enzyme less substituted than the plant enzyme. The three-dimensional structures of several laccases have been reported. All fungal laccases show a similar architecture consisting of three sequentially arranged domains of a β-barrel type structure. [92, 94]

The active site is well conserved. It contains four copper ions that mediate the redox process: one type 1 copper ion, one type 2 copper ion, and two type 3 copper ions, based on the copper's coordination and spectroscopic properties. The type 2 and type 3 copper centers are close together forming a trinuclear copper cluster, and the type 1 site contains a single type 1 copper ion (**Figure 6.23**). [95]

The oxidation reactions catalyzed by laccases lead to the formation of free radicals that act as intermediate substrates for the enzymes (**Figure 6.22**, b). [81]

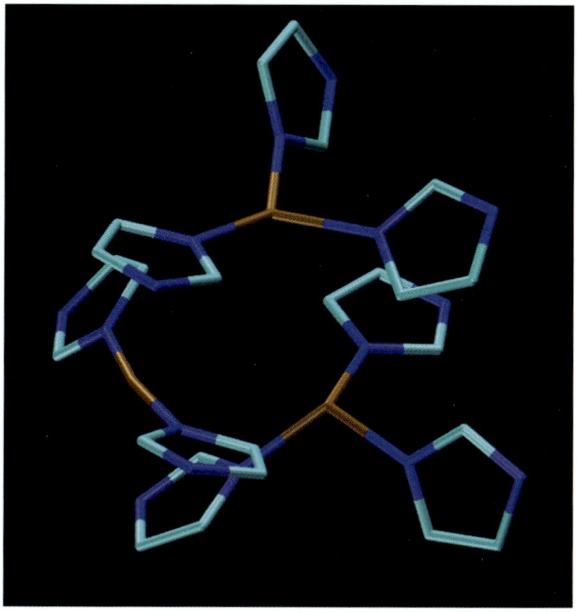

Figure 6.23 The tri-copper site found in many laccases. Each copper center is bound to imidazole (copper in brown, nitrogen in blue). [96]

These mediators can leave the enzyme site and react with high-redox potential substrates and, thus, create non-enzymatic routes of oxidative polymerizing or depolymerising reactions. As laccases work efficiently on many substrates without cofactors, they may be valuable in many biotechnological applications, such as pulp bio-bleaching, biosensors, food industries, textile industries, soil remediation, and in the production of complex polymers. [81]

Heme peroxidases

Lignin peroxidases

Lignin peroxidases (also called diarylpropane peroxidases, EC 1.11.1.14) are heme-containing glycoproteins and play a central role in the biodegradation of lignin. [81] The enzymes catalyze the overall reaction (**Equation 6.2**): [92]

$$\text{1,2-bis(3,4-dimethoxyphenyl)propane-1,3-diol} + H_2O_2 \rightarrow \text{3,4-dimethoxybenzaldehyde} + \text{1–(3,4-dimethoxyphenyl) ethane-1,2-diol} + H_2O \quad (6.2)$$

Lignin peroxidases catalyze the H_2O_2-dependent oxidative depolymerization of non-phenolic lignin compounds (diarylpropane), *β-O*-4 non-phenolic lignin model compounds and many phenolic compounds with redox potentials up to 1.4 V. Lignin peroxidases oxidize the substrates in multi-step electron transfers and form intermediate radicals, such as phenoxy radicals and non-phenolic veratryl alcohol (**Figure 6.24**) radical cations (**Figure 6.22, e**).

These intermediate radicals undergo non-enzymatic reactions, such as radical coupling and polymerization, side-chain cleavage, demethylation, and intramolecular addition and rearrangement (**Figure 6.22, f**). Unlike the other peroxidases, lignin peroxidase is able to oxidize non-phenolic aromatic substrates and does not require the participation of mediators due to its high redox potential (**Figure 6.22, g**).

Lignin peroxidase was first discovered in *P. Chrysosporium*, and various isoforms are known to exist with this microorganism and other white-rot fungi. [92] The lignin peroxidase isozymes are glycoproteins of 38–46 kDa.

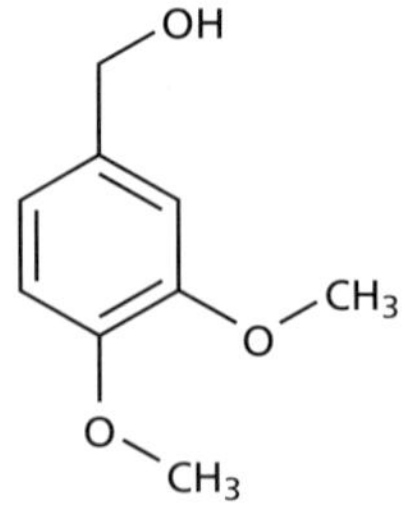

Figure 6.24 Veratryl alcohol (3,4-dimethoxybenzyl alcohol).

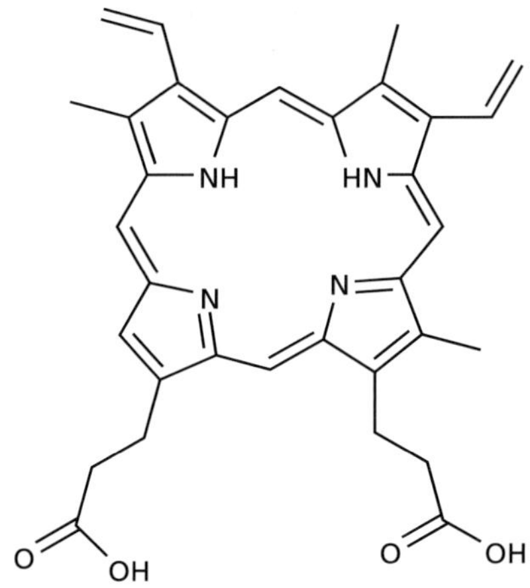

Figure 6.25 Protoporphyrin IX, a porphyrin derivative that combines with iron.

Lignin peroxidase has a distinctive property of an unusually low pH optimum near pH 3. The enzyme contains one mol of iron protoporphyrin IX (**Figure 6.25**) per mole of protein.

The crystal structure of *P. Chrysosporium* lignin peroxidase has been described. [97] The enzyme is globular with a dimension of 50 × 40 × 40 Å comprised of a proximal (C-terminal) and distal (N-terminal) domain (**Figure 6.26**).

The heme is buried in the interior of the protein and has access to the outer medium through a channel. [98] Although the size of the channel is not sufficient to allow the large lignin polymer to access the heme group, small molecule substrates can find a suitable binding site. [81] The enzyme molecule contains eight major and eight minor α-helices and a limited β-structure in the proximal domain. [92] Several N- and O-glycosylation sites can be identified in the crystal structure. The lignin peroxidase has eight cysteine residues, all forming disulfide bridges. There are two calcium-binding sites, one in each domain. The heme iron is predominantly pentacoordinated with histidine at the proximal side as an axial fifth ligand, and stabilized by hydrogen bonding to the distal histidine (**Figure 6.27**).

Lignin peroxidase has a typical peroxidase catalytic cycle. The mechanism of lignin peroxidase-catalyzed reaction includes two steps: [92]

1. A 2 electron oxidation of the native ferric enzyme [Fe(III), LiP] to yield compound I intermediate that exists as a Fe(IV) porphyrin radical cation [Fe(IV)=O$^{\cdot+}$, LiP-I], with the peroxide substrate cleaved at the O-O bond (**Equation 6.3**):

$$\mathrm{LiP[Fe(III)] + H_2O_2 \rightarrow LiP\text{-}I[Fe(IV){=}O^{\cdot+}] + H_2O} \qquad (6.3)$$

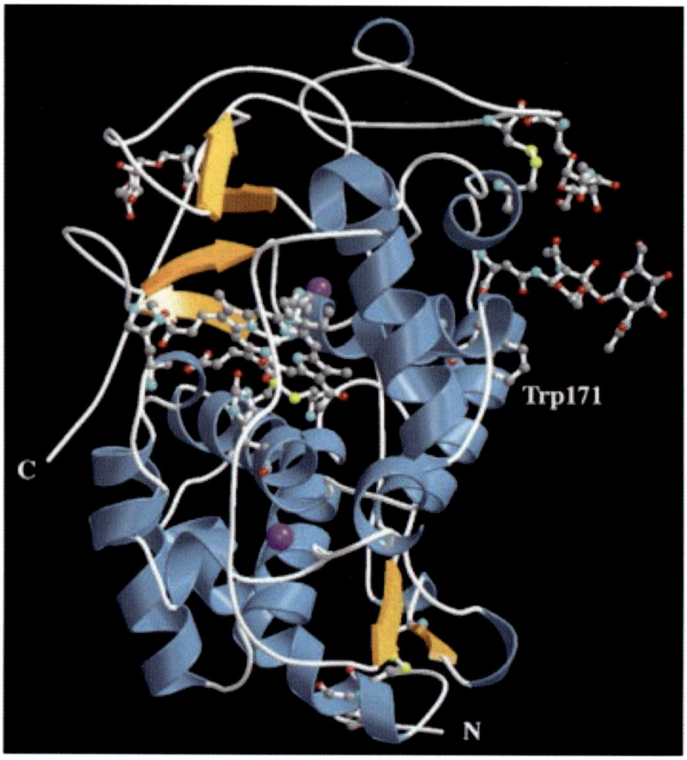

Figure 6.26 Structure of a fungal lignin peroxidase. The heme with the proximal and distal histidine and the two distal water molecules, the two calcium ions (purple spheres), His82 at the entrance of the active-site channel (left site), four carbohydrate molecules, Trp171 (a radical center formed during the redox cycle), and the disulfide bridges (S atoms as yellow spheres) are represented as ball and stick models. [97]

2. A two-consecutive single electron reduction of LiP-I by electron donor substrates to the native enzyme. The first reduction of LiP-I by a reducing substrate (AH), such as veratryl alcohol, yields compound II [Fe(IV)=O, LiP-II] and a radical cation. A second reduction returns the enzyme to the ferric oxidation state (**Equations 6.4** and **6.5**).

$$\text{LiP-I} + \text{AH} \rightarrow \text{LiP-II[Fe(IV)=O]} + \text{A}^{\cdot+} \tag{6.4}$$

$$\text{LiP-II} + \text{AH} \rightarrow \text{LiP} + \text{A}^{\cdot+} \tag{6.5}$$

Manganese peroxidases

Manganese peroxidases (Mn(II):hydrogen-peroxide peroxidases, EC 1.11.1.13) are extracellular glycoproteins and are secreted in multiple isoforms which con-

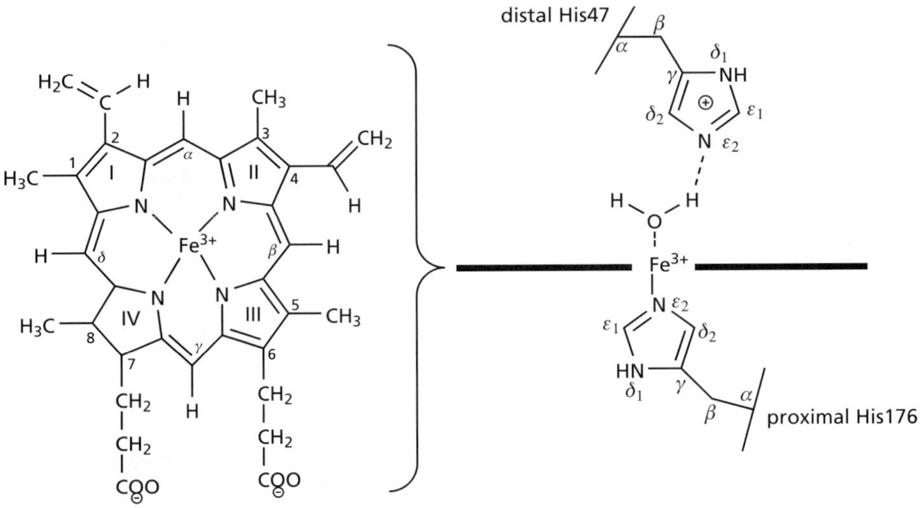

Figure 6.27 The heme iron environment in lignin peroxidase. Copyright © Humana Press 2008. [92]

tain one molecule of heme as iron protoporphyrin IX. [81] The enzymes catalyze the peroxide-dependent oxidation of Mn(II) to Mn(III) (**Equation 6.6**):

$$2\ Mn(II) + 2\ H^+ + H_2O_2 \leftrightarrow 2\ Mn(III) + 2\ H_2O \tag{6.6}$$

The heme iron in the native protein is in the pentacoordinate, ferric state with a histidine residue coordinated as the fifth ligand. [92] The overall structure of manganese peroxidases is similar to lignin peroxidases, consisting of two domains with the heme sandwiched in-between (**Figure 6.28**).

The protein molecule contains ten major helices and one minor helix, as found in lignin peroxidase. The Mn(II) is located in a cation-binding site at the surface of the protein.

Mn-dependent peroxidases are unique in utilizing Mn(II) as the reducing substrate. [92] The enzyme oxidizes Mn(II) to Mn(III) (**Figure 6.22, h**), which is then released from the enzyme surface in complex with oxalate or with other chelators (**Figure 6.22, i**). [81] Chelated Mn(III) complex acts as a reactive diffusible redox-mediator (**Figure 22, j**) of phenolic substrates including simple phenols, amines dyes, phenolic lignin substructures and dimers. The oxidation potential of Mn(III) complex is only limited to phenolic lignin structures. However, for the oxidation of non-phenolic substrates by Mn(III), reactive radicals must be formed in the presence of a second mediator. Organic acids, such as oxalate and malonate, are the primary compounds that act as second mediators in the production of reactive radicals. In the absence of hydrogen peroxide, these radicals can be used by manganese peroxidase as a source of peroxides and increase the lignin-degrading efficiency of the fungi. The cata-

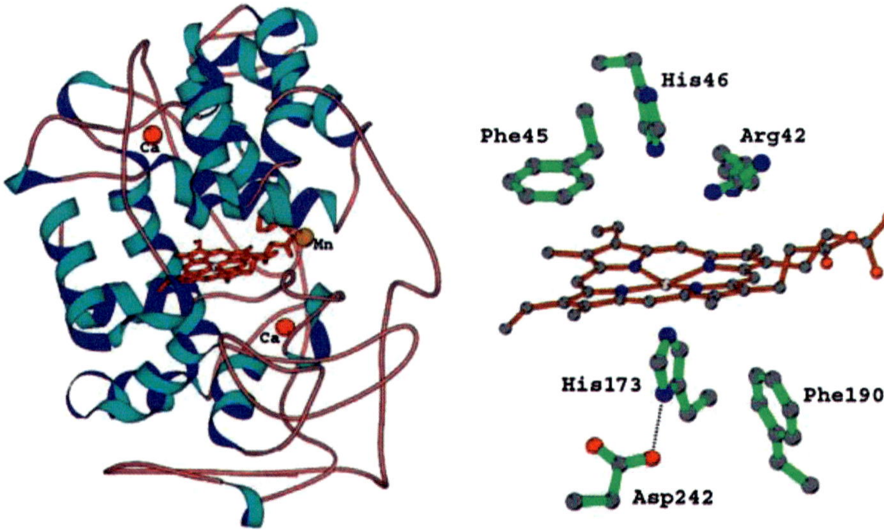

Figure 6.28 **(A)** Structure of manganese peroxidase from *Phanerochaete chrysosporium*. The red spheres are structural Ca(II) ions conserved in extracellular heme peroxidases. The location of the substrate Mn(II), near the heme, is indicated. **(B)** The active site structure of MnP This architecture is highly conserved in heme peroxidases. The main variations are the Phe residues which are Trp in the intercellular peroxidases, cytochrome c and ascorbate peroxidase. The Asp242-His173 pair is conserved. [99]

lytic cycle, thus, includes the oxidation of Mn(II) by manganese peroxidase compound I (MnP-I) and manganese peroxidase compound II (MnP-II) to yield Mn(III) (**Equations 6.7** to **6.9**): [92]

$$MnP + H_2O_2 \rightarrow MnP\text{-}I + H_2O \quad (6.7)$$

$$MnP\text{-}I + Mn^{2+} \rightarrow MnP\text{-}II + Mn^{3+} \quad (6.8)$$

$$MnP\text{-}II + Mn^{2+} \rightarrow MnP + Mn^{3+} + H_2O \quad (6.9)$$

Mn(III) in turn mediates the oxidation of organic substrates (**Equation 10**):

$$Mn^{3+} + RH \rightarrow Mn^{2+} + R^{\cdot} + H^+ \quad (6.10)$$

The features of the cycle are very similar to that of lignin peroxidase. [92] Addition of 1 equivalent of H_2O_2 to the native enzyme yields MnP-I, which is a Fe(IV)-oxo-porphyrin radical cation [$Fe(IV){=}O^{\cdot+}$]. The peroxide bond of H_2O_2 is cleaved subsequent to a $2e^-$ transfer from the enzyme heme-porphyrin. Addition of 1 equivalent of Mn^{2+} reduces compound I to compound II.

Manganese peroxidases may be capable of rivaling the potential applications of laccases in biotechnology. [81] Studies have shown that the presence of manganese peroxidase can increase the degree of dye decolorization. In addition, manganese peroxidase from white-rot fungi is considered as the primary enzyme responsible for biobleaching of kraft pulps.

Versatile peroxidases

It has been reported that some manganese peroxidases exhibit activities on aromatic substrates similar to those of lignin peroxidases. [92] This group of enzymes, known as versatile peroxidases (EC 1.11.1.16), is not only capable of oxidizing Mn(II) as in manganese peroxidase, but also phenolic and non-phenolic substrates that are typical for lignin peroxidase (**Figure 6.22**).

Molecular characterization of versatile peroxidases reveals structures that are closer to the lignin peroxidases than to the manganese peroxidase isozymes. [92] A Mn(II) binding site containing Glu/Glu/Asp residues typical of manganese peroxidases is found near the heme internal propionate. Furthermore, residues involved in lignin peroxidase interaction with veratryl alcohol and other aromatic substrates are also found in the protein structure. The molecular model of the *Pleurotus eryngii* versatile peroxidase reveals 12 helices, 4 disulfide bonds, a heme pocket containing the characteristic proximal histidine, and the distal histidine, 2 structural Ca^{2+} sites and a Mn(II) binding site (**Figure 6.29**).

Versatile peroxidases form an attractive ligninolytic enzyme group due to their catalytic versatility. [81] This makes versatile peroxidases more efficient than both lignin peroxidases and manganese peroxidases, which are not able to efficiently oxidize phenolic compounds in the absence of veratryl alcohol or oxidize phenols in the absence of Mn(II), respectively. Similar to the manganese peroxidase mechanism, Mn(III) is released from versatile peroxidases and acts as a diffusible oxidizer of phenolic lignin and free phenol substrates. Like other members of heme peroxidases, heme is buried in the interior of the protein and has access to the outer medium through two channels. The function of the first channel is similar to that described for lignin peroxidase and is conserved among all heme peroxidases. Conversely, the second channel is found to be specific to versatile peroxidase and manganese peroxidase, and is where the oxidation of Mn(II) to Mn(III) takes place.

Accessory enzymes involved in lignin degradation

Besides ligninases, other fungal extracellular enzymes, which act as accessory enzymes, are involved in lignin degradation. [81] These include oxidases generating hydrogen peroxide required by peroxidases, and mycelium-associated dehydrogenases, which reduce lignin-derived compounds (**Figure 6.22,1**). Oxidases generating hydrogen peroxide include aryl-alcohol oxidase (AAO) and glyoxal oxidase (GLOX). Furthermore, aryl-alcohol dehydrogenases (AAD) and

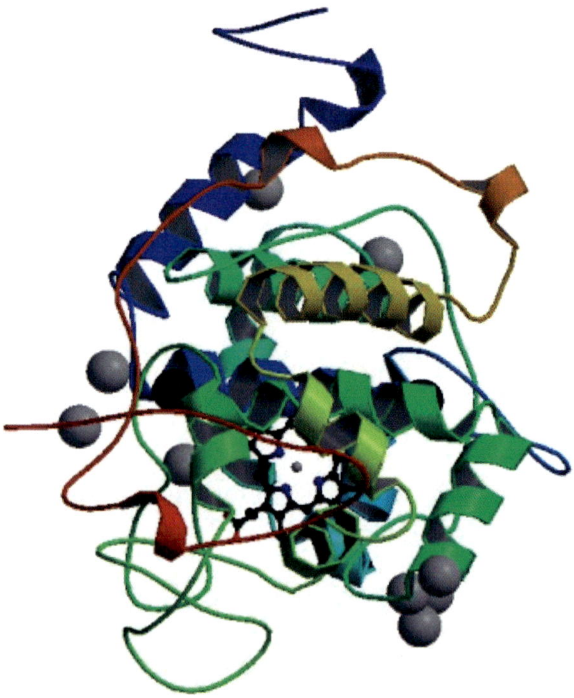

Figure 6.29 Crystal structure of *Pleurotus eryngii* versatile peroxidase. [100, 101]

quinone reductases (QR) are also involved in lignin degradation. Moreover, it has been shown that cellulose dehydrogenase (CDH) is also involved in lignin degradation in the presence of hydrogen peroxide and chelated iron ions.

Oxidative (non-lignocellulolytic) lignocellulose degradation

Studies over the last few decades have provided evidence in support of the involvement of non-enzymatic mechanisms in plant cell-wall polysaccharide degradation. [81] These mechanisms are mostly assisted by oxidation through the production of free hydroxyl radicals (**Figure 6.22, o**). Many white- and brown-rot fungi produce hydrogen peroxide which enters the Fenton reaction, resulting in the release of hydroxyl radicals (**Equation 6.11**):

$$Fe^{2+} + H_2O_2 \rightarrow Fe^{3+} + OH^- + OH^{\cdot} \qquad (6.11)$$

By attacking polysaccharides and lignin in a non-specific manner, these radicals create a number of cleavages that facilitate penetration of the cell wall by lignocellulolytic enzymes. The pathways by which fungi generate free hydroxyl radicals are: cellobiose dehydrogenase catalyzed reactions, low molecular weight peptides/quinone redox cycling and glycopeptide-catalyzed Fenton reactions (**Figure 6.22**).

Cellobiose dehydrogenase, an extracellular monomeric protein, has been shown to be produced in a number of basidiomycete fungi (mostly white-rot fungi) and in the culture filtrates of non-ligninolytic, cellulolytic microorganisms. [81, 102] This enzyme is able to oxidize cellobiose, higher cellodextrins and other disaccharides or oligosaccharides with β-1,4 linkages. Furthermore, cellobiose dehydrogenase has been found with cellulose binding module (CBM) (in ascomycetes) and without CBM (in basidiomycetes). In the absence of CBM, cellobiose dehydrogenase is able to bind to cellulose through hydrophobic interactions. It is now accepted that cellobiose dehydrogenase is able to degrade and modify the three major components of biomass by producing free hydroxyl radicals in a Fenton-type reaction.

White- and brown-rot fungi have been shown to produce low molecular weight chelators which are able to penetrate into the cell wall. [81] For example, a low molecular weight peptide, known as short fiber generating factor, can degrade cellulose into short fibers by an oxidative reaction. Some of these low molecular weight compounds are quinones, which must first be converted to hydroquinones before free hydroxyl radicals can be produced by the Fenton reaction.

Glycopeptides have been found in many brow- and white-rot fungi. Similar to other mechanisms, glycopeptides are able to catalyze redox reactions and thus to produce free hydroxyl radicals. [81]

6.3.5 Properties and applications

As a biopolymer, lignin is unusual because of its heterogeneity and lack of a defined primary structure. Lignin is the glue that holds plant cell walls together and is the only aromatic carbon source generated in nature. [103]

Lignin is a three-dimensional amorphous polymer mainly consisting of methoxylated phenylpropane structures. [104, 105, 106] It is a thermoplastic polymer and can be compounded with most thermoplastics for making composites. [107] Generally, softwood lignins have a slightly higher T_g (glass transition temperature) than hardwood lignins, which is probably due to a more-cross-linked structure in softwoods. Unmodified lignin in wood has the lowest T_g (65-105°C), and for modified lignins, T_g is reported to be higher: milled wood lignin (lignin extracted from finely ground wood) 110–160 °C and kraft lignin (lignin found in black liquor at kraft pulp mills) 124–174 °C. [108] Isolated lignins generally have a molecular weight of around 2000 to 5000.

Lignin is one of the most abundant renewable resources. Worldwide, about 50 million tonnes of lignin are produced annually as residue in paper production processes. [109] World production of lignosulfonates, which are water-soluble anionic polyelectrolyte polymers derived from the production of wood pulp using sulfite pulping, is estimated to be close to one million tonnes. [110]

World production of kraft lignins not used as fuel is about 0.1 million tonnes, whereas that of sulfur-free lignins is limited. Most lignin waste is burned to generate energy for the pulp mills. However, based on its interesting functionalities and properties, lignin offers perspective for higher added value applications in biobased products.

Lignin has several relatively low-value applications such as: [110]

- fuel, yielding more energy when burned than cellulose;
- cement additive, especially as cement set retarder;
- asphalt additive, especially due to its anti oxidant characteristics;
- feed binder to plasticize and hold the pellet together; and
- additive in biomass-based fuel pellets. [111]

High-value applications for lignin and modified lignin include: [110]

- raw material for vanillin production;
- lignin-based composites, especially lignin matrix composites; [112]
- macromolecular binder for durable wood, panel, and board (medium density fiberboard, plywood) and design products, involving partial substitution of phenol-formaldehyde resins by modified lignins due to their structural similarity, as well as the development of emission-free renewable binders such as lignin-furan binders; [113]
- depolymerized lignin for aromatic base chemicals (e.g. phenols), the main routes for lignin depolymerization being base catalyzed depolymerization followed by hydrodeoxygenation, pyrolysis and fractionation, catalytic liquefaction, hydrocracking, and supercritical depolymerization;
- key building block to make carbon fibers, soil conditioners, nitrogenous fertilizers, and pulping catalysts; [108, 114]
- component of polymeric materials such as starch film, conducting polymers, polyurethanes, and thermoplastics; [114]
- photostabilization and upgrading of lignin-rich mechanical wood pulp and paper; [114]
- UV stabilizer and coloring agent; [109]
- surfactant; [109] and
- raw material for aromatic synthon production. [103]

As an illustration of the commercial valorization of lignin, the Canadian company Lignol (see § 8.5.3) produces high-purity lignin as an ethanol co-product through an ***organosolv*** process (see § 7.3.4). A clear benefit of any organosolv-based process is that it fractionates biomass into its three components at relatively high purity. Organosolv lignin from Lignol is suitable for the following demonstrated applications: [115]

- phenol-formaldehyde resin and wood adhesive substrate;
- printed circuit board encapsulated resins;
- foundry resins and molding compounds;
- friction materials, green strength binders, organic particles;
- antioxidants in rubber, lubricants, feed additives;
- rubber tackifiers;
- renewable surfactants, concrete admixtures, air-entrainers, super-plasticizers;
- carbon fiber and activated carbon production; and
- animal feed applications.

In Europe, Borregaard LignoTech, a division of the Norwegian company Borregaard (see § 8.5.3), is the world's leading producer of lignin and lignosulfonate based products, with over 60 years supplying binding and dispersing agents to the chemical industry. [116] Applications and functionalities of Lignotech's lignin-based products include:

- binding agent;
- crystal growth modifier;
- dispersing agent;
- dust suppressant;
- emulsion stabilizers;
- extrusion aids;
- retarders; and
- rheology control.

These products are used in areas such as concrete additives, textile dyes, agrochemicals, batteries, oil drilling, ceramic materials and animal feed.

The LignoBoost process (see § 7.3.3) is based on a research collaboration between Innventia, a Swedish R&D company, and Swedish Chalmers University of Technology, who developed a process for extracting high quality lignin from sulfate pulp mills. [117] In 2008, Innventia sold the process to the Finnish company Metso. In 2010, the Swedish Energy Agency granted support to a demonstration plant at a Swedish pulp mill.

In France, the CIMV organosolv process (see § 8.5.3) has been developed with the objective of isolating sulfur-free linear lignin for applications such as a binding agent for phenolic powder resins, a polyol component in polyurethane formulations, and a phenolic component of epoxy resins. [118, 119] The process has been designed for the manufacture of whitened pulp, non-degraded linear lignin and C5 sugar syrup from annual fiber crops and hardwood. [120]

Potential for commercial lignin applications is estimated to become enormous over the upcoming decades. Carbon fibers and composite materials are amongst the most attractive examples of exploitable markets.

6.4 References

[1] N. Fernandez-Garcia, M. Carvajal and E. Olmos, *Ann. Bot. 93*, 53, 2004 in http://aob.oxfordjournals.org/cgi/content/full/93/1/53

[2] R. M. McCourt, R. L. Chapman, M. Buchheim, and B. D. Mishler in the Tree of Life Web Project, http://tolweb.org/tree?group=Green_plants&contgroup=Eukariotes

[3] J.L. Wertz, O. Bedue and J.P. Mercier, *Cellulose Science and Technology*, EPFL Press, Lausanne, 2010

[4] Complex Carbohydrate Research Center, The University of Georgia, *Plant Cell Walls,Hemicelluloses*, 2007 in http://www.ccrc.uga.edu/~mao/xyloglc/Xtext.htm

[5] H.V. Scheller and P. Ulvskov, Ann. Rev. Plant Biol. **61**, 263, 2010 in http://arjournals.annualreviews.org/doi/full/10.1146/annurev-arplant-042809-112315?amp;searchHistoryKey=%24%7BsearchHistoryKey%7D

[6] B. Alberts, A. Johnson, J. Lewis, M. Raff, K. Roberts and P. Walter, *Molecular Biology of the Cell*, 4th Edition, New York, 2002 in http://www.ncbi.nlm.nih.gov/bookshelf/br.fcgi?book=cell&part=A5208

[7] D. Hildebrand, *Plant Biochemistry*, BCH/PPA/PLS 609, 2010 in www.uky.edu/~dhild/biochem/11B/lect11B.html

[8] http://www.drugbank.ca/drugs/DB04303

[9] MetaCyc Pathway: *xylan biosynthesis*, 2008 in http://metacyc.org/META/new-image?type=PATHWAY&object=PWY-5800

[10] R.H. Atalla and J.M. Hackney, I. Uhlin and N.S. Thompson, *Int J. Biol. Macromol. 15*, 109, 1993 in http://www.fpl.fs.fed.us/documnts/pdf1993/atall93a.pdf

[11] M.G. Handford, T.C. Baldwin, F. Goubet, T.A. Prime, J. Miles, Y. Xiaolan and P. Dupree, *Planta 218*, 27, 2003 in http://cat.inist.fr/?aModele=afficheN&cpsidt=15433283

[12] J.J. Ordaz-Ortiz, S.E. Marcus and J.P. Knox, *Mol. Plant 2*, 910, 2009 in http://mplant.oxfordjournals.org/cgi/reprint/2/5/910.pdf

[13] Z.A. Popper and M.G. Tuohy, *Plant Physiol. 153*, 373, 2010 in http://www.plantphysiol.org/cgi/content/full/153/2/373

[14] S. C. Fry, *J. Exp. Bot. 40*, 1, 1989 in http://jxb.oxfordjournals.org/cgi/content/abstract/40/1/1

[15] N. Raikhel, Center for Plant Cell Biology at UC Riverside in www.cepceb.ucr.edu/members/raikhel.htm

[16] Complex Carbohydrate Research Center, The University of Georgia http://www.ccrc.uga.edu/~mao/intro/ouline.htm

[17] A. Varki, R. D. Cummings, J. D. Esko, H. H. Freeze, P. Stanley, C. R. Bertozzi, G. W. Hart, M. E. Etzler, *Essentials of Glycobiology*, Second Edition, 2009 in www.ncbi.nlm.nih.gov/bookshelf/br.fcgi?book=glyco2&part=ch22

[18] S.E. Marcus, Y. Verhertbruggen, C. Herve, J.J. Ordaz-Ortis, V. Farkas, H.L. Pedersen, W.G.T. Willats and J.P. Knox, *BMC Plant Biol. 8*, 60, 2008 in http://www.biomedcentral.com/1471-2229/8/60

[19] J.P. Vincken, W.S. York, G. Beldman and A.G.J. Voragen, *Plant Physiol. 114*, 9, 1997 in http://www.plantphysiol.org/cgi/reprint/114/1/9.pdf

[20] L. Von Schantz, F. Gullfot, S. Scheer, L. Filonova, L. Cicortas Gunnarsson, J.E. Flint, G. Daniel, E. Nordberg Karlsson, H. Brumer and M. Ohlin, BMC *Biotechnol. 9*, 92, 2009 in http://www.biomedcentral.com/1472-6750/9/92

[21] M. Hoffman, Z. Jia, M.J. Pena, M. Cash, A. Harper, A.R. Blackburn, A. Darvill and W.S. York, *Carbohydr. Res. 340*, 1826, 2005 in http://www.ncbi.nlm.nih.gov/pubmed/15975566

[22] W. Zeng, M. Chatterjee and A. Faik, *Plant Physiol. 147*, 78, 2008 in http://www.ncbi.nlm.nih.gov/pmc/articles/PMC2330321/

[23] S. Persson, K.H. Caffall, G. Freshour, M. T. Hilley, S. Bauer, P. Poindexter, M.G. Hahn, D. Mohnen and C. Somerville, *Plant Cell 19*, 237, 2007 in http://www.ncbi.nlm.nih.gov/pmc/articles/PMC1820957

[24] C. Lee, Q. Teng, W. Huang, R. Zhong and Z.H. Ye, *Plant Physiol. 153*, 526, 2010 in http://www.plantphysiol.org/cgi/content/full/153/2/526

[25] Complex Carbohydrate Research Center, The University of Georgia, *Plant Cell Walls, Galacturonans*, in www.ccrc.uga.edu/~mao/galact/gala.htm

[26] M.E. Etzler and D. Mohnen, *Essentials of Glycobiology*, Second Edition, The Consortium of Glycobiology Editors, La Jolla, California, 2009 in www.ncbi.nlm.nih.gov/bookshelf/br.fcgi?book=glyco2&part=ch22

[27] E.M. Lord and J.C. Mollet, *Proc. Natl Acad Sci USA 99*, 15843, 2002 in http://ukpmc.ac.uk/articles/PMC138523;jsessionid=C80D46E0CCD0440BB60A411E4FA8B8BA.jvm4

[28] www.ccrc.uga.edu/~mao/rg2/intro.htm

[29] http://www.ccrc.uga.edu/~lwang/bcmb8020/plant-carbo.pdf

[30] F. Delmas, M. Seveno, J.G.B. Northet, M. Hernould, P. Lerouge, P. Mccourt and C. Chevalier, *J. Exp. Bot. 59*, 2639, 2008 in http://jxb.oxfordjournals.org/cgi/content/full/ern118

[31] M. Irshad, Thesis, *Dynamics of Cell Wall Proteins During Cell Elongation in Etiolated Hypocotyls of Arabidopsis thaliana as Shown by Proteomic and Transcriptomic Surveys*, Université de Toulouse, 2008 in http://tel.archives-ouvertes.fr/docs/00/32/32/17/PDF/Complete_thesis.pdf

[32] C. Somerville, S. Bauer, G. Brininstool, M. Facette, T. Hamann, J. Milne, E. Osborne, A. Paredez, S. Persson, T. Raab, S. Vorwerk, H. Youngs, *Science 306*, 2206, 2004

[33] E.M. Kerr and S.C. Fry, *Planta 217*, 327, 2003

[34] G.M. Cooper, *The Cell: A Molecular Approach*, 2nd Edition, Boston University, Sinauer Associates, 2000 in http://www.ncbi.nlm.nih.gov/bookshelf/br.fcgi?book=cooper&part=A1498

[35] O. Lerouxel, D.M. Cavalier, A.H. Liepman and K. Keegstra, *Curr. Opin. Plant Biol. 9*, 621, 2006 in http://www.ncbi.nlm.nih.gov//pubmed/17011813

[36] G. J. Seifert, *Curr. Opin. Plant Biol. 7*, 277, 2004 in http://www.ncbi.nlm.nih.gov/pubmed/15134748

[37] A.S. Karthikeyan, The Arabidopsis Information Resource (TAIR), *MetaCyc Pathway: xyloglucan biosynthesis*, 2008 in http://metacyc.org/META/new-image?type=PATHWAY&object=PWY-5936

[38] R.M. Perrin, A.E. Derocher, M. Bar-Peled, W. Zeng, L. Norambuena, L. Orellana, N.V. Raikhel, and K. Keegstra, *Science 284*, 1976, 1999 in http://www.ncbi.nlm.nih.gov/pubmed/10373113

[39] G.F. Vanzin, M. Madson, N.C. Carpita, N.V. Raikhel, K. Keegstra and W.D. Reiter, *Proc. Natl. Acad. Sci. USA 99*, 3340, 2002 in http://www.ncbi.nlm.nih.gov/pmc/articles/PMC122520

[40] M. Madson, C. Dunand, X. Li, R. Verma, G.F. Vanzin, J. Caplan, D.A. Shoue, N.C. Carpita and W.D. Reiter, *Plant Cell 15*, 1662, 2003 in http://www.plantcell.org/cgi/content/full/15/7/1662

[41] X.M. Li, I. Cordero, J. Caplan. M. Molhoj and W.D. Reiter, *Plant Physiol. 134*, 940, 2004 in http://www.plantphysiol.org/cgi/content/full/134/3/940?ijkey=c2c19c81525f47696cc2006861fa28875a124e8d

[42] A. Faik, N.J. Price, N.V. Raikhel and K. Keegstra, *Proc. Natl. Acad. Sci. USA 99*, 7797, 2002 in http://www.pnas.org/content/99/11/7797.long

[43] D.M. Cavalier and K. Keegstra, *J. Biol. Chem. 281*, 34197, 2006 in http://www.ncbi.nlm.nih.gov/pubmed/16982611

[44] Y. Verhertbruggen, L. Yin, A. Oikawa and H.V. Scheller, *Plant Signal. Behav. 6*, 1620, 2011 in http://www.ncbi.nlm.nih.gov/pmc/articles/PMC3256401

[45] R. Zhong, M.J. Pena, G.K. Zhou, C.J. Nairn, A. Wood-Jones, E.A. Richardson, W.H. Morrison, A.G. Darvill, W.S. York, Z.H. Ye, *Plant Cell 17*, 3390, 2005 in http://www.ncbi.nlm.nih.gov/pubmed/16272433

[46] S. PERSSON, K.H. CAFFALL, G. FRESHOUR, M T. HILLEY, S. BAUER, P. POINDEXTER, M.G. HAHN, D. MOHNEN and C SOMERVILLE, *Plant Cell 19*, 237, 2007 in http://www.ncbi.nlm.nih.gov/pmc/articles/PMC1820957

[47] M.J. PENA, R. ZHONG, GK. ZHOU, E.A. RICHARDSON, M.A. O'NEILL, A.G. DARVILL, W.S. YORK and Z.H. HE, *Plant Cell 19*, 549, 2007 in http://www.plantcell.org/cgi/content/short/19/2/549

[48] A.M. Wu, E. Hornblad, A. Voxeur, L. Gerber, C. Rihouey, P. Lerouge and A. Marchant, *Plant Physiol. 153*, 542, 2010 in http://www.plantphysiol.org/cgi/content/full/153/2/542

293

[49] C. Lee, R. Zhong and Z.H. Ye, *Plant Cell Physiol. 53*, 135, 2012 in http://pcp.oxfordjournals.org/content/53/1/135.abstract

[50] C. Lee, R. Zhong and Z.H. Ye, *Plant Signal. Behav. 7*, 332, 2012 in http://www.landesbioscience.com/journals/psb/article/19269/?nocache=1087295047

[51] O.O. Obembe, E. Jacobsen, R.G.F. Visser and J.P. Vincken, *Biotechnol. Mol. Biol. Rev. 1*, 76, 2006 in http://www.academicjournals.org/bmbr/PDF/Pdf2006/Sep/Obembe%20et%20al.pdf

[52] CAZY, Carbohydrare Active enzymes, 2010 in http://www.cazy.org/Glycoside-Hydrolases.html

[53] S.C Fry, R.C Smith, K.F. Renwick, D.J. Martin, S.K. Hodge and K.J. Matthews, Biochem J. **282**, 821, 1992 in http://www.ncbi.nlm.nih.gov/pmc/articles/PMC1130861

[54] P. Campbell and J. BRAAM, Trends Plant Sci; 4, 361, 1999 in http://www.ncbi.nlm.nih.gov/pubmed/10462769

[55] J.K.C. Rose, Cornell University, 2006 in http://labs.plantbio.cornell.edu/XTH/overview.htm

[56] J.K.C. Rose, J. Braam, S.C. Fry and K. Nishitani, *Plant Cell Physiol. 43*, 1421, 2002 in http://pcp.oxfordjournals.org/cgi/content/full/43/12/1421

[57] Y.R. Soro, Purification et caractérisation de l'alpha-glucosidase du suc digestif de Archachatina ventricosa (Achatinidae) ; Application à la synthèse de polyglucosylfructosides, Thesis, Toulouse, 2007 in http://eprint.insa-toulouse.fr/archive/00000198/01/Soro.pdf

[58] J.M. Eklof and H. Brumer, *Plant Physiol. 153*, 456, 2010 in http://www.plantphysiol.org/cgi/content/full/153/2/456

[59] P. Johansson, H. Brumer, Iii, M.J. Baumann, A.M. Kallas, H. Henriksson, S.E. Denman, T.T. Teeri and T.A. Jones, *Plant Cell 16*, 874, 2004 in http://www.plantcell.org/cgi/content/full/16/4/874

[60] P. Johansson, H. Brumer, III, M.J. Baumann, A.M. Kallas, H. Henriksson, S.E. Denman, T.T. Teeri and T.A. Jones, *Plant Cell 16*, 874, 2004 in http://www.plantcell.org/cgi/content/full/16/4/874

[61] N.A. Eckardt, *Plant Cell, 16*, 792, 2004 in http://www.plantcell.org/cgi/content/full/16/4/792

[62] M.J. Baumann, JM Eklof, G. Michel, A.M. Kallas, T.T. Teeri, M. Czjzek and H. Brumer, III, *Plant Cell 19*, 1947, 2007 in http://www.plantcell.org/cgi/content/full/19/6/1947?ijkey=eb13d280656f78b1561142d8f8f7350e65429869

[63] http://www.ccrc.uga.edu/~mao/intro/ouline.htm

[64] R. Brunecky, T.B. Vinzant, S.E. Porter, B.S. Donohoe, D.K. Johnson and M.E Himmel, *Biotechnol. Bioeng. 102*, 1537, 2009 in http://www3.interscience.wiley.com/cgi-bin/fulltext/121542373/PDFSTART

[65] M.G. Handford, T.C. Baldwin, F. Goubet, T.A. Prime, J. Miles, X. Yu and P. Dupree, *Planta, 218*, 27, 2003 in http://cat.inist.fr/?aModele=afficheN&cpsidt=15433283

[66] A. Bacic, G.B. Fincher and B.A. Stone, Eds., *Chemistry, Biochemistry and Biology of (1,3)-β-Glucans and Related Polysaccharides*, Academic Press, Amsterdam, 2009

[67] D. Shallom and Y. Shoham, Curr. Opin. *Microbiol. 6*, 219, 2003 in http://biotech.technion.ac.il/Media/Uploads/Documents/Yuval/Publications/Yuval_67.pdf

[68] T.W. Jeffries, C. Ratledge (ed.), *Biochemistry of Microbial Degradation*, 233, Kluwer Academic Publishers, the Netherlands, 1994 in http://www.fpl.fs.fed.us/documnts/pdf1994/jeffr94b.pdf

[69] ACS Symposium #864 *Hemicelluloses: Science and Technology*, P. Gatenholm and M. Tenkanen, Eds. Washington, DC, 2003 in http://pubs.acs.org/isbn/9780841238428

[70] S.E. Marcus, Y. Verhertbruggen, C. Herve, J.J. Ordaz-Ortiz, V. Farkas, H.L. Pedersen, W.G.T. Willats and J.P. Knox, BMC *Plant Biology 8*, 60, 2008 in http://www.biomedcentral.com/1471-2229/8/60

[71] N.M.L. Hansen and D. Plackett, *Biomacromolecules. 9*, 1493, 2008 in http://pubs.acs.org/doi/full/10.1021/bm800053z

[72] http://en.wikipedia.org/wiki/Guar_gum

[73] http://en.wikipedia.org/wiki/Locust_bean_gum

[74] http://www.glucomannan.com/gum.htm

[75] S.V. PATIL, D.R. JADGE and S.C. DHAWALE, Pharmainfo.net, *Tamarind Gum: A Pharmaceutical Overview, Latest reviews 6*, Issue 4, 2008 in http://www.pharmainfo.net/reviews/tamarind-gum-pharmaceutical-overview

[76] N.D. BONAWITZ and C. CHAPPLE, *Annu. Rev. Genet. 44*, 337, 2010 in http://www.annualreviews.org/doi/pdf/10.1146/annurev-genet-102209-163508

[77] F.R.D. VAN PARIJS, K. MOREEL, J. RALPH, W. BOERJAN and R.M.H. MERKS, *Plant Physiol. 153*, 1332, 2010 in http://www.plantphysiol.org/cgi/content/full/153/3/1332

[78] R. VANHOLME, B. DEMEDTS, K. MORREEL, J. RALPH and W. BOERJAN, *Plant Physiol. 153*, 895, 2010 in http://www.plantphysiol.org/cgi/content/full/153/3/895

[79] J. RALPH, K. LUNDQUIST, G. BRUNOW, F. LU, H. KIM, P.F. SCHATZ, J.M. MARITA, R.D. HATFIELD, S.A. RALPH, J.H. CHRISTENSEN and W BOERJAN, *Lignins: Natural polymers from oxidative coupling of 4-hydroxyphenyl-propanoids, Phytochem. Rev. 3*, 29, 2004 in http://www.springerlink.com/content/lx20h1488802t565

[80] J. RALPH, *Lignin Structure: Recent Developments*, Proceedings of the 6th Brazilian Symposium Chemistry of Lignins and Other Wood Components, 1999 in http://www.dfrc.wisc.edu/DFRCWebPDFs/JR_Brazil99_Paper.pdf

[81] M. DASHTBAN, H. SCHRAFT, T.A. SYED and W. Qin, Int. J. Biochem. Mol. Biol. **1**, 36, 2010 in http://www.ijbmb.org/files/IJBMB1004005.pdf

[82] W. BOERJAN, J. RALPH and M. BAUCHER, *Annu. Rev. Plant Biol., 54*, 519, 2003 in http://www.dfrc.wisc.edu/DFRCWebPDFs/2003-Boerjan-ARPB-54-519.pdf

[83] X. LI and C. CHAPPLE, *Plant Physiol. 154*, 449, 2010 in http://www.plantphysiol.org/cgi/content/full/154/2/449/FIG1

[84] University of Maine, Department of Chemistry, 2011 in http://chemistry.umeche.maine.edu/CHY431/Shikimate.html

[85] P. RIPPERT, J. PUYAUBERT, D. GRISOLLET, L. DERRIER and M. MATRINGE, *Plant Physiol. 149*, 1251, 2009 in http://www.plantphysiol.org/content/149/3/1251.full

[86] A. HOLMGREN, G. BRUNOW, G. HENRIKSSON, L. ZHANG and J. RALPH, *Org. Biom. Chem. 4*, 3456, 2006 in http://ddr.nal.usda.gov/bitstream/10113/14222/1/IND44051453.pdf

[87] A. BARAKAT, *Etude de la lignification de parois végétales de graminées par des assemblages modèles : Réactivité, organisation et structure supramoléculaire*, Thesis INRA, Université de Reims Champagne-Ardenne, 2007 in http://ebureau.univ-reims.fr/slide/files/quotas/SCD/theses/exl-doc/GED00000531.pdf

[88] M.T.T. NGUYEN, *Synthesis and Oxidation of Lignin-Carbohydrate Model Compounds*, Thesis, University of Maine, 2008 in http://www.library.umaine.edu/theses/pdf/NguyenMTT2008.pdf

[89] T. KOSHIJIMA and T. WATANABE, *Association Between Lignin and Carbohydrates in Wood and Other Plant Tissues*, Springer, 2003

[90] MicrobeWiki, Kenyon College, *Phanerochaete chrysosporium* 2008 in http://microbewiki.kenyon.edu/index.php/Phanerochaete_chrysosporium

[91] S. Shary, S.A. Ralph and K.E. Hammel, *Appl. Environ. Microbiol. 73*, 6691, 2007 in http://aem.asm.org/cgi/content/full/73/20/6691

[92] D.W.S. Wong, *Appl. Biochem. Biotechnol. 157*, 174, 2009 in http://www.udea.edu.co/portal/page/portal/bibliotecaSedesDependencias/unidadesAcademicas/CorporacionAcademicaAmbiental/BilbiotecaDiseno/Archivos/Structure-ActionMechanismLigninolyticEnzymes.pdf

[93] P.M. Hanna, D.R. Mcmillin, M. Pasenkiewicz-Gierula, W.E. Antholine and B. Reinhammar, *Biochem. J. 253*, 561, 1988 in http://www.ncbi.nlm.nih.gov/pmc/articles/PMC1149334

[94] K. Piontek, M. Antorini and T. Choinowski, *J. Biol. Chem. 277*, 37663, 2002 in http://www.jbc.org/content/277/40/37663.full

[95] UC Davis, ChemWiki, by University of California, *Laccase*, 2008 in http://chemwiki.ucdavis.edu/Wikitexts/UCD_Chem_124A%3A_Berben/Laccase/Laccase_1

[96] http://en.wikipedia.org/wiki/Laccase

[97] T. Choinowski, W. Blodig, K.H. Winterhalter and K. Piontek, *J. Mol. Biol. 286*, 809, 1999 in http://www.ncbi.nlm.nih.gov/pubmed/10024453

[98] K. Piontek, A.T. Smith and W. Blodig, *Biochem. Soc. Trans 29*, 111, 2001 in http://www.biochemsoctrans.org/bst/029/0111/bst0290111.htm

[99] M. Sundaramoorthy, M.H. Gold and T.L. Poulos, *J. Inorg. Biochem. 104*, 683, 2010 in http://www.ncbi.nlm.nih.gov/pubmed/20356630

[100] http://www.pdb.org/pdb/explore.do?structureId=2BOQ

[101] M. Perez-Boada, F.J. Ruiz-Duenas, R. Pogni, R. Basosi, T. Choinowski, M.J. Martinez, K. Piontek and A.T. Martinez, *J. Mol. Biol. 354*, 385, 2005 in http://www.ncbi.nlm.nih.gov/pubmed/16246366?dopt=Abstract and http://www.cib.csic.es/repositorio_bd/grupo/3/html/2005-Perez-Boada-JMB-354-385.pdf

[102] S.D. Mansfield, E. De Jong and J.N. Saddler, *Appl. Environ. Microbiol. 63*, 3804, 1997 in http://www.ncbi.nlm.nih.gov/pmc/articles/PMC1389261/pdf/hw3804.pdf

[103] D. Meier, *Catalytic Hydrocracking of Lignins to Useful Aromatic Feedstocks*, DGMK Conference, Berlin, 2008 in http://www.dgmk.de/petrochemistry/abstracts_content16/Meier.pdf

[104] J. Zakzeski, P.C.A. Bruijnincx, A.L. Jongerius and B.M. Weckhuysen, *Chem. Rev. 110*, 3552, 2010 in http://pubs.acs.org/doi/abs/10.1021/cr900354u

[105] P. Dole and F. Bouxin, *Macromolecular and Molecular Uses of Lignin*, GFP 2008, Lyon, 2008 in http://gfp2008.ccsd.cnrs.fr/docs/00/32/05/99/PDF/abstract gfp2008_DOLE.pdf

[106] L. JOUANIN, *Lignines*, Académie d'Agriculture de France, 2010 in http://www.academie-agriculture.fr/mediatheque/seances/2010/20100217resume2.pdf

[107] T. LIU and Q.W. WANG, *Advanced Materials Research 113-114*, 606, 2010 in http://www.scientific.net/AMR.113-116.606

[108] I. BRODIN, *Chemical Properties and Thermal Behaviour of Kraft Lignins*, KTH Royal Institute of Technology, Stockholm, 2009 in http://kth.diva-portal.org/smash/get/diva2:234300/FULLTEXT01

[109] J. VAN DAM, R. GOSSELINK and E. DE JONG, *Lignin Applications*, Wageningen UR, Agrotechnology & Food Innovations in http://www.biomassandbioenergy.nl/infoflyers/LigninApplications.pdf

[110] R. GOSSELINK et al. *Valorization of Lignin Resulting from Biorefineries, Agrotechnology & Food Sciences Group*, Wageningen UR, 2008 in http://www.rrbconference.com/bestanden/downloads/145.pdf

[111] M. KLUKO, Densified Fuel Pellets, United States Patent Application 20090205546, 2009 in http://www.freepatentsonline.com/y2009/0205546.html

[112] N. EISENREICH, EU project: BIOCOMP, *New Classes of Engineering Composites Materials from Renewable Resources*, 2008 in http://www.biocomp.eu.com/uploads/Final_Summary_Report.pdf

[113] Wageningen UR, 2010 in http://www.biobasedproducts.wur.nl/UK/projects/fibres

[114] H. HATAKEYAMA, *Chemical Modification, Properties and Usage of Lignin*, T.Q. HU Ed., 2002 in http://www.springer.com/life+sciences/forestry/book/978-0-306-46769-1

[115] Lignol Energy, *Cellulosic Ethanol – The Sustainable Fuel*, TAPPI, International Conference on Renewable Energy, 2007 in http://www.tappi.org/content/Events/07renew/07ren06.pdf

[116] Borregaard LignoTech in http://www.lignotech.no

[117] Innventia, Press release: *EU approves Swedish support to LignoBoost demonstration* in http://www.innventia.com/templates/STFIPage____9421.aspx

[118] CIMV, The Biorefinery Concept, *CIMV Technology* in http://www.cimv.fr/cimv-technology/cimv-technology/5-.html?lang=en

[119] Biocore Project, Chemical and thermochemical transformations in http://www.biocore-europe.org/page.php?optim=Chemical-and-thermochemical-transformations

[120] M. DELMAS, *Chem. Eng. Technol. 31*, 792, 2008 in http://www.cimv.fr/uploads/vegetal-refining-and-agrichemistry-chemical-engineering-and-technology-2008.pdf

Chapter 7

Pretreatments of Lignocellulosic Biomass

7.1 Introduction

Biofuels and biobased chemicals produced from various lignocellulosic materials have the potential to be valuable substitutes for fossil-based products. [1] Many structural and compositional factors hinder the enzymatic digestibility of cellulose present in lignocellulosic biomass. [2] The goal of any pretreatment is to alter the structure of the lignocellulosic materials to make cellulose and hemicelluloses more accessible to the enzymes that convert the polysaccharides into fermentable sugars. This goal is reached by (1) degrading and removing hemicelluloses and lignin, (2) reducing the crystallinity of cellulose, and (3) increasing the porosity of the lignocellulosic materials (**Figure 7.1**). [3]

Pretreatment must meet the following requirements: (1) improve the formation of sugars or the ability to subsequently form sugars by hydrolysis, (2) avoid the degradation or loss of carbohydrates, (3) avoid the formation of byproducts that are inhibitory to the subsequent hydrolysis and fermentation processes, and (4) be cost-effective. [1]

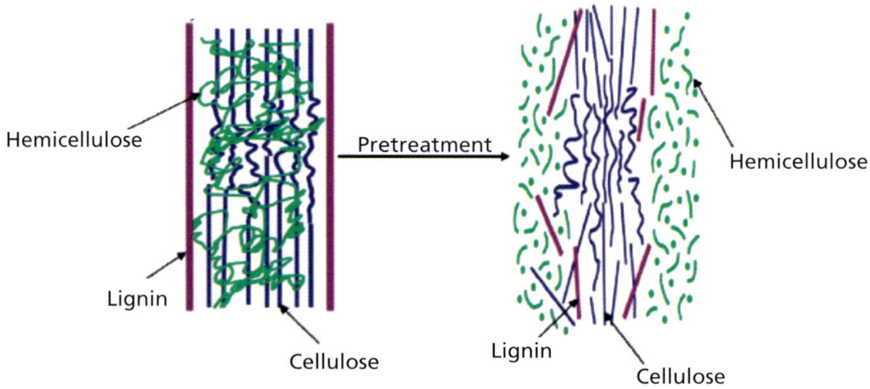

Figure 7.1 Schematic of the role of pretreatment in the conversion of lignocellulosic biomass to biofuels and biobased chemicals. [1]

During the last years, a variety of pretreatment methods that change the physical and chemical structure of the lignocellulosic biomass and improve hydrolysis rates have been developed. Many methods have been shown to result in high sugar yields, above 90% of the theoretical yield for lignocellulosic biomasses such as woods, grasses, corn, and so on. In this chapter, we give an overview of the various available pretreatment technologies and we review the recent literature on the use of these technologies for pretreatment of lignocellulosic biomass.

The pretreatment methods can be divided into different categories, namely: (1) physical pretreatments of biomass involving milling and high-energy radiation; (2) chemical methods involving liquid hot water, dilute and concentrated acids, alkali, organic solvents (***organosolv***), oxidizing agents and ***ionic liquids***; (3) physicochemical methods involving steam explosion, ammonia pretreatments, CO_2 explosion, mechanical/alkaline pretreatment; and (4) biological methods. [3-4]

7.2 Physical pretreatments

7.2.1 Mechanical comminution

Reduction of particle size is often needed to make material handling easier and to increase surface/volume ratio. [3] This can be done by chipping, milling or grinding. Mechanical pretreatment is usually carried out before a following processing step, and the desired particle size is dependent on these subsequent steps. Factors like capital costs, operating costs, scale-up possibilities, and depreciation of equipment are important in mechanical pretreatment.

7.2.2 High energy radiation

Digestibility of cellulosic biomass has been enhanced by the use of high-energy radiation methods, including γ-rays, ultrasound, electron beam, pulsed electrical field, UV, and microwave. [5] The action mode behind high energy radiation could be one or more changes in biomass structure, including increase in specific surface area, decrease in the polymerization degree and crystallinity of cellulose, hydrolysis of hemicelluloses, and partial depolymerization of lignin. However, these high-energy radiation methods are usually slow, energy-intensive, and expensive. They also appear to be strongly substrate-specific.

The Iowa State University investigated the application of ultrasound to pretreat corn or biomass for ethanol production. [6] They found that (1) ultrasonics can enhance fermentable sugar yield when integrated with other pretreatment methods and (2) ultrasonics are effective in enhancing the dissolution time of switchgrass in ionic liquids. A recent patent application assigned to Bio-Sep Limited and Brooks claimed a method of processing biomass comprising the digestion of biomass in an aqueous solvent system subjected to ultrasonic irradiation and separating the biomass into its constituents. [7]

7.3 Chemical pretreatments

Here are described the pretreatments that are purely initiated by chemical reactions for disruption of lignocellulosic biomass. Combinations of chemical pretreatment and other types of pretreatments will be reviewed in other sections.

7.3.1 Liquid hot water

Liquid hot water (LHW) processes are biomass pretreatments with water at high temperature and pressure. [3] Flow-through processes pass water maintained in the liquid state at elevated temperatures through lignocellulosic materials. [2] This type of pretreatment has been termed ***autohydrolysis***, ***hydrothermolysis***, hydrothermal pretreatment, aqueous or steam/aqueous fractionation, ***solvolysis***, or ***aquasolv***. [8]

Solvolysis by hot compressed liquid water contacts water with biomass for up to 15 minutes at temperatures of 200–230 °C. [2-3] Between 40 and 60% of the total biomass is dissolved in the process, with 4–22% of the cellulose, 35–60% of the lignin and all of the hemicelluloses being removed. Over 90% of the hemicelluloses is recovered as monomeric sugars when acid was used to hydrolyze the resulting liquid. Variability in results was related to the biomass type with high lignin solubilization impeding recovery of hemicellulose sugars.

Liquid hot water pretreatments are both helped and hindered by the cleavage of *O*-acetyl and ***uronic acid*** substitutions from hemicelluloses to generate

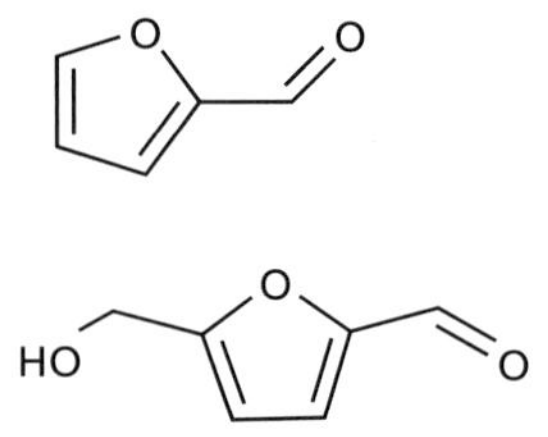

Figure 7.2 Furfural structure (top) and 5-hydroxymethylfurfural structure (bottom).

acetic acid and other organic acids. [2] The release of these acids helps to catalyze formation and removal of oligosaccharides. However, the polysaccharides and especially hemicelluloses may be further hydrolyzed to monomeric sugars which are subsequently partially degraded to aldehydes if acid is used. These aldehydes, mainly ***furfural*** from pentoses and 5-hydroxymethylfurfural from hexoses, are inhibitory to microbial fermentation (**Figure 7.2**).

There are three types of liquid hot water reactor configurations: co-current, countercurrent, and flow-through (**Figure 7.3**).

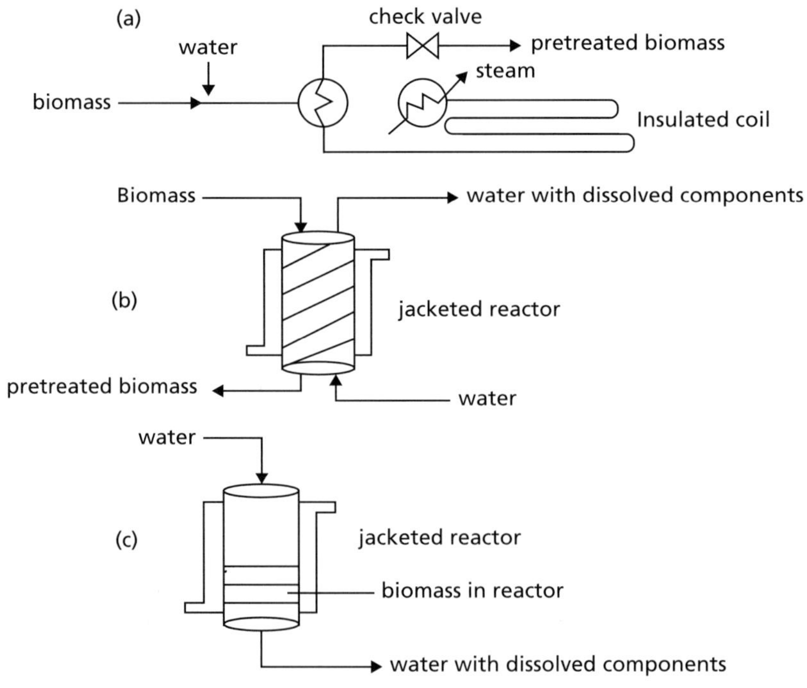

Figure 7.3 Schematic illustration of co-current, counter-current and flow-through pretreatment methods: (a) Co current liquid hot water pretreatment, (b) counter-current reactor, (c) flow-through reactor. [2]

In co-current pretreatment, slurry of biomass and water is heated to the desired temperature and held at the pretreatment conditions for a controlled residence time before being cooled. Counter-current pretreatment is designed to move water and lignocellulose in opposite directions through the pretreatment reactor. In a flow-through system, hot water passes over a stationary bed of lignocellulose hydrolyzes and dissolves lignocellulose components and carries them out of the reactor.

The Danish company Inbicon has developed a patented hydrothermal pretreatment, which is integrated in a three-stage process: mechanical conditioning of the biomass, hydrothermal pretreatment, and enzymatic hydrolysis. [9] During pretreatment the biomass is continuously mixed with water and heated to ~180–200 °C for 5–15 min in order to destroy the lignin and make the cellulose accessible to enzymes.

Another important feature of the Inbicon's pretreatment is to separate the alkali content (particularly Na and K) and a portion of the inhibitors formed through partial degradation of the hemicellulose. The process operates at high (30–40%) dry matter content, minimizing water and energy consumption. The biomass is fed into the pressure reactor with proprietary particle pumps.

7.3.2 Acid hydrolysis

General

Acid pretreatment of lignocellulosic biomass has received considerable research attention. [2, 10] Hemicelluloses are the first of the constituents of biomass to break down during acid hydrolysis. For example, when sulfuric acid is used, 0.01M is generally enough to break down the hemicelluloses into their monomers.

Dilute sulfuric acid has been added to lignocellulosic materials for some years to commercially produce furfural. [11] Dilute sulfuric acid is mixed with biomass to hydrolyze hemicelluloses to xylose and other sugars, and then continue to break xylose down to form furfural. The furfural is recovered by distillation. The biomass/acid mixture is held at temperatures of 160–220 °C for periods ranging from seconds to minutes.

Addition of sulfuric acid has been initially applied to remove hemicelluloses either in combination with breakdown of cellulose to glucose or prior to acid hydrolysis of cellulose. [2] Hemicelluloses are removed when sulfuric acid is added and this enhances digestibility of cellulose in the residual solids. Most approaches are based on dilute sulfuric acid, but nitric acid, hydrochloric acid, and phosphoric acid have also been tested. Furthermore, it is worth mentioning the autocatalyzed acid hydrolysis, a process where no acid is added. The process solely depends on the acetate in the biomass to form acetic acid, which becomes the acid during hydrolysis.

Process description

Dilute acid hydrolysis

Dilute acid treatment is one of the most effective pretreatment methods for lignocellulosic biomass. [1, 3] Two types of dilute-acid treatment processes are typically used:

1. a high-temperature ($T>160$ °C), continuous-flow process for low solids loadings (5–10 wt% substrate concentration);

2. a low- temperature ($T<160$ °C) batch process for high solids loadings (10–40 wt% substrate concentration).

Dilute (mostly sulfuric) is mixed with lignocellulosic material and the mixture is held at the proper temperature for short periods up to a few minutes. [3] Hydrolysis of hemicelluloses then occurs, releasing monomeric sugars and soluble oligomers from the cell wall into the hydrolysate. Hemicellulose removal increases porosity and enhances enzymatic digestibility, with maximum digestibility usually coinciding with complete hemicellulose removal. [12] As an alternative to inorganic acids, organic acids, such as maleic acid and fumaric acid, can be used for dilute acid pretreatment.

The treatment offers good performance in terms of recovering hemicellulose sugars, but it has some important limitations, including corrosion, that mandate expensive material of construction. [2-3] The acid must be neutralized before the sugars proceed to fermentation. Furthermore, formation of degradation products and release of biomass fermentation inhibitors are other characteristics of acid treatment. The pretreatment is especially suitable for biomass with low lignin content, as almost no lignin is removed from biomass.

Concentrated acid hydrolysis

Concentrated acids (mostly sulfuric acid and hydrochloric acid) have also been used to treat lignocellulosic materials because they are powerful agents for cellulose hydrolysis, and no enzymes are needed after the acid hydrolysis. [1, 3] However, concentrated acids are toxic, corrosive, hazardous, and, thus, require reactors that are resistant to corrosion. In addition, the concentrated acid must be recovered after hydrolysis to make the process economically viable.

Advantages of concentrated acid hydrolysis are the flexibility in terms of feedstock choice, high monomeric sugar yield, as well as mild temperature conditions. [3] Today, several companies are in the process of commercializing concentrated acid hydrolysis for fermentation to ethanol (BlueFire, [13] Biosulfurol Energy [14]).

Mode of action

Acid hydrolysis releases oligomers and monosaccharides and has historically been modeled as a homogeneous reaction in which acid catalyzes breakdown of cellulose to glucose followed by breakdown of the glucose to form hydroxymethyl furfural and other degradation products. [2, 15] This reflects the approximately equal reactivity of glycosidic bonds with respect to hydrolysis. The model was adapted to describe the hydrolysis of hemicelluloses and formation of furfural and other decomposition products.

Oligomers are generally ignored in many models since they are viewed as being too short-lived to be important. [2] Other studies have shown oligomers to be present in batch hydrolysis systems and to be a significant fraction of the product for very dilute acid and water only flow-through systems. Kinetics models have been modified to include hemicellulose hydrolysis to oligomers followed by their breakdown to sugars where the sugars degrade to furfural and other compounds. Modeling of hemicellulose hydrolysis as a biphasic reaction incorporates fast-and slow-hydrolyzing hemicellulose fractions.

Specific effects on lignocellulosic substrates

Acid pretreatment has been used on a wide variety of feedstocks ranging from hardwoods to grasses and agricultural residues. [1]

Ishizawa et al. [16] evaluated whether porosity was one of the factors governing the overall enzymatic digestibility of the cellulose in dilute-acid-pretreated biomass. [1] Corn stover was subjected to dilute sulfuric acid pretreatments with a range of severities. All the pretreated samples showed higher pore volumes than untreated corn stover, but no significant differences in pore volume were measured for the different pretreated samples. The authors presented evidence that porosity may distinguish lignocellulosic substrates of low digestibility from those of high digestibility; however the results also indicated that porosity may not be the unique variable affecting the overall yields of ethanol production from lignocellulosic materials. [17]

Lu et al. [18] emphasized the use of different conditions to attain the highest yields of xylose and glucose, from pretreatment and enzymatic hydrolysis. [1] They performed pretreatment of corn stover for sulfuric concentrations of 2%, 4%, and 6% at 80 °C, 100 °C and 120 °C. The optimum conditions were a sulfuric acid concentration of 2% and a reaction time of 43 minutes at 120 °C. Up to 77% xylose yield was obtained, while the glucose yield was only 8.4%. The corresponding solid phase showed good susceptibility towards enzymatic hydrolysis, leading to solutions containing up to 42.11 g glucose/100 g substrate, equivalent to a conversion yield of 70% under the optimum conditions. The total sugar from the two stages was 62.8 g/100 g substrate.

Cara et al. [19] studied the production of fermentable sugars from olive-tree biomass by dilute-acid pretreatment and further saccharification of the pretreated solid residues. Pretreatment was performed at 0.2%, 0.6%, 1.0% and 1.4% (w/w) sulfuric acid concentrations while temperature was in the range of 170–210 °C. Attention was paid to sugar recovery both in the liquid fraction issued from pretreatment (prehydrolysate) and that in the water-insoluble solid. A maximum of 83% of hemicellulosic sugars in the raw material was recovered in the prehydrolysate obtained at 170 °C., 1% sulfuric acid concentration, but the enzyme accessibility of the corresponding pretreated solid was not very high. In turn, a maximum enzymatic hydrolysis yield of 76.5% was attained from a pretreated solid at 210 °C and 1.4% acid concentration, in which cellulose solubilization was detected, but sugar recovery in the hydrolysate was the poorest among all the experiments performed. To take into account the fermentable sugars generated by pretreatment and the glucose released by enzymatic hydrolysis, an overall sugar yield was calculated. The maximum value of 36.3 g sugar/ 100 g raw material was obtained when pretreating olive-tree biomass at 180 °C and 1% sulfuric acid concentration, representing 75% of all sugars in the raw material.

Yat et al. [20] reported dilute-acid pretreatment of four timber species (aspen, balsam fir, basswood and red maple) and switchgrass using dilute sulfuric acid for 50 g dry biomass/L under similar conditions. The primary goal was to obtain kinetic data of xylose formation and degradation from a match between a first-order reaction model and the experimental at various final reactor temperatures (160–190 °C), sulfuric acid concentrations (0.25–1.0% w/v), and particle sizes (28–10/20 mesh). Reaction rates for the generation of xylose from hemicelluloses and the generation of furfural from xylose were strongly dependent on both temperature and acid concentration. However, no effect was observed for the particle sizes studied. Oligomer sugars were observed early in the reaction period for all sugars, but were reduced to low concentrations at later times. Maximum yields for xylose ranged from 70% (balsam) to 94% (switchgrass), for glucose from 10.6% to 13.6%, and for other minor sugars from 8.6% to 58.9%. There was a near linear dependence on acid concentration observed for xylose degradation.

Addition of very dilute sulfuric acid (~0.1% versus the 0.7–3.0% typical for batch dilute-acid technology) in a flow-through reactor configuration is effective at acid levels lower than 0.1%. [1] Despite achieving excellent hemicellulose sugar yields and highly digestible cellulose with low acid loadings, the equipment configurations and the high ratio of water to solids employed in flow-through systems require significant energy for pretreatment and product recovery. Although dilute-acid pretreatment can significantly improve cellulose hydrolysis, its cost is usually higher than those of physicochemical pretreatment processes such as steam explosion or AFEX. Neutralization is necessary for the downstream enzymatic hydrolysis or fermentation processes. Dilute-acid pretreatment is also known to have the potential to adversely affect the efficiency of enzymatic hydrolysis of biomass. Selig et al. [21] reported the formation of spherical droplets on the surface of residual corn stover following dilute-acid pretreatment at high temperature. They suggested that these droplet formations were composed of lignin and possible lignin-carbohydrate complexes. It was demonstrated that these droplets were produced from corn stover during pretreatment under neutral or acidic conditions at and above 130 °C and that they can deposit back onto the surface of residual biomass. The deposition of droplets produced under certain pretreatment conditions (acidic pH; T > 150 °C) and captured onto pure cellulose was shown to have a negative effect on the enzymatic saccharification of this substrate. Therefore, it is important to carefully examine the appropriate dilute-acid pretreatment of lignocellulosic biomass species. It has been reported that materials that have been subjected to acid hydrolysis can be harder to ferment due to the presence of toxic substances. [1]

7.3.3 Alkaline hydrolysis

The general principle behind alkaline pretreatment methods is the removal of lignin whereas cellulose and a portion of hemicelluloses remain in the solid fraction. Alkaline treatment of woody biomass is well known in the pulp and paper industry as kraft pulping. After grinding, and sometimes steam activation, wood chips are cooked with sodium hydroxide and sodium sulfide at temperatures above 150 °C. [22-23] Kraft treatment conditions degrade and dissolve readily lignins molecules by alkaline cleavage of ether bonds. [24] The reaction starts with an alkaline rearrangement of a phenolate to a quinone methide with cleavage of α-aryl ether bonds (**Figure 7.4**). [25] A nucleophile adds to the quinone methide with the subsequent cleavage of the adjacent β-O-4 ether bond (**Figure 7.5**).

However, the cooking conditions for global solubilization of lignin are also favorable to cellulose and glucomannan degradation, mainly by peeling oxidation reaction. [23, 26] Therefore, delignification goes together with carbohydrate decomposition and the final yield of the reaction is significantly lower than the theoretical target. [23, 27] Additives like anthraquinone (AQ) are known to favor lignin solubilization.

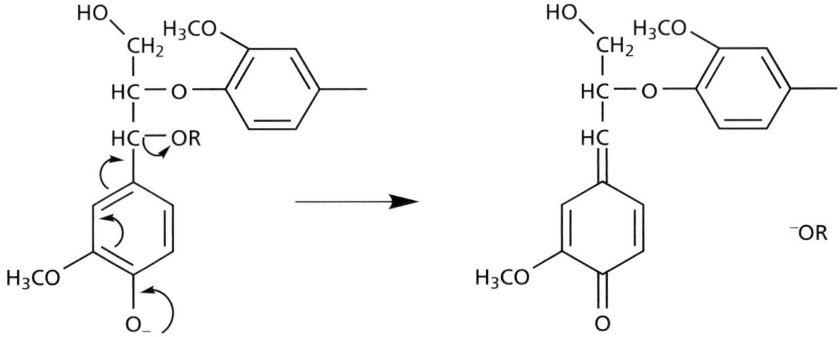

Figure 7.4 Alkaline cleavage of α-aryl ether bonds in phenolic phenylpropane units. R = H, alkyl, aryl. [24]

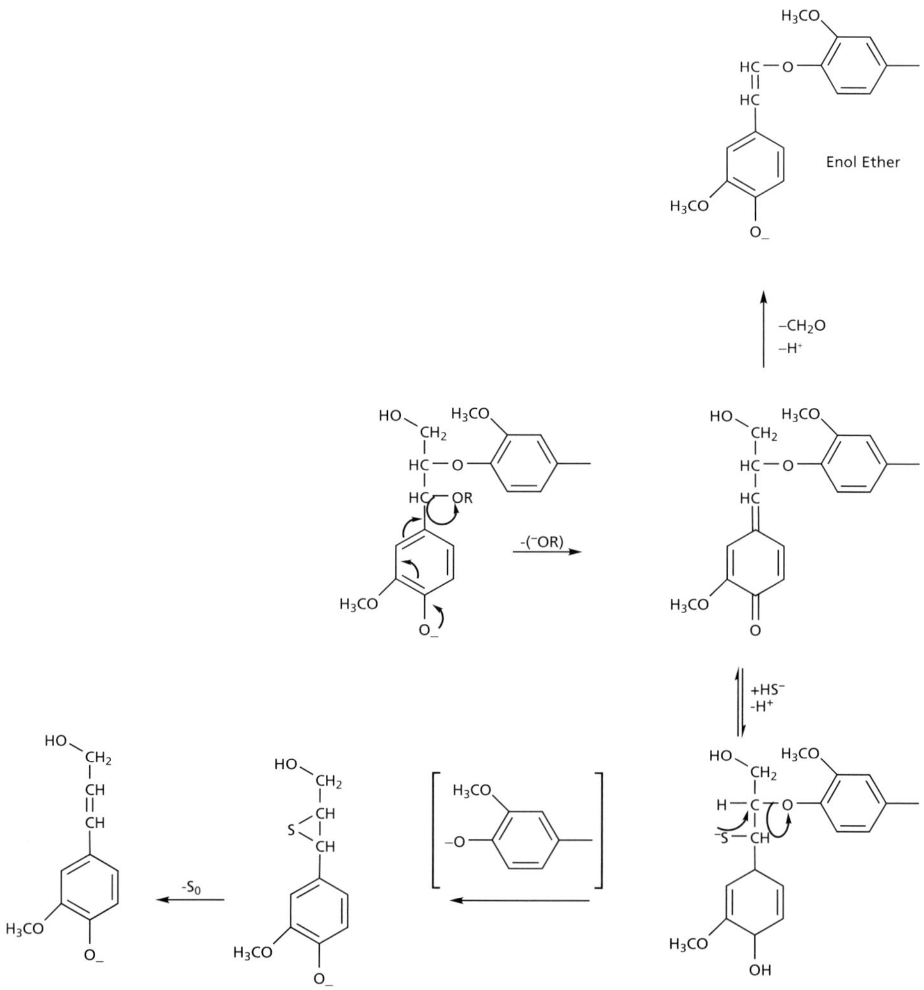

Figure 7.5 Sulfidolytic cleavage of β-aryl ether bonds in phenolic phenylpropane units and conversion into enol-ether units. [24]

To avoid environmental impact and reduce cost due to the use of expensive chemicals, the kraft process needs to be a closed loop with recovery steps for lime and soda. These recovery steps can be strongly perturbed by the precipitation of sodium silicate (amphoteric salt) obtained from silica present in lignocellulosic raw materials. As silica content goes from 1 kg/t in wood, to about 20 kg/t in bamboo, and 50–150 kg/t in wheat or rice straw, kraft process is not applicable for most non-woody lignocellulosic residues.

The idea of transforming a standard kraft pulping mill into an integrated biorefinery becomes a reality. [28-29] An advantage is that the conversion of

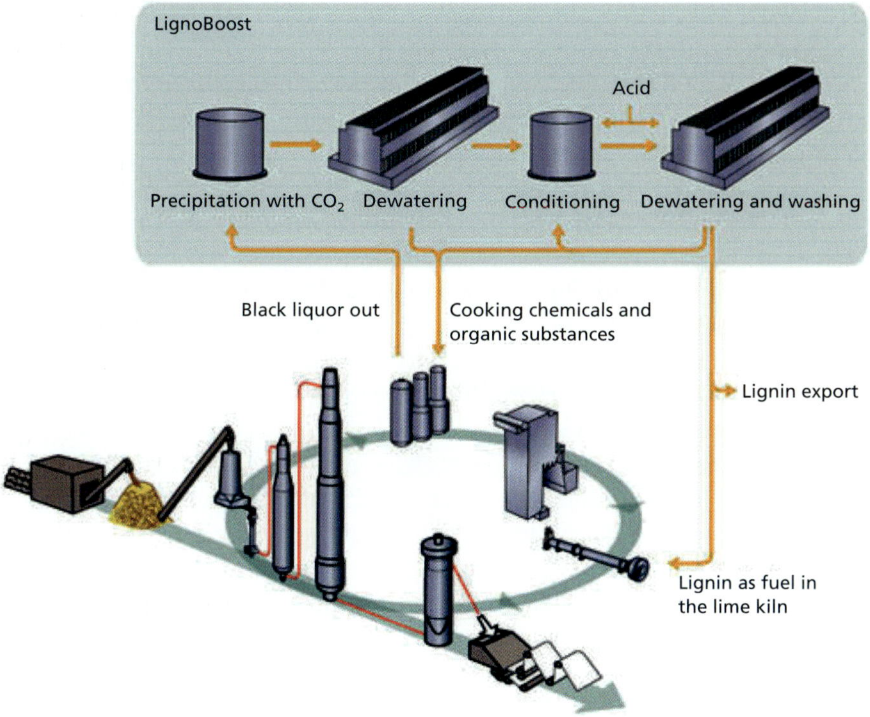

Figure 7.6 Extraction of lignin from kraft pulp mill black liquor by the LignoBoost process. © 2012 Metso. [31]

a kraft mill from 100% pulp making to 100% biorefinery can be done in a stepwise fashion. One biorefinery concept involves extracting hemicelluloses, which normally end up in the black liquors of kraft pulp mills, prior to pulping, and using the extract for the production of acetic acid and ethanol. [30] The extracted liquor undergoes hydrolysis, separation of lignin, liquid-liquid extraction to separate acetic acid and furfural from monosaccharides, fermentation, and distillation for the production of ethanol, acetic acid, and furfural.

Developed recently, the LignoBoost process is designed to extract lignin from pulp mill black liquor (**Figure 7.6**). [31]

It starts with lignin being precipitated from the black liquor by lowering the pH with CO_2 gas. [32] Precipitation is followed by a separation step in which the lignin is dewatered. This is done by using the Metso Vertical Plate Pressure Filter. The first filter press produces a lignin cake with a solid lignin content of ~50% by weight. The unwashed lignin cake is dissolved in a re-suspension step together with spent wash water (from the downstream washing step). This slurry is then acidified with sulfuric acid in order to lower the pH to the same level as in the washing step. By doing so a large part of the remaining sodium is transferred from the lignin to the water phase. The re-slurried lignin is once again dewatered using a Vertical Plate Pressure Filter and finally washed with wash water, i.e. diluted sulfuric acid. The lignin can be exported or used as fuel. The demonstration plant in Bäckhammar (Sweden) owned by Metso Corporation produces ~4000 t/year of kraft lignin.

Several authors evaluated the ability of sodium hydroxide solutions in different concentrations to enhance enzymatic digestibility of corn stover, sorghum straw, wheat straw, cotton stalks, or switch grass. [33-35] Treatment is applied alone or with addition of irradiation, such as microwave and ultra high frequency (UHF). [36-37] Pedersen et al. [38] assessed the influence of pretreatment pH, temperature and time, and their interactions on the enzymatic glucose and xylose yields from mildly pretreated wheat straw. They concluded that pretreatment pH exerted significant effects and factor interactions on the enzymatic glucose and xylose releases. Quite extreme pH values (pH 1 or pH 13) were necessary with mild thermal pretreatment conditions ($T<140$ °C, $t<10$ min). Alkaline pretreatments generally induced higher sugar release and did so at lower temperatures than required with acidic pretreatments. Mild conditions in terms of temperature and alkaline concentration reduce the endwise degradation of cellulose and hemicelluloses, and thus the quantity of acidic monomers, cyclic enols and phenolic compounds released during degradation. [23]

As glucomannans are much less stable in alkaline media than xylans, the proportion of remaining hemicelluloses in the solid part may vary significantly, especially with the composition of lignocellulosic raw material. Alkaline treatments are well known to be more efficient on hardwoods and straw than on softwood. In any case, further enzymatic treatment requires hemicellulases both acting on dissolved and solid hemicelluloses fractions. [39]

By using sodium hydroxide or calcium hydroxide (slaked lime), salts are formed that may be incorporated in the biomass and that need to be removed or recycled. [3] Slake lime has the additional benefits of low reagent cost and safety, and being recoverable from water as insoluble calcium carbonate by reaction with carbon dioxide. [2] The carbonate can then be converted to lime (CaO or burnt lime) using established lime-kiln technology.

Slaked lime treatments have been applied to different cellulosic materials (poplar, straw, bagasse, corn stover, switchgrass, etc.), showing good activation yields and enzymatic digestion similar to or even better than after acidic treatments. [40-44] As with soda, treatment efficiently solubilizes lignins, and removes acetyl and uronic acid substitutions on hemicellulosic compounds.

Essentially developed at the laboratory scale by Holtzapple et al. [45-47] the lime activation process is currently under industrialization by the US company Terrabon (MixAlco process). [48] The process converts biomass such as municipal solid waste, sewage sludge, forest product residues and energy crops into valuable chemicals such as acetic acid, ketones and alcohols that can be processed into renewable hydrocarbon fuels (**Figure 7.7**).

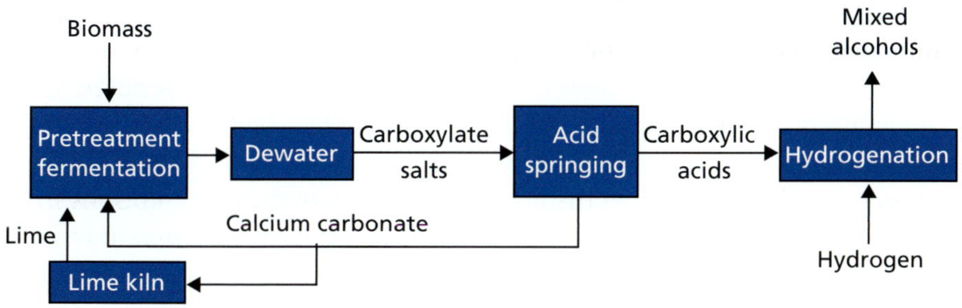

Figure 7.7 Schematic of the MixAlco process. [49, 50]

The biomass is treated with lime and air to enhance digestibility. [49] This material is fed to a mixture culture of microorganisms that produce carboxylic acids, which are transformed into salts by addition of calcium carbonate. After drying, the salts are acidified and thermally converted into ketones. Finally the ketones are hydrogenated to alcohols. Two different versions of the MixAlco process are available. Version one is the original process which produces mixed alcohol fuels. Version two produces carboxylic acids and primary alcohols (ethanol). Terrabon expects to achieve economic viability and a fully operational biorefinery for 2013.

During the pretreatment phase, lignocellulosic biomass is contacted with lime, which removes lignin to enhance reactivity. [51] Low-lignin (<10%) biomass does not require pretreatment. Medium-lignin (10–18%) biomass responds well to 1-2 hours of contact with boiling lime water. High-lignin (18–24%) biomass responds well to long-term lime plus air (~55 °C, 1 month, atmospheric pressure). Very high-lignin (>24%) biomass requires lime plus oxygen (~150 °C, 2 hours, ~20 atm). Biomass, lime and calcium carbonate are blended and stored in a large pile. [49] Air is blown up through the pile while water is trickled down through the pile. The combination of air and lime removes lignin from the biomass rendering the biomass more digestible. The pile is then inoculated with acid-forming anaerobic microorganisms. The microorganisms digest the biomass producing carboxylic acids. The carboxylic acids combine with calcium carbonate to form carboxylate salts, which are extracted from the pile with water.

7.3.4 Organosolv/solvent fractionation

Organosolv pretreatment processes use an organic solvent or mixtures of organic solvents with water to remove lignin before enzymatic hydrolysis of the cellulosic fraction. [3] In addition to lignin removal, hemicellulose hydrolysis occurs resulting in improved enzymatic digestibility of cellulose. Common solvents for the processes include ethanol, methanol, acetone and ethylene glycol. Temperatures used for the processes can be as high as 200 °C, but lower temperatures can be sufficient depending on e.g. the type of biomass and the use of a catalyst. Possible catalysts include inorganic and organic acids.

Originally developed in the years 1980–1990 as an "environmentally friendly" alternative to kraft and sulfite pulping, this process received regain of interest recently with the development of biorefineries. Main advantages are absence of odoriferous sulfur compounds, increase of yield compared to kraft pulp process, and possibility to apply to all feedstocks; main drawbacks are energy consumption and complicated recovery systems for solvent.

Main reactions during organosolv cooking are hydrolysis of ether linkages in lignin and hemicellulose-lignin bonds. [23, 52] Easily hydrolyzable α-ether bonds are more readily broken, but it is likely that β-aryl ether bonds are also broken under the conditions of many processes. Most of the organosolv processes use either a neutral solvent, with or without an acid added catalyst, or an acidic solvent. In cases where no acid is added, the liquor becomes acidic as a

result of release of acetic acid from the wood. Under these conditions neutral or acidic solvolysis reactions of lignin may be expected to occur.

Model compound have shown that α-aryl ether linkages are more easily split that β-aryl ether linkages, especially when they occur in a lignin structural unit containing a free hydroxyl group in the para position. [52] In this case, the formation of a quinone methide intermediate is possible (**Figure 7.8**). Otherwise, a nucleophilic substitution reaction occurs at the benzilic position (**Figure 7.9**). In the presence of added acid, the hydrolysis of benzyl ether linkages is facilitated by acid catalysis as shown in **Figure 7.9**.

The likelihood of β-ether cleavage is greater in more acidic systems. [52] Formation of either a resonance-stabilized benzyl carbocation or a similarly stabilized transition state is followed by elimination of water to form the readily hydrolyzed enol ether (**Figure 7.10**). This releases guaiacol and forms β-hydroxyconiferyl alcohol, which exist in equilibrium with its keto form. In general, β-ether cleavage is likely to be more important in the pulping of hardwoods than in the pulping of softwoods. Hardwoods are delignified more easily

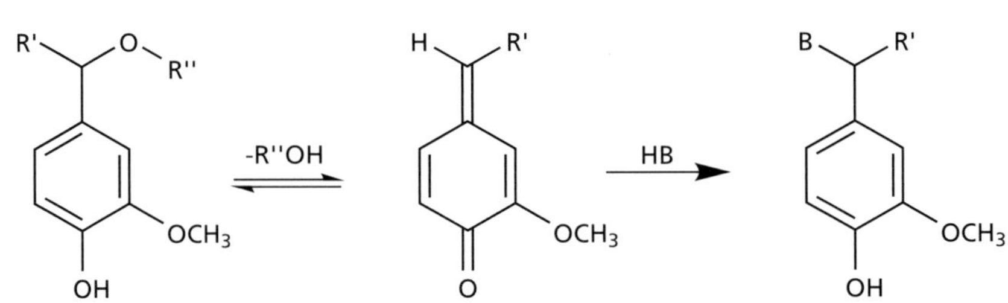

Figure 7.8 Solvolytic cleavage of a phenolic α-aryl ether linkage via a quinone methide intermediate. B = OH, OCH_3. [52]

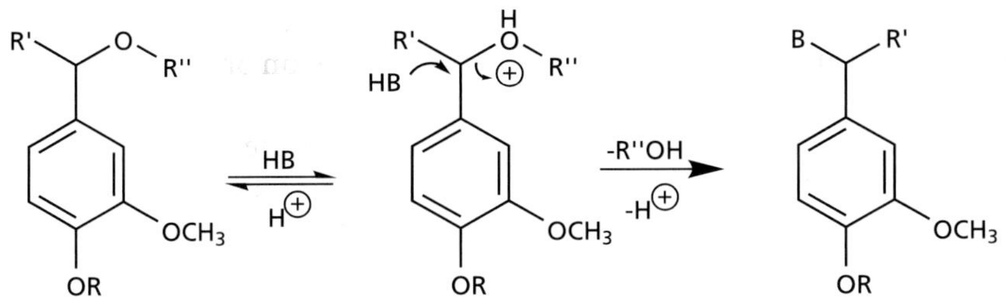

Figure 7.9 Solvolytic cleavage of an α-aryl ether linkage by nucleophilic substitution. R = H or CH_3; B = OH, OCH3, etc. [52]

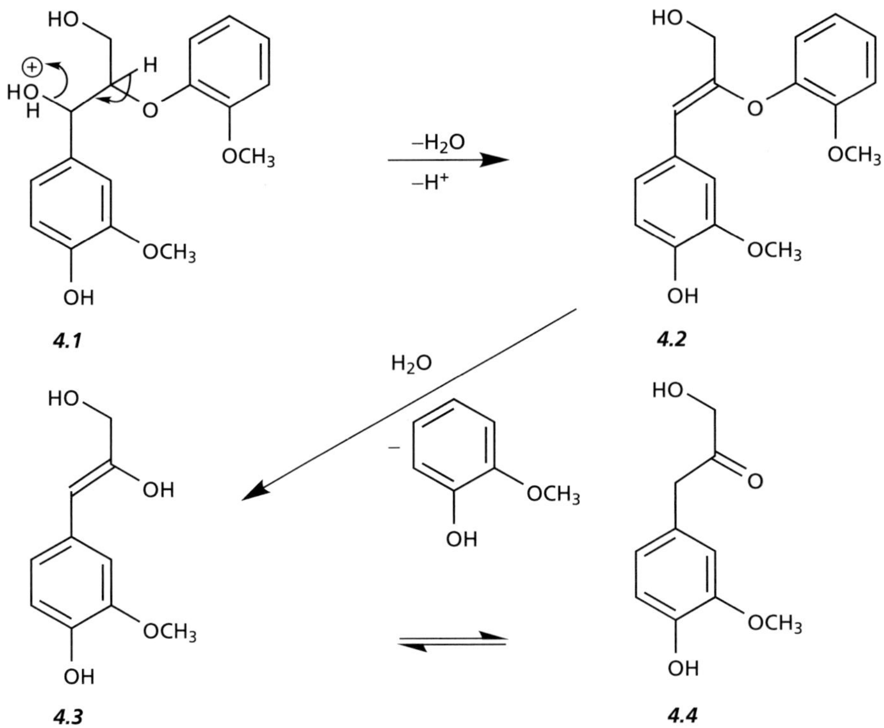

Figure 7.10 Solvolytic cleavage of a β-aryl ether linkage to form β-hydroxyconiferyl alcohol, which is in equilibrium with its keto form. [52]

than softwoods, mainly due to differences in β-ether reactivity, α-ether concentration, lignin concentration and propensity to undergo condensation reactions.

Organosolv pulping in alkaline systems is less distinct from conventional processes than are the autocatalyzed and acidic organosolv processes. [52] Unlike the latter, the alkaline systems would remain viable pulping processes if the nonaqueous component were omitted. In alkaline systems, α-ether are readily cleaved if they occur in units containing free phenolic groups, but not otherwise. Alkaline cleavage of β-ethers also occurs, especially in the presence of nucleophilic additives such as hydrosulfide ion or an anthraquinone derivative.

In a review related to organosolv pulping, Muurinen [53] concluded in 2000 that organosolv pulping processes are still in a developing stage and are not yet ready to seriously threat the leading position of the kraft process. Distillation seems to be the main option as the process for recovering the solvent in organosolv pulping. The main organosolv processes until 2000 are summarized in **Table 7.1**. Some of them were exploited industrially in the 1990s.

Recent reviews show the growing interest of organosolv processes for pretreatment of biomass. [54-56] The organosolv processes with either acetic acid (Acetosolv), or formic acid (Formosolv), or peroxyformic acid (Milox) have been tested on miscanthus, showing efficient routes for fractionation. [57]

Some studies utilize high-boiling-point organic solvents, such as glycerol, but most of literature is based on ethanol-water mixes, eventually with acidic catalysis and sometimes coupled with steam explosion or mechanical treatments. [58] Ethanol-water treatment at high temperature (~200°C) produces spontaneously carboxylic acids, but addition of sulfuric or hydrochloric acid increases delignification rate and xylan degradation. [59] Studies using the ethanol organosolv process show potential for conversion to biofuel and valuable coproducts. [60-61]

Table 7.1 Main organosolv processes developed for pulp production until 2000. [53]

Process Name	Solvent system
Asam	Water + alkaline sulfide + anthraquinone + methanol
Organocell	Water + sodium hydroxide + methanol
Alcell (APR)	Water+ low aliphatic alcohol
Milox	Water + formic acid + hydrogen peroxide (forming peroxyformic acid)
Acetosolv	Water + acetic acid + hydrochloric acid
Acetocell	Water + acetic acid
Formacell	Water + acetic acid + formic acid
Formosolv	Water + formic acid + hydrochloric acid
Battelle-Geneva	Water + phenol + acid catalyst
Acos	Water + acetone + mineral acid
MEA	Water + monoethanolamine

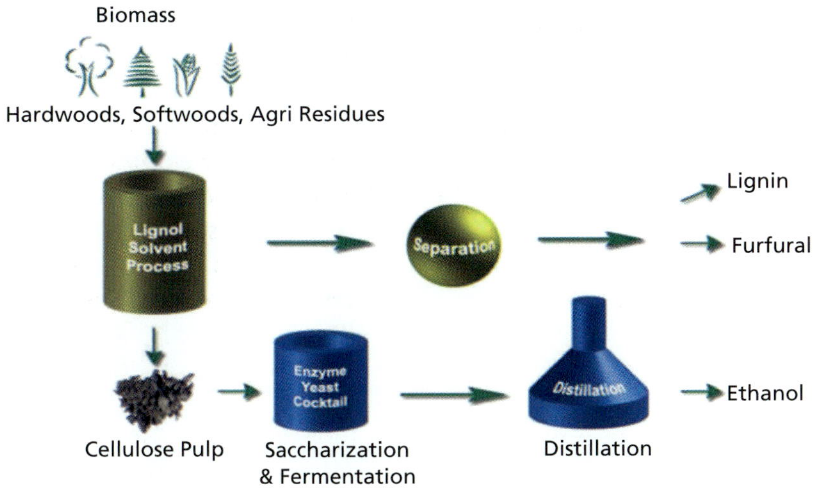

Figure 7.11 Lignol's process and products. [63]

Lignol, a Canadian company undertaking the development of biorefining technologies, and Novozymes, a Danish enzyme company, are associated to build a biorefinery based on former Alcell (ethanol/water) process (see **Table 7.1**) improved as Alcell-Plus. US Department of Energy (DOE) is funding the project at $30 million for a global amount estimated at $100 million. The objective is to build a completely flexible biorefinery, able to treat different feedstocks and produce fuel ethanol and valuable chemicals (**Figure 7.11**). Construction completion is scheduled for October 2013 and project completion for June 2014. [62]

From the former Formacell process (see **Table 7.1**), some new initiatives are developed that are now close to pilot step or industrialization. The CIMV company has patented acetic acid/formic acid/water process for the fractionation of lignocellulosic biomass. [64-65] The CIMV organosolv process cogenerates partially depolymerized hemicelluloses and sulfur-free, low molecular weight lignins. In preliminary studies, cellulose extracted from the CIMV process appeared to be highly amenable to cellulase liquefaction/saccharification, with more than 90% of glucose being recovered in monomeric form. The CIMV process, including biomass fractionation, sugar and lignin extraction, and separation of sugars and lignin, is illustrated in **Figure 7.12**. [66-67]

CIMV biomass fractionation process will be the heart of the biorefinery concept being developed by the EU Biocore project. [69]

7.3.5 Oxidative delignification

Delignification of lignocellulose can be achieved by a treatment with an oxidizing agent such as hydrogen peroxide, ozone, oxygen, or air. [3] The effec-

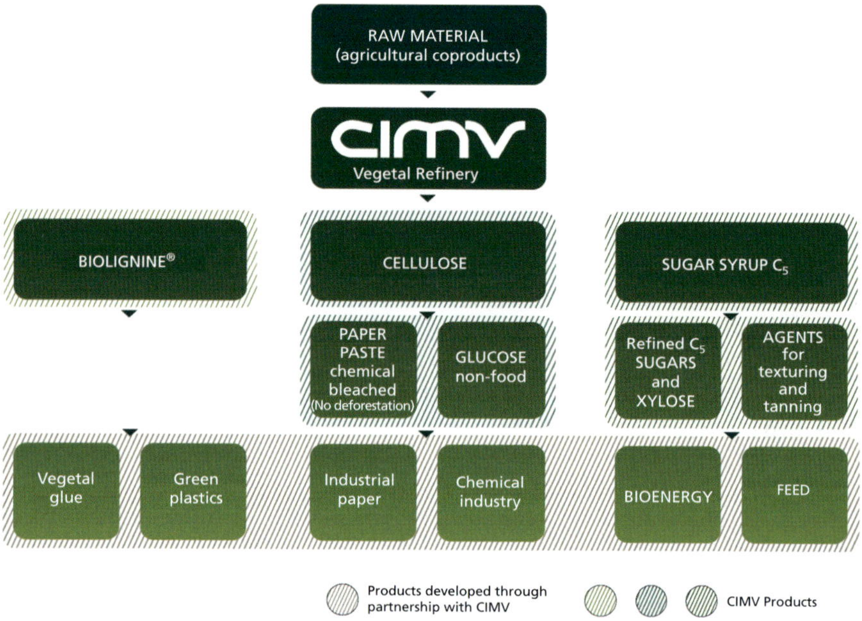

Figure 7.12 CIMV organosolv process. [68]

tiveness of delignification can be attributed to the high reactivity of oxidizing agents with the aromatic ring. Thus, the lignin polymer will be converted into compounds such as carboxylic acids. Since these acids will act as inhibitors in the fermentation step, they have to be neutralized or removed. In addition to an effect on lignin, oxidative treatment also affects the hemicellulose component of the lignocellulosic biomass. A significant part of the hemicelluloses might be degraded and cannot be used anymore for sugar production.

Hydrogen peroxide

An oxidative agent commonly used is hydrogen peroxide. [3] Dissolution of ~50% of lignin and most of the hemicelluloses has been achieved in a solution of 2% hydrogen peroxide at 30 °C. The yield of subsequent enzymatic hydrolysis can be as high as 95%.

There is not much literature on peroxide treatment, but it is generally applied in alkaline conditions (pH 10–14). Alkaline hydrogen peroxide (AHP) treatment has been mostly studied on non-woody materials (corn stover [70], different types of grass and straw, [71-72] miscanthus [73]). Conditions may change slightly but are generally mild: temperature of 20–25 °C, biomass loading of 1-15%, hydrogen peroxide loading of 0.1–0.5g/g and durations of 24–48h. Authors report good results concerning enzymatic digestibility improvement, even compared to dilute acid, lime or AFEX pretreatments. [74-75]

Ozonolysis

Ozone treatment degrades preferentially lignin by attacking and cleaving aromatic ring structures. [3] It can be used to disrupt the structure of many lignocellulosic materials, such as wheat straw, bagasse, pine, peanut, cotton straw, and poplar sawdust.

Wet oxidation

Wet oxidation operates with oxygen or air in combination with water at elevated temperature and pressure. [3] It was presented as an alternative to steam explosion, dilute acid hydrolysis and hydrothermal pretreatment, operating in the same temperature range (170–220 °C) and reaction time (1–30 min). [76] However, steam explosion pretreatment operates with saturated steam with or without the addition of sulfur dioxide or sulfuric acid as catalyst whereas wet oxidation operates with water and requires addition of oxygen or air. Bjerre et al. [77] discovered that combination of alkali and wet oxidation (alkaline wet oxidation) prevented the formation of furfural and 5-hydroxymethylfurfural. [78] Industrially, wet air oxidation processes have been used for the treatment of wastes with a high organic content by oxidation of soluble or suspended materials by using oxygen in aqueous phase at high temperature and high pressure. [3]

Palonen et al. [79] investigated the wet oxidation pretreatment of softwood for enhancing enzymatic hydrolysis. Six different combinations of reaction time, temperature and pH were applied, and the compositions of solid and liquid fractions were analyzed. The solid fraction after wet oxidation pretreatment contained 58–64% cellulose, 2–16% hemicelluloses and 24–30% lignin. Important factors affecting enzymatic hydrolysis are the temperature of the pretreatment and the residual hemicellulose content of the substrate. The highest sugar yield, 79% of theoretical, was obtained using a pretreatment at 200 °C for 10 min at neutral pH.

Lissens et al. [80] investigated the wet oxidation pretreatment of woody yard waste with high lignin content (22% of dry matter). The effects of temperature (185–200 °C), oxygen pressure (3–12 bar) and addition of sodium carbonate (0–3.3 g per 100 g dry matter biomass) on enzymatic cellulose and hemicellulose (xylan) convertibility were studied. The enzymatic cellulose conversion was highest after wet oxidation for 15 min at 185 °C with addition of 12 bars of oxygen and 3.3 g sodium carbonate per 100 g waste. Total carbohydrate recoveries were high (91–100% for cellulose and 72–100% for hemicelluloses) and up to 49% of the original lignin and 79% of the hemicelluloses could be solubilized during wet oxidation pretreatment and converted into carboxylic acids mainly.

Klinke et al. [76] investigated the alkaline wet oxidation pretreatment (water, sodium carbonate, oxygen, high temperature and pressure) of wheat straw. They found that the pretreatment resulted in solid fractions with high cellulose recovery (96%) and high enzymatic convertibility to glucose (67%).

Carbonate and temperature were the most important factors for fractionation of wheat straw by wet oxidation. Optimal conditions were 10 min at 195 °C with addition of 12 bar oxygen and 6.5 g/l sodium carbonate.

7.3.6 Room temperature ionic liquids (RTIL)

The two terms that are the most employed in literature to describe ionic liquids (ILs) are "tunable properties" and "green solvents"; they are presented for about 10 years as very promising solvents for catalysis, analysis, and organic synthesis. [81-85] The first term comes from the very high number of possible combinations between an organic cation and an organic/inorganic anion: the main cations mentioned in literature include imidazolium, pyridinium, quaternary ammonium, quaternary phosphonium, and pyrazolium; the main anions of generally smaller size include chloride, bromide, iodide, nitrate, and phosphate (**Figure 7.13**). [82]

The second term emphasizes the fact that they are liquid at room temperature (typically below 100 °C). Their vapor pressure is extremely low, so that they are non-volatile, do not evaporate and are non-flammable. They are an environmentally friendly alternative to classical organic solvents. [86] The solvent properties are mainly linked to polarity, but also to abundant different interactions that can take place in the presence of these compounds and make them very complex systems (**Figure 7.14**). [82]

Densities observed for ionic liquids are typically above 1g.cm^{-3}. Viscosities can be in the range of 100–1000 cPs at room temperature, which is considered as a drawback in terms of processing compared to standard organic solvents. Most ionic liquids are barely biodegradable, [87] and some can be toxic [88] or corrosive. [89]

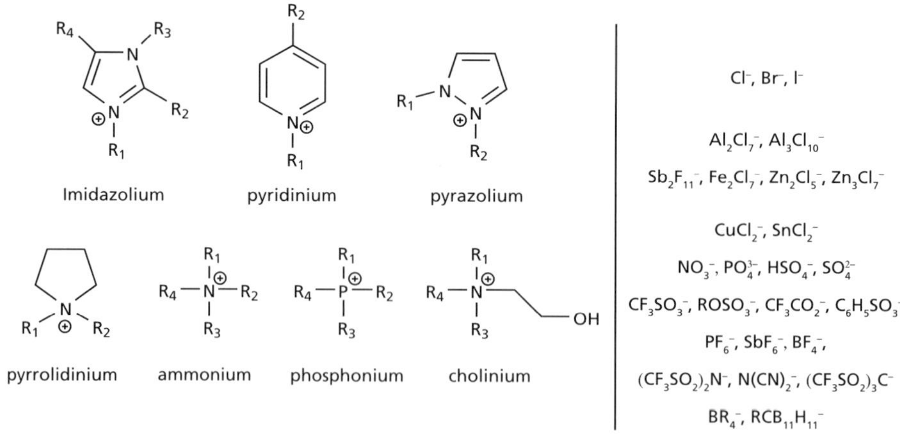

Figure 7.13 Main cations and ions in ionic liquids. [82]

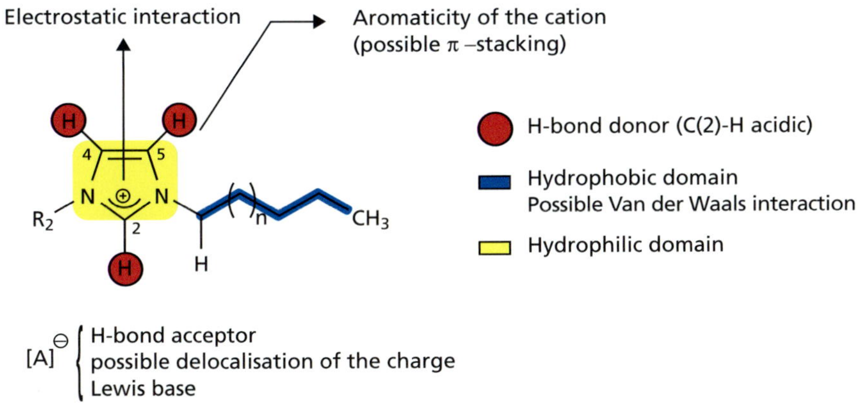

Figure 7.14 Different types of interactions present in imidazolinium-based ionic liquids ($[A]^-$, anion). [82]

Concerning the ability of ionic liquids to dissolve cellulose, it is generally accepted that the results obtained in University of Alabama by Rogers et al. [90] were the starting point: high DP cellulose solutions (up to 25%) were obtained in 1-butyl-3-methylimidazolium chloride (BmimCl) at temperatures of 70–110 °C. Zhang et al [91] studied 1-allyl-3-methylimidazolium chloride (AmimCl) and made solutions at ~15% concentration at 80°C, also claiming the easy recycling of the solvent after regeneration of cellulose with water. 1-Ethyl-3-methylimidazolium chloride (EmimCl) has been used successfully by several authors, [92-93] confirming that chloride-based ionic liquids generally give the best results in terms of cellulose solubilization. It is to be noticed that a large part of the literature about ionic liquids and cellulose is oriented toward cellulose grafting and chemical modification. [94-95] Microwave activation has been shown to raise dissolution rates but also causes increased degradation (mainly by temperature increase). [90, 93, 96] Water content in ionic liquids has clearly been identified as a key factor for solubilization potential with the less water content giving the best results. [93, 97] As chloride-based ionic liquids are most hygroscopic and corrosive, interest has raised for ionic groups like formate, phosphonate and especially acetate. [98-100] Amongst acetate-based ionic liquids, 1-ethyl-3-methylimidazolium acetate (EmimAc) seems to be a good candidate for future pilot/industrial exploitation, including biomass pretreatment. [101-102]

As lignocellulose is also composed of lignin and hemicelluloses, it is of interest to study the action of ionic liquids on all these components. Several reviews have been recently published describing the use of ionic liquids for lignocellulosic biomass pretreatment. [102-104] Wood pretreatment has been extensively studied: partial to complete dissolution has been obtained on different species of soft and hard woods, at temperatures of 80–150 °C and for durations of 8–48 hours (**Table 7.2**). [103]

Table 7.2 Literature review for dissolution of different wood species by ionic liquids. [103]

Ionic liquid	Wood	Dissolution	Conditions
[MMIM][$MeSO_4$]	Maple wood flour	Very low	80 °C, 24h
[MMIM][$(MeO)_2PO_2$]	Spruce chips	5 wt% Partial	90 °C
[EMIM][Cl]	Spruce chips	5 wt% Partial	90 °C
[EMIM][OAc]	Spruce chips	5 wt% Complete	90 °C
	Southern yellow pine, chips 0.25-0.5 mm	92.6%	110 °C, 16h
	Poplar chips	92% conversion to glucan after enz. hydrolysis	120 °C, 24h
[AMIM][Cl]	Maple wood flour	>30g/kg	90 °C 24h
	Norway spruce saw dust	8 wt%	80 °C, 8h
	Norway spruce TMP	7 wt%	130 °C, 8h
	Spruce chips	5 wt% Complete	90 °C
[BMIM][Cl]	Pine	67 wt%	100 °C, 24h, cosolvent DMSO-d_6
	Wood chips	Partially soluble	130 °C, 15h
	Norway spruce TMP	7 wt%	130 °C, 8h
	Spruce chips	5 wt% Partial	90 °C
	Southern yellow pine, chips 0.35-0.5 mm	26%	110 °C, 16h
	Softwood	Partial	150 °C
	Maple wood flour	<0.1 g/kg	90 °C, 24h
[BMIM][BF_4]	Maple wood flour	<0.1 g/kg	90 °C, 24h
[BMIM][PF_6]	Maple wood flour	>30 g/kg	90 °C, 24h
[BMIM][OTf]	Maple wood flour	>10 g/kg	90 °C, 24h
	Southern pine TMP	5 wt%	130 °C, 8h
[BnMIM][DCA]	Aspen wood chips or powder	Dissolved	150 °C, 24h
[3MeOBnMIM][Cl]	Aspen wood chips or powder	Almost completely dissolved	100 °C, 24h
[HDBU][Cl]	Aspen wood chips or powder	Partly dissolved	100 °C, 24h

Table 7.2 (cont.)

Ionic liquid	Wood	Dissolution	Conditions
[HDBU][HCOO]	Aspen wood chips or powder	Partly dissolved	100 °C, 48h
[HDBU][OAc]	Aspen wood chips or powder	Partly dissolved	100 °C, 48h
[BDBU][Cl]	Aspen wood chips or powder	Partly dissolved	100 °C, 48h
[C_8DBU][Cl]	Aspen wood chips or powder	Partly dissolved	100 °C, 48h

As pointed out by Kilpelainen et al., [105] low water content was crucial for complete dissolution of ground and dried wood in BmimCl or AmimCl for ~8 hours at 80–130 °C. Like other authors [106-107], they claimed no degradation after treatment, and recovery of an amorphous-rich cellulose structure by addition of a non-solvent (water, acetone, etc.); this cellulose structure was more reactive toward enzymatic hydrolysis. Even if patents have been issued in this area, it is still not very clear how the different dissolved components can be efficiently separated and recycled. [108-109]

It has also been found that complete dissolution in ionic liquids was not always necessary to significantly enhance the enzymatic depolymerization of biomass polysaccharides. Varanasi et al. [110] claimed a method for conversion of the lignocellulosic carbohydrates to sugars with improvement in yield and rate of sugar production by using short time contact pretreatment (**Table 7.3**). The pretreatment has as effect to mainly disrupt lignin sheathing and swell the remaining structure.

Table 7.3 Percent corn stover glucan conversion to glucose with varying pretreatment time and temperature as a function of hydrolysis (Hyd) time; pretreatment with BmimCl. [110]

Hyd Time (hr)	Conversion to glucose (%) Incubation Temperature °C and Time in hr						
	Untreated	130 °C 10 min	130 °C 1 hr	130 °C 3 hr	150 °C 10 min	150 °C 1 hr	150 °C 3 hr
6	14	41	31	44	42	61	75
28	27	48*	58	65	52*	85	91
36	26	-	68	70	-	100	100

* Glucose conversion at 24 hours hydrolysis.

Partial dissolution is mentioned in several articles, some insisting on cellulose dissolution mainly with chloride and acetate imidazolium salts, [111-112] others on lignin dissolution especially with sulfate and sulfonate salts. [113-115] As a matter of fact, it appears that, generally, delignification and disruption of cellulose crystalline structure are simultaneous during pretreatment, and that both are equally important for biomass activation. [116]

As shown by several studies, most ionic liquids tend to inactivate cellulases, imposing cumbersome precipitation/purification steps before enzymatic treatment. [117] Therefore, some research teams have been exploring the possibility of direct enzymatic hydrolysis of cellulose in specific mixtures of cellulases and ionic liquids (**Figure 7.15**). [118-119]

Zhao et al. [120] and Kamiya et al. [121] demonstrated that anion structure and solvent viscosity were crucial in terms of cellulase activity. To avoid this inhibition of cellulases, another route was to mix ionic liquids and acids to combine their effects. Catalytic amounts of hydrochloric acid or sulfuric acid were sufficient to drive the hydrolysis reaction and obtain good yields of reducing sugars and glucose. [122-123] Solid catalysts were also studied with interesting results, but the mechanisms involved remained mysterious. [124-125] As a conclusion, there is up to now no efficient and industrial process able to (1) achieve efficient separation/recovery of final glucosidic monomers and (2) achieve efficient recovery of the solvent: this is the key of the development of ionic liquids technology in the future. For example, Brennan et al. [126-127] claimed 100% sugar recovery by solvent extraction and complexation with boronic acid. Solvent/boronic acid mix

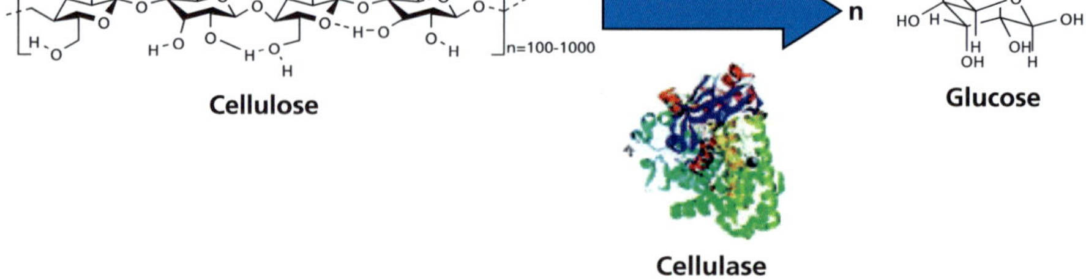

Figure 7.15 Hydrolysis of cellulose in a mixture of cellulases and tris-(2-hydroxyethyl)methylammonium methylsulfate (HEMA). Reproduced with permission from S. BOSE, D.W. ARMSTRONG and J.W. PETRICH, Journal of Physical Chemistry B 114, 8221, 2010. Copyright © 2010 American Chemical Society. [119]

can be applied on ionic liquid after biomass solubilization or on ionic liquid/water after enzymatic treatment.

7.4 Physicochemical treatments

7.4.1 Steam explosion

General

In general terms, a steam explosion is a violent boiling or flashing of water into steam. [128] Pressure vessels that operate at above atmospheric pressure can also provide the conditions for a rapid boiling event that can be characterized as steam explosion.

The process of steam explosion was developed by W.H. Mason in 1925 for hardboard production. [2, 129] Wood chips are conveyed into large vessels, and high-pressure steam is applied for a few minutes without addition of chemicals. At a set time, some steam is rapidly vented from the reactor to reduce the pressure, and the contents are discharged into a large vessel to flash cool the biomass. Use of steam explosion for biomass pretreatment was introduced in early 1980s on aspen wood. [130]

Extensive research has been conducted on steam explosion for biomass pretreatment. [2] In 1984, Schultz et al. [131] compared effectiveness of steam explosion pretreatment on mixed hardwood chips, rice hulls, corn stalks and sugarcane bagasse. [129] Some steam explosion processes have been mostly used as pretreatment methods for kraft pulp bleaching in paper and pulp industry. Well known steam-explosion-based pretreatments reported include Masonite, Siropulper, Iotech, and Stake/SunOpta processes. Most steam treatments yield high hemicellulose solubility (producing mainly oligosaccharides) along with low lignin solubility. [129] Studies without added catalyst report sugar recoveries between 45–65%.

Process description

Steam explosion refers to a pretreatment technique in which lignocellulosic biomass is rapidly heated with high-pressure saturated steam. [2] The biomass/steam mixture is held for a period of time to promote hemicellulose hydrolysis and other chemical and physical changes, and terminated by an explosive decompression. Steam explosion is typically initiated at a temperature of 160–260 °C for several seconds to a few minutes before the material is exposed to atmospheric pressure. [4]

The apparatus for steam explosion consists of a boiler (steam generator) and a reactor that is subjected to a rapid decompression. A schematic of the steam explosion process is shown in **Figure 7.16**. [132] The reactor is filled with biomass through valve 1. Valve 1 is then closed and steam is let into the

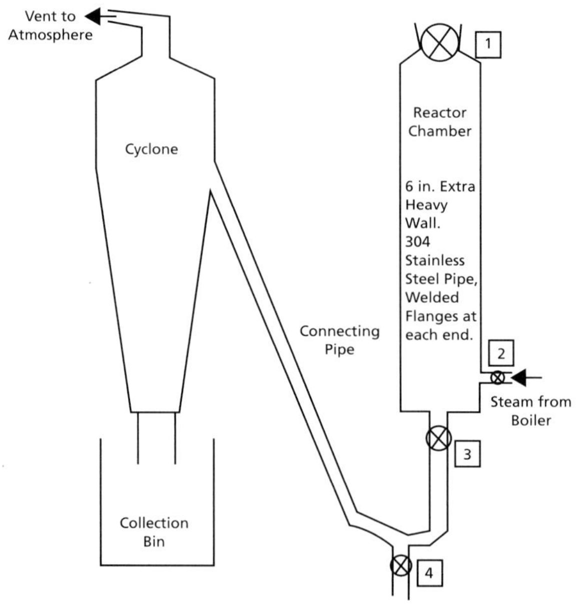

Figure 7.16 A schematic of the steam explosion process. Valve 1: sample charging valve. Valve 2: Saturated steam supply valve. Valve 3: discharge valve. Valve 4: condensate drain valve. Source: T. JEOH, Thesis, Virginia Polytechnic Institute and State University, 1998. [132]

reactor through valve 2. The reactor is allowed to reach target temperature before time begins. Typically, ~20 seconds are required to attain the desired temperature. At the end of the allotted steaming time, valve 3 is opened for the explosive decompression to occur. The steam-exploded material shoots through the connecting piping and collected in the collection bin.

Another example of a steam explosion pilot plant, as developed by Gembloux Agro-Bio Tech (Belgium), is shown in **Figure 7.17**. During the decompression step, the material is ejected from the reactor and recovered by a burster. [133]

Both batch and continuous systems are used to carry out steam explosion. [134] Batch reactors are cheap, versatile and easy to manage. Relatively small amounts of biomass (0.1–1 kg) are sufficient in each experiment. On the other side, continuous plants can work only with feeding rates above 100 kg/h of dry biomass, so they are more expensive, require additional plant facilities and are work-demanding.

Mode of action

According to Chornet and Overend, [135] steam explosion for biomass pretreatment is a physicochemical process where breakdown of structural components of lignocellulosic materials is aided by heat in the form of steam, shear forces due

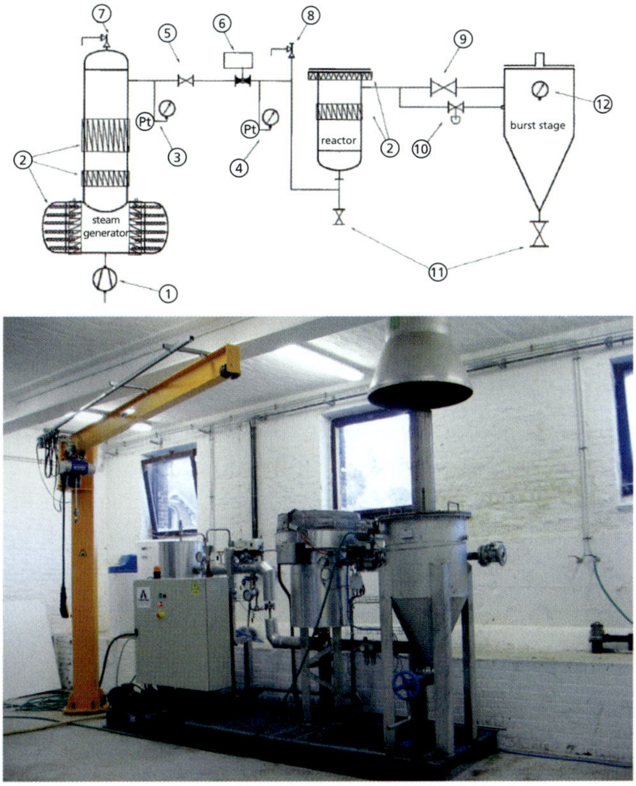

Figure 7.17 TOP Steam explosion pretreatment process flow diagram including a steam generator, a reactor and a burster as developed by Gembloux Agro-Bio Tech. 1. High-pressure pump. 2. Heating collars. 3. Generator pressure and temperature probes. 4. Reactor pressure and temperature probes. 5. Isolation valve. 6. Loading valve. 7. Generator safety valve. 8. Reactor safety valve. 9. Bursting disk. 10. Purge valve; 11. Product recovery valves. 12. Burst manometer [133] BOTTOM Photograph of the steam explosion equipment at Gembloux Agro-Bio Tech. Reproduced by courtesy of Nicolas JACQUET, GxABT.

expansion of moisture, and hydrolysis of glycosidic bonds. [129, 133] The process can be described as being composed of two distinct stages: steam cracking (i.e. the breakdown of complex molecules into smaller ones under the effect of steam) and explosive decompression. [133]

In the first stage, steam under high pressure penetrates within the structure of the material. [133] Consequently, steam condenses and wets the material. The condensed water, together with high temperature, initiates the hydrolysis of acetyl and methylglucuronic acid groups attached to hemicelluloses. The released organic acids catalyze the depolymerization of hemicelluloses.

In the second stage, the explosive decompression results in the instantaneous evaporation of the condensed moisture within the structure. [133] This expansion of the water vapor exerts a shear force on the surrounding structure. If the shear force is high enough, the vapor will cause the mechanical breakdown of the lignocellulosic structures. Combined effects of both stages include modification of the physical properties of the material (specific surface area, water retention, coloration, cellulose crystallinity, etc.), hydrolysis of hemicellulosic components, and modification of the chemical structure of lignin, facilitating their extraction. These phenomena permit the opening of the lignocellulosic material.

The two governing parameters controlling the steam explosion technique are: reaction temperature and retention time. [133, 136] The amount of time the biomass spends in the reactor helps to determine the extent of hemicellulose hydrolysis by the organic acids. [132] However, long retention times will also increase the production of degradation products, which must be minimized in a subsequent fermentation process. Temperature governs the steam pressure within the reactor. Higher temperatures translate to higher pressures, therefore

increasing the difference between reactor pressure and atmospheric pressure. The pressure difference is in turn proportional to the shear force of the evaporating pressure.

Severity factor

With the numerous studies using different biomass structures came a need to standardize the process parameters to facilitate comparisons. [132] For example, it is important to be able to relate the net product yields to the pretreatment severity. From the observation that treatment temperatures and times are interchangeable, Overend and Chomet defined the severity of a steam explosion pretreatment in terms of the combined effect of both temperature and residence time. The model is based on the assumptions that the process kinetics is first order and obeys Arrhenius' law. In doing so, they were able to develop the reaction ordinate (R0) (**Equation 7.1**):

$$\mathrm{R0} = \int_0^t \exp\left[(\mathrm{Tr} - \mathrm{Tb})/14.75\right] \mathrm{d}t \qquad (7.1)$$

where Tr is the reaction temperature (°C), Tb is the base temperature (boiling point of water at atmospheric pressure; 100 °C), t is the retention time (min) and 14.75 is the conventional energy of activation assuming that the overall process is hydrolytic and the overall conversion is first order. The $\log_{10}$ value of the reaction ordinate gives the severity factor (or severity) that is used to map the effects of steam explosion pretreatment on biomass (**Equation 7.2**):

$$\mathrm{Severity} = \log_{10}(\mathrm{R0}) \qquad (7.2)$$

Chornet and Overend demonstrated the application of the reaction ordinate model using previously documented steam explosion data. [132-133] The data used by the authors were based on wood feedstocks. A study using steam-exploded sugarcane bagasse, however, concluded that the reaction ordinate model does not apply universally. In particular, the authors found that glucose yields from enzyme hydrolysis of steam-exploded sugarcane bagasse were not constant at a given severity over a range of temperatures.

Specific effects on lignocellulosic substrates

Uncatalyzed steam explosion

Tanahashi et al. [137] studied the effects of steam explosion on the morphology and physical properties of wood. [132] They found that at pressures greater than 28 kg/cm2 (230 °C) and 16-minute residence time, the microfibrils become completely separated from each other. The microfibrils were found to be thicker and shorter with increased steaming time. [138] Furthermore, the cellulose crystallinity was found to increase. This led the authors to conclude that amorphous cellulose becomes crystalline during the steaming process. Steam explosion at moderate severities was also shown to promote delignification. In another study, Tanahashi et al. [139] observed the chemical effects of steam explosion on wood. The hemicelluloses were found to be readily hydrolyzed to oligosaccharides by steaming, at lower severities. Higher severities further hydrolyzed the hemicelluloses to monosaccharides, but also increased the concentration of furfural and 5-hydroxymethyl furfural.

Similarly, Excoffier et al. [140] found that the degree of crystallinity of cellulose increases due to the steam treatment. [132] This observation was attributed to the crystallization of amorphous regions during the heat treatment. The authors also found that while the hemicelluloses are removed by hydrolysis, lignin softens, and depolymerizes.

Attala [141] investigated the effects of steam explosion on cellulose itself. [132] He found that higher temperature treatments resulted in increasing order of the cellulose lattice structure, thereby increasing crystallinity. The author also stated that the mechanical action during the explosive decompression similarly increased structural order in cellulose. Treatment at higher temperature was also found to result in enhanced delignification.

Focher et al. [142] studied steam exploded wheat straw by scanning electron microscopy (SEM) and observed the formation of droplets on the fibers at high severities believed to be a physically modified form of lignin. Marchessault et al. [143] observed similar globular deposits on steam-exploded pulp.

Wu et al. [144] optimized steam explosion to enhance hemicellulose recovery and enzymatic hydrolysis of cellulose in softwoods.

Recent studies have shown that steam explosion causes modifications in the lignin structure. Sun et al. [145] studied the characteristics of degraded lignins obtained from steam exploded wheat straw. Li et al. [146] revealed the competition between lignin depolymerization and repolymerization and identified the conditions required for these two types of reaction. Addition of a reactive phenol, 2-naphtol was shown to strongly inhibit the repolymerization reaction, resulting in an improved delignification. In another paper, Li et al. [147] studied steam explosion lignins, their extraction, structure and potential as feedstock for biodiesel and chemicals. Bonini et al. [148] characterized lignins from steam explosion of corn stalk and pine, and investigated the role of superoxide ion and ozone.

The effects of steam explosion on lignocellulose can be summarized as follows:

- Lignocellulosic materials are defibrillated.
- Crystallinity of cellulose is increased due to crystallization of amorphous portions.
- Hemicelluloses are easily hydrolyzed.
- Delignification is promoted due to modifications in lignin structure.

Catalyzed steam explosion

Numerous studies have also been carried out to investigate the possibilities of combining steam explosion pretreatment with catalysts.

Morjanoff et al. [149] were the first to show that the addition of sulfuric acid during the pretreatment improves the enzymatic hydrolysis and decreases the production of inhibitors such as furfurals and hydroxymethylfurfurals. [133] Optimum conditions for pretreatment of sugarcane bagasse were treatment for 30 seconds with saturated steam at 220 °C and the addition of 1 g sulfuric acid/100 g dry bagasse.

Later, Vignon et al. [150] investigated hemp core samples impregnated in acid solution (0.1% w/w sulfuric acid) and steamed at temperatures between 200 and 240 °C for 180 seconds. Treatment at 200 and 210 °C led to samples that were difficult to delignify because the destructuring and disintegration of lignocellulosic materials were insufficient. A temperature of ~220–230 °C is required to obtain well-separated fibers. However, at a temperature of 240 °C, degradation and fiber damage were noted.

Viola et al. [151] pretreated the marine eelgrass *Zostera marina* by steam explosion, and selected temperature, time of pretreatment and oxalic acid load as experimental factors. The best results were obtained at 180 °C, 300 seconds and 2 wt% of oxalic acid.

7.4.2 Ammonia pretreatments

Ammonia fiber explosion (AFEX)

In the ammonia fiber explosion/expansion (AFEX) pretreatment, lignocellulosic biomass is exposed to liquid ammonia at high temperature and pressure for a period of time, and then the pressure is suddenly reduced (**Figure 7.18**). [1, 3, 4]

The AFEX process is very similar to steam explosion. The operating conditions of a typical AFEX process are a dosage of liquid ammonia of 1–2 kg of ammonia/kg of dry biomass, a temperature of 90 °C, and a residence time of 30 min. AFEX pretreatment can significantly improve the fermentation rate of various herbaceous crops and grasses (**Figure 7.19**). [1]

The technology has been used for the pretreatment of many lignocellulosic materials including aspen wood, wheat straw, wheat chaff and alfalfa stems. [152] During pretreatment only a small amount of the solid

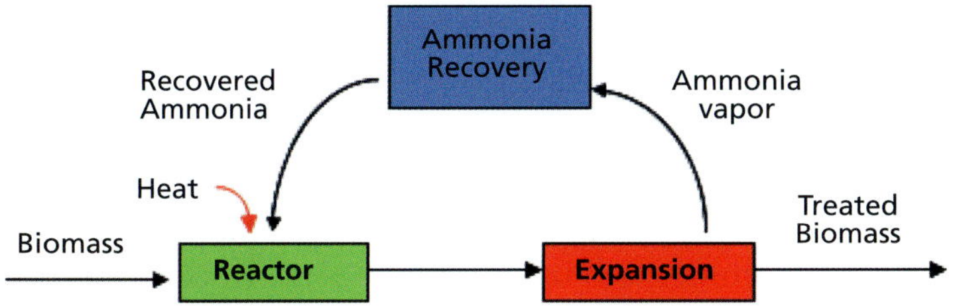

Figure 7.18 AFEX process. AFEX™ is a trademark of MBI. Source: MBI and Prof. Bruce DALE of Michigan State University.

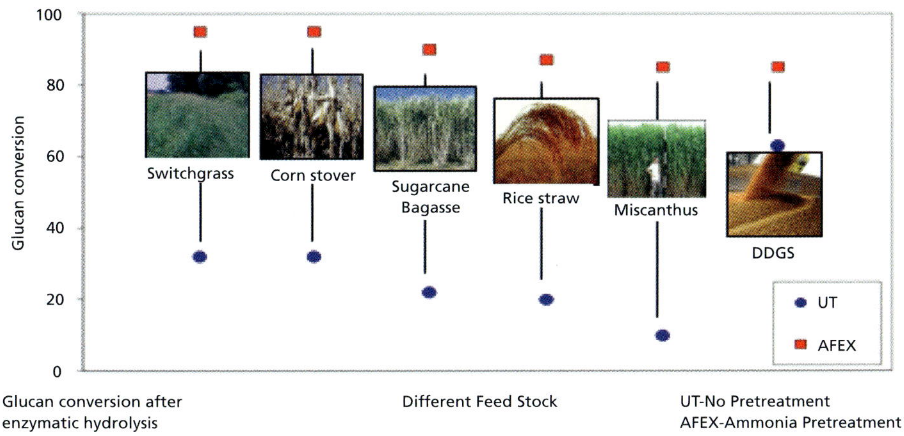

Figure 7.19 Enzymatic hydrolysis of different feedstocks before and after AFEX. Source: MBI and Prof. Bruce DALE of Michigan State University.

material is solubilized. [1] The hemicelluloses are degraded to oligomeric sugars and deacetylated. [153] The structure of the material is changed, resulting in higher digestibility. Over 90% hydrolysis of cellulose and hemicelluloses was obtained after AFEX treatment of biomass with low lignin content (<15% lignin). [154] However, the AFEX process was not very effective for biomass with higher lignin content such as newspaper (18–30% lignin) and aspen chips (25% lignin). [155] Hydrolysis yields of AFEX-pretreated newspaper and aspen chips were reported to be 40% and below 50%, respectively. [156]

To reduce the cost and protect environment, ammonia must be recycled after the pretreatment. In an ammonia recovery process, a superheated ammonia vapor with a temperature up to 200 °C was used to vaporize and strip the residual ammonia in the pretreated biomass and the evaporated ammonia was then withdrawn from the system by a pressure controller for recovery. [155]

 AFEX pretreatment does not require small particle size for efficacy. [1, 155]

Ammonia recycle percolation (ARP)

Another type of process using ammonia is the ammonia recycle percolation (ARP). [1, 157] In this process, aqueous ammonia (10–15 wt%) passes through biomass at elevated temperatures (150–170 °C) with a fluid velocity of 1 cm/min and a residence time of 14 min, after which the ammonia is recovered. In the ARP method, the ammonia is separated and recycled (**Figure 7.20**).

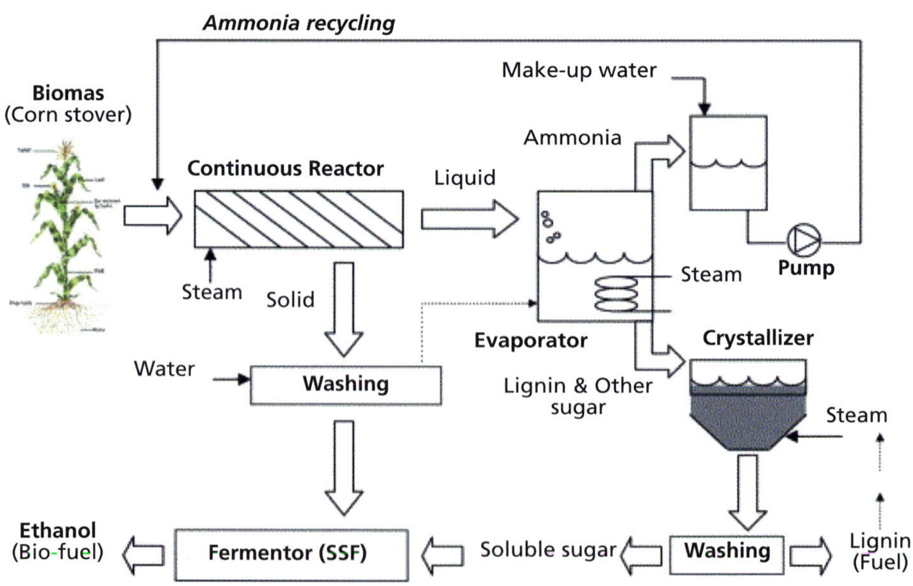

Figure 7.20 ARP (Ammonia Recycle Percolation) process diagram. [158]

Under these conditions, aqueous ammonia reacts primarily with lignin and causes depolymerization of lignin and cleavage of lignin-carbohydrate linkages. [1-2] The ammonia pretreatment does not produce inhibitors for the downstream biological processes.

Specific effects on lignocellulosic biomass

Generally, AFEX and ARP processes are not differentiated in the literature, although AFEX is carried out with liquid ammonia and ARP with an aqueous ammonia solution. [1] The AFEX pretreatment simultaneously reduces lignin content and removes some hemicelluloses while decrystallizing cellulose. It can have a profound effect on the rate of cellulose hydrolysis. The cost of ammonia, and especially of ammonia recovery, drives the cost of the AFEX pretreatment.

Recently, a significant amount of research has been carried out to determine the optimum conditions for ammonia pretreatment of lignocellulosic materials. [1] Teymouri et al. [159] evaluated the optimum process conditions such as ammonia loading, moisture content of biomass, temperature and residence time necessary for maximum effectiveness of the AFEX process on corn stover (~17% lignin). A comparison of enzymatic digestibility of corn stover with AFEX at different ratios of ammonia to biomass is shown in **Figure 7.21**.

The optimum pretreatment conditions were found to be a temperature of 90 °C, an ammonia/dry corn stover mass ratio of 1:1, a moisture content of 60% (dry weight basis), and a residence time of 5 min. About 98% of the theoretical glucose yield was obtained during enzymatic hydrolysis of the opti-

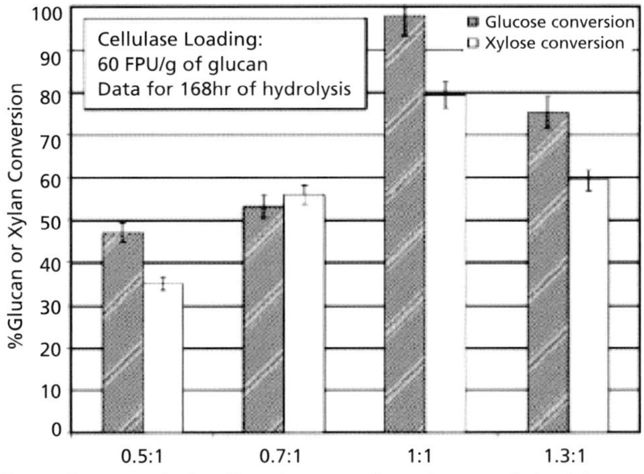

Figure 7.21 Effect of ammonia loading (grams of NH_3/gram of dry biomass) on the enzymatic conversion of glucan and xylan for APEX pretreatment of corn stover at 90 °C and 60% moisture content. [159]

mally treated corn stover. The ethanol yield of this sample was increased up to 2.2 times over that of untreated sample.

Alizadeh et al. [160] reported the optimal process conditions for the pretreatment of switchgrass. The optimal pretreatment conditions were found to be a temperature of ~100 °C, an ammonia loading of 1 kg of ammonia per kg of dry matter, a moisture content of 80% (dry weight basis) and a residence time of 5 min. Hydrolysis results of AFEX-treated and untreated samples showed 93% vs 16% glucan conversion, respectively. The ethanol yield of optimized AFEX-treated switchgrass was measured to be ~0.2 g ethanol/g dry biomass, which is 2.5 times more than that of the untreated sample.

In another article, Murnen et al. [161] reported the optimization of AFEX pretreatment and enzymatic hydrolysis of miscanthus. Pretreatment conditions including temperature, moisture, ammonia loading, residence time, and enzyme loadings were varied to maximize hydrolysis yields. The optimal pretreatment conditions determined were 160 °C, 2:1 (w/w) ammonia to biomass loading, 233% moisture (dry weight basis), and 5 min residence time for water-soaked miscanthus. About 96% glucan and 81% xylan conversions were achieved after 168 h enzymatic hydrolysis at 1% glucan loading.

7.4.3 Carbon dioxide explosion

Another pretreatment that utilizes fiber separation via a rapid pressure drop is CO_2 explosion and particularly supercritical CO_2 explosion. [162] The method is similar to steam explosion and ammonia fiber explosion; high pressure carbon dioxide is injected into the batch reactor and then liberated by an explosive decompression. [3] It was hoped that, because CO_2 forms carbonic acid when dissolved in water, the acid increases the hydrolysis rate. [1, 163] Carbonic acid may offer the benefits of acid catalysts without the use of an acid like sulfuric acid. [3] The pH of carbonic acid is determined by the partial pressure of CO_2 in water, and can be neutralized by releasing the reactor pressure. Carbon dioxide molecules are comparable in size to water and ammonia and should be able to penetrate small pores accessible to water and ammonia molecules. Compared to steam explosion, CO_2 explosion offers the benefit of lower operating temperatures (35–80 °C) which reduces sugar decomposition. [162] Dale et al. [164] used the method to pretreat alfalfa (4 kg of CO_2/kg of fiber at a pressure of 5.62 MPa) and obtained 75% of the theoretical glucose released during 24 h of the enzymatic hydrolysis. [1] The yields were relatively low compared to those of steam or ammonia explosion pretreatments but high compared to that of enzymatic hydrolysis without pretreatment.

An alternative use of CO_2 in pretreatment is the extraction with ***supercritical*** CO_2 (SC-CO_2). CO_2 becomes supercritical under relatively mild conditions (critical temperature = 304 °K and critical pressure = 7.38 MPa).

SC-CO_2 has been widely used as an extraction solvent. For several years, it has been considered as possible pretreatment process for lignocellulosic biomass. Sahle Demessie et al. [165] reported the increased permeability of Douglas-fir heartwood by SC-CO_2. Zheng and al. [166] compared SC-CO_2 explosion with steam and ammonia explosion for pretreatment of cellulosic materials such as recycled paper mix, sugarcane bagasse and repulping waste of recycled paper and found that CO_2 explosion was more cost-efficient than ammonia explosion. Kim et al. [167] investigated SC-CO_2 as possible pretreatment for enzymatic hydrolysis of hardwood and softwood. They found a positive effect of the SC-CO_2 pretreatments for both hardwood and softwood with moisture contents higher than 40% (w/w). Recently, Srinivasan and al. [168] studied the pretreatment of waste biomass using SC-CO2-based method. The pretreatment used involved: adding water to the bagasse, raising system temperature, pressurizing using SC-CO_2, holding the system for a period of time, and exploding the bagasse. The yields of sugars produced from enzymatic hydrolysis of the treated biomass were used as pretreatment efficiency indicators.
They found that supercritical method outperformed other methods and gave much higher overall sugar yields. Alinia et al. [169] designed an experimental apparatus to investigate the effect of pretreatment of dry and wet wheat straw using SC-CO_2 alone and the combination of SC-CO_2 with steam by varying temperature (160–200 °C) and the residence time (between 10 and 70 min). The methods were evaluated with the aim of achieving the highest yield of reducing sugar. Finally, Luterbacher et al. [170] developed a high pressure (200 bar) CO_2–water process for pretreating lignocellulosic biomass at high-solid contents, while minimizing chemical inputs.

7.4.4 Mechanical/alkaline pretreatment

Combined mechanical/alkaline pretreatment consists of a continuous mechanical pretreatment (such as milling, extrusion and refining) of lignocellulosic materials with the aid of an alkali. [3] The resulting fractions include a soluble fraction (containing lignin, hemicelluloses, and inorganic components) and a cellulose-enriched solid fraction.

As opposed to the acid-catalyzed pretreatment methods, the general principle behind alkaline pretreatment methods is the removal of lignin whereas cellulose and a portion of the hemicelluloses remain in the solid fraction. [3] The solid fraction is subjected to enzyme hydrolysis for the production of C6- and C5 sugars. Therefore, this pretreatment process is especially suitable in combination with fermentation routes in which both C6- and C5 sugars can be converted to valuable products.

By performing mechanical and chemical pretreatment in one step the accessibility of cellulose to enzymes is improved, resulting in higher delignification and improved enzymatic hydrolysis. [3] Furthermore, the moderate operating temperatures of this process prevent the formation of degradation and oxidation products.

7.4.5 Torrefaction

Biomass torrefaction is gaining attention as an important preprocessing step to improve the quality of biomass in terms of physical properties and chemical composition. [171] Torrefaction is a slow heating of biomass in an inert or reduced environment to a maximum temperature of ~300 °C. Torrefaction can also be defined as a group of products resulting from the partially controlled and isothermal pyrolysis of biomass occurring in a temperature range of 200–280 °C. Thus, the process can be called a mild pyrolysis as it occurs at the lower temperature range of the pyrolysis process. At the end of the torrefaction process, a solid uniform product with lower moisture content and higher energy content is obtained. Most of the volatiles are removed during torrefaction, which produces a final product that will have a lower mass but a higher heating value.

The initial heating of biomass during torrefaction removes unbound water. [171] Further heating results in the removal of bound water through chemical reactions. Further heating between 180–270 °C results in an exothermic reaction and initiates the decomposition of hemicelluloses, which causes the biomass to change color due to loss of water, CO_2, acetic acid and phenols. The energy value of these compounds is relatively low, resulting in an increase in the energy density of the biomass. The process becomes completely exothermic at temperatures greater than 280 °C, resulting in significant increases in the production of CO_2, acetic acid, phenols and other higher hydrocarbons. Torrefaction of lignocellulosic materials results in decomposition of more hemicelluloses than lignin and cellulose. These decomposition reactions lead in destruction of hydroxyl groups, creating a hydrophobic product.

Advantages of torrefaction are:

- The hygroscopic raw biomass is converted to hydrophobic material.
- Easier storage and delivery.
- Easier grindability.
- No inhibition in successive enzymatic hydrolysis and fermentation. [172-173]

7.5 Biological pretreatment

In biological pretreatments, microorganisms such as brown- white- and soft-rot fungi are employed to degrade components of biomass based on their enzymatic systems. [1] Brown-rot and soft-rot fungi attack polysaccharides predominantly, and do not degrade lignin extensively, whereas white-rot fungi

Figure 7.22 White-rot in a Birch tree. Reproduced by permission of Sten under the Creative Commons Attribution-Share Alike 3.0 Unported, 2.5 Generic, 2.0 Generic and 1.0 Generic license. [174]

are the most efficient in causing lignin degradation. Lignin degradation by white-rot fungi takes place through the action of lignin-degrading enzymes such as peroxidases and laccases. These enzymes are regulated by carbon and nitrogen sources. White-rot fungi are the most effective for biological pretreatment of lignocellulosic biomass (**Figure 7.22**).

Hatakka et al. [175] studied the pretreatment of wheat straw with 19 white-rot fungi for enzymatic saccharification of cellulose. They found that 35% of the straw was converted to reducing sugars by *Pleurotus ostreatus* after 5 weeks, while only 12% of the untreated control straw was converted to reducing sugars. They concluded that the economic viability of a non-optimized biological pretreatment process is still poor due to the long cultivation times required. Akin et al. [176] also reported the delignification of Bermuda grass by white-rot fungi.

Biological treatment in combination with other pretreatment technologies has also been studied. [1] Itoh et al. [177] studied bioorganosolve pretreatments for simultaneous saccharification and fermentation of beech wood by ethanolysis and white-rot fungi. The biological pretreatments saved 15% of the electricity needed for the ethanolysis. In another article, Balan et al. [178] studied the effect of the fungal conditioning of rice straw followed by AFEX pretreatment and enzymatic hydrolysis. They found that the pretreatment of rice straw with white-rot fungi followed by AFEX gave higher glucan and xylan conversions and less-severe AFEX conditions than the pretreatment of rice straw with AFEX alone.

The advantages of biological pretreatments include low energy requirement and mild environmental conditions. [1, 3] Nevertheless, the rate of biological hydrolysis is usually low, resulting in long residence times. Recently, Tanjore [179] studied biological pretreatments of corn stover through aerobic and anaerobic solid substrate fermentation. Shi et al. [180] investigated cotton stalks degradation by solid state cultivation using *Phanerochaete chrysosporium* They showed that moisture content significantly affected lignin degradation, with 75% and 80% moisture content degrading ~6% more lignin than 65% moisture content after 14 days. Singh et al. [181] screened different fungi species, mentioning best degradation results after treatment of sugarcane trash with *Aspergillus terreus* and by *Cellulomonas uda*. A recent paper reviewed biological pretreatment of lignocellulosics, showing potential, progress and challenges. [182]

7.6 Summary

The mode of action of various important pretreatments on the structure of lignocellulosic biomass and the generation of fermentation inhibitors is summarized in **Table 7.4**. [1-3]

Table 7.4 shows the following:

- Wet oxidation, AFEX, ARP and CO_2 explosion reduce cellulose crystallinity.
- All pretreatments partially or totally remove hemicelluloses.
- Alkaline, organosolv, wet oxidation, AFEX, ARP, mechanical/alkaline and biological partially or totally remove lignin.
- Fermentation inhibitors are formed with liquid hot water, acid and steam explosion.
- No or small amounts of fermentation inhibitors are formed with alkali, organosolv, wet oxidation, AFEX, ARP, CO_2 explosion, mechanical/alkaline, and biological. [183]

Specific advantages and disadvantages for various pretreatment methods are summarized below. [1]

1. Mechanical comminution reduces crystallinity but more importantly gives reduction of particle size, making material handling easier and increasing surface/volume ratio. [3] These mechanical processes usually have high energy and capital costs.

2. Liquid hot water may generate fermentation inhibitors, but using water as a solvent is an economic and environmental advantage. [2, 3, 184] Chemical pretreatments that use alkalis, organic solvents or oxidants are largely focused on lignin removal, thereby enhancing enzymatic degradability of cellulose. [3] Alkaline hydrolysis requires long residence times and leads to the formation of irrecoverable salts and their incorporation into biomass. Organosolv requires solvents to be drained from the reactor, evaporated,

Table 7.4 Effect of various pretreatments on the structure of lignocellulosic biomass (in addition to increasing surface area) and the formation of fermentation inhibitors. [1, 2, 3]

Pretreatment	Effect on cellulose	Effect on hemicelluloses	Effect on lignin	Inhibitor formation
Mechanical comminution	Decrystallization			
Liquid hot water1)		Removal (major)	Alteration	Yes
Dilute acid1)		Removal (major)	Alteration (minor)	Yes
Strong acid	Hydrolysis	Hydrolysis	Alteration	Yes
Alkaline		Removal (minor)	Removal (major)	
Organosolv		Removal (minor) [3]	Removal (major)	
Wet oxidation	Decrystallization	Removal (minor)	Removal (major)	
Ionic liquids	Dissolution	Dissolution	Dissolution	
Steam explosion 1)		Degradation (major)	Alteration	Yes
AFEX	Decrystallization	Removal	Removal	No
ARP	Decrystallization	Removal (minor)	Removal (major)	No
CO_2 explosion	Decrystallization	Removal		No
Mechanical/ alkaline		Removal (minor)	Removal (major)	
Torrefaction		Removal		No
Biological		Removal	Removal	

condensed and recycled, resulting in high cost. [1] Acid hydrolysis generates fermentation inhibitors and requires corrosion-resistant equipment, resulting in high cost. [1]

3. Steam explosion is an attractive, potentially cost-effective process because it makes limited use of chemicals, it does not result in excessive dilution of the resulting sugars, and it requires low energy input with no recycling or environmental cost. [184] Limmitations of steam explosion include destruction of a portion of the xylan fraction, incomplete dis-

ruption of the lignin-carbohydrate matrix and generation of fermentation inhibitors. AFEX/ARP do not produce fermentation inhibitors but are somewhat not efficient for biomass with high lignin content. [1, 184] The low cost of CO_2 as a pretreatment solvent, no generation of toxins, the use of low temperatures and high solids capacity are attractive features of CO_2 explosion. [184] However, the high cost of equipment that can withstand high-pressure conditions of the treatment is a limitation. Torrefaction does not generate downstream inhibition and produces an upgraded biomass.

4. The main disadvantage of biological pretreatment is that the rate of hydrolysis is very low. [1, 3]

Biomass pretreatment remains a key bottleneck in the processing of lignocellulosics for biofuels and other biobased products. [184] Although some pretreatment methods show apparent advantages, it is unlikely that one method will become the method of choice for the diversity of biomass feedstocks, which can react differently to a given technology.

7.7 References

[1] P. Kumar, D.M. Barrett, M.J. Delwiche and P. Stroeve, *Ind. Eng. Chem. Res. 48*, 3713, 2009 in http://pubs.acs.org/doi/abs/10.1021/ie801542g

[2] N. Mosier, C. Wyman, B. Dale, R. Elander, Y.Y. Lee, M. Holtzapple and M. Ladisch, *Biores. Technol. 96*, 673, 2005 in http://www.sciencedirect.com/science/article/pii/S0960852404002536

[3] P. Harmsen, W. Huijgen, L. Bermudez and R. Bakker, *Literature Review of Physical and Chemical Pretreatment Processes for Lignocellulosic Biomass*, Wageningen UR Food & Biobased T-Research, 2010 in http://www.biomassand-bioenergy.nl/filesdwnld/Literature%20review_FBR.pdf

[4] N. Kaushik, S. Biswas and P.R. Basak, *New Generation Biofuels-Technology and Economic Perspectives, 2010* in http://www.tifac.org.in/index.php?view=article&catid=120%3Apublication-bioprocess-a-bioroducts&id=773%3Anew-generation-biofuels-technology-a-economic-perspectives&tmpl=component&print=1&page=&option=com_content&Itemid=1380

[5] Y. Zheng, Z. Pan and R. Zhang, *Int. J. Agric. Biol. Eng 2*, 51, 2009 in http://www.ijabe.org/index.php/ijabe/article/viewFile/168/83

[6] Iowa State University, *Application of ultrasound to pretreat corn or biomass for ethanol production* in http://dgrewell.public.iastate.edu/research/biofuels/ultrasound_pretreatment.pdf

[7] S. Brooks, Wo 2010/058185 A1 in http://www.wipo.int/patentscope/search/fr/WO2010058185

[8] H.H. Yoon, *Korean J. Chem. Eng. 15*, 631, 1998 in http://www.springerlink.com/content/800662041m04q366

[9] Inbicon, *Hydrothermal Pretreatment* http://www.inbicon.com/Technologies/Hydrothermal_pretreatment/Pages/Hydrothermal%20pretreatment.aspx

[10] Y.Y. Lee, P. Iyer and R.W. Torget, *Adv. Biochem. Eng./Biotechnol. 65*, 93, 1999 in http://www.springerlink.com/content/tf1ttedvjwqlm9df/

[11] K.J. Zeitsch, *The Chemistry and Technology of Furfural and its Many By-Products*, Sugar Series 13, Elsevier, New York, 2000 in http://www.sciencedirect.com/science/bookseries/01677675

[12] Y. Chen, R.R. Sharma-Shivappa, D. Keshwani and C. Chen, *Appl. Biochem. Biotechnol. Part A 142*, 276, 2007 in http://www.springerlink.com/content/a2542710357w3u12

[13] BlueFire in http://bfreinc.com/our-technology

[14] Biosulfurol Energy in http://www.biosulfurol-energy.com/wp/technology

[15] J.F. Saeman, *Ind. Eng. Chem. 37*, 43, 1945 in http://pubs.acs.org/doi/abs/10.1021/ie50421a009

[16] C.I. Ishizawa, M.F. Davis, D.F. Schell and D.K. Johnson, *J. Agric. Food Chem. 55*, 2575, 2007 in http://www.ncbi.nlm.nih.gov/pubmed/17335219

[17] M.S. Buckeridge and G.H. Goldman (Eds.), *Routes to Cellulosic Ethanol*, Springer, 2011 in http://www.amazon.com/Routes-Cellulosic-Ethanol-Silveira-Buckeridge/dp/0387927395

[18] X.B. Lu, Y.M. Zhang, J. Yang and Y. Liang, *Chem. Eng. Technol. 30*, 938, 2007 in http://onlinelibrary.wiley.com/doi/10.1002/ceat.200700035/abstract

[19] C. Cara, E. Ruiz, J.M. Oliva, F. Saez and E. Castro, *Bioresour. Technol. 99*, 1869, 2008 in http://www.sciencedirect.com/science/article/pii/S0960852407002891

[20] S.C. Yat, A. Berger and D.R. Shonnard, *Bioresour. Technol. 99*, 3855, 2008 in http://www.sciencedirect.com/science/article/pii/S0960852407005536

[21] M.J. Selig, S. Viamajala, S.R. Decker, M.P. Tucker, M.E. Himmel and T.B. Vinzant, *Biotechnol. Prog. 23*, 1333, 2007 in http://cat.inist.fr/?aModele=afficheN&cpsidt=19919409

[22] S.A. Rhydolm, *Pulping Processes*, Wiley Interscience, New York, 1966

[23] D.N.S. Hon and N. Shiraishi, *Wood & Cellulosic Chemistry*, Marcel Dekker Inc., New-York, 1991

[24] Institute of Paper Science and Technology, *Basics of Kraft Pulping* in http://www.ipst.gatech.edu/faculty/ragauskas_art/technical_reviews/Basics%20of%20Kraft%20Pulping.pdf

[25] H. Xu, *Biobleaching of Kraft Pulp with Recombinant Manganese Peroxidase*, Proquest Information and Learning Company, 2007

[26] J.L. WERTZ, O. BEDUE and J.P. MERCIER, *Cellulose Science and Technology*, EPFL Press, 2010 in http://www.epflpress.com/livres/EPFL978-2-940222-41-4.html

[27] M. MCLEOD, *Paperi ja Puu-Paper and Timber 89*, 1, 2007 in http://kraftpulping-course.knowledgefirstwebsites.com/f/Top_Ten.pdf

[28] M. MOSHKELANI, M. MARINOVA and J. PARIS, *The forest biorefinery and its implementation in the pulp&paper industry, 2nd European Conference on Polygeneration, March30-April 1*, Taragona, 2011 in http://six6.region-stuttgart.de/sixcms/media.php/773/ABS-03-Jean-Paris.pdf

[29] A. VAN HEININGEN, *Forest Biorefinery Research at the University of Maine, NRBP Steering Committee Meeting, Latham, NY, May 25*, 2006 in http://www.nrbp.org/events/meetings/060524/presentations/heiningen.pdf

[30] H. MAO, J.M. GENCO, A. VAN HEININGEN and H. PENDSE, *BioRes. 5*, 525, 2010 in http://www.ncsu.edu/bioresources/BioRes_05/BioRes_05_2_0525_Mao_GVP_Kraft_Mill_Biorefinery_Acetic_Acid_Ethanol_723.pdf

[31] Metso, *New process extracts lignin from pulp mill black liquor: LignoBoost*, 2010 in http://www.metso.com/pulpandpaper/recovery_boiler_prod.nsf/WebWID/WTB-090518-22575-328A1?OpenDocument

[32] H. WALLMO, M. WIMBY and A. LARSSON, Tappi Engineering Pulping & Environmental Conference, Memphis, 2009 in http://www.tappi.org/Downloads/Conference-Papers/2009/09EPE/09epe98.aspx

[33] S. MCINTOSH and T. VANCOV, *Bioresour Technol. 101*, 6718, 2010 in http://www.sciencedirect.com/science/article/pii/S0960852410005936

[34] K.C. NLEWEM, M.E. THRASH, *Bioresour. Technol. 101*, 5426, 2010 in http://www.sciencedirect.com/science/article/pii/S0960852410003020

[35] Y. LI, R. RUAN, P.L. CHEN, Z. LIU, X. PAN, X. LIN, Y. LIU, C.K. MOK and T. YANG, *Trans. ASAE 47*, 821, 2004 in http://asae.frymulti.com/abstract.asp?aid=16078&t=2

[36] D.R. KESHWANI, J.J. CHENG, JC. BURNS, L. LI and V CHIANG, *Microwave Pretreatment of switchgrass to enhance enzymatic hydrolysis*, ASABE Annual International Meeting, 2007 in http://asae.frymulti.com/abstract.asp?aid=23472&t=2

[37] Z.H. HU and Z.Y. WEN, *Biochem. Eng. J. 38*, 369, 2008 in http://www.sciencedirect.com/science/article/pii/S1369703X07002756

[38] M. PEDERSEN, K.S. JOHANSEN and A.S. MEYER, *Biotechnol Biofuels 4*, 11, 2011 in http://www.ncbi.nlm.nih.gov/pubmed/21569460

[39] G. ZACCHI, *Pretreatment of biomass for ethanol production*, International Conference on Lignocellulosic Ethanol, Copenhagen, 13-15 October 2010 in http://ec.europa.eu/energy/renewables/events/doc/2010_10_13/1_pretreatment_guido_zacchi.pdf

[40] R.T. ELANDER, B.E. DALE, M. HOLTZAPPLE, M.R. LADISCH, Y.Y. LEE, C. MITCHINSON, J.N. SADDLER and C.E. WYMAN, *Cellulose 16*, 649, 2009 in http://www.springerlink.com/content/d47121610u58n877

[41] B.C. Saha and M.A. Cotta *Biomass Bioenerg. 32*, 971, 2008 in http://www.sciencedirect.com/science/article/pii/S0961953408000184

[42] S.C. Rabelo, R.M. Filho and A.C. Costa, *Appl. Biochem. Biotechnol. 153*, 139, 2009 in http://www.springerlink.com/content/m746r7x236707242

[43] R. Sierra, C. Granda and M.T. Holzapple, *Biotechnol. Prog. 25*, 323, 2009 in http://onlinelibrary.wiley.com/doi/10.1002/btpr.83/abstract

[44] P. Kumar, D.M. Barett, M.J. Delwiche and P. Stroeve, *Ind. Eng. Chem. Res. 48*, 3713, 2009 in http://www.che.ncsu.edu/ILEET/CHE596web_Spr2010/resources/biomass-biofuels/Methods-for-Pretreatment.pdf

[45] S. Kim and M.T. Holtzapple, *Bioresour. Technol. 97*, 778, 2006 in http://www.ncbi.nlm.nih.gov/pubmed/15961306

[46] F.K. Agbogbo and M.T Holzapple, Recent advances in the MixAlco process for the production of mixed alcohol fuel in http://www.eri.ucr.edu/ISAFXVCD/ISAFXVAF/RcAMAP.pdf

[47] V.S. Chang, W.E. Kaar, B. Burr and M.T. Holtzapple, *Biotechnol. Lett., 23*, 1327, 2001 in http://www.springerlink.com/content/pg75137v4k838580/

[48] Terrabon, MixAlco – Overview, 2008 in http://www.terrabon.com/mixalco_overview.html

[49] M.H. Lau, J.W. Richardson, J.L. Outlaw, M.T. Holtzapple, R.F. Ochoa, *The Economics of Ethanol from Sweet Sorghum Using the MixAlco Process*, AFPC research report 06-2, 2006 in www.afpc.tamu.edu/pubs/2/446/RR%2006-2.pdf

[50] Bioref-Integ, *Development of advanced biorefinery schemes to be integrated into existing industrial fuel producing complexes*, EU 7th Framework Programme, 2009 in http://www.bioref-integ.eu/fileadmin/bioref-integ/user/documents/D3total_and_D6.3.1_and_D6.6.pdf

[51] M.T.Holtzapple, *Biological Hydrogen Carriers: Harnessing Fossil Fuels in a Carbon-Constrained World*, rev.ing. no.29, p.121-128 Jan.-June 2009, ISSN 01214993 in http://revistaing.uniandes.edu.co/pdf/A13%2029.pdf

[52] T.J. McDonough, *The chemistry of organosolv delignification,* IPST Technical Paper Series Number 455, 1992 in http://smartech.gatech.edu/bitstream/handle/1853/2069/tps-455.pdf?sequence=1

[53] E. Muurinen, *Organosolv pulping, a review and distillation study related to peroxyacid pulping,* Doctoral thesis, Oulo University Library, Finland, 2000 in http://herkules.oulu.fi/isbn9514256611/isbn9514256611.pdf

[54] X. Zhao, K. Cheng and D. Liu, *Appl. Microbiol. Biotechnol. 82*, 815, 2009 in http://www.ncbi.nlm.nih.gov/pubmed/19214499

[55] P. Alvira, E. Tomas-pejo, M. Ballesteros and M.J. Negro, *Bioresour. Technol., 101*, 4851, 2010 in http://www.scribd.com/doc/53790135/Pre-Treatment-Technologies-for-an-Efficient-Bio-Ethanol-Production-Process

[56] H. CHEN and W. QIU, *Biotechnol. Adv. 28*, 556, 2010 in http://www.ncbi.nlm.nih.gov/pubmed/20546879

[57] J.J. VILLAVERDE, P. LIGERO and A. DE VEGA, *Open Agri. J. 4*, 102, 2010 in http://www.benthamscience.com/open/toasj/articles/V004/SI0085TOASJ/102TOASJ.pdf

[58] F.B. SUN and H.Z. CHEN, *Bioresour. Technol. 99*, 5474, 2008 in http://www.sciencedirect.com/science/article/pii/S0960852407009285

[59] R. EL HAGE, *Prétraitement du Miscanthus x giganteus. Vers une valorisation optimale de la biomasse cellulosique.* Thèse, Université de Nancy (France), 2010 in http://www.scd.uhp-nancy.fr/docnum/SCD_T_2010_0063_EL_HAGE.pdf

[60] N. BROSSE, R. EL HAGE, P. SANNIGRAHI and A. RAGAUSKA, *Cellulose Chem. Technol. 44*, 71, 2010 in http://www.cellulosechemtechnol.ro/pdf/CCT44,1-3%20(2010)/P.71-78.pdf

[61] X. PAN, D. XIE, R. YU and J. SADDLER, *Biotechnol. Bioeng. 101*, 39, 2008 in http://onlinelibrary.wiley.com/doi/10.1002/bit.21883/pdf

[62] M. RUSHTON, *Lignol Biorefinery Demonstration Plant*, Peer Review, DOE Biomass Program, 2011 in http://www.investorvillage.com/smbd.asp?mb=237&mid=10039600&pt=msg

[63] http://www.greencarcongress.com/2007/08/canadian-wood-c.html

[64] European Biocore project on http://www.biocore-europe.org/page.php?optim=An-innovative-fractionation and http://cordis.europa.eu/fetch?CALLER=FP7_PROJ_EN&ACTION=D&DOC=1&CAT=PROJ&RCN=94207

[65] G.AVIGNON, M.DELMAS , EP 1180171, 2002 and WO 0068494, 2000

[66] H.QUOC LAM, Y. LE BIGOT, M. DELMAS and G. AVIGNON, *Ind. Crops Prod. 14*, 139, 2001 in http://www.cimv.fr/uploads/new-procedure-for-the-destructuring-vegetable-matter-industrial-crops-2001.pdf

[67] R. BAKKER, *Advanced chemical/physical fractionation,* Workshop on the BioSynergy project, Reims, 2010 in http://www.biosynergy.eu/fileadmin/biosynergy/user/docs/WP1-Advanced_Physical__chemical_fractionation-RBakker-BIOSYNERGY_workshop_Reims__FR__17-11-2010.pdf

[68] CIMV in http://www.cimv.fr/technologie/technologie-cimv/technologie-cimv.html

[69] Biocore Project : "an innovative concept to valorize the biomass" in http://www.biocore-europe.org/page.php?optim=An-innovative-concept

[70] M.J. SELIG, T.B. VINZANT, M.E.HIMMEL, S.R.DECKER, *Appl.Biochem.Biotechnol. 155*, 397, 2009 in http://www.springerlink.com/content/b7157368g1021435/

[71] B.C. SAHA, M.A. COTTA, *Biotechnol. Prog. 22*, 449, 2006 in http://ddr.nal.usda.gov/bitstream/10113/25800/1/IND43819470.pdf

[72] B. QI, X. CHEN, Y. SU, Y. WAN, *Ind. Eng. Chem. Res. 48*, 7346, 2009 in http://pubs.acs.org/doi/abs/10.1021/ie8016863

[73] B. WANG, X. WANG, H. FENG, *Bioresour. Technol. 101*, 752, 2010 in http://www.ncbi.nlm.nih.gov/pubmed/19762230

[74] B.C. SAHA, M.A. COTTA, *New. Biotechnol. 27*, 10, 2010 in http://www.sciencedirect.com/science/article/pii/S1871678409012540

[75] G. BANERJEE, S. CAR, J.S. SCOTT-CRAIG, D.B. HODGE, J.D. WALTON, *Biotechnology for Biofuels, 4*, 16, 2011 in http://www.biotechnologyforbiofuels.com/content/4/1/16/abstract

[76] H.B. KLINKE, B.K. AHRING, A.S. Schimdt and A.B. THOMSEN, *Bioresour. Technol. 82*, 15, 2002 in http://www.sciencedirect.com/science/article/pii/S0960852401001523

[77] A.B. BJERRE, A.B. OLESEN, T. Fernqvist, A. PLOGER and A.S. SCHMIDT, *Biotechnol. Bioeng.. 49*, 568, 1996 in http://www.ncbi.nlm.nih.gov/pubmed/18623619

[78] C. MARTIN, M. MARCET and A.B. THOMSEN, *Bioresour. 3, 670*, 2008 in http://www.ncsu.edu/bioresources/BioRes_03/BioRes_03_3_0670_Martin_MT_Bagasse_Pretreatment_WetOx_SteamExp.pdf

[79] H. PALONEN, A.B. THOMSEN, M. TENKANEN, A.S. SCHMIDT and L. VIIKARI, *Appl. Biochem. Biotechnol. 117*, 1, 2004 in http://www.springerlink.com/content/q3v545878317m682

[80] G. LISSENS, H. KLINKE, W. VERSTRAETE, B. AHRING and A.B. THOMSEN, *J. Chem. Technol. Biotechnol. 79*, 889, 2004 in http://onlinelibrary.wiley.com/doi/10.1002/jctb.1068/abstract

[81] H. OLIVIER-BOURBIGOU and Y. CHAUVIN, *J. Mol. Catal. A: Chem. 214*, 9, 2004 in http://www.sciencedirect.com/science/article/pii/S1381116904001402

[82] H. OLIVIER-BOURBIGOU, L. MAGNA, D. MORVAN, *Applied Catalysis A: General 373*, 1, 2010 in http://www.sciencedirect.com/science/article/pii/S0926860X09007030

[83] J. LU, J.YAN, J. TEXTER, *Prog. Polym. Sci. 34*, 431, 2009 in http://www.sciencedirect.com/science/article/pii/S0079670008001226

[84] N.V. PLECHKOVA, K.R. SEDDON, *Chem. Soc. Rev 37*, 123, 2008 in http://pubs.rsc.org/en/Content/ArticleLanding/2008/CS/b006677j

[85] M. J. EARLE, K.R. SEDDON, *Pure Appl. Chem. 72*, 1391, 2000 in http://www.iupac.org/publications/pac/72/7/1391/

[86] Katholieke Universiteit Leuven, *Ionic Liquids Info*, 2011 in http://www.kuleuven.be/ionic-liquids/ionic_liquid.php

[87] "Ionic Liquid Environmental Fate and Toxicity", Oral session at The 231st ACS National Meeting, Atlanta, GA, March 26-30, 2006 in http://www.polyacs.org/355.html

[88] BASF *Basionic Ionic Liquids* in http://www.intermediates.basf.com/chemicals/ionic-liquids/products

[89] M.F. ARENAS, R.G. REDDY, *Journal of Mining and Metallurgy 39*, 81, 2003 in http://www.jmmab.com/index.php?option=com_content&task=view&id=156&Itemid=50

[90] R. P. SWATLOWSKI, S.K. SPEAR, D. JOHN, J.D. HOLBREY, R.D. ROGERS, *J. Am. Chem. Soc. 124*, 4974, 2002 in http://pubs.acs.org/doi/abs/10.1021/ja025790m

[91] H. ZHANG, J. WU, J. ZHANG, J. HE, *Macromolecules 38*, 8272, 2005 in http://pubs.acs.org/doi/abs/10.1021/ma0505676

[92] B. KOSAN, C. MICHELS, C. MEISTER, *Cellulose 15*, 59, 2008 in http://www.springerlink.com/content/l827521240874537/

[93] J. VITZ, T. Erdmenger, C. HAENSCH, U.S. SCHUBERT, Green. *Chem. 11*, 417, 2009 in http://pubs.rsc.org/en/content/articlelanding/2009/gc/b818061j

[94] S. BARTEL, T. HEINZE, *Green Chem. 8*, 301, 2006 in http://pubs.rsc.org/en/Content/ArticleLanding/2006/GC/b513157j

[95] J. WU, J. ZHANG, H. ZHANG, J. HE, Q.R EN, M. GUO, *Biomacromolecules 5*, 266, 2004 in http://pubs.acs.org/doi/abs/10.1021/bm034398d

[96] Z. ZHANG, Z. ZHAO, *Carbohydrate Research 344*, 2069, 2009 in http://www.sciencedirect.com/science/article/pii/S0008621509003450

[97] M. MAZZA, D-A. CATANA, C. VACA-GARCIA, C. CECUTTI, *Cellulose 16*, 207, 2009

[98] Y. FUKAYA, A. SUGIMOTO, H. OHNO, *Biomacromolecules 7*, 3295, 2006 in http://pubs.acs.org/doi/abs/10.1021/bm060327d

[99] Y. FUKAYA, K. HAYASHI, M. WADA and H. OHNO, *Green Chem. 10*, 44, 2008 in http://pubs.rsc.org/en/Content/ArticleLanding/2008/GC/b713289a

[100] T. LIEBERT, *Cellulose solvents – Remarkable History, Bright Future* in *ACS Symposium Series 1033*, 3, 2010 in http://pubs.acs.org/doi/abs/10.1021/bk-2010-1033.ch001

[101] J. ZHANG, H. ZHANG, J. WU, J. ZHANG, J. HE and J. XIANG, *Phys. Chem. Chem. Phys 12*, 1941., 2010 in http://pubs.rsc.org/en/Content/ArticleLanding/2010/CP/b920446f

[102] M. ZAVREL, D. BROSS, M. FUNKE, J. BUCHS and A.C. SPIESS, *Bioresour.Technol. 100*, 2580, 2009 in http://www.ncbi.nlm.nih.gov/pubmed/19157872

[103] P. MAKI-ARVELA, I. ANUGWOM, P. VIRTANEN, R. SJOHOLM and J.P. MIKKOLA, *Ind. Crops Prod. 32*, 175, 2010 in http://www.mendeley.com/research/dissolution-lignocellulosic-materials-constituents-using-ionic-liquidsa-review

[104] G. BRODEUR, E. YAU, K. BADAL, J. COLLIER, K.B. RAMACHANDRAN and S. RAMAKRISHNAN, *Enzyme Research 2011*, 17 pages, 2011 in http://www.hindawi.com/journals/er/2011/787532/abs/

[105] I. KILPELAINEN, H. XIE, A. KING, M. GRANSTROM, S. HEIKKINEN and D. ARGYROPOULOS, *J. Agric. Food Chem. 55*, 9142, 2007 in http://pubs.acs.org/doi/abs/10.1021/jf071692e

[106] N. Sun, M. Rahman, Y. Qin, M.L. Maxim, H. Rodriguez and R.D. Rogers, *Green Chem. 11*, 646, 2009 in http://pubs.rsc.org/en/content/articlelanding/2009/gc/b822702k

[107] D.A. Fort, R.C. Remsing, R.P. Svatloski, P. Moyna, G. Moyna and R.D. Rogers, *Green Chem. 9*, 63, 2007 in http://pubs.rsc.org/en/content/articlelanding/2007/gc/b607614a

[108] V. Millimaky, WO2005017001, 2005

[109] D. Argyropoulos, WO2008098032, 2008

[110] V. Varanasi, C.A. Schall, A.P. Dadi, J. Anderson, K. Rao and P. Paripati, WO2008112291, 2008

[111] A.P. Dadi, S. Varanasi and P.A. Schall, *Biotechnol. Bioeng. 95*, 904, 2006 in http://onlinelibrary.wiley.com/doi/10.1002/bit.21047/abstract

[112] B. Li, J. Asikkala,I. Filpponen and D. Argyropoulos, *Ind.Eng.Chem.Res. 49*, 2477, 2010 in http://www4.ncsu.edu/~dsargyro/Publications/documents/FactorsAffectingWoodDissolutionandRegenerationsinILsASAPPaper.pdf

[113] Y. Pu, N. Jiang and A.J. Ragauskas, *Wood Chem.Technol. 27*, 23, 2007 in http://www.tandfonline.com/doi/abs/10.1080/02773810701282330

[114] S.H. Lee, T.V. Doherty and J.S. Linhardt, *Biotechnol. Bioeng. 102*, 1368, 2009 in http://onlinelibrary.wiley.com/doi/10.1002/bit.22179/abstract

[115] S.S.Y. Tan, D.R. Macfarlane, J. Upfal, L.A. Edye, W.O.S. Doherty, A.F. Patti, J.M. Pringle and J.L. Scott, *Green Chem. 11*, 339, 2009 http://pubs.rsc.org/en/content/articlelanding/2009/gc/b815310h

[116] H. Xie and Z.K. Zhao, "Selective breakdown of (Ligno)cellulose in ionic liquids" in "Ionic Liquids Applications and Perspectives", A.Kokorin Ed., Intech, 674pp, 2011

[117] M.B. Turner, S.K. Spear, J.G. Huddleston, J.D. Holbrey and R.D. Rogers, *Green Chem. 5*, 443, 2003 in http://pubs.rsc.org/en/content/articlelanding/2003/gc/b302570e

[118] H. Zhao, *Journal of Chemical Technology and Biotechnology 101*, 891, 2010 in http://onlinelibrary.wiley.com/doi/10.1002/jctb.2375/full

[119] S. Bose, D.W. Armstrong and J.W. Petrich, *Journal of Physical Chemistry B 114*, 8221, 2010 in http://pubs.acs.org/doi/abs/10.1021/jp9120518

[120] H. Zhao, G.A. Baker, Z. Song, O. Olubajo, T. Crittle and D. Peters, *Green Chem. 10*, 696, 2008 in http://pubs.rsc.org/en/content/articlelanding/2008/gc/b801489b

[121] N. Kamiya, Y. Matsushita, M. Hanaki, K. Nakashima, M. Narita, M. Goto and H. Takahashi, *Biotechnology Letters 30*, 1037, 2008 in http://www.springerlink.com/content/q5g232n94326785q/

[122] C. Li and Z.K. Zhao, *Adv. Synth. Catal. 349*, 1847, 2007 in http://onlinelibrary.wiley.com/doi/10.1002/adsc.200700259/abstract

[123] C.Z. Li, Q. Wang and Z.K. Zhao, *Green Chem. 10*, 177, 2008 in http://bioconversion.dicp.ac.cn/EWEB/07%20publ.html

[124] R. Rinaldi, R. Palkovits and F. Schuth, *Angewandte Chemie-International Edition 47*, 8047, 2008 in http://onlinelibrary.wiley.com/doi/10.1002/anie.200802879/suppinfo

[125] R. Rinaldi, J.Vom Stein, R. Palkovits and F. Schuth, *ChemSusChem 3*, 266, 2010 in http://onlinelibrary.wiley.com/doi/10.1002/cssc.200900281/abstract

[126] T.C. Brennan, S. Datta, H.W. Blanch, B.A. Simmons and B.M. Holmes, *Recovery of Sugars from Ionic Liquid Biomass Liquor by Solvent Extraction*, Postprints, Multi-Campus, eScholarship, University of California, 2010 and *Bioenerg. Res. 3*, 123, 2010 in http://escholarship.org/uc/item/0nw0d6xv#page-1

[127] T.C. Brennan, B.M. Holmes, B.A. Simmons and H.W. Blanch, WO2011041455, 2011 in http://www.sumobrain.com/patents/wipo/Recovery-sugars-from-ionic- liquid/WO2011041455.html

[128] http://en.wikipedia.org/wiki/Steam_explosion

[129] F. Carvalheiro, L.C. Duarte and F.M. Girio, *J. Sci. Ind. Res. 67, 849*, 2008 in http://nopr.niscair.res.in/bitstream/123456789/2429/1/JSIR%2067(11)%20849-864.pdf

[130] P. Foody, *Optimization of Steam Explosion Pretreatment*, Final Report to DOE, Contract AC02-79ET23050, 1980 in http://www.osti.gov/energycitations/product.biblio.jsp?osti_id=5170904

[131] T.P. Schultz, M.C. Templeton, C.J. Biermann and G.D. McGinnis, *J. Agri. Food Chem. 32*, 1166, 1984 in http://pubs.acs.org/doi/abs/10.1021/jf00125a058

[132] T. Jeoh, Thesis, Virginia Polytechnic Institute and State University, 1998 in http://scholar.lib.vt.edu/theses/available/etd-011499-120138/unrestricted/ETD.PDF

[133] N. Jacquet, C. Vanderghem, C. Blecker and M. Paquot, *Biotechnol. Agron. Soc. Environ. 14*, 561, 2010 in http://popups.ulg.ac.be/Base/document.php?id=6226

[134] http://www.omcmsrl.com/project/steam-explosion-plant.htm

[135] E. Chornet and R.P. Overend, Proc. Int. Workshop on Steam Explosion Techniques: Fundamentals and Industrial Applications, 21, 1988 in http://www.amazon.fr/Steam-Explosion-Techniques-Fundamentals-International/dp/2881244572

[136] M. Ibrahim, Thesis, Virginia Tech, 1998 in http://scholar.lib.vt.edu/theses/available/etd-2998-114756/unrestricted/e-body1.pdf

[137] M. TANAHASHI, S. TAKADA, T. AOKI, T. GOTO, T. HIGUCHI and S. HANAI, *Wood Res. 69*, 36, 1983 in http://ci.nii.ac.jp/els/110000012836.pdf?id=ART0000336830&type=pdf&lang=en&host=cinii&order_no=&ppv_type=0&lang_sw=&no=1310116852&cp=

[138] M. TANAHASHI, *Characterization and Degradation Mechanisms of Wood Components by Steam Explosion and Utilization of Exploded Wood*, Abstract of the PhD Thesis by the author, Kyoto University, 1989 in http://repository.kulib.kyoto-u.ac.jp/dspace/bitstream/2433/53271/1/KJ00000017898.pdf

[139] M. TANAHASHI, K. TAMABUCHI, T. GOTO, T. AOKI, M. KARINA and T. HIGUCHI, *Wood Res. 75*, 1, 1998 in http://ci.nii.ac.jp/els/110000012875.pdf?id=ART0000336908&type=pdf&lang=en&host=cinii&order_no=&ppv_type=0&lang_sw=&no=1310116757&cp=

[140] G. EXCOFFIER, A. PEGUY, M. RINAUDO and M.R. VIGNON, *Evolution of Lignocellulosic Components during Steam Explosion. Potential Applications*, B. Focher, A. Marzetti and V. Crescenzi (eds.), Proceedings of the International Workshop on Steam Explosion Techniques : Fundamentals and Industrial Applications, 83, 1998 in http://www.amazon.fr/Steam-Explosion-Techniques-Fundamentals-International/dp/2881244572

[141] R.H. ATALLA, *Structural transformations in Celluloses*, B. Focher, A. Marzetti and V. Crescenzi (eds.), Proceedings of the International Workshop on Steam Explosion Techniques : Fundamentals and Industrial Applications, 97, 1988 in http://www.amazon.fr/Steam-Explosion-Techniques-Fundamentals-International/dp/2881244572

[142] B. FOCHER, A. MARZETTI, P.L. BELTRAME, P. CARNITI and A. VISCIGLIO, *Steam Explosion of Wheat Straw. Product Fractionation and Enzymatic Hydrolysis of the Cellulosic Component*, B. Focher, A. Marzetti and V. Crescenzi (eds.), Proceedings of the International Workshop on Steam Explosion Techniques : Fundamentals and Industrial Applications, 331, 1998

[143] R.H. MARCHESSAULT and J.M. ST-PIERRE, *A New Understanding of the Carbohydrate System*, I. Chemrawn, L.E. St-Pierre and G.R. Brown (Eds.), Future Sources of Organic Raw Materials, 613, Pergamon Press, Oxford, 1980

[144] M.M. WU, K. CHANG, D.J. GREGG, A. BOUSSAID, R.P. BEATSON and J.N. SADDLER, *Appl. Biochem Biotechnol. 77*, 47, 1999 in http://cat.inist.fr/?aModele=afficheN&cpsidt=1904987

[145] X.F. SUN, F. XU, R.C. SUN, Y.X. WANG, P. FOWLER and M.S. BAIRD, *Polym. Degrad. Stabil. 86*, 245, 2004

[146] J. LI, G. HENRIKSSON and G. GELLERSTEDT, *Bioresour. Technol. 98*, 3061, 2007

[147] J. LI, G. GELLERSTEDT and K. TOVEN, *Bioresour. Technol. 100*, 2556, 2009 in http://www.ncbi.nlm.nih.gov/pubmed/19157871

[148] C. BONINI, M. D'AURIA, P. DI MAGGIO and R. FERRI, *Ind. Crops Prod. 27*, 182, 2008 in http://www.sciencedirect.com/science/article/pii/S0926669007001069

[149] P.J. Morjanoff and P.P. Gray, *Biotechnol. Bioeng. 29*, 733, 1987 in http://onlinelibrary.wiley.com/doi/10.1002/bit.260290610/abstract

[150] M.R. Vignon, C. Garcia-Jaldon and D. Dupeyre, *Int. J. Biol. Macromol. 17*, 395, 1995 in http://www.ncbi.nlm.nih.gov/pubmed/8789346

[151] E. Viola, M. Cardinale, R. Santarcangelo, A. Villone and F. Zimbardi, *Biomass Bioenerg. 32*, 613, 2008 in http://www.sciencedirect.com/science/article/pii/S0961953407002346

[152] M. Mes-Hartree, B.E Dale and W.K. Craig, *Appl. Microbiol. Biotechnol. 29*, 462, 1988 in http://www.springerlink.com/content/m653710m565251gt

[153] L.E. Gollapalli, B.E. Dale and D.M. Rivers, *Appl. Biochem. Biotechnol. 98*, 23, 2002 in http://www.ncbi.nlm.nih.gov/pubmed/12018251

[154] M.T. Holtzapple, J.H. Jun, G. Ashok, S.L. Patibandla and B.E. Dale, *Appl. Biochem. Biotechnol. 28/29*, 59, 1991 in http://www.springerlink.com/content/x462833787910938/

[155] Y. Sun and J. Cheng, *Bioresour. Technol. 83*, 1, 2002 in http://www.sciencedirect.com/science/article/pii/S0960852401002127

[156] J.D. McMillan, *Pretreatment of Lignocellulosic Biomass in Enzymatic Conversion of Biomass For Fuels Production*, M.E. Himmel, J.O. Baker and R.P. Overend, Eds., American Chemical Society, Washington D.C., 292, 1994

[157] H.H. Yoon, Z.W. Wu and Y.Y. Lee, *Appl. Biochem. Biotechnol. 51/52*, 5, 1995 in http://cat.inist.fr/?aModele=afficheN&cpsidt=3554674

[158] T.H. Kim and Y.Y. Lee, *Bioresourc. Technol. 96*, 2007, 2005 in http://www.sciencedirect.com/science/article/pii/S0960852405000684

[159] F. Teymouri, L. Laureano-Perez, H. Alizadeh and B.E. Dale, *Appl. Biochem. Biotechnol. 115*, 951, 2004 in http://www.ncbi.nlm.nih.gov/pubmed/15054244

[160] H. Alizadeh, F. Teymouri, T.I. Gilbert and B.E. Dale, *Appl. Biochem. Biotechnol. 121-124*, 1133, 2005 in http://www.ncbi.nlm.nih.gov/pubmed/15930586

[161] H.K. Murnen, V. Balan, S.P.S. Chundawat, B. Bals, L. Da Costa Sousa and B.E. Dale, *Biotechnol. Prog. 23*, 846, 2007 in http://cat.inist.fr/?aModele=afficheN&cpsidt=18990934

[162] J. Ruffell, *Pretreatment and Hydrolysis of Recovered Fibre for Ethanol Production*, Thesis, The University of British Columbia, 2008 in http://opac.tistr.or.th/Multimedia/Web/0051/wb0051907.pdf

[163] P.G. Walsum and H. Shi, *Bioresour. Technol. 93*, 217, 2004 in http://www.sciencedirect.com/science/article/pii/S0960852403003353

[164] B.E. Dale and M.J. Moreira, *Biotechnol. Bioeng. Symp. 12*, 31, 1982 in http://www.osti.gov/energycitations/product.biblio.jsp?osti_id=5893198

[165] E. SAHLE DEMESSIE, A. HASSAN, K.L. LEVIEN, S. KUMAR and J.J. MORRELL, *Wood Fiber Sci. 27*, 296, 1995 in http://swst.metapress.com/content/bg25qk0780343206

[166] Y. ZHENG, H. LIN and G.T. TSAO, *Biotechnol. Prog. 14*, 890, 1998 in http://www.ncbi.nlm.nih.gov/pubmed/9841652

[167] K.H. KIM and J. HONG, *Bioresour. Technol. 77*, 139, 2001 in http://www.sciencedirect.com/science/article/pii/S0960852400001474

[168] N. SRINIVASAN and L.K. JU, *Bioresour. Technol. 101*, 9785, 2010 in http://www.ncbi.nlm.nih.gov/pubmed/20696574

[169] R. ALINIA, S. ZABIHI, F. ESMAEILZADEH and J.F. KALAJAHI, *Biosyst. Eng 107*, 61, 2010 in http://www.sciencedirect.com/science/article/pii/S153751101000139X

[170] J.S. LUTERBACHER, J.W. TESTER and L.P. WALKER, *Biotechnol. Bioeng. 107*, 451, 2010 in http://onlinelibrary.wiley.com/doi/10.1002/bit.22823/abstract?systemMessage=Wiley+Online+Library+will+be+disrupted+2+July+from+10-12+BST+for+monthly+maintenance

[171] J.S. TUMULURU, S. SOKHANSANJ, C.T. WRIGHT, R.D. BOARDMAN and J.R. HESS, *Review on Biomass Torrefaction Process and Product Properties and Design of Moving Bed Torrefaction System Model Development*, 2011 ASABE Annual International Meeting in http://www.inl.gov/technicalpublications/Documents/5094548.pdf

[172] A.M. RASPOLLI GALLETTI and C. ANTONETTI, *Biomass pre-treatment: separation of cellulose, hemicellulose and lignin. Existing technologies and perspectives*, Eurobioref Project, 2011 in http://www.eurobioref.org/Summer_School/Lectures_Slides/day2/Lectures/L04_AG%20Raspolli.pdf

[173] D. CHIARAMONTI, A.M. RIZZO, M. PRUSSI, S. TEDESCHI, F. ZIMBARDI, G. BRACCIO, E. VIOLA and P.D. PARDELLI, *Biomass Conv. Bioref.* DOI 10.1007/s13399-010-0001-z in http://www.spikerenewables.com/Pagine/pubblicazioni/Springer1.pdf

[174] http://en.wikipedia.org/wiki/Wood-decay_fungus

[175] A.I. HATAKKA, Appl. Microbiol. *Biotechnol. 18*, 350, 1983 in http://www.springerlink.com/content/l651h46ru63h0674/

[176] D.E. AKIN, L.L. RIGSBY, A. SETHURAMAN, W.H. MORRISON III, G.R. GAMBLE and K.E.L. ERIKSSON, *Appl. Environ. Microbiol. 61*, 1591, 1995 in http://www.ncbi.nlm.nih.gov/pmc/articles/PMC167414/pdf/611591.pdf

[177] H. ITOH, M. WADA, Y. HONDA, M. KUWAHARA and T. WATANABE, *J. Biotechnol. 103*, 273, 2003 in http://www.sciencedirect.com/science/article/pii/S0168165603001238

[178] V. BALAN, L. DA COSTA SOUSA, S.P.S. CHUNDAWAT, R. VISMEH, A.D. JONES and B.E. DALE, *J. Ind. Microbiol. Biotechnol. 35*, 293, 2008 in http://www.springerlink.com/content/8684g07830563p57

[179] D. TANJORE, *Biological pretreatments of corn stover biomass through aerobic and anaerobic solid substrate fermentation*, Ph.D., The Pennsylvania State University, 2009 in http://gradworks.umi.com/33/99/3399715.html

[180] J. SHI, M.S. CHINN, R.R. SHARMA-SHIVAPPA, *Bioresour. Technol. 99*, 6556, 2008 in http://www.ncbi.nlm.nih.gov/pubmed/18242083

[181] P. SINGH, A. SUMAN, P.TIWARI, N.ARYA, A.GAUR, A.K. SHRIVASTAVA, WORLD J. *Microbiol. Biotechnol. 24*, 667, 2008 in http://www.springerlink.com/content/c0022255t624805q

[182] S. CHEN, X. ZHANG; D.SINGH, Y. HONGBO, X. YANG, *Biofuels, 1*, 177, 2010 in http://www.ingentaconnect.com/content/fs/bfs/2010/00000001/00000001/art00019

[183] M.J. RAY, D.J. LEAK, P.D. SPANU and R.J. MURPHY, *Biomass Bioener. 34*, 1257, 2010 in http://www.sciencedirect.com/science/article/pii/S0961953410001108

[184] V.B. AGBOR, N. CICEK, R. SPARLING, A. BERLIN and D.B. LEVIN, *Biotechnol. Adv. 29*, 675, 2011 in http://www.ncbi.nlm.nih.gov/pubmed/21624451

Chapter 8

Biochemical Conversion of Biomass

8.1 Introduction

Lignocellulosic biomass feedstock can be converted to biofuels and chemicals through multiple processes. Conversion technologies fall into two main categories: biochemical conversion, which also includes in this book direct catalytic conversion of biomass, [1] and thermochemical conversion (Chapter 9). [2] Thermochemical conversion includes both gasification and pyrolysis, which use heat and pressure to convert the biomass into synthesis gas and bio-oils, respectively. [3] The advantages of thermochemical conversion are low residence time and the ability to handle various feedstocks in a continuous manner; however, the conversion process is not selective. Thermochemical conversion typically works best with dry feedstocks. [4] Biochemical conversion, alternatively, offers high selectivity in deconstructing biomass to valuable end products. Higher moisture feedstocks are generally good candidates for biochemical conversion.

Biochemical conversion proceeds at lower temperatures and lower reaction rates than thermochemical conversion. [5] Biochemical conversion generally first uses low-severity thermochemical pretreatment at temperatures between 100 and 300 °C to break down the lignocellulosic matter and improve enzymatic or chemical accessibility (Chapter 7). The pretreatment step is usually followed by hydrolysis of cellulose to fermentable sugars. There are two main types of hydrolysis processes: enzymatic/microbial (the most investigated route) and acid (either concentrated or dilute). The fermentable sugars obtained from hydrolysis can then be fermented into ethanol or chemicals using native or genetically modified microorganisms. Hexoses, such as glucose, galactose and mannose, can readily be fermented by many naturally occurring organisms, but the pentoses, such as xylose and arabinose, are fermented by few native strains, and usually at relatively low yields. [6] The development of recombinant strains resulted in bacteria and yeasts capable of co-fermenting hexoses and pentoses into ethanol and other value-added products at high yields. [6] Biomass processing schemes involving enzymatic or microbial hydrolysis involve four biologically mediated transformations: cellulase production, cellulose hydrolysis, hexose fermentation and pentose fermentation. [7] Four process configurations proposed for the biological transformations differ in their degree of integration: Separated Hydrolysis and Fermentation (SHF), Simultaneous Saccharification and Fermentation (SSF), Simultaneous Saccharification and Co-Fermentation (SSCF) and Consolidated Bioprocessing (CBP) – featuring cellulase production, hydrolysis and fermentation in one step.

Direct catalytic conversion of biomass into chemicals aims to bypass the processes involving biological fermentation, reducing the need for large and expensive biorefineries and expanding the range of biomolecules that can be produced. [8] The catalytic conversion of plant-based sugars into either furan derivatives as developed by Avantium or into hydrocarbon molecules as developed by Virent enters in this category.

This chapter is subdivided into three main sections:

- review of the processes involved in biochemical conversion,
- production of lignocellulosic biomass-derived ethanol (also called cellulosic ethanol) and other fuels, including case studies, and
- production of lignocellulosic biomass-derived chemicals, including case studies.

8.2 Processes: pretreatment, hydrolysis and fermentation

Production of ethanol and other chemicals from lignocellulosic biomass contains three major processes, including pretreatment (see Chapter 7), hydrolysis and fermentation. Pretreatment can fractionate the cellulose, hemicellulose and lignin components. The main advantage of most pretreatments is their

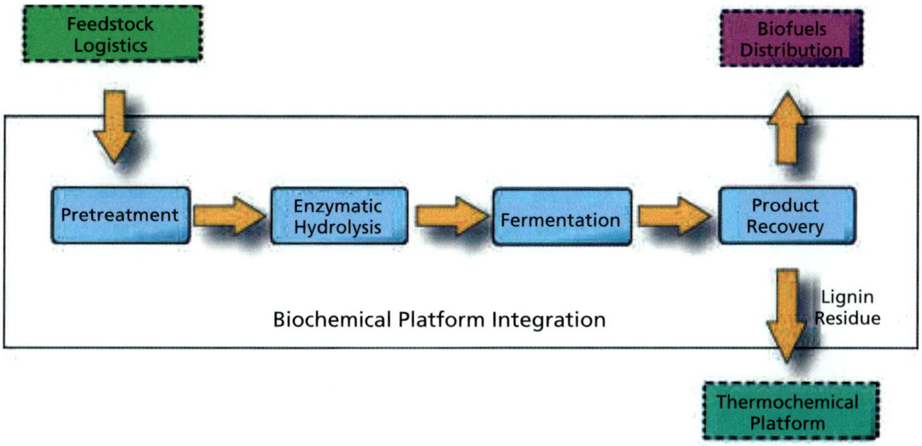

Figure 8.1 Very simplified version of the biochemical conversion process. [10]

ability to produce an aqueous slurry that includes a solid fraction (water-insoluble fraction) composed of mainly cellulose and lignin and a liquid fraction (water-soluble fraction) composed of mainly partially hydrolyzed hemicellulosic sugars. [9]

A very simplified version of the biochemical conversion process from lignocellulosic biomass to ethanol is shown in **Figure 8.1**.

The schematic illustrates the four key main steps of the process: pretreatment and conditioning, enzymatic hydrolysis, fermentation and distillation and dehydration. The generic process diagram for the biochemical conversion is shown in **Figure 8.2**. [11-12] It emphasizes three different levels of integration for the biological steps: SHF, SSF, and SSCF. When enzymatic hydrolysis is applied, the integration of cellulase production and SSCF in a single step gives rise to a CBP configuration.

8.2.1 Hydrolysis

The second step after biomass pretreatment in biochemical conversion is hydrolysis to depolymerise the carbohydrate polymers to soluble sugars. [3] Currently, there are two major ways of converting cellulose to sugars: enzymatic versus chemical. Considerable research efforts on both ways have been made for decades worldwide.

Enzymatic hydrolysis

Cellulase systems

Enzymatic hydrolysis of carbohydrate polymers is carried out by synergistic enzyme cocktails that are highly specific (see Chapter 5). [3, 13] These cocktails

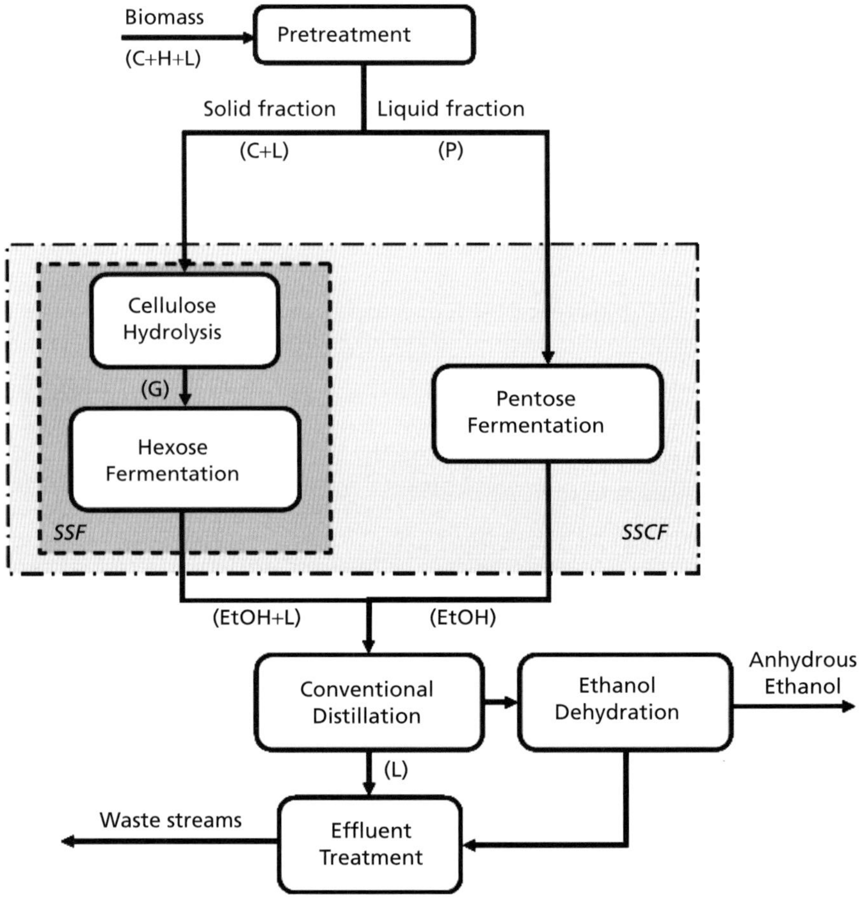

Figure 8.2 Generic block diagram of ethanol production from lignocellulosic biomass. Possibilities for reaction-reaction integration are shown inside the shaded boxes: SSF, simultaneous saccharification and hexose fermentation; SSCF, simultaneous saccharification and co-fermentation. C, cellulose; H, hemicelluloses, L, lignin; G, glucose; P, pentoses; Eth, ethanol. Copyright © 2011 – Instituto de Biotecnologia Agropecuaria. [11]

include exoglucanases (processive enzymes), endoglucanases (non-processive enzymes), and β-glucosidases (cellobiases) for depolymerizing cellulose, as well as several classes of hemicellulases and accessory enzymes for depolymerising hemicelluloses. The products of the hydrolysis are usually reducing sugars including glucose. Fungi and aerobic bacteria produce noncomplexed cellulase systems, whereas anaerobic bacteria produce multienzyme complexes. Many authors have studied the differences between the free and complexed enzyme systems.

Although many cellulolytic bacteria, particularly the cellulolytic anaerobes such as *Clostridium thermocellum*, produce cellulases with high specific activity, they do not produce high enzyme titers. [13] Because the anaerobes have a very low growth rate and require anaerobic growth conditions, most research for commercial cellulase production has focused on fungi, such as *Trichoderma reesei* (recently reclassified as *Hypocrea jecorina*). Fungi are of interest commercially because they can secrete proteins to titers more than 100 g/liter and their enzyme mixtures are naturally quite effective at biomass deconstruction. [3]

The *T reesei* exocellulase Cel7A is amongst the best-characterized cellulases and a major component of fungal enzyme cocktails. It consists of a small Family 1 carbohydrate-binding module (CBM), an *O*-glycosylated linker, and a large catalytic domain (CD) containing a 50 Å-long tunnel and three sites for N-glycosylation. [3]

Several strategies for improving cellulase activity exist. [3] They are categorized as follows:

- directed evolution, such as improving the thermal stability of fungal cellulases via high-throughput screening,
- rational engineering of cellulases, such as improving the binding affinity of the CBM, changing cellulase glycosylation, and
- addition of hemicellulases, accessory enzymes and other helper proteins.

Industrial production of enzymes

Enzymes have found a variety of applications in sectors such as medicine, food industries (baking, alcohol, brewing), textile, leather, detergent, starch, and animal feed. [14-15] The most important industrially produced microbial enzymes are α-amylase, proteases and lipases.

Enzymes are derived from natural resources such as fungi and bacteria, and then are genetically engineered or "molecularly evolved" in the laboratory to impart specific new properties. [16] The new genetic material is inserted into the nucleus of a recombinant host (such as the fungus *Aspergillus* or the bacteria *Bacillus*) that is easy to scale up via fermentation. Today, 95% of all industrial enzymes are made by manufacturing genetically modified organisms.

Most commercialized microbial enzymes come from a small number of fungi and bacteria. *Aspergillus*, *Fusarium*, *Trichoderma* and *Humicola* (all ascomycetes), *Mucor* and *Rhyzomucor* (both zygomycetes) are the main fungi from which biocatalysts are retrieved. [17] *Bacillus* and *Pseudomonas* are the main bacterial strains employed for the production of enzymes.

Three main different molecular techniques are used for the large-scale production of microbial enzymes:

- ***Expression cloning***, in which cDNAs are first translated in prokaryotic or eukaryotic cells and then assayed for a specific biologic activity; [18-19]
- ***Molecular screening***, which is based on similarities among enzyme-encoding gene sequences; [18]

- ***Protein engineering***, which is the process of constructing enzymes with altered properties using two general strategies: ***rational design*** and ***directed evolution.***

The process of developing and producing an industrial enzyme can be summarized as follows: [20]

- creating biological diversity from Nature's diversity and molecular evolution,
- screenings,
- creating expression system,
- upscaling process including fermentation, purification and formulation, and
- production.

Commercial cellulases

Today, cellulases are the third largest industrial enzyme worldwide because of their use in cotton processing, paper recycling, as detergent enzymes, in juice extraction and as animal feed processing. [21] Industrial cellulases are almost all produced from aerobic cellulolytic fungi, such as *Trichoderma reesei* (*Hypocrea jecorina*) or *Humicola insolens*. This is due to the ability of engineered strains of these organisms to produce very large amounts of crude cellulase (over 100 g/liter), the relatively high specific activity of their crude cellulase on crystalline cellulose, and the ability to genetically modify these strains to tailor the set of enzymes they produce. The cost of cellulases is generally estimated at ~50 cents per gallon of ethanol. [21-22]

Two major manufacturers of cellulases are Novozymes (Denmark), the world's largest producer of industrial enzymes and Genencor (now subsidiary of DuPont), the world's second largest producer of industrial enzymes.

Novozymes is active in cellulosic ethanol through broad partnerships that employ many feedstocks and processes. [23] Novoymes' recent Cellic® CTec2 contains a unique protein complex that is known to improve the performance of the enzyme and reduce the dosage required to access and convert the cellulose into fermentable sugars. [24] Cellic CTec2 has also been proven to work on a wide variety of feedstock types, including corn cobs and stalks, wheat straw, sugarcane bagasse, woodchips and even municipal waste. Novozymes has partnered with leading companies in the biofuel industry, such as POET, Greenfield Ethanol, Inbicon, Lignol, ICM, M&G, CTC, COFCO, Siopec and PRAJ to help accelerate process technology development and implementation. [25]

In 2011, DuPont acquired Danisco (Denmark), corporate parent of Genencor. Genencor is a pioneer in enzyme innovation with more than 400 patented enzyme technologies. [26] Genencor's recent Accellerase® TRIO product provides a combined cocktail of enzymes in one product to breakdown the glucan and xylan in the biomass feedstock into fermentable sugars. With Accelerase TRIO, Genencor enables biofuel producers to more cost-effectively manufacture cellulosic ethanol from a wide range of renewable non-food feedstocks such as switchgrass, wheat straw, and corn stover, as well as from municipal solid waste.

Research on cellulases

One possible route to reducing cellulase cost is to produce cellulases in the crop biomass itself, rather than producing them in microbial tanks. Along this line, the gene coding for the catalytic domain of the thermostable *Acidothermus cellulolyticus* endoglucanase E1 enzyme was successfully expressed in a broad array of plant species, and the enzyme accumulation had no apparent deleterious effects on plant growth and development (**Figure 8.3**). [27]

Green plants are the lowest cost source of proteins, so that if cellulases could be produced at a high level without hurting the yield of a productive crop, their cost could be reduced. [21] A challenge with plant production of cellulases is the multiplicity of enzymes that are needed to efficiently degrade plant cell walls. Fungi produce large numbers of enzymes, and it is sometimes not clear which enzymes are required to degrade a given substrate; hence, it is difficult to engineer a plant to produce a mixture with equal activity to that produced by a cellulolytic fungus.

Extensive research is being carried out to try to obtain cellulases with higher activity on pretreated biomass substrates by screening and sequencing new organisms, engineering cellulases with improved properties and by identifying proteins that can stimulate cellulases. [21]

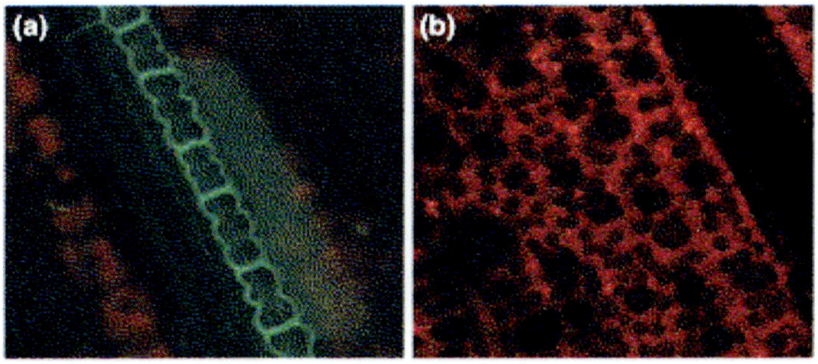

Figure 8.3 The production of the *Acidothermus cellulolyticus* endoglucanase (E1) gene product in corn leaf apoplast. **(a)** E1 transgenic maize leaf tissue shows apparent storage of E1 in the plant apoplast (green areas around each of the cells). Image produced with immunofluorescent confocal laser microscopy. **(b)** Leaf tissue from an untransformed control maize leaf, showing the characteristics of no expression of E1 enzyme. [27]

Acid hydrolysis

Dilute acid hydrolysis

The dilute acid hydrolysis process is the oldest technology for converting cellulosic biomass to ethanol. [28] Typically, a solution of 1% sulfuric acid is used. The first attempt to commercialize a wood-to-ethanol process was done in Germany in 1898. It involved the use of dilute acid, and was able to produce 7.6 liters of ethanol per 100 kg of wood waste (18 gallons per ton). An industrial process was promptly developed by the Germans with an improved yield of 50 gallons per ton of biomass. This process was then introduced to the United States, where two plants were operating during World War I. These plants used what was called the American Process, a one-stage dilute sulfuric acid hydrolysis.

In 1932, the Germans developed an improved percolation process, known as the Scholler Process, in which dilute sulfuric acid was pumped through a bed of wood chips. The subsequent US work resulted in the Madison Wood Sugar process, which showed improvements in productivity and yields over its German predecessor. Tennessee Valley Authority's pilot plant studies further refined the process. The dilute acid hydrolysis percolation reactor is still one of the simplest means of producing sugars from biomass, and is used as a benchmark.

By 1985, it was generally accepted that, while the dilute acid percolation designs were well understood, these systems had reached the limits of their potential. [28] Their relatively high glucose yields (~70%) were obtainable only with highly dilute sugar streams. Kinetic models and process design work showed that both high solids concentration and some form of countercurrent flow would be required for the most efficient systems.

After more than 100 years of R&D, the dilute acid hydrolysis process has evolved into the general scheme shown in **Figure 8.4**. [28] The process is conducted under high temperature and pressure, and has reaction times in the range of seconds or minutes, which facilitates continuous processing. [29]

The hydrolysis occurs in two stages to accommodate the differences between hemicelluloses and cellulose. [28] Hemicelluloses, composed primarily of 5-carbon sugars (xylose), are indeed more easily hydrolyzable than cellulose, composed of 6-carbon sugars (glucose). Furthermore, under the hydrolysis conditions, 5-carbon sugars degrade more rapidly into other chemicals, such as furfural, than cellulose sugars. [29] Not only does sugar degradation reduce sugar yield, but also the furfural and other by-products can be poisonous to the fermentation microorganisms. With the two-stage process, the first stage can be operated under milder conditions (e.g. 190 °C), which maximize yield from hemicelluloses, whereas the second stage can be optimized (e.g. 215 °C) for hydrolysis of the more resistant cellulose fraction. The liquid sugars are recovered at each stage and fermented to ethanol. [31] Resid-

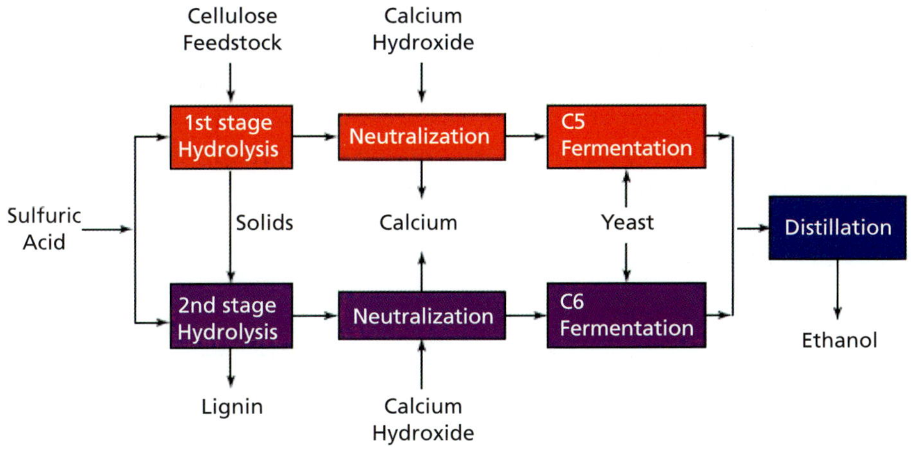

Figure 8.4 General scheme of two-stage dilute acid hydrolysis. [30]

ual cellulose and lignin serve as boiler fuel for electricity or steam production. However, in the two-stage process, sugar degradation still occurs and yields are limited to around 80 gallons of ethanol per ton (330 l/t) of dry wood. [29]

To our knowledge, there is no industrial plant existing on full dilute acid hydrolysis for fuel production. Dilute acid is essentially applied as "pretreatment" and combined with other technologies to yield improvements in production processes.

Wood-pulp producers like Tembec, USA, and Georgia Pacific, USA, operate sulfite pulp mills, which utilize a dilute acid hydrolysis process to dissolve hemicelluloses and lignin from wood, and produce specialty cellulose pulp. [28] The hexose sugars in the spent sulfite liquor are fermented to ethanol. The lignin is either burnt to generate process steam or converted to added-value products.

Concentrated acid hydrolysis

The concentrated acid hydrolysis process for producing ethanol from cellulosic biomass has a long history. [32] The ability to dissolve and hydrolyze native cellulose using concentrated sulfuric acid followed by dilution with water was reported as early as in 1883. The concentrated acid disrupts the intermolecular hydrogen bonds, converting the cellulose into an amorphous material. Once decrystallized, the cellulose is highly susceptible to hydrolysis. Thus, dilution with water at relatively mild temperatures gives complete and rapid hydrolysis to glucose with little degradation.

In 1918, researchers from the US Department of Agriculture (USDA) introduced the idea of using dilute acid pretreatment of agricultural residues to remove hemicelluloses before decrystallization and hydrolysis of the cellulose fraction. [32] The benefit was the ability to isolate the hemicelluloses (primarily 5-carbon) sugars from the cellulose (6-carbon) sugars. In 1937, commercial concentrated acid plants based on hydrochloric acid were operated in Germany. The Japanese developed a concentrated sulfuric acid process using membranes to separate the sugar and acid in the product stream. The Japanese process was commercialized in 1948. In the 1980s, US research on the USDA concentrated sulfuric acid process led to improvements in the recycling of the acid. The studies were performed particularly at Purdue University and at Tennessee Valley Authorities.

Currently, the concentrated acid hydrolysis process is illustrated by the general scheme shown in **Figure 8.5**. [33] The diagram, which gives an example of how such a process might be configured, is based on information from Arkenol, Inc. (USA). The process uses relatively mild temperatures and the only pressures involved are usually those created by pumping materials from vessel. [29]

The core of the process is the decrystallization followed by the dilute acid hydrolysis. [32] To make the process economically viable, Arkenol claims to have solved key steps, such as:

- efficient acid recovery and reconcentration;
- high sugar concentration at high purity;
- the ability to handle silica in biomass feedstocks;
- all by-products are usable and marketable.

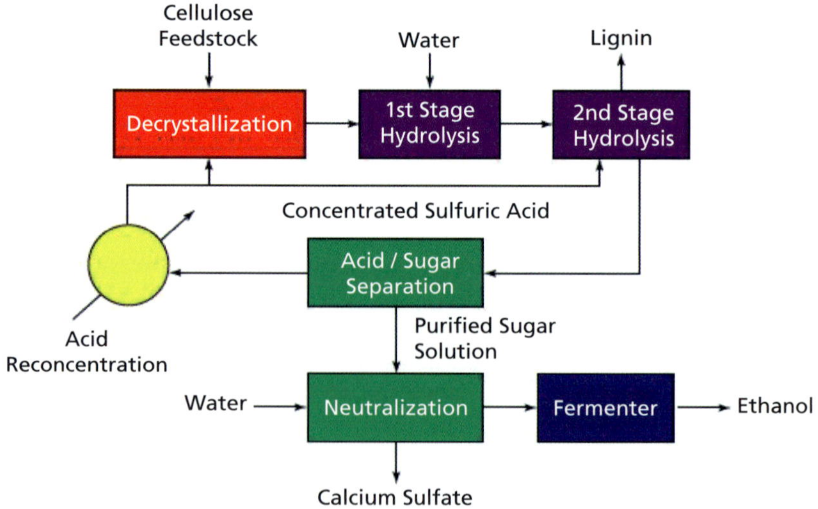

Figure 8.5 General scheme of the concentrated acid hydrolysis. [30]

BlueFire Renewables was established to deploy the proven Arkenol process for the conversion of cellulosic waste materials into renewable liquid fuels and other products. [34] BlueFire is the exclusive North America licensee of the technology. The company is currently in the process of developing two cellulosic facilities in Lancaster, California and Fulton, Mississippi. [35] Production from the Mississippi ethanol facility will be totally procured by Tenaska BioFuels under a 15-year off-take agreement. [36]

Masada Resource Group based in Alabama, USA has developed a similar acid hydrolysis process that can utilize municipal solid waste (MSW) as a feedstock. [37] The process is known as the OxyNol™ Process. [38]

8.2.2 Process configurations for ethanol production

The third step after biomass pretreatment and hydrolysis in biochemical conversion is fermentation, in which microorganisms convert sugars into ethanol or other chemicals. Microorganisms are capable of growing on a wide range of substrates (such as glucose) and can produce a remarkable spectrum of products. [39] Genetic engineering has extended the range of products that may be produced by microorganisms and has provided new methods for increasing the yields of existing ones. Fermentation is the term used by microbiologists to describe any process for the production of a product by means of the culture of microorganisms or cells. [40-41]

Two modes of fermentation exist:

- Batch fermentation is carried out in a closed-system bioreactor. The sterilized nutrient medium is inoculated with living cells. The mixture is incubated with stirring under specific conditions, and the culture grows. The composition of the medium and the concentration of biomass and nutrients change with time.
- Continuous fermentation is considered to be an open system where nutrients are continuously added to the system and product is continuously removed. The cell growth is kept at steady state. [41]

There are different strategies for carrying out hydrolysis and fermentation, including separate (sequential) enzymatic hydrolysis and fermentation (SHF), simultaneous saccharification and fermentation (SSF), simultaneous saccharification and co-fermentation (SSCF) and consolidated bioprocessing (CBP) (**Figure 8.6**).

Separate hydrolysis and fermentation (SHF)

In the SHF process, pretreated lignocellulosic materials are hydrolyzed to glucose and subsequently fermented to ethanol in separate units. [43-47] SHF can be designed in various ways. A typical SHF configuration is shown in

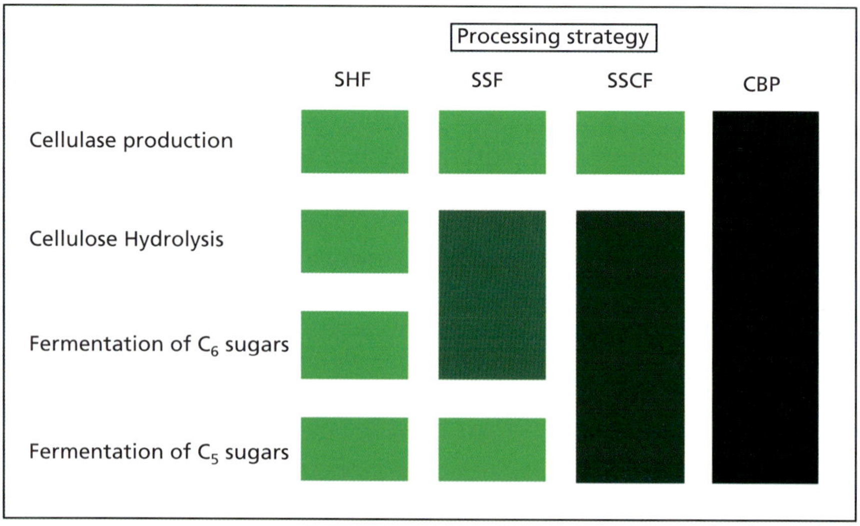

Figure 8.6 The four process configurations and their degree of consolidation in converting lignocellulosic biomass to ethanol: separate hydrolysis and fermentation (SHF); simultaneous saccharification and fermentation (SSF); simultaneous saccharification end co-fermentation (SSCF) and consolidated Bioprocessing (CBP) (adapted from Lynd, 1996). Lisa du Plessis. [42]

Figure 8.7. [48] Prior to ethanol distillation the hemicellulosic sugars released from the pretreatment and glucose released in the enzymatic hydrolysis can either be fermented together or separately. When hexose and pentose fermentation is preformed together, the term separate hydrolysis and co-fermentation (SHCF) is also used to describe this SHF variant. [49-50]

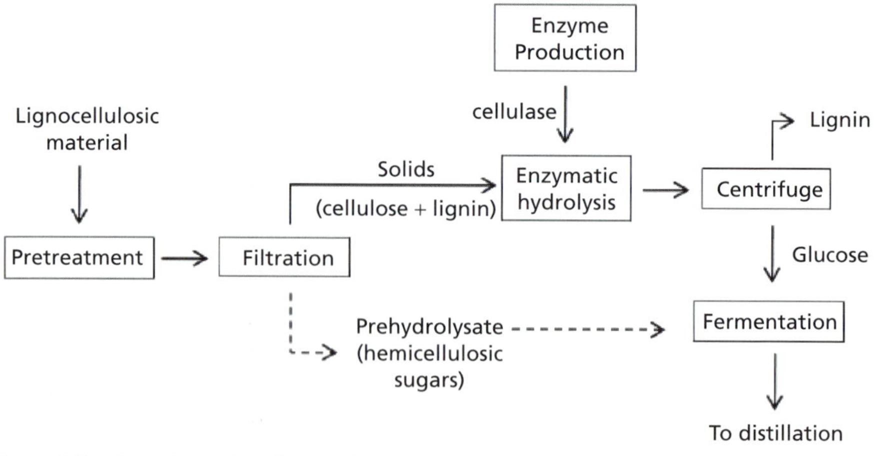

Figure 8.7 Flow sheet showing a schematic picture of a possible bioethanol process using separate enzymatic hydrolysis and fermentation (SHF). J.Axelsson/Master's Thesis/Linköping University/2011. [48]

In another SHF configuration, the joint liquid flow from both hemicellulose and cellulose hydrolysis first enters the glucose fermentation reactor; the mixture is then distilled to remove the ethanol leaving the unconverted pentose behind. In a second reactor, pentose is fermented to ethanol, and the ethanol again distilled. [51]

The main advantage of this process is it allows for each step to be performed at its optimum conditions for temperature and pH. [52] Cellulases are typically maximally active at ~50 °C and a pH of 4.8, whereas most microorganisms ferment better at temperatures near ~30 °C and pH near 6.0. However, a major disadvantage of SHF is that the accumulation of glucose and cellobiose during hydrolysis can lead to end-product inhibition. To maintain acceptable ethanol yields, the hydrolysis step in SHF must often be carried out at low solid loadings, resulting in a relatively dilute ethanol stream. Economically, this can increase costs for downstream processing, namely for distillation.

Simultaneous saccharification and fermentation (SSF)

One of the most successful methods for ethanol production from lignocellulosic biomass is combination of the enzymatic hydrolysis of pretreated biomass and fermentation in one step (**Figure 8.8**). [43] SSF has been established as a promising option for ethanol production since the overall ethanol yield has been reported to be higher than if the enzymatic hydrolysis and fermentation are carried out separately (SHF). Furthermore, it has also been established that xylose consumption increases in SSF. [53-54]

In this process, the glucose produced by the hydrolyzing enzymes is consumed immediately by the fermenting microorganism present in the culture. [43] This is a great advantage for SSF compared to SHF, since the inhibition effects of glucose and cellobiose to the enzymes are minimized by keeping a low concentration of these sugars in the media. SSF gives higher reported ethanol yields from cellulose than SHF and requires lower amounts of enzyme. The risk of contamination in SSF is lower than in SHF, since the presence of ethanol reduces the possibility of contamination. Furthermore, the number of vessels required for SSF is reduced compared to SHF, resulting in lower capital cost of the process.

Nevertheless, SSF has the disadvantage of operating at temperature and pH conditions that represent a compromise between the optima of theses parameters for both enzymatic activity and the growth of the fermenting organisms. [55] The temperature for SSF is typically kept at ~37 °C, which is a compromise between the optimal temperatures for hydrolysis and fermentation. Furthermore, the yeasts cannot be reused in an SSF process due to the problems of separating the yeast from the lignin after fermentation. [53] Therefore, the yeast will necessarily represent a yield loss in an SSF process, if the yeast is produced from carbohydrates within the process or a running cost

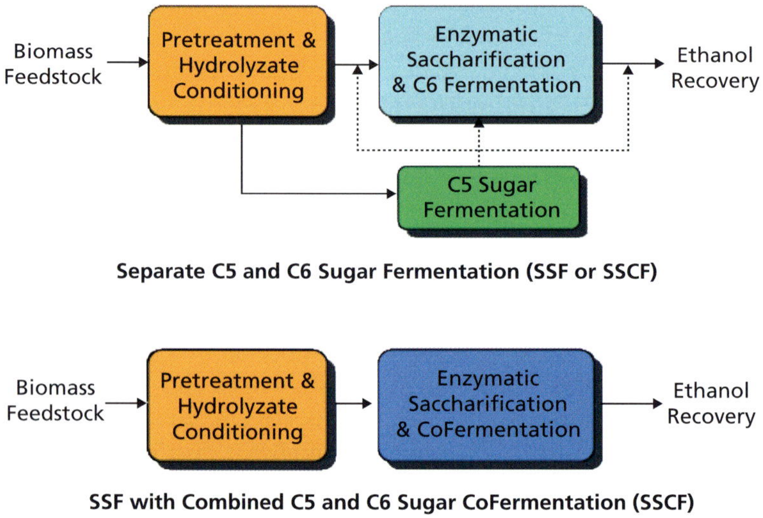

Figure 8.8 Simultaneous saccharification and fermentation with separate pentose and hexose fermentation (SSF) or combined pentose and hexose co-fermentation (SSCF). [49]

if it is externally supplied. The enzymes are equally difficult to reuse, also in an SHF process. The enzymes are either produced within the process, thereby representing a loss of substrate, or are externally supplied and thereby add to the costs. Recirculation of enzymes is equally difficult since the enzymes bind to the substrate.

Hydrolysis is usually the rate-limiting process in SSF. [55] Several thermotolerant yeasts, such as *Candida acidothermophilum* and *Kluyveromyces marxianus*, as well as various species of bacteria, have been used in SSF to raise the temperature to the optimal hydrolysis temperature. Another disadvantage of SSF is that cellulases are inhibited by ethanol.

Simultaneous saccharification and co-fermentation

SSCF refers to simultaneous saccharification and co-fermentation (SSCF), in which co-fermentation refers to the fermentation of both 5-carbon and 6-carbon sugars to ethanol. [43] The hydrolyzed hemicelluloses and the solid cellulose are not separated after pretreatment, allowing the conversion of hemicellulose sugars to occur simultaneously with the SSF of cellulose. In the SSCF process, both hexoses and pentoses are fermented in a single reactor with a single microorganism. The SSCF process is considered to be an improvement to SSF.

Progress is rapid in the field of xylose fermentation to ethanol, but few industrial yeast strains have yet the demonstrated capability of fermenting xylose in lignocellulosic hydrolyzates efficiently. [54] Recently, Hahn-Hägerdal et al. [56] summarized the performance of industrial yeast strains with the ability to ferment all lignocellulose-derived sugars. All strains covered in their review were xylose reductase and xylitol dehydrogenase expressing strains, which also overexpressed xylulokinase. TMB 3400 was the only industrial pentose- and hexose-fermenting *S. cerevisiae* strain for which results on SSF have also been reported. Ethanol concentrations reaching 40 g/L and yields up to 80% of the theoretical based on xylose and glucose have been achieved. In SSCF studies using TMB 3400, substantial xylose conversions have been reported for various materials such as spruce, corn stover, wheat straw, and sugar cane bagasse. [57]

Consolidated Bioprocessing (CBP)

Biomass processing schemes involving enzymatic or microbial hydrolysis commonly involve four biological transformations: the production of saccharolytic enzymes, the hydrolysis of carbohydrate components present in pretreated biomass to sugars, the fermentation of hexose sugars, and the fermentation of pentose sugars (see **Figure 8.6**). [58] These four transformations occur in a single step in CBP, which is distinguished from the other process configurations in the absence of a dedicated step for enzyme production.

CBP, which carries out the four biological events in one bioreactor, should reduce the enzyme production cost and, at the same time, has all the advantages of SSCF. [59] A strong case has been made that CBP is the ultimate low-cost industrial configuration to produce cellulosic ethanol. This results from avoided costs associated with enzyme production. In addition, several factors support the possibility of realizing higher hydrolysis rates, and, hence, reduced reactor volume and capital investment. [58]

However, the crucial problem with CBP is to develop a microorganism which can degrade lignocellulosic biomass efficiently and at the same time utilize all the sugars released to produce ethanol. [50, 59] Researchers are focusing on either engineering an ethanologen, such as *S. cerevisiae*, to be able to produce cellulases/hemicellulases (recombinant strategy) or engineering a lignocellulose degrader, such as *Clostridium thermocellum*, to be an efficient ethanol producer (native strategy).

Native occurring cellulolytic microorganisms are starting points for CBP organism development via the native cellulolytic strategy, with anaerobes being of particular interest. [58] The primary objective of such developments is to engineer product yields and titers to satisfy the requirements of an industrial process. Among cellulolytic anaerobes, gene-transfer systems have been described for *Clostridium cellulolyticum* and for *Clostridium thermocellum*. Product tolerance is a key property impacting the feasibility of CBP through the native strategy.

Non-cellulolytic microorganisms with desired product formation are starting points for CBP organism development by the recombinant cellulolytic strategy. [58] The primary objective of such developments is to engineer a heterologous cellulase system that enables growth and fermentation on pretreated lignocellulose. The heterologous production of cellulases has been pursued primarily with bacterial hosts producing ethanol at high yield (engineered strains of *E. coli*, *Klebsiella oxytoca* and *Zymomonas mobilis*), and the yeast *S. cerevisiae*.

8.2.3 Fermentation

General

From a biochemical viewpoint, fermentation can be defined as an anaerobic process that derives energy from the oxidation of sugars and other organic molecules, using an endogeneous electron acceptor, which is usually an organic compound. [60-61] Fermentation is a major source of energy for those organisms that can survive only in the absence of air. Other fermentative organisms that can grow in the presence or absence of air use fermentation as a source of energy only when oxygen is absent. Like cellular respiration, fermentation starts with a cellular process called ***glycolysis*** that oxidizes the glucose molecule into two 3-carbon molecules called pyruvic acid ($CH_3COCOOH$). [62] The most common type of glycolysis is the ***Embden Meyerhof Parnas pathway*** (EMP pathway), which is a sequence of ten reactions characterized by a net gain of two ATP molecules and two NADH molecules (**Figure 8.9** and **Equation 8.1**).

$$\text{Glucose} + 2\,\text{NAD}^+ + 2\,\text{ADP} + 2\,\text{P}_i \rightarrow 2\,\text{Pyruvate} + 2\,\text{NADH} + 2\,\text{ATP} + 2\,\text{H}^+ + 2\,\text{H}_2\text{O} \quad (8.1)$$

The energy of other sugars is also harvested using this process. A variety of D-hexoses including fructose, mannose and galactose can be funnelled into glycolysis. [64]

The amount of ATP in fermentation is largely restricted to the amount formed during glycolysis. [60]

If glycolysis were to continue indefinitely, all of the NAD^+ would be used up, and glycolysis would stop. To allow glycolysis to continue, organisms must be able to oxidize NADH back to NAD^+. One method of doing that is to simply have the pyruvate do the oxidation. In fermentation reactions, pyruvate is converted to a variety of organic compounds. [60] These reactions involve the transfer of electrons and hydrogen from NADH to organic compounds. The end products of fermentations and the pathway by which they are formed are specific for microbial species or groups. Generally, fermentation pathways are named after the characteristic end products that are formed. Some of the

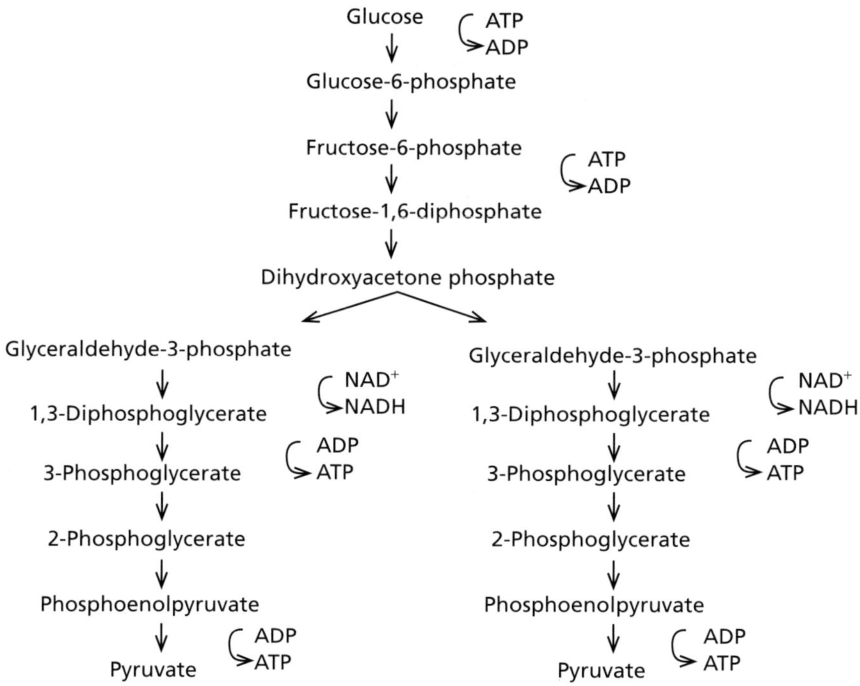

Figure 8.9 Glycolysis as a biochemical pathway. © Copyright by Gretchen Kirchner 1996, 2001. [63]

important fermentation pathways are lactic acid fermentation, ethanol fermentation, succinic acid fermentation, propionic acid fermentation, butyric acid fermentation, and mixed acid fermentation.

Not long ago, the term fermentation was strictly confined to the context of anaerobic respiration. [65] However, the situation is complicated by the ability of many microorganisms, including many widely-used yeasts, to switch between aerobic and anaerobic respiration, according to the availability of oxygen. Further, where several species of microorganism are simultaneously involved in a process, aerobic and anaerobic pathways may well occur side by side. Therefore, the term fermentation is currently used in a less precise sense, to describe any process, anaerobic or aerobic which includes the breakdown of food by microorganisms.

Ethanol fermentation

Hexose fermentation to ethanol

Ethanol fermentation, also referred to as alcoholic fermentation, is a process in which sugars such as glucose, fructose and sucrose (which has to be hydrolyzed first into glucose and fructose) are converted into cellular energy

Figure 8.10 Grapes fermenting during the production of wine. Reproduced by permission of AGNE27 under the Creative Commons Attribution-Share Alike 3.0 Unported License. [65]

and, thereby, produce ethanol and carbon dioxide as metabolic waste products. [66] Because yeasts perform this conversion in the absence of oxygen, ethanol fermentation is classified as anaerobic (**Figure 8.10**).

Ethanol fermentation has considerable economic importance. [60] For example, it is used to make bread, beer, wine, and distilled spirits. Ethanol fermentation is carried out by many types of yeast, such as *Saccharomyces cerevisiae*, and a few bacteria such as *Zymomonas mobilis* and *Pseudomonas saccharophila*.

In this fermentation pathway, pyruvate is converted to ethanol and carbon dioxide in a two-step process (**Figure 8.11**). [67] In the first step, pyruvate undergoes decarboxylation in the irreversible reaction catalyzed by pyruvate decarboxylase. This reaction is a simple decarboxylation and does not involve the net oxidation of pyruvate. Pyruvate decarboxylase requires Mg^{2+} and has a tightly bound coenzyme, thiamine pyrophosphate (TPP, a coenzyme derived from vitamin B1). In the second step, acetaldehyde is reduced to ethanol, with NADH furnishing the reducing power. Ethanol and CO_2 are thus the end products of alcoholic fermentation.

The overall equation for alcoholic fermentation can be expressed as follows (**Equation 8.2**):

$$\text{Glucose} + 2\ \text{ADP} + 2\text{P}_i \rightarrow 2\ \text{Ethanol} + 2\ \text{CO}_2 + 2\ \text{ATP} \qquad (8.2)$$

There is no net change in the ratio of hydrogen to carbon atoms when glucose (H/C ratio = 2) is fermented to two ethanol and two CO_2 (combined H/C ratio = 12/6 = 2). [67] In all fermentations, the H/C ratio of the reactants and products remains the same.

Pentose fermentation to ethanol by engineered S. cerevisiae

Due to the very attractive properties of *S. cerevisiae* in industrial fermentation, there have been considerable efforts made in the past decades to design

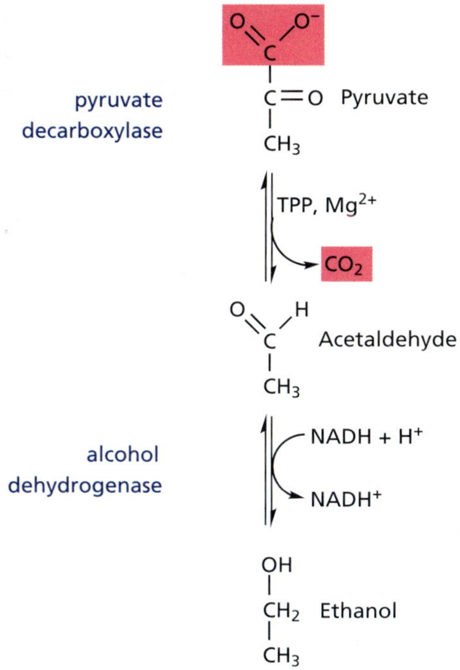

Figure 8.11 Ethanol fermentation pathway. [67]

recombinant pentose-fermenting strains of this yeast. [54] Since *S. cerevisiae* cannot utilize xylose, but does utilize and ferment its isomer D-xylulose, the obvious way to allow xylose metabolism is to introduce a heterogeneous pathway converting xylose to xylulose. [68] Several approaches have been explored to express a pentose utilization pathway from naturally pentose-utilizing bacteria and fungi in *S. cerevisiae*. **Figure 8.12** summarizes the initial pathways for D-xylose utilization in bacteria and fungi.

In naturally D-xylose-utilizing bacteria, xylose is isomerized to D-xylulose by xylose-isomerase (XI). [68] Xylulose is then phosphorylated to xylulose 5-phosphate, which is an intermediate of the ***pentose phosphate pathway*** (PPP; secondary pathway for glucose metabolism). A similar pathway has been found in an anaerobic fungus; however, most naturally xylose-utilizing fungi contain a more complex pathway consisting of reduction-oxidation reactions involving the cofactors NAD(P)H and NAD(P)$^+$. Xylose is reduced to xylitol by a xylose reductase (XR), and xylitol is then oxidized to D-xylulose by a xylitol dehydrogenase (XDH). As in bacteria, xylulose is phosphorylated to D-xylulose 5-phosphate by a xylulokinase (XK). The conversion of L-arabinose into intermediates of the PPP requires more enzymatic reactions than the conversion of xylose (**Figure 8.13**).

Xylose-fermenting strains of *S. cerevisiae* can be constructed by introducing genes encoding xylose isomerase from bacteria and fungi, or genes encod-

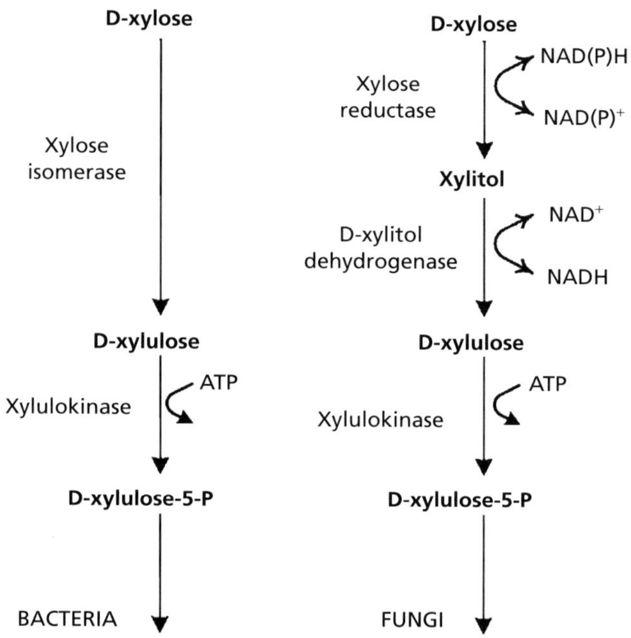

Figure 8.12 The initial xylose utilization pathways in bacteria and fungi. [56]

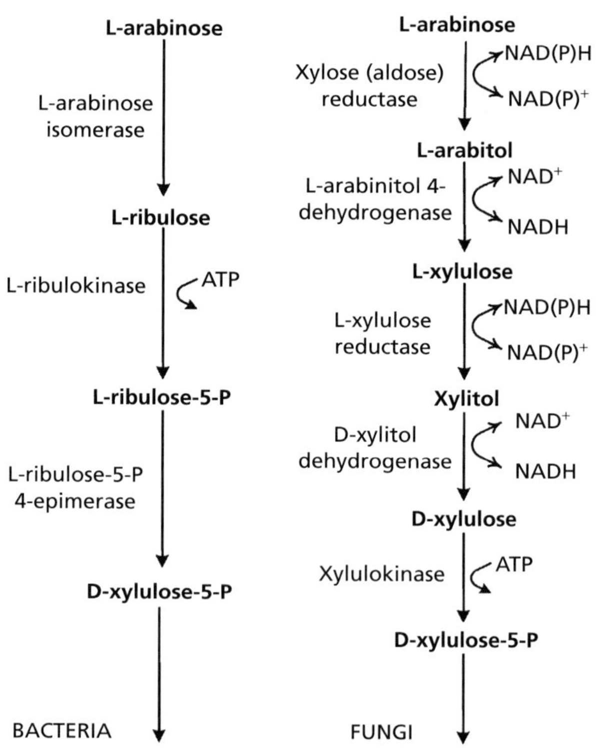

Figure 8.13 The initial L-arabinose pathways in bacteria and fungi. [56]

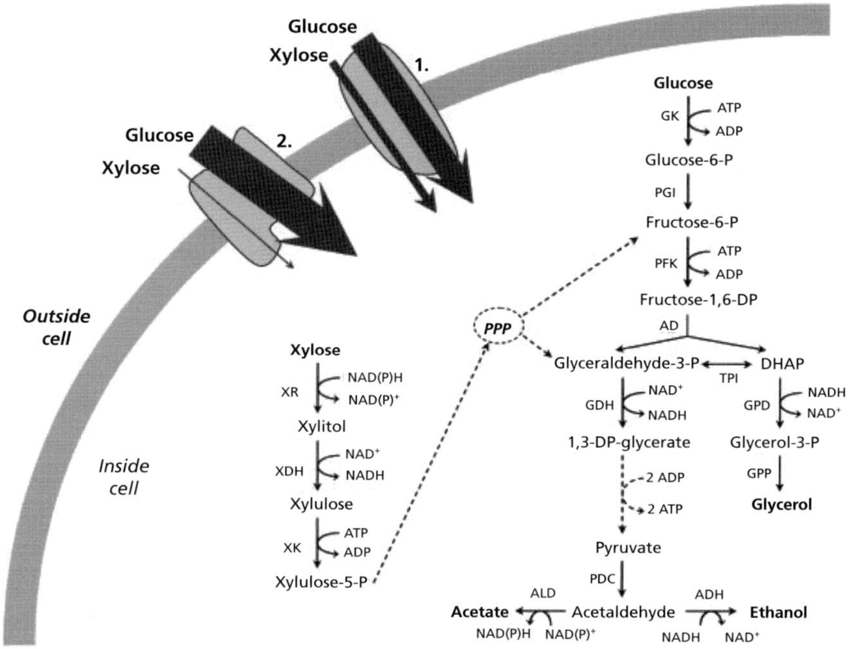

Figure 8.14 Simplified scheme of sugar transport and metabolism in *S. cerevisiae*. 1 Low- and intermediate-affinity hexose transporters. 2. High-affinity hexose transporters. PPP, pentose phosphate pathway; XR, xylose reductase; XDH, xylitol dehydrogenase; XK, xylulokinase, GK, Glucokinase; PGI, phosphoglucose isomerase; PFK, phosphofructokinase; AD, aldolase; TPI, triose phosphateisomerase; GDH, glyceraldehyde-3-P dehydrogenase; GPD, glycerol-3-P dehydrogenase; GPP, glycerol-3-phosphotase; PDC, pyruvate decarboxylase; ALD, acetaldehyde dehydrogenase; ADH, alcohol dehydrogenase. © 2008 Olofsson et al.; licensee BioMed Central Ltd./ CREATIVE COMMONS. [54]

ing xylose reductase and xylitol dehydrogenase from fungi. [54] Also, the endogeneous gene encoding xylulokinase has to be overexpressed to obtain significant xylose fermentation. Transport proteins are needed for uptake of xylose, as well as other sugars in yeast. In *S. cerevisiae*, it has been found that xylose is transported by the hexose transporters, but the affinity for xylose is ~200-fold lower than for glucose (**Figure 8.14**). Hence, xylose uptake is competitively inhibited by glucose. It has been suggested that xylose is taken up by both high- and low-affinity systems of glucose transporters, but the uptake is increased in the presence of low glucose concentrations. Consequently, in order to obtain efficient co-fermentation of xylose and glucose in SSCF with recombinant *S. cerevisiae*, it is necessary to keep the glucose concentration low.

Lactic acid fermentation

Important genera of lactic acid bacteria are *Streptococcus, Lactococcus* and *Lactobacillus*. [60, 69] Lactic acid fermentations are of two types: homolactic and heterolactic. When the EMP pathway of glycolysis is used in the lactic acid

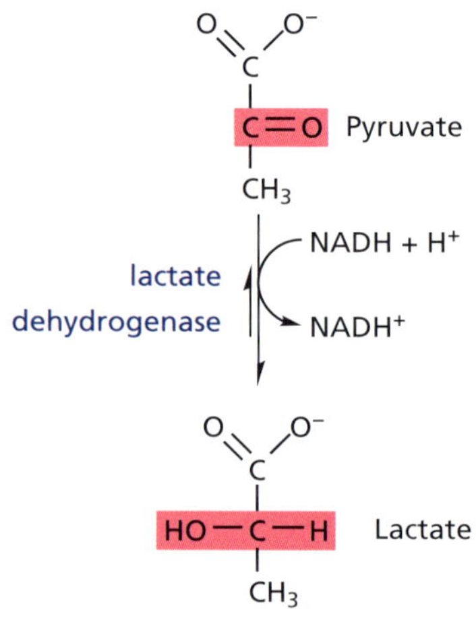

$\Delta G^{o'} = -25.1$ kJ/mol

Figure 8.15 Homolactic acid fermentation pathway. Copyright: Worth Publishers. [67]

fermentation, the overall pathway is homolactic fermentation because the only end product formed is lactic acid (**Figure 8.15**).

In homolactic acid fermentation, the reduction of pyruvate is catalyzed by lactate dehydrogenase, which forms the L-isomer of lactic acid. [67] The overall equilibrium of this reduction reaction strongly favors lactate formation, as shown by the large negative standard free-energy change. Because the reduction of two molecules of pyruvate to two of lactate regenerates two molecules of NAD^+, the overall process is balanced and can continue indefinitely: one molecule of glucose is converted to two of lactate, with the generation of two ATP molecules from one of glucose, and NAD^+ and NADH are continuously interconverted with no net gain or loss in the amount of either (**Equation 8.3**).

$$\text{Glucose} + 2\ \text{ADP} + 2\ \text{P}_i \rightarrow 2\ \text{Lactic acid} + 2\ \text{ATP} \tag{8.3}$$

Although there are two oxidation-reduction steps as glucose is converted into lactate, there is no net change in the oxidation state of carbon; in glucose ($C_6H_{12}O_6$) and lactic acid ($C_3H_6O_3$), the H/C ratio is the same. [67] Nevertheless, some of the energy of the glucose molecule has been extracted by its conversion to lactate, enough to give a net yield of two molecules of ATP for every one of glucose consumed.

In heterolactic acid fermentation, ethanol and carbon dioxide are produced in addition to lactic acid. [60, 70] The overall reaction for the heterolactic acid fermentation can be expressed as follows (**Equation 8.4**):

$$\text{Glucose} + \text{ADP} + \text{P}_i \rightarrow \text{Lactic acid} + \text{Ethanol} + CO_2 + \text{ATP} \tag{8.4}$$

Succinic acid fermentation

Succinic acid, also known as amber acid or butanedioic acid, is a dicarboxylic acid of molecular formula HOOC-C_2H_4-COOH.

Succinic acid produced by microbial fermentation of carbohydrates can be used as a precursor of many industrially important chemicals in food, chemical and pharmaceutical industries (**Figure 8.16**). [71]

The processes for fermentative succinic acid production are carried out by several promising succinic acid producers including *Actinobacillus succinogenes* (anaerobic bacterium), *Anaerobiospirillum succiniciproducens* (anaerobic bacterium), *Mannheimia succiniciproducens* (***capnophilic*** bacterium) and recombinant *Escherichia coli*. [71]

Succinic acid is synthesized as one of the mixed-acid fermentation product. Mixed acid fermentation produces a mixture of products including succinic acid, ethanol, lactic acid, formic acid and acetic acid, from which the succinic acid must be isolated. [71-72] The major metabolic pathways involved in the fermentative production of mixed acids is shown in **Figure 8.17**. Interestingly, phosphoenolpyruvate carboxylation is a major CO_2-fixing step.

Process developments are focussed on reducing side-product formation through organism selection.

Several companies are currently involved in pilot plants and large scale plants for producing biobased succinic acid: [73]

- Bioamber, initially a joint venture between DNP Green Technology (USA) and French Agro-Industrie Recherches et Développements (ARD)

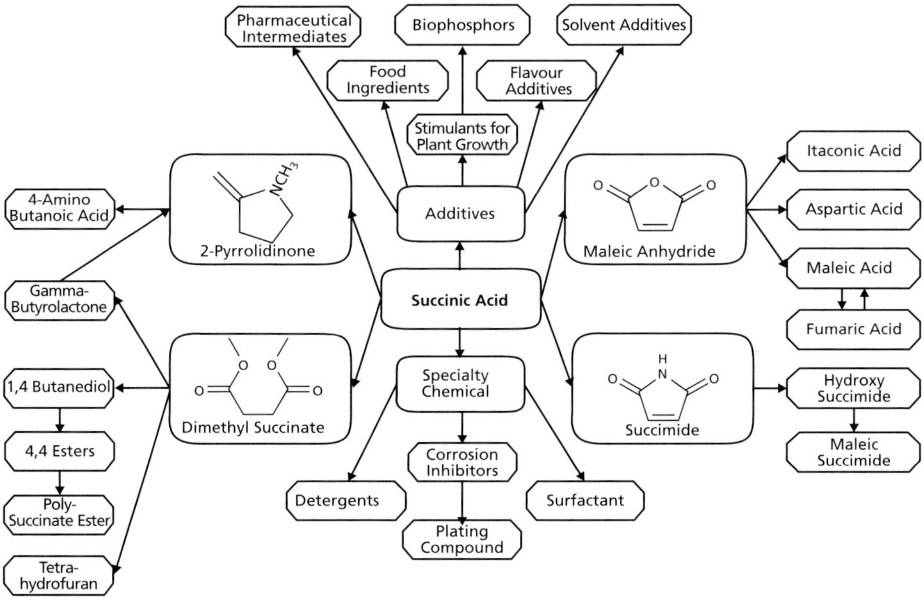

Figure 8.16 Various chemicals that can be synthesized from succinic acid. [71]

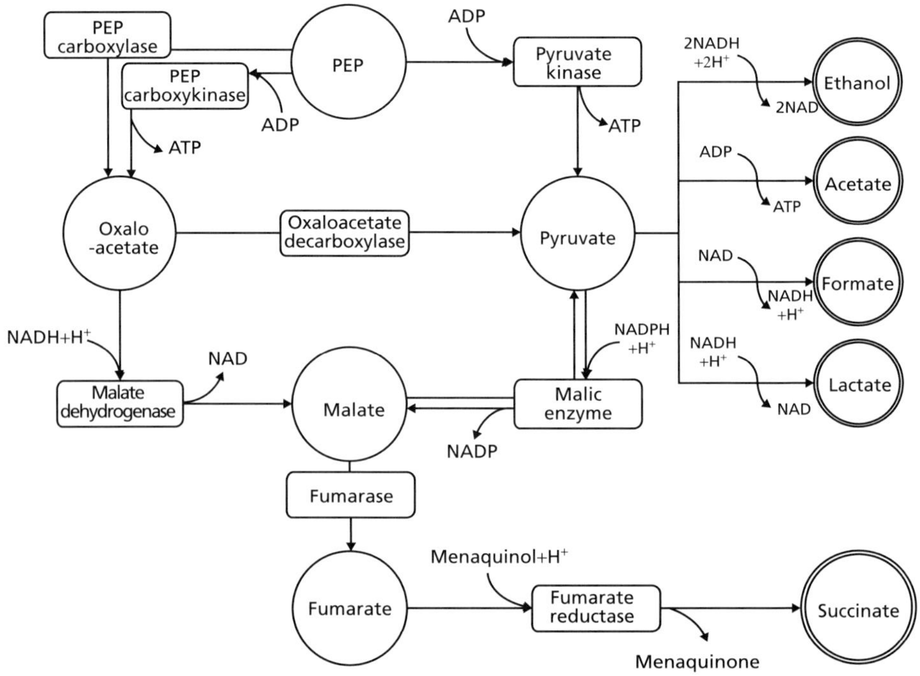

Figure 8.17 Major metabolic pathways leading to the formation of succinic acid and byproducts in *M. succiniciproducens*. Intracellular metabolites are shown in single circles while the excretory metabolic products are shown in double circles. Enzymes catalyzing the reactions are shown in rectangles. PEP, phosphoenolpyruvate. [71]

recently commissioned a demonstration plant in Pomacle, France with a capacity of 2000 tonnes per year. [73] Now, BioAmber is a private US company. [74] Based on an engineered *Escherichia coli* licensed from the US Department of Energy, BioAmber has developed an innovative fermentation technology that delivers cost-effective succinic acid. [75] An important feature of the technology is that the bacterium sequesters CO_2 during the production process. Therefore, succinate salts are produced from sugar and CO_2. With 25% of the carbon in the product coming from CO_2, overall the biosynthesis consumes CO_2 rather than generating it. Early 2011, BioAmber announced that it is developing a new generation microorganism for the production of succinic acid derived from renewable feedstocks including lignocellulosic material, under exclusive license from Cargill. [76] Also in 2011, BioAmber partnered with Mitsubishi Chemical in succinic acid. [77]

- Early 2010, DSM and Roquette opened a demonstration plant in Lestrem, France. [78] In 2011, the companies announced their intention to build a commercial scale plant in 2012 with a capacity of ~10000 tonnes per year.

The new commercial production facility in Italy will use a proprietary yeast-based fermentation process, which operates at much lower pH than competing processes, allowing succinic acid to be produced with a higher energy efficiency compared to the traditional method. It is also one of the first biobased processes that sequesters carbon dioxide in the production process.

- Myriant Technologies as successor to BioEnergy International, has been awarded USD 50 million by the US Department of Energy to help fund the construction of a succinic acid plant in Louisiana, USA. [73] The project will produce up to ~13600 tonnes per year using sorghum and carbon dioxide as feedstocks.
- BASF and Purac, the Dutch worldwide market leader in lactic acid and lactic acid derivatives, have announced their collaboration and intention to begin commercial production of succinic acid in 2010. Succinic acid will be used as a building block in a variety of biopolymers.

- Mitsubishi Chemical and Thailand's PTT are investigating the feasibility of the manufacture of biobased polybutylene succinate (PBS), a biodegradable polymer, in Thailand. Mitsubishi Chemical has developed a process for the production of succinic acid, a precursor to PBS, from biomass.

8.3 Life cycle assessments

8.3.1 Introduction

Lignocellulosic biorefineries are supposed to contribute to a more sustainable resource supply and to a reduction in greenhouse gas emissions. [79] However, biobased products and fuels may also be associated with environmental disadvantages due to e.g. land use or eutrophication of water.

Water consumption for cellulosic ethanol is expected to be significantly less than for corn ethanol. [80] Many cellulose sources require no irrigation, and those that do are expected to require less water than does corn. Technology of 2007 for cellulosic ethanol uses ~6 gal water/gal ethanol produced. Technology improvements are expected to reduce this demand to less than 2 gal water/gal ethanol.

According to the U.S. Environmental Protection Agency's Renewable Fuel Standard Program (RFS2) Regulatory Impact Analysis of February 2010, first-generation corn-to-ethanol plants have the potential to reduce GHG emissions by as much as 52% over petroleum-based fuels. [81-82] Second-generation ethanol made from cellulosic feedstocks (e.g., switchgrass), agricultural residues (e.g. corn stover), or wood forest residues may reduce GHG emissions by as much as 129%, compared to gasoline.

8.3.2 LCA of a lignocellulosic biorefinery

A life cycle assessment of a lignocellulosic biorefinery assumed to use straw as input, hydrochloric acid for hydrolysis, and to produce ethanol, lignin, and sugar substitute xylite (or xylitol) has been performed and compared to conventional product alternatives (**Figure 8.18**). [79]

The biorefinery was found to have the greatest environmental impacts in the three categories: fossil fuel use, respiratory effects, and carcinogenics. The environmental impacts predominantly result from the provision of hydrochloric acid and to a smaller extent from the provision of process heat. Various variants of the biorefinery were analyzed. The optimum variant (acid and heat recoveries) yields better results than the fossil alterna-

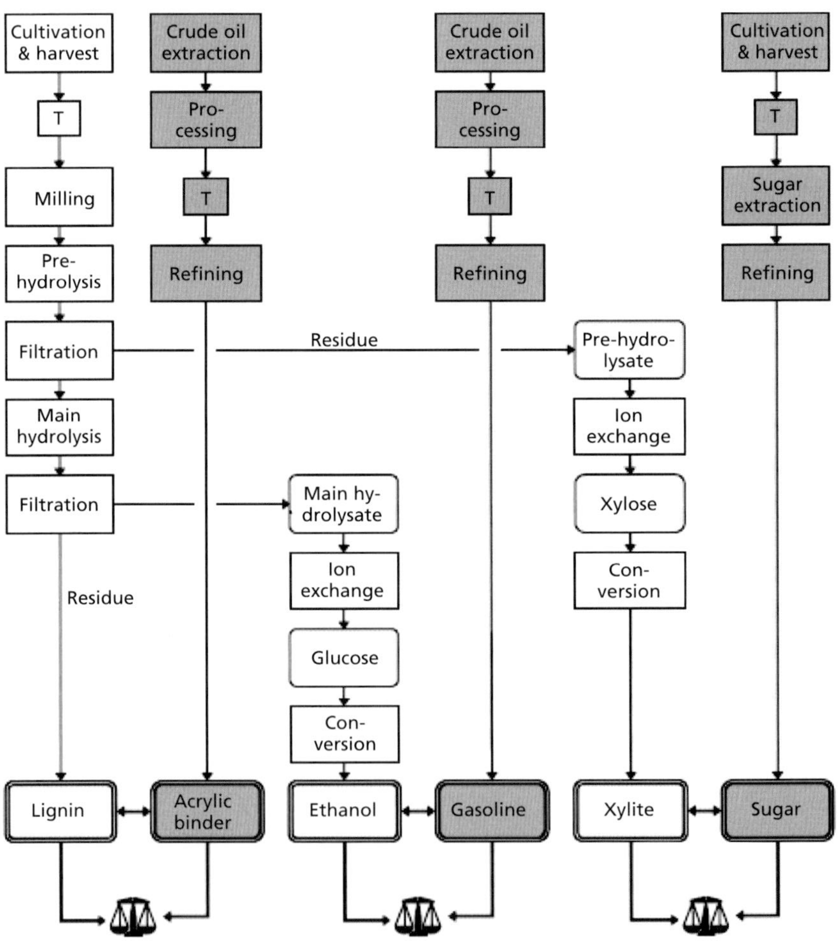

Figure 8.18 Schematic life cycle comparison between biorefinery and conventional products. T, transport. Copyright © 2009, Elsevier. [79]

tives, with the total environmental impacts being ~41% lower than those of the fossil counterparts. For most biorefinery variants analyzed, the environmental performance in some impact categories is better than that of the fossil counterparts while disadvantages can be seen in other categories. [79] Nevertheless, the overall results indicate that future lignocellulosic biorefineries will be competitive with the existing fossil alternatives under the environmental point of view, especially when some process technologies are improved. [79]

8.3.3 LCA of a biorefinery concept producing ethanol, bioenergy and chemicals from switchgrass

An LCA of a biorefinery concept which produces ethanol, bioenergy and chemicals from switchgrass has also been performed. [83] Results are compared with a fossil reference system producing the same products/services from fossil sources (**Figure 8.19**). The paper investigates key issues like land use change and soil N_2O emissions. Since climate change mitigation and energy security are the two most important driving forces for biorefinery development, the assessment had a focus on greenhouse gas (GHG) emissions and cumulative primary energy demand (distinguished into fossil and renewable), but other environmental impact categories such as abiotic depletion, eutrophication, are assessed as well.

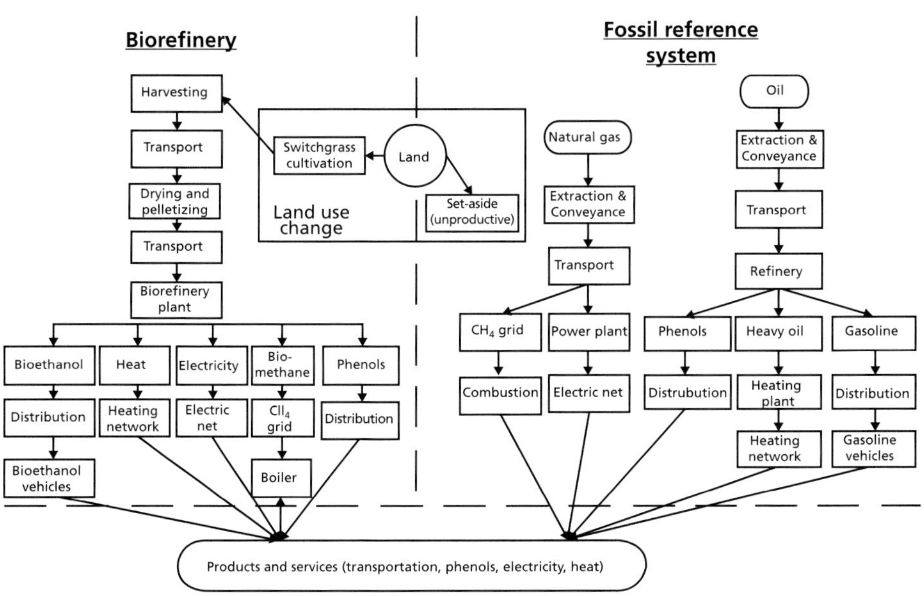

Figure 8.19 Comparison between the production chains of the biorefinery and the fossil reference systems. Copyright © 2009, Springer-Verlag. [83]

The use of switchgrass in a biorefinery saves GHG emissions and reduces fossil energy demand: GHG emissions are decreased by 79% and ~80% of non-renewable energy is saved. [83] Soil C sequestration is responsible for a large GHG benefit, while switchgrass production is the most important contributor to total GHG emissions. If compared with the fossil reference system, the biorefinery system releases more N_2O emissions, while both CO_2 and CH_4 emissions are reduced. The investigation of the other impact categories revealed that the biorefinery has higher impacts in two categories: acidification and eutrophication. Results are mainly affected by raw material production and land use change effects. The provision of biomass with sustainable practices is then a crucial point to ensure a renewable energy supply to biorefineries.

It can be concluded that the biorefinery system is an effective option for mitigating climate change, reducing dependence on fossil fuels, and enhancing cleaner production chains based on local and renewable resources. [83] However, determination of the environmental impacts is complex and a certain degree of uncertainty remains in final results. The study demonstrates that the perennial switchgrass enhances carbon sequestration in soils if established on set-aside land, thus, considerably increasing the GHG savings of the system for the first 20 years after crop establishment.

8.3.4 LCA of a biorefinery using agricultural residues as feedstocks

Another LCA has been performed on a biorefinery concept which produces ethanol, bioenergy and chemicals from two types of agricultural residues, corn stover and wheat straw. [84] The analysis investigates land-use change aspects, i.e. the effects of crop residue removal (like decrease in grain yields, change in soil N_2O emissions and decrease of soil organic carbon). The biorefinery systems are compared with the respective fossil reference systems producing the same amount of products/services from fossil instead of biomass (**Figure 8.20**).

Results show that the use of crop residues in a biorefinery saves GHG emissions and reduces fossil energy demand: GHG emissions are reduced by ~50% and more than 80% of non-renewable energy is saved. [84] Land-use change effects have a strong influence in the final GHG balance (~50%). Concerning the other impact categories, biorefinery systems have higher eutrophication potential than fossil reference systems.

It can be concluded that when using agricultural residues as feedstocks, best management practices and harvest rates need to be carefully established. [84] Rotation, tillage, fertilization management, soil properties, and climate can play an important role in the determination of the amount of crop residue that can be removed minimizing soil carbon losses.

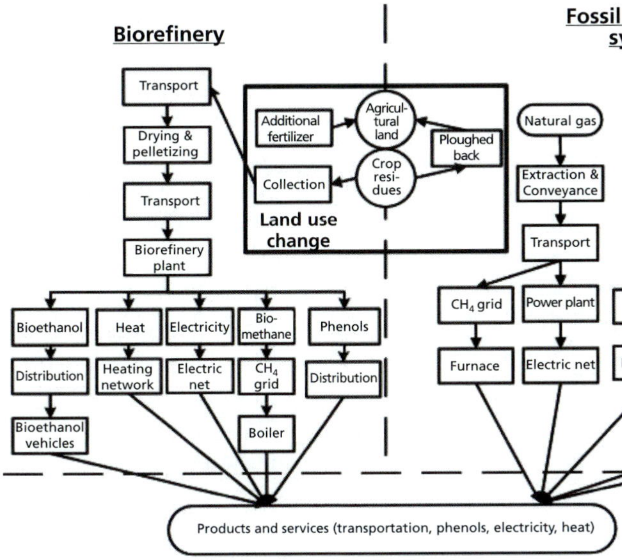

Figure 8.20 Comparison between the production chains of the biorefinery and the fossil reference systems. [84]

8.4. Production of cellulosic ethanol: case studies

Due to the complexity of the process, the production cost for biochemical lignocellulosic ethanol is still higher than for corn ethanol, both in capital and operating cost.

Like it used to be during the last 20 years, the major barrier for the development and cost reduction of lignocellulosic ethanol is to overcome the recalcitrance of the cell-wall structure. Research strategies to address this include the development of specific energy crops with maximum yield, optimized structure, and composition, as well as the development of improved biological systems to hydrolyze the cell-wall components. [85]

There are plenty of routes for improvement and reach the Holy Grail, and as there was no plant producing biochemical lignocellulosic ethanol in 2005, a few demonstration ones are running now, together with well advanced technology pilots.

Recent reports and articles mention that cellulosic biofuel output is likely to be 6.6 million gallons in 2011, far below the RFS (Renewable Fuels Standards) target of 250 million gallons [86] And in January 2011, the Rand Corp. released a report stating that the U.S. military should back off its research and investment in alternative fuels because "there is little promised benefit over using fossil fuels." In the same time USDA and DOE announced several huge loan guarantees for major biofuel plants. This is now obviously a crossroad for advanced biofuels technology, [87] and hopefully case studies below and developing technologies described will show us the way out.

8.4.1 Abengoa

This multidisciplinary (energy, telecommunication, transport) Spanish company plays an important role in the development of alternative energies (bio and solar alternatives) for several years now. The company owns ~15 sites producing first generation bioethanol, located in USA, Europe and Brazil. [88]

Abengoa Bioenergía Nuevas Tecnologías (ABNT) began its operations in 2003 with the purpose of positioning Abengoa Bioenergy as the leading innovator in the bioenergy industry (**Figure 8.21**).

Abengoa has a biorefinery pilot plant in York, Nebraska, sponsored by DOE, which converts 1.5 tons of agricultural residues per day through an enzymatic hydrolysis process. [90] There is also a biomass ethanol commercial demonstration plant in Salamanca, Spain, supported by the European Commission, which uses 70 tons/day wheat straw as a feedstock and produces 5 million L/year ethanol. Abengoa has partnerships with enzyme partners like Dyadic and Novozymes to help develop efficient degradation process steps.

Abengoa has various gasification, catalyst development, and ethanol reforming projects. One such project is a hybrid starch and biomass commercial plant in a conceptual design phase. Its output will be 700 tons/day, integrated with a cereal ethanol plant. [91] Abengoa also develops the concept of a biorefinery integrating both thermochemical and biochemical processing and syngas conversion to ethanol and chemicals.

Abengoa Bioenergy has signed an agreement to acquire 100% of the capital of the Dedini Agro group of companies for €216 million. Dedini Agro is one of the major companies in the Brazilian bioethanol and sugar market. The acquisition is highly important for the global biofuel sector because Abengoa's

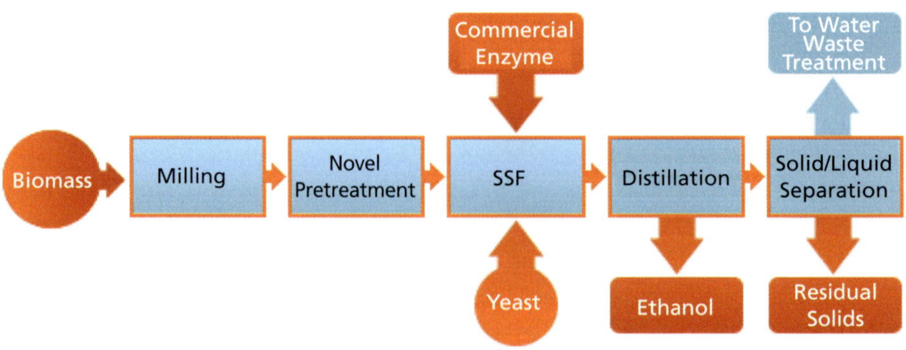

Figure 8.21 Abengoa's typical enzyme-based bioethanol process. © 2011 Abengoa Bioenergy, S.A. [89]

cellulosic ethanol technologies will now be applied to sugar cane husks and to the processing byproduct bagasse, to yield a fuel the energy balance of which may come close to or even surpass that of petroleum based fuels. [92]

In September 2011, U.S. Energy Secretary Steven Chu announced that DOE finalized a $132.4 million loan guarantee to Abengoa Bioenergy Biomass of Kansas to support the development of a 23 million gallon cellulosic ethanol plant. [93] The project will be located in Hugoton, Kansas, and will convert approximately 300,000 tons of agricultural crop residues, including corn stover (stalks and leaves), into ethanol using an enzymatic hydrolysis process. The facility will use unconverted biomass to generate 20 megawatts of electricity to power the cellulosic ethanol plant. This is now one of the most important projects funded by DOE in the view of restoring its image after the big Range Fuels failure. Construction of the facility is expected to be completed by the end of 2013 with some ethanol production starting the same year, though ramp up will be in 2014. [94]

8.4.2 Shell/Iogen

Despite breaking its partnership with Choren a few years ago, the Royal Dutch Shell remains clearly at the head of the top ten strategic investors in biofuels. [95]

The $12 billion joint venture unveiled in 2011 by Shell and Cosan, a big Brazilian producer of ethanol, could be interpreted as a reversal compared to previous years of efforts to develop second-generation biofuel process. [96-97]

However, the new entity named Raizen is not only in charge of production of over 2.2 billion liters per year of ethanol based on sugar cane, it also includes part of Shell's interest in Canadian Iogen Energy, a 50-50 joint venture between Shell and Iogen Corporation since 2002, as well as Shell's interest in US Codexis, a biocatalysis company, in order to enable Raizen to deploy next generation biofuels technologies in the future. [98-99]

Iogen Energy is a world leading biotechnology firm specializing in cellulosic ethanol. The company has built and operated since 2004 a demonstration scale facility in Ottawa to convert wheat straw to cellulosic ethanol using enzyme technology. [100]

Iogen process combines modified steam explosion process as pretreatment, enzyme production, enzymatic hydrolysis, ethanol fermentation, and process integration (**Figure 8.22**). The pilot plant has a capacity production of 5,000–6,000 liters of cellulosic ethanol per day. [101] A plant using Iogen technology is in early development in Brazil at one of Cosan's facilities. [102]

In April 2012, Shell and Iogen announced that they will not pursue the jointly owned 23 million gallon Iogen Energy cellulosic ethanol project in Manitoba, Canada. [104] As a result of this decision, Iogen Energy will refocus its strategy and activities towards a smaller development program. [105] Shell

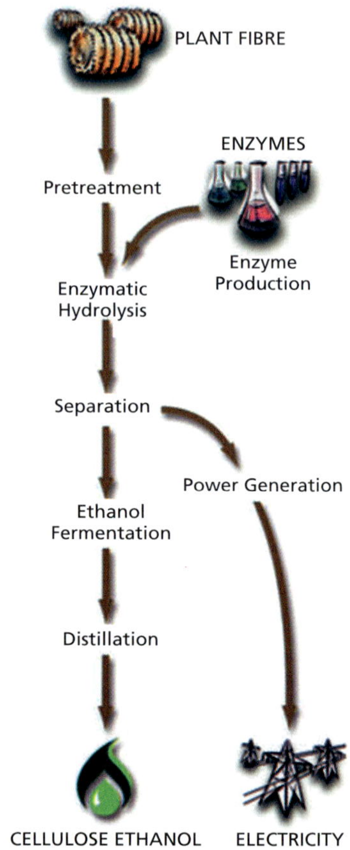

Figure 8.22 Iogen's process description. Source: Iogen [103]

said it will continue to explore pathways to find a commercial solution for the production of advanced biofuels on an industrial scale.

In January 2013, Iogen announced that Novozymes will acquire its wholly owned industrial enzyme business, Iogen Bio-Products. [106] The acquisition does not include the purchase of process technology assets that relate to Iogen Corporation's second generation biofuels business.

8.4.3 DuPont Cellulosic Ethanol

In 2007, Genencor, the biotechnology division of Danisco (Denmark), launched first-ever commercial enzyme product for cellulosic ethanol. [107] Danisco had acquired Genencor in 2005 and with this acquisition had become the second largest supplier of industrial enzymes in the world. [108]

In May 2011, DuPont finalized the acquisition of Danisco for $6.4 billion, becoming number one in the biotechnologies market. DuPont Danisco

Cellulosic Ethanol LLC (DDCE), now DuPont Cellulosic Ethanol, aims to be the world's leading cellulosic ethanol company by delivering comprehensive, integrated technology, engineering, and service solutions. [109]

DDCE process can afford different types of biomass feedstocks but has demonstrated efficiency on corncob and switchgrass. It includes mechanical milling, dilute ammonia pretreatment, enzymatic hydrolysis, mixed-sugar fermentation based on the bacterium *Zymomonas mobilis* that has high ethanol yield and high tolerance to ethanol.

The DDCE demonstration plant in Vonore (Tennessee), which opened in 2010, is based on the Integrated Corn-Based Biorefinery (ICBR) technology. It was partly funded by DOE. DDCE has assessed the sustainability of the process through comparative life cycle analysis. [110] This demonstration conversion facility is being utilized to develop a commercial scale facility by 2013 that has a production capacity of 25–50 million gallons of ethanol. [111]

In November 2012, DuPont started construction of a commercial-scale cellulosic ethanol biorefinery in Nevada, Iowa, with completion expected in mid-2014. [112] The conversion process deployed by DuPont includes five major steps: milling, pretreatment, saccharification, fermentation and separation. [113] The more than $200 million facility in Iowa will be among the first and largest cellulosic biorefineries in the world. This new facility is expected to generate 30 million gallons annually from corn stover residues. The stover will be collected from farms in a 30 mile radius around the facility. The commercial biorefinery will require a capital investment of ~$7 per gallon of annual capacity.

8.4.4 BP/Verenium

Until 2010, US Verenium was a pioneer in the development of cellulosic ethanol and high-performance specialty enzymes. [114] In 2008, Verenium entered the start-up phase at its 1.4 million-gallon-per-year demonstration facility in Jennings, Louisiana. [115]

In 2010, BP announced acquisition of Verenium's cellulosic biofuels business, including Verenium's facilities in Jennings, Louisiana, and San Diego, California, for $98.3M. [116] Verenium, which was struggling badly and possibly approaching bankruptcy, will continue as an enzyme company, trying to develop industrial enzymes. [117]

In October 2012, BP announced that it is cancelling plans to build a commercial-scale cellulosic ethanol plant in Highlands County, Florida and refocusing its US biofuels strategy on research and development as well as licensing its industry-leading biofuels technology. [118] BP originally announced plans to build the Florida facility in 2008 with the intention of turning thousands of acres of energy crops into 36 million gallons per year of cellulosic ethanol.

8.4.5 POET-DSM Advanced Biofuels

POET, formerly known as Broin Companies, is the world's largest producer of biofuels. Its initial partnership with DuPont in an Integrated Corn-Based Biorefinery project failed in mid-2008, and subsequently POET opened its own pilot plant in Scotland (South Dakota) end 2009. The pilot is integrated on the site of a 9 million gallon per year starch ethanol production facility. Strategically, POET is pursuing an integrated starch- and cellulose-to-ethanol biorefinery model that could see cellulosic production capacity added to its 26 plants that currently produce 1.5 billion gallons of ethanol from corn per year. [119]

By working on the optimization of the process, POET has been able to reduce the production cost from \$4/gallon to approximately \$2.3/gallon. The pilot plant's production costs were cut in many ways. First, the plant used alternative energy (wind/solar) to provide its electricity, and the energy used to produce the fuel was cut in half. Second, its chemical raw materials costs were cut, reducing the cost per gallon by \$0.20. Finally, the biggest cuts came from reducing the cost of the enzymes that POET uses, working with Novozymes. The cost of enzymes has already been cut in half, and POET expects that number to drop lower. [120]

Project Liberty will be POET's first commercial-scale, cellulosic ethanol plant. Scheduled to begin operations in Iowa in 2013, it will produce 25 million gallons of ethanol per year from corncobs, leaves and husks, provided by farmers in the area around Emmetsburg, Iowa. The process leaves behind excess materials, which is sent to anaerobic digesters, generating power for both the cellulosic ethanol plant and the adjacent grain-based ethanol plant. In January 2012, DSM and POET announced a joint venture to commercially demonstrate and license cellulosic ethanol, based on their proprietary and complementary technologies. [121] POET-DSM Advanced Biofuels is scheduled to start production in the second half of 2013 at one of the first commercial-scale cellulosic ethanol plants in the United States. In October 2012, POET-DSM contracted with Andritz to supply a two-step biomass treatment for the cellulosic ethanol plant in Emmetsburg. [122] The Andritz technology is a two-stage process that includes a vertical reacter, an inter-stage washer and then a continuous steam explosion technology.

8.4.6 Inbicon/Dong

Inbicon, which operates the largest cellulosic biofuel demonstration plant in Europe, received the award "Bioethanol Corporation of the Year" by the World Refining Association (WRA) at the Biofuels 2011 Conference in Amsterdam. [123] Inbicon is a fully owned subsidiary of Dong Energy, Denmark's largest energy group.

Since beginning of the 2000s, the Danish company has developed and improved a cellulosic biofuel process based on mechanical conditioning of the biomass, hydrothermal pretreatment and enzymatic hydrolysis.

During the pretreatment, the biomass is continuously mixed with water and heated to around 180–200°C for 5–15 minutes in order to destroy the protecting lignin structure and make the cellulose available for enzymes. Another important function of the pretreatment is to separate the alkali content (mainly sodium and potassium) and a portion of the inhibitors formed through partial degradation of the hemicelluloses.

After hydrothermal pretreatment, enzymes are added to the fiber mass, which consists mainly of cellulose and lignin. The fiber mass is liquefied as part of the cellulose is converted to lower carbohydrates. The process takes place in a horizontal reactor with paddles, a so-called free-fall mixer. [124-125] The process is continuous. The fiber mass leaves the pretreatment with 30–40% dry matter content, enters the reactor at one end, and, after a couple of hours, exits the other end of the reactor. [125]

The pilot plant for cellulosic biofuel has been raised in Kalundborg, Denmark, and integrated to a power plant station. Started in 2009, the demonstration plant converts 4 tonnes of straw per hour, equivalent to 30,000 tonnes of straw per year. Genencor or Novozymes enzymes are mentioned for biomass degradation. The output will be 4,300 tonnes/5,400 m^3 of ethanol per year. [126-128]

Dong has signed a contract with Norwegian Statoil, who will purchase the first five million liters of Inbicon's bioethanol. [126, 129]

In June 2012, Inbicon and Leifmark, an independent Inbicon partner, announced a new partnership to market Inbicon cellulosic ethanol technology in North America. [130]

8.4.7 Mascoma/Valero

In January 2011, Valero, the petrochemical giant based in San Antonio, Texas, invested in the US renewable fuels company Mascoma, which develops microbes that convert wood chips and grasses into cellulosic ethanol. [131] Valero also signed a non-binding letter of intent to support the construction of Mascoma's plant in Michigan that would be capable of generating 40 million gallons of ethanol a year.

In February 2009, Mascoma started producing cellulosic ethanol from wood chips at a demonstration-scale plant in Rome, New York, USA. Mascoma has made its name by developing proprietary yeasts that, according to the firm, eliminate the need for a number of expensive enzymes used in biomass conversion.

The pilot plant in Rome uses the consolidated bioprocessing (CBP) process configuration. In 2009, Mascoma announced that the company has made major research advances in CBP, leading to production cost reductions. [132-134]

In 2010, Mascoma acquired SunOpta BioProcess Inc. (SBI), a world-leading fiber preparation and pretreatment company, creating a company with comprehensive capabilities for converting non-food cellulose (wood chips, energy crops, and organic solid waste) into ethanol and high value co-products. With the addition of SBI and with Valero, Mascoma has now covered the entire process of commercializing cellulosic ethanol, from raw materials supply, to pre-processing, through Mascoma's CBP process and into final distribution. [135]

In June 2012, Lallemand Ethanol Technology, a global provider of fermentation ingredients to the fuel ethanol industry, and Mascoma announced that the Mascoma Grain Technology, or MGT™, yeast product will be marketed under the commercial name TransFerm™ for use in the fuel ethanol industry. [136] TransFerm™, which is the first commercial application of Mascoma's proprietaty CBP, is a bioengineered drop-in substitute for conventional fermenting yeast that lowers costs for corn ethanol producers.

8.4.8 BlueFire

BlueFire Renewables, based in California, USA, has developed proprietary improvements to concentrated acid hydrolysis such that the process is ready for commercial implementation. [35] The technology enables widely available cellulosic materials to be converted into sugars for subsequent fermentation or chemical conversion into ethanol fuel and chemicals.

BlueFire's first cellulosic ethanol plant in the US will be located in Lancaster, California. [35-36] The facility will use post-sorted cellulosic wastes diverted from Southern California's landfills to produce ~3.9 million gallons of ethanol per year. BlueFire's second commercial plant will be located in Fulton, Mississippi. It will produce ~19 million gallons of ethanol per year from woody biomass, mill residue and other cellulosic waste.

In February 2012, SucreSource, a wholly owned subsidiary of BlueFire, signed agreements with GS Caltex, a Koran petroleum company, to build cellulose to sugar plant in Korea. [137]

8.5 Production of lignocellulose-derived chemicals: case studies

8.5.1 Introduction

Biorefinery development has two strategic goals: the displacement of imported petroleum in favour of renewable raw materials and the establishment of a robust biobased industry. [138] Large-scale displacement of petroleum is expected to come from the production of biofuels. But despite its high volume, fuel is a low value product. As a result, biofuel-only biorefineries are not easily economically viable. High value, lower volume biobased chemicals provide the necessary financial incentive to stimulate expansion of the biorefining industry.

However, the choice of appropriate chemicals for addition to the biorefinery's portfolio is challenged by a lack of broad-based conversion technology coupled with a plethora of potential targets. [138] In 2004, the US Department of Energy (DOE) identified twelve top value-added building blocks that can be produced from sugars via chemical or biological conversions (see §3.4.1). [139] Since DOE's original report, considerable progress has been made in the use of carbohydrates for chemical production. In 2010, Bozell et al. [123] presented an overview of the revised top chemical opportunities from biorefinery carbohydrates. The revisited list included ethanol, furans, glycerol and derivatives, hydrocarbons, lactic acid, succinic acid, hydroxypropionic acid/aldehyde, levulinic acid, sorbitol, and xylitol.

As an additional biobased product and feedstock, lignin can afford the entire family of aromatic compounds that are difficult to produce via sugars or oils. [139]

This section will be devoted to (1) carbohydrate-derived chemicals and (2) extracted lignin and lignin-derived chemicals.

8.5.2 Carbohydrate-derived chemicals

ZeaChem

ZeaChem, Inc. has developed a cellulose-based biorefinery platform capable of producing advanced ethanol, fuels and chemicals. [140] The company uses a hybrid combination of biochemical and thermochemical processing steps (**Figure 8.23**).

After fractioning the biomass, the C5 and C6 sugar stream is sent to fermentation where an ***acetogenic*** process is utilized to ferment the sugars to acetic acid without carbon dioxide as a by-product. In comparison, traditional yeast fermentation produces one molecule of CO_2 for every molecule of ethanol. The acetic acid is converted to an ester, which can then be reacted with hydrogen to make ethanol. To get the hydrogen necessary to convert the ester to ethanol, ZeaChem takes the lignin residue from the fractionation process and gasifies it to create a hydrogen-rich syngas stream. The hydrogen is separated from the syngas and used for ester hydrogenation and the remainder of the syngas is burned to create steam and power for the process.

ZeaChem's approach can produce many chemicals and fuels within the carbon chain product groups: [141]

- C2 chain: acetic acid produced by its current C2 organism, ethyl acetate, ethanol, ethylene, ethylene glycol;

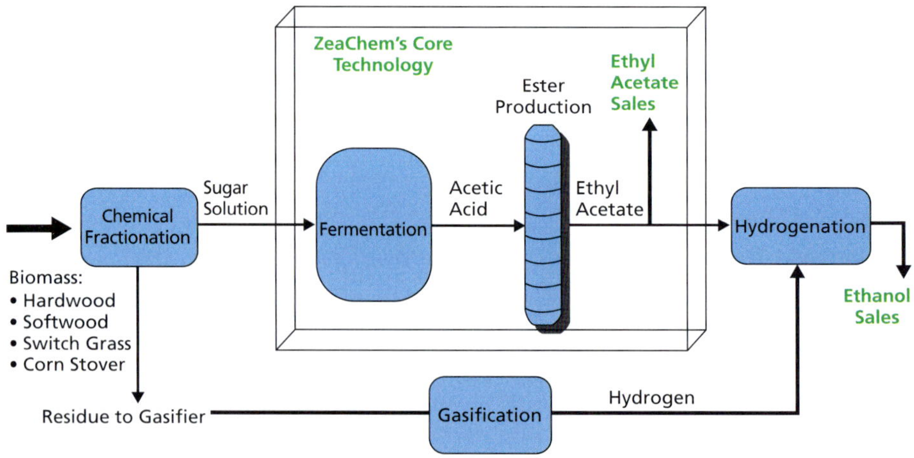

Figure 8.23 ZeaChem's process. Source: ZeaChem [140]

- C3 chain: lactic acid produced by its C3 organism, propylene glycol, acrylic acid and esters, propionic acid, propylene, methacrylic acid and esters, diluent;
- C4 chain: butanol;
- C6 chain: hexanol and hexane.

In June 2011, ZeaChem announced a multi-year joint development agreement with Procter & Gamble. The agreement will accelerate development of ZeaChem's product platform beyond C2 through the commercialization of "drop in" biobased chemicals and other products. [142] It also aims to boost P&G's commitment to sustainable consumer products.

In October 2012, ZeaChem completed construction of its 250,000 gallons per year cellulosic ethanol biorefinery in Boardman, Oregon. [143] The project was completed on budget and ZeaChem expects to begin production of cellulosic ethanol by the end 2012. In the technology, the acetic acid is processed into an acetate ester, which is then converted into ethanol via hydrogenolysis. [144]

Virent

Virent is commercializing an innovative advanced biofuel technology that catalytically transforms a wide range of soluble plant sugars into hydrocarbon molecules like those produced in a petroleum refinery. [145] Virent's renewable hydrocarbons can be blended seamlessly to make gasoline, jet fuel and diesel.

Virent made its groundbreaking discoveries in 2006 and is advancing its innovative BioForming® platform for the production of biofuels (**Figure 8.24**). [146]

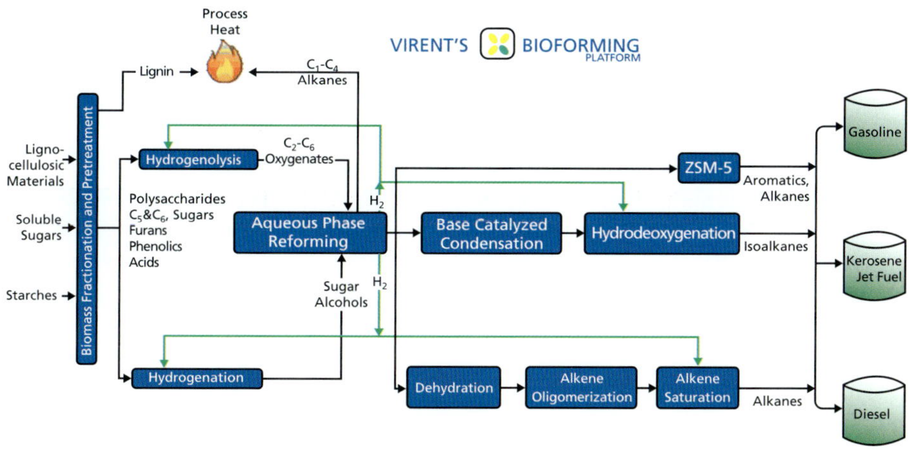

Figure 8.24 Virent BioForming® process to produce conventional liquid transportation fuels from biomass feedstocks. Aqueous Phase Reforming enables to partially defunctionalize carbohydrate feedstocks for further catalytic upgrading. Source: Virent. [146-147]

The BioForming® process is based on the novel combination of Virent's core ***aqueous phase reforming*** (APR) technology with conventional catalytic processing technologies such as catalytic hydrotreating and catalytic ***condensation*** processes, including ***ZSM-5*** (a zeolite) acid condensation, base catalyzed condensation, acid catalyzed dehydration, and alkylation. Each of these process steps can be optimized and modified to produce a particular slate of desired hydrocarbon products. For example, a gasoline product can be produced from ZSM-5 based process, jet fuel and diesel can be produced from using a base catalyzed condensation route, and a high octane fuel can be produced using a dehydration/oligomerization route.

Depending on the particular application, the water soluble carbohydrates may undergo an initial hydrotreating step to convert the sugars and organic acids resulting from biomass pretreatment into polyhydric alcohols. [147] This step consists of either ***hydrogenation*** to form ***polyhydric alcohols*** or ***hydrogenolysis*** to form shorter-chain oxygenated compounds such as glycerol. After the hydrotreating step is the APR. The aqueous product stream from the hydrotreating step is fed to the APR reactor where the oxygenated hydrocarbons react with water over a catalyst to form hydrogen, CO_2, alcohols, ketones, aldehydes, and by-products alkanes, organic acids, and furans. It is believed that the reactions include: (1) reforming to generate hydrogen; (2) dehydrogenation of alcohols/hydrogenation of carbonyls; (3) deoxygenation reactions; (4) hydrogenolysis; and (5) cyclization. One key feature of this process is that the hydrogen is generated in situ and used to further de-functionalize the highly reactive carbohydrate to the less reactive mono-oxygenated species such as alcohols, ketones and aldehydes. The formation of hydrogen and condensable intermediates by APR is the critical step in the BioForming process. While oxygenates generated by APR are generally limited to six or fewer carbon atoms, their chemistry support the formation of carbon-carbon bonds and the production of hydrocarbons appropriate for liquid fuels.

The BioForming process can convert alcohols, ketones, aldehydes and organic acids to gasoline-range hydrocarbons, including aromatics and iso-alkanes, via direct catalytic condensation over acid catalysts, such as solid acids and zeolites. [147]

In the case of a zeolite ZMS-5 catalyst, a series of reactions occurs, including: (1) the dehydration of oxygenates to alkenes; (2) oligomerization of the alkenes; (3) cracking; (4) cyclization and dehydrogenation of larger alkenes to form aromatics; (5) alkane isomerization; and (6) hydrogen-transfer to form alkanes. [147]

The BioForming process is also able to convert alcohols, ketones and aldehydes into jet fuel range hydrocarbons via direct catalytic condensation over a base catalyst. In this application, oxygenates are passed over multi-functional solid-base catalysts to promote condensation reactions to occur through aldol and decarboxylative condensation. The process enables the generation of longer branched-chain hydrocarbons from shorter-chain oxygenated compounds. In addition, by coupling base condensation with hydrogenation of the resulting condensed product, high yields of hydrocarbons appropriate for gasoline and jet fuels can be obtained.

Dehydration of alcohols produced by APR can also generate alkenes as feedstocks for alkylation or condensation. Alkenes may also be dimerized to produce gasoline-range product or oligomerized for kerosene or diesel fuel using catalysts. In this case, the oxygenates derived from the APR process are converted into paraffins and olefins through dehydration reactions and/or successive hydrogenation-dehydration reactions.

In June 2011, Virent announced that it has successfully made paraxylene (PX, a direct precursor of terephthalic acid) from 100% renewable plant sugars. [148-149] The PX molecule, when combined with existing PET technology, allows manufacturers to offer customers 100% natural, renewable, plant-based PET and packaging. Biobased ethylene glycol is available and replaces 30% of the fossil fuel needed to produce PET, whereas PX is the missing 70%. This announcement is the culmination of Virent's mixed-xylenes development, which started in 2010. [150] In its gasoline process, Virent generates indeed an aromatic-rich stream that can be processed to generate PX. [151] The company has made PX in its 10,000 gallon per year demonstration plant in Madison, Wisconsin. Virent's corporate partners include Cargill, Coca-Cola, Honda and Shell. [152]

In December 2011, Coca-Cola announced partnerships with Virent, Gevo and Avantium to develop commercial solutions for plastic PET bottles made entirely from plants. [153] In June 2012, Coca-Cola, Ford Motor, Heinz, Nike, and Procter & Gamble formed collaborative to accelerate the development and use of 100% plant-based PET materials and fiber in their products. [154]

In February 2012, Shell announced that it has built a next generation pilot plant at Shell's Westhollow Technology Center in Houston, USA to produce drop-in biofuels rather than ethanol. [155] It uses a thermo-catalytic process technology licensed from its commercial partner Virent, which is similar to the process being used at the Virent pilot plant in Madison, Wisconsin.

In March 2012, Virent converted Virdia's cellulosic pine tree sugars to drop-in hydrocarbon fuels within the BIRD (Israel-US Binational Industrial Research and Development) Energy project. [156] The project demonstrated that Virdia's deconstruction process generated high-quality sugars from cellulosic biomass, which were converted to fuel via Virent's BioForming process.

Gevo

Gevo, based in Colorado, USA, is a renewable chemicals and advanced fuels company. [157] Gevo is developing biobased alternatives to petroleum-based products using a combination of synthetic biology and chemistry. The company plans to produce isobutanol, a versatile platform chemical for the liquid fuels and petrochemicals markets. Isobutanol has broad market applications as a solvent and a gasoline blendstock that can help refiners meet their renewable fuel and clean air obligations. It can also be further processed using well-known chemical processes into jet fuels and feedstocks for the production of synthetic rubber, plastics and polyesters.

Gevo has pioneered a platform technology based on a proprietary fermentation method that relies on an innovative biocatalyst and the efficient separation of isobutanol. [158] The technology is similar to ethanol production, except that Gevo replaces the ethanol-producing biocatalyst with its isobutanol-producing biocatalyst and then adds a separation unit.

The combination of these two proprietary innovations, Gevo's Integrated Fermentation Technology® (GIFT®), was designed to enable the low cost retrofit of existing ethanol capacity for isobutanol production.

In 2009, Total acquired an interest in Gevo. [159-160] Total's investment is part of a new share issue. After the capital increase, which will finance the next stages in the company's growth, Total will be a core shareholder. This investment is part of Total's recently initiated corporate venture activity.

Gevo's commercialization efforts are focused on isobutanol, a naturally occurring C4 alcohol. [161] Isobutanol is an important platform chemical with broad applications in large chemicals and fuels markets and a "drop-in" product that should allow customers to replace petroleum-derived raw materials with isobutanol-derived raw materials without modification to their equipment or production processes (**Figure 8.25**).

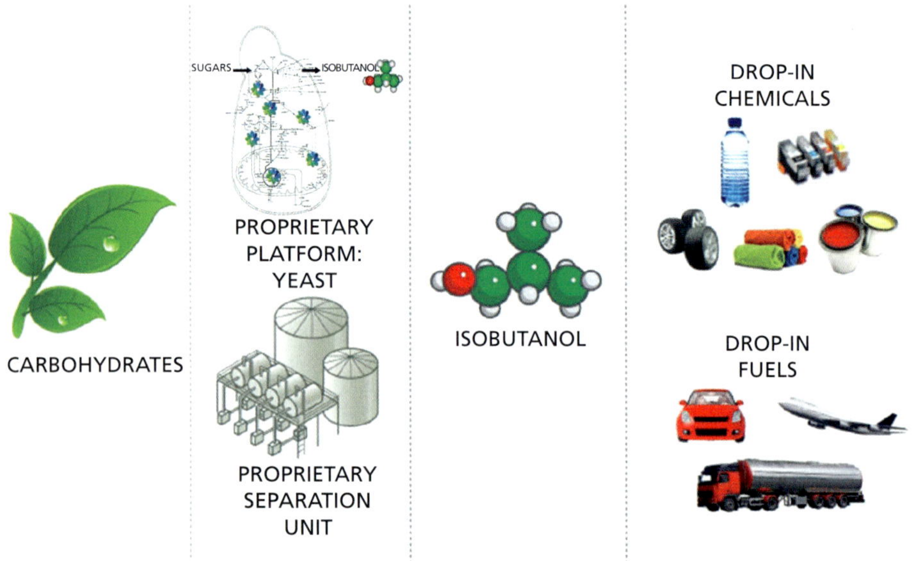

Figure 8.25 Gevo's isobutanol. Reproduced by permission of Gevo. [161]

Gevo's isobutanol can be produced from a variety of starch and sugar feedstocks. In the future, inedible sources, such as corn stover, bagasse, and wood residues, could also be used as feedstocks. [162] In 2011, Gevo received a $5 million grant from the US Department of Agriculture for the development of biojet fuel from woody biomass and forest product residues.

In December 2011, Gevo announced an agreement with Coca-Cola to create renewable para-xylene from plant based isobutanol, which will accelerate the development of Coca-Cola's second-generation PlantBottle™ packaging made from 100% plant-based materials. [163]

In May 2012, Gevo started up the world's first commercial biobased isobutanol plant located in Luverne, Minnesota. [164] Gevo retrofitted the Luverne plant to incorporate its proprietary yeast and Gevo Integrated Fermentation Technology® system to produce biobased isobutanol. In September 2012, Gevo decided to optimize certain specific parts of its technology to further enhance bio-isobutanol production rates, and to temporarily shift the plant back to ethanol production. [165]

Avantium

Avantium, based in Amsterdam, the Netherlands, is a technology company specialized in high-throughput R&D for applications in the energy, chemicals and pharmaceuticals industries. [166]

In 2005, Avantium initiated its program to convert biomass into materials and fuels using its world-leading high-throughput catalyst development technology. [167] Using small-scale reactor systems (0.1–1 ml of catalyst per reactor) simultaneously allows for broader screening and quicker catalyst identification than conventional methods. [168-169] Avantium focused its efforts on furanics (i.e. furan derivatives), a novel class of building blocks to make materials and fuels (**Figure 8.26**).

Furanics have been referred as "sleeping giants" because of their enormous market potential, as evidenced by its position in the US Department of Energy top 12 of high-potential biobased products. Furanics were never commercialized because they could not be produced in an economic manner. Avantium has discovered a revolutionary chemical, catalytic process that enables the economic production of furanics on basis of a wide range of carbohydrates (**Figure 8.26**). XYX is the brand name for the furanics technology to produce green building blocks.

Avantium has initiated several collaborations to develop YXY for polyesters, polyamides, polyurethanes and a number of chemical applications. Its partners are, among others, NatureWorks, Solvay, and Teijin Aramids. Target polymers for 2,5-furandicarboxylic acid (FDCA) are polyesters, polyamides, and copolymers. The collaboration with NatureWorks is for the development of YXY polyesters. This collaboration has demonstrated that YXY polyesters have properties that are very similar to PET, making it possible to develop a green version of this bulk polymer. The companies will further test its properties to evaluate applications in packaging, automotive, fibers, coatings and engineering plastics.

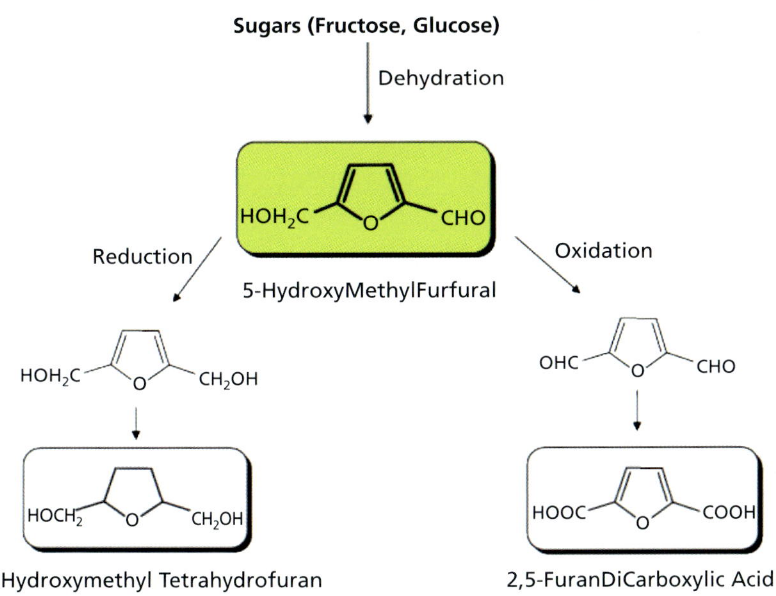

Figure 8.26 Avantium's furan-based molecules with potential for use as monomers for biopolymer applications. [169]

Avantium is also working on the development of YXY for fuels. The company has demonstrated the use of YXY as a green component to blend with diesel, gasoline, and jet fuel.

In December 2011, Avantium and Coca-Cola signed partnership agreement to develop next generation 100% plant based plastic: polyethylene furanoate (PEF), a polyester based on ethylene glycol and FDCA. [170] In March 2012, Avantium and Danone signed development partnership for bio-based PEF. [171]

Aemetis

Headquartered in Cupertino, California, Aemetis (formerly AE Biofuels) is an industrial biotechnology company producing renewable chemicals and advanced fuels using patented microbes and processes. [172] Aemetis operates a 55 million gallon biofuels plant in California acquired from Cilion in 2012, and built in 2007 and operates a nameplate 50 million gallon per year renewable chemicals and advanced fuels production facility on the east coast of India. [173-174] In 2011, Aemetis received a California Energy Commission grant to commercialize technology that enables the production of advanced biofuels from both non-food and traditional feedstocks. [175]

Aemetis patented technology platform is based on a unique patented aerobic marine organism (*Saccharophagus degradans* 2–40 trademarked Z-Microbe [176]) and propagation process that enables scientists at Aemetis to manufacture a variety of specialty biochemical and renewable fuels products for large markets using renewable feedstocks. [173-174] Aemetis holds 4 granted patents on the Z-Microbe technology and has 14 pending patents on its microbial and enzymatic "Ambient Temperature Starch/Cellulose Hydrolysis (ATSCH)" technology and production processes. [177]

The Z-Microbe, which naturally generates ~90 enzymes, [178] is a patented organism that converts a variety of renewable feedstocks such as sugar, starch and cellulose directly into renewable chemicals and fuels. [179] It exhibits unique and powerful degradative abilities on a broad spectrum of feedstocks. These characteristics enable the Z-Microbe to be deployed at existing chemical and fuel facilities, thereby reducing capital expenditures, construction time and start-up costs associated with building plants for processing biomass into chemicals and fuels. The Z-Microbe technology utilizes consolidated bio processing (CBP) technology to convert multiple feedstocks into chemicals and fuels in one, low-cost process. Aemetis' plans are to convert existing biofuels plants (such as their California ethanol facility) into biorefineries through joint ventures.

Ambient Temperature Starch/Cellulose Hydrolysis (ATSCH) is a biorefining process acquired and developed by Aemetis. [180] In 2008, Aemetis built and commissioned a 9,000-square foot integrated cellulose and starch ethanol commercial demonstration facility in Butte, Montana where Aemetis scientists evaluated a variety of feedstocks to optimize the ATSCH technology. [180]

The Z-Microbe will enable Aemetis to produce isoprene (synthetic rubber), glycerine, and fuels (jet fuel and green diesel) and in the pipeline, adipic acid (nylon precursor) and butanediols (plastics precursors). [176, 181]

In October 2012, Aemetis granted fifth US patent for cellulosic biochemical and biofuel technology. [182] This patent represents another significant milestone for Aemetis Z-microbe development.

In November 2012, Aemetis partnered with Edeniq, a biomaterials and sustainable fuels company, to implement Edeniq technology at the Aemetis plant in California. [182] The agreement will improve ethanol yield and pave the way for commercial cellulosic ethanol production. [183]

Verdezyne

California-based Verdezyne is an industrial biotechnology company using proven and proprietary metabolic pathway engineering tools to develop unique yeast strains for cost effective production of biobased chemicals and fuels. [184] Current investors in Verdezyne include BP Alternative Energy Ventures, DSM Venturing B.V., OVP Venture Partners, and Monitor Ventures. [185]

Verdezyne has taken a novel combinatorial approach to pathway engineering. [186] The company uses synthetic gene libraries to introduce diversity into each pathway gene, combinatorial assembly of multiple pathway genes to introduce enzymatic diversity into a metabolic pathway, and biological selection or high-throughput screening to rapidly identify the most productive combination of pathway genes.

The company is applying these tools to develop a versatile yeast production platform that optimizes the conversion of sugars from a variety of feedstocks, including starch, sugarcane and lignocellulosic biomass, to renewable fuels and chemicals.

The two major targets in Verdezyne's pipeline include ethanol (from hexoses and pentoses) and adipic acid with the formula $(CH_2)_4(COOH)_2$. [187] At the end of 2011, the company opened its first pilot plant to produce adipic acid, a key component of nylon 6,6, from renewable sources, such as non-food based vegetable oils. [185, 188]

In December 2012, Verdezyne entered in a strategic relationship with Universal Fiber Systems, a global multi-polymer fiber producer. [189] Verdezyne will supply its bio-based adipic acid to Universal Fiber Systems for production of nylon 6,6 fibers.

In December 2012 also, Verdezyne announced the sale of its proprietary xylose isomerase technology, enabling the metabolism of 5-carbon sugars for use in the production of biofuels and biochemicals, to DuPont Industrial Biosciences. [190]

Amyris

Amyris, a California-based firm, is building an integrated renewable products company. [191] Amyris' initial portfolio of products is based on Biofene® (trans-β-farnesene), a C15 hydrocarbon molecule (**Figure 8.27**). Amyris uses genetically modified microorganisms to ferment plant sugars into farnesene that can be processed into fuels and chemicals. [192]

Using Biofene® as a building block, Amyris is developing a slate of different products for the fuels and the specialty chemicals markets. [192] Biofene is the first of many isoprenoid molecules (i.e. compounds derived from isoprene) made possible by its microbial platform.

In Amyris fermentation technology, Acetyl-CoA is converted through the mevalonate pathway into isopentenyl pyrophosphate (IPP), which is further converted into farnesenyl pyrophosphate (FPP) and into C15 isoprenoids (**Figure 8.28**). [194] Amyris has developed this technology to pilot stage for sugar streams from corn or cane sugars.

In February 2011, the National Advanced Biofuels Consortium's (NABC) fermentation of lignocellulosic sugars process strategy has extended the diesel fuel fermentation process developed at Amyris to be able to use complex lignocellulosic streams. [194-197] The project team will develop an economic, feasible, integrated process between lignocellulosic feedstocks, Amyris hydrocarbon isoprenoid production, and refiners and end-users of hydrocarbon fuels. The fermentation of lignocellulosic sugars process strategy will adapt the current Amyris process to incorporate sugar streams from lignocellulosic feedstocks (**Figure 8.28**).

In October 2011, Amyris acquired failing renewable chemicals and products company Draths (Michigan, USA). [198] Draths has the lab-scale fermentation technology to produce biobased monomers from muconic acid that can lead to production of 100% biobased polymers such as PET and nylon. [199] Draths' product portfolio includes bio-based terephthalic acid (PTA), caprolactam, adipic acid, and hexamethylene diamine (HMDA).

In October 2011 also, Amyris and ETH Bioenergia, a leading producer of ethanol, electric energy and sugar, signed an agreement for the formation of a joint venture to produce Biofene® in Brazil. [200] The joint venture would be able to access up to two million tons of sugarcane crush capacity per year at one of ETH's mills in Brazil.

In November 2011, Total and Amyris have agreed to expand their ongoing research and development collaboration to accelerate the deployment of

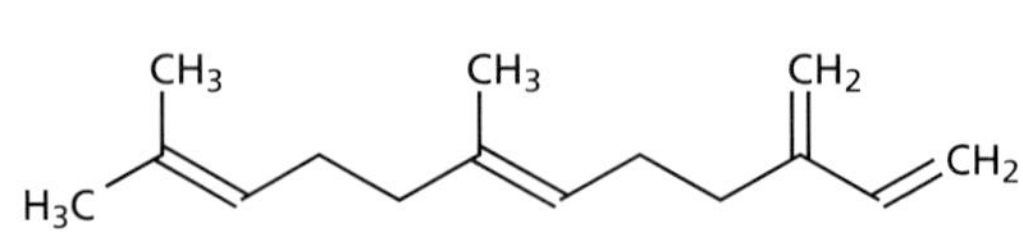

Figure 8.27 Trans- β-farnesene. [193]

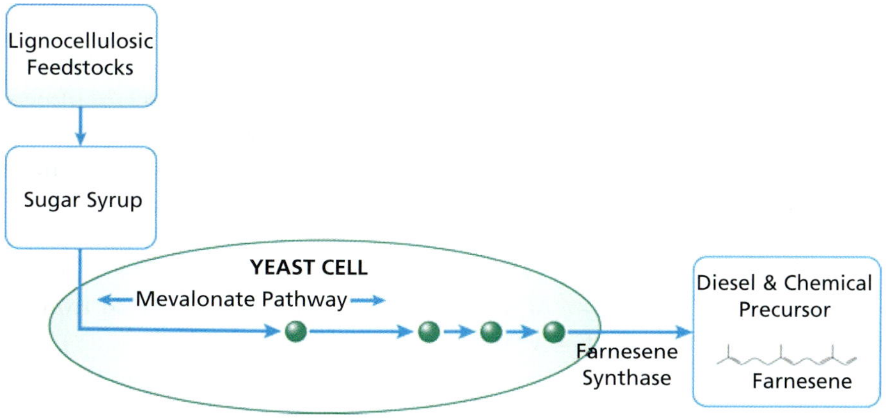

Figure 8.28 Amyris fermentation of lignocellulosic sugars process. [194]

biofene and develop renewable diesel based on this molecule. [201] The R&D program aims to develop the necessary stages to bring the next generation renewable fuels to market.

In February 2013, Amyris shipped the first commercial order of Biofene® from its new plant in Brazil. [202] The farnesene molecule, produced from sugar cane, will be used for a variety of specialty chemical and fuel applications.

8.5.3 Lignin and lignin-derived products

Lignin is present (15–30%) in all lignocellulosic biomass and any cellulosic ethanol production will have lignin as a residue. [203] Lignin can be combusted to provide heat and/or power for the ethanol process; however, increasing the value of this residue by converting it to higher value compounds can significantly increase the competitiveness of cellulosic ethanol technology.

US Department of Energy

The Department of Energy (DOE) project aims at developing a process for making fuel additives from lignin. [203]

The first step to converting lignin into a product compatible with the transportation fuel market is to reduce its molecular weight. [203] Currently, base catalyzed depolymerization (BCD) is the process being studied to achieve this. Further processing of depolymerized lignin is necessary before the product is suitable for fuel applications. The product must be either partially or completely deoxygenated depending on whether an oxygenate or hydrocarbon product is the final target. Furthermore, a conceptual design of the process was constructed to convert lignin from a cellulosic ethanol plant into a high-octane hydrocarbon fuel additive.

CIMV

The vegetal refining concept developed by CIMV (Compagnie Industrielle de la Matière Végétale) provides a technology that can cleanly valorize the three plant components and meet the challenges raised by the end of all oil. [204]

The CIMV organosolv process promotes the three main components of lignocellulosic biomass into intermediary products for both chemical and biotechnology industries (see **Figure 7.12**):

- Biolignin™ (phenolic polymer with a linear structure and a low molecular weight).
- Cellulose and glucose.
- Sugars syrup (monomers, dimers of xylose and other C5 sugars). [205]

The CIMV process uses a mixture of acetic acid, formic acid and water as the cooking chemical. [206]

In the first stage, the vegetal matter is treated at atmospheric pressure by the acids, which dissolve lignin and hydrolyze the hemicelluloses into oligo- and monosaccharides with high xylose content. [207-208] The raw pulp is then filtered, the solvent is removed, and the pulp is bleached with hydrogen peroxide. The commercial value of the raw pulp is close to that of eucalyptus chemical pulp. Organic acids are then recycled by evaporation from the organic solution. The remaining syrup is then treated with water to precipitate lignin, which is easily separated. The obtained lignin shows a linear structure, which can be used as a reactant for chemical production of new polymers and composite synthesis. The raw syrup, which, without purification, resembles molasses, can be used directly for numerous industrial applications (**Figure 8.29**).

The European FP7 (Seventh Framework Programme) project Biocore will create and demonstrate a lignocellulosic biorefinery based on the CIMV frac-

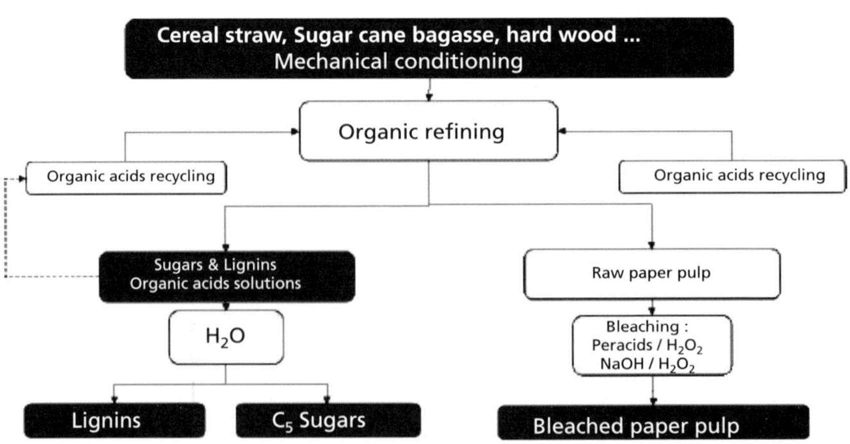

Figure 8.29 CIMV process. [207]

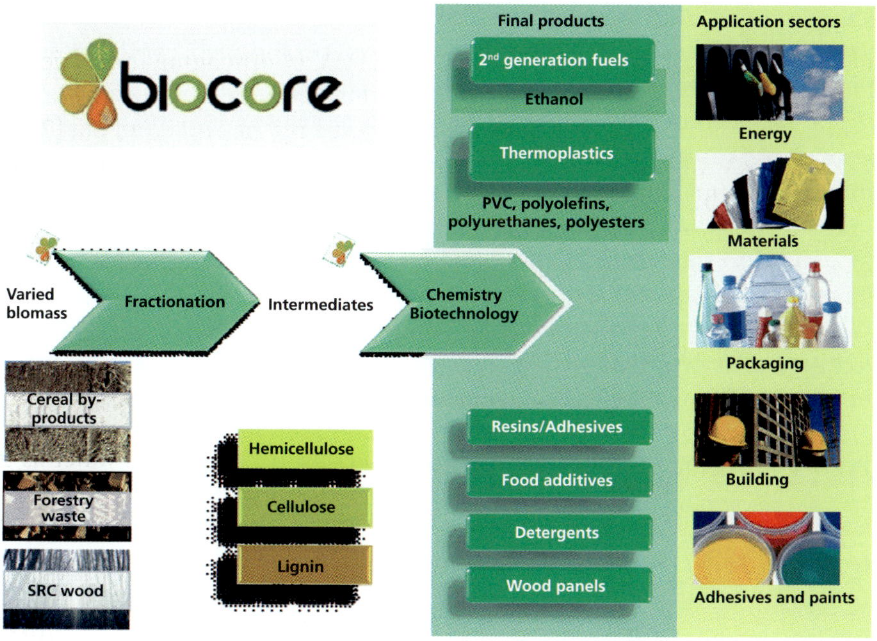

Figure 8.30 Biocore in brief. Biocore project FP7 GA-241566. [188]

tionation process. [209-210] Biocore, which benefits from a budget of EUR 20.3 million for four years, involves 24 partners from 13 different countries and is coordinated by the Institut National de la Recherche Agronomique, France. The products will be second-generation biofuels, bulk chemicals, polymers, speciality molecules, heat and power (**Figure 8.30**).

In April 2012, Technip, a specialist in engineering and construction for the energy industry, signed a cooperation agreement in the green chemistry business with CIMV to industrialize CIMV's process. [211]

Lignol

Lignol is a Canadian company that is undertaking to construct biorefineries for the production of ethanol, pure lignin, and other valuable biochemicals from non-food cellulosic biomass feedstocks. [212] Lignol has acquired and since modified, a solvent based pretreatment technology that was originally developed by a subsidiary of General Electric. The pretreatment technology is based on aqueous ethanol organosolv extraction sometimes known as "Alcell." Lignol's pretreatment technology facilitates the rapid, high-yield conversion of cellulose to ethanol and the production of renewable chemicals, including high purity HP-L™ lignins. [213] HP-L™ lignin represents a new class of high purity lignin extractives (and their subsequent derivatives) which can be engineered to meet the chemical properties and functional requirements of a range of industrial applications that until now has not been possible with traditional lignin by-products generated from other processes.

The Lignol process (see **Figure 7.11**) uses a blend of ethanol and water in the range of 50:50(w/w) at ~200 °C, and 400 psi (28 bar) to extract most of the lignin from wood chips or other lignocellulosic materials. [214]

Lignin is recovered as a fine precipitate by flashing the pulping liquor to atmospheric pressure, followed by rapid dilution with water. Other co-products such as hemicellulose sugars and furfural are recovered from the water-soluble stream.

In April 2011, Lignol announced that it has recently begun to sell tonnage quantities of its proprietary High Performance Lignin (HP-L™) to its development partners for industrial production trials. [213]

In January 2013, Lignol announced first commercial supply agreement for HP-L™ lignin in the thermoplastics sector. [215]

Wageningen UR Lignin Platform

The mission of Wageningen UR (University & Research centre) is to explore the potential of nature to improve the quality of life. [216] Wageningen UR Food & Biobased Research Unit is both founder and coordinator of the Wageningen UR Lignin Platform launched in June 2011. [217]

The objective of the Wageningen UR Lignin Platform is to promote interdisciplinary, scientifically challenging, and precompetitive research and to create a network on the valorization of lignin for the industrial production of lignin-derived chemicals. [218] Lignin valorization will be an essential part of the integral biorefinery concepts and will be an important driver for development of economic and sustainable biorefineries. The Wageningen UR Lignin Platform brings together unique competences covering the whole chain from cultivating lignocellulosic biomass to lignin derived biobased products.

Wageningen UR Food &Biobased Research has been the coordinator of the project LignoValue devoted to value-added valorization of lignin for optimal biorefinery of lignocellulose towards energy carriers and products (**Figure 8.31**). [219]

This project focuses on:

1. Primary biorefinery involving pretreatment and fractionation for production of streams of cellulose, hemicelluloses, and lignin.

2. Secondary biorefinery in which lignin is converted into phenols, performance products like resins, fuel additives, electricity, and heat via thermochemical routes such as pyrolysis and depolymerization under supercritical conditions.

3. Development of an integral biorefinery in which the technologies have been optimally integrated and all biomass fractions consist of an optimal quality for further refining and valorization.

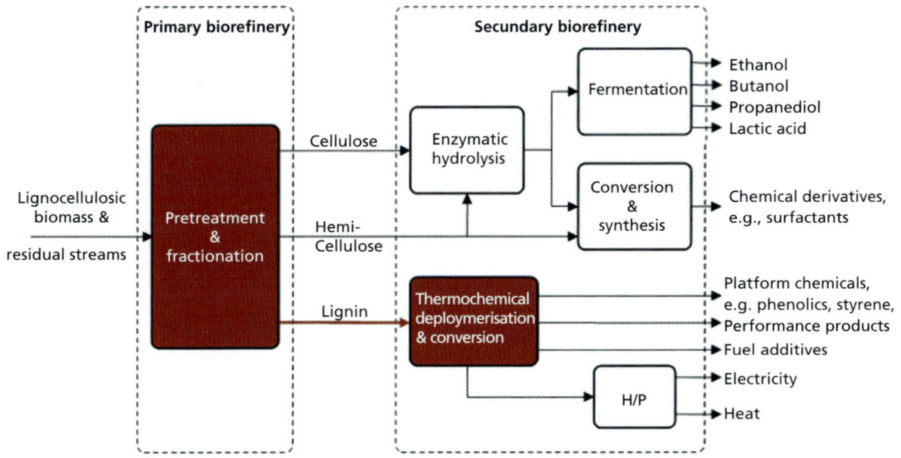

Figure 8.31 Multi-product biorefinery including lignin upgrading. [219]

Borregaard

Since the beginning of the 20th century, Norwegian company Borregaard is in the business of lignocellulosic products. Now part of Orkla Group (32,000 people and EUR 7.5 billion turnover), Borregaard is divided in five divisions according to market and product type: Borregaard ChemCell is responsible for specialty cellulose and bioethanol, Borregaard LignoTech for lignin-based products and trading activities, Borregaard Synthesis for fine chemicals/pharmaceuticals, Borregaard Ingredients for vanillin and omega-3 for foodstuffs, and Borregaard Energy for electrical power. [220]

Borregaard aims to use as much as possible of their raw materials (wood) in the production process (**Figure 8.32**). [221-222] The company utilizes the different component parts of timber in a range of different products.

The wood pulp production plant in Sarpsborg (Norway) is a good example of the biorefinery concept developed by Borregaard. [223] The plant consumes spruce wood and transforms it into specialty cellulosic pulp and chemicals. The biorefinery in Sarpsborg uses every year 650,000 m3 of solid wood and produces 20 million liters of ethanol, 160,000 tonnes of specialty cellulose (for production of cellulose acetate, nitrocellulose, viscose, etc.), ~150,000 tonnes of lignosulfonates, and 1,000 tonnes of vanillin.

The process includes several steps:

- The bark is removed and the log is cut into chips. The bark is sent to incineration.
- The chips are cooked in acidic conditions (also called sulfite pulping process). In these conditions, lignin residues are oxidized and transformed almost completely in lignosulfonates, [224] while considerable hydrolysis of hemicelluloses occurs, which make the process particularly valuable to

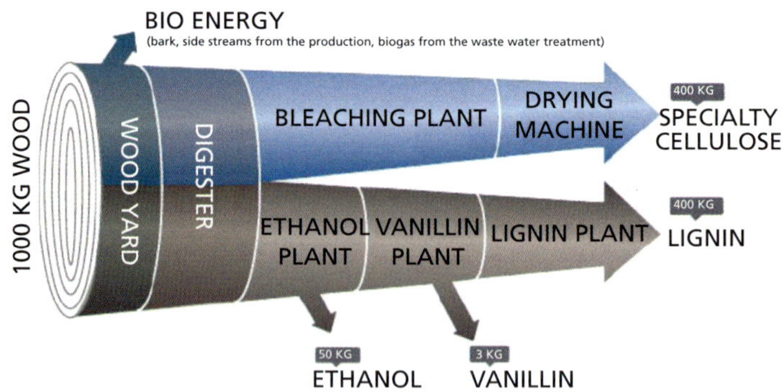

Cellulose	Lignin	Vanillin	Ethanol
Construction materials	Concrete additives	Food	Pharmaceutical industry
Cosmetics	Animal Feed	Perfumes	Bio Fuel
Food	Mining	Pharmaceuticals	Paint / varnish
Tablets	Batteries		Car care
Paint / varnish	Briquetting		
Filters	Soil Conditioner		
Textiles			

Figure 8.32 Borregaard's biorefinery concept. [222]

obtain high purity pulps. The cellulose is then separated and goes through a range of bleaching and final purification stages. [223]

- The remaining parts of the timber, which includes binding agents and sugar compounds, are separated in the pulping process. This flow of raw materials (waste liquor) is sent to the ethanol factory, where the sugar is converted into ethanol through fermentation.
- The sugar-free waste liquor contains lignosulfonates, which are a starting point for the production of valuable products produced in the lignin and vanillin (**Figure 8.33**) factories. Lignosulfonates are polydisperse (20–80 kDa) and contain 0.6–1.2 sulfonate groups per monomer. [223] They can be used for various applications like corrosion inhibition, concrete formulation and rheology modification. Vanillin (used in food, perfumes, etc.) is obtained by catalytic oxidation of lignosulfonates.
- The residual organic material from the timber that is not included in any products goes to a biological treatment plant where the material is broken down with the help of bacteria. In this process biogas (methane) is produced, which is used as an energy source. The biological mud from the breakdown process is burned in the bark incineration plant.

Borregaard is also participating to several research and development projects on biorefineries, including two important projects financed by European Union Seventh Framework Programme, namely EuroBioRef and Suprabio.

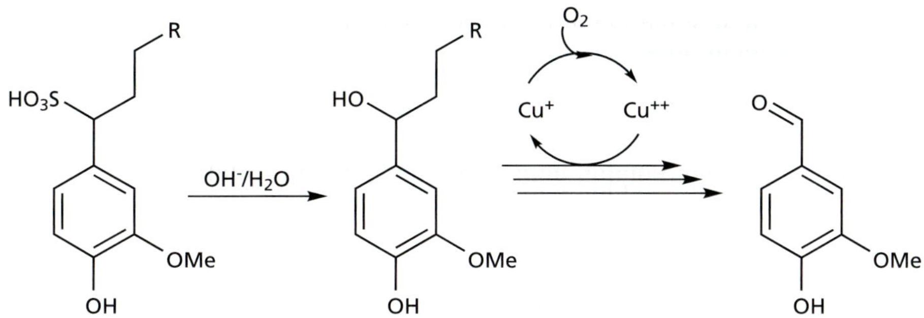

Figure 8.33 Oxidation of lignosulfonate to vanillin. [223]

8.6 Overview of selected companies

An overview of selected companies using biochemical technology to convert lignocellulosic biomass into ethanol as primary product is shown in **Table 8.1**.

The group of companies converting biomass into fuels presents the following features or raises the following remarks:

- No company has yet reached the commercial scale but several commercial projects are underway.
- There is much restructuration in the sector, such as Danisco acquired by DuPont, Verenium acquired by BP, Shell abandoning the project with Iogen.
- It is not yet clear which technology will be the first commercially operational and in which delays, but commercial solutions should be available in the near future.

An overview of selected companies using biochemical technology to convert lignocellulosic biomass into ethanol and/or chemicals is shown in **Table 8.2**.

The group of companies converting biomass into fuels and/or chemicals presents the following features:

- Some companies have reached the commercial stage.
- The technologies in the group are generally very innovative and attractive.
- The technologies from Virent, Gevo, and Avantium have been selected by Coca-Cola to reach global scale. The three companies are targeting commercial production of 100% biobased PET or PEF in the next few years.
- The various emerging technologies in the group produce a wide variety of building blocks including acetic acid, isobutanol, furanics, isoprene,

Table 8.1 Selected companies using biochemical technologies for converting lignocellulosic biomass into ethanol as primary products. [225-226]

Company (country of the parent company)	Technology	Technology scale (capacity in gallon per year; location)
Abengoa (ES)	Enzymatic hydrolysis	Commercial project (23 million, Kansas)
Iogen (CA)	Enzymatic hydrolysis	Demonstration (0.66 million, Ottawa) Commercial project with Shell cancelled
DuPont (US)	Enzymatic hydrolysis	Commercial project (30 million, Iowa)
BP (GB)	Enzymatic hydrolysis	R&D and demonstration (1.4 million, Louisiana) Commercial project cancelled
POET-DSM (US)	Enzymatic hydrolysis	Commercial project (25 million, Iowa)
Inbicon (DK)	Hydrothermal pretreatment and enzymatic hydrolysis	Demonstration (1.4 million, DK)
Mascoma (US)	Consolidated bioprocessing	Commercial project (20 million, Michigan)
BlueFire (US)	Concentrated acid hydrolysis	Commercial project (19 million, Mississipi)

farnesene, and adipic acid. This variety in technologies and products is synonymous with promises for the future.

An overview of selected companies using biochemical technology to convert lignocellulosic biomass into lignin and lignin derivatives as primary products is shown in **Table 8.3**.

Features related to the group of companies converting biomass into lignin and lignin derivatives as primary products are the following:

- There is a growing awareness in the world of the need to valorize lignin.
- Only Borregaard is operating at the industrial scale, using the sulfite process for making wood pulp.
- There is a wide variety of short-, medium-, and long-term technologies in extracting lignin and developing lignin derivatives.
- The prevailing technologies of tomorrow remain still obscure as well as time-to-market for most projects.

Table 8.2 Selected companies using biochemical technologies for converting lignocellulosic biomass into fuels and/or chemicals as primary products. [225-226]

Company (country of the parent company)	Technology	Primary products	Technology scale (capacity in gallon per year; location)
Zeachem (US)	Biochemical (acetogenic process) and thermochemical	Ethanol, fuels and chemicals	Commercial plant (0.25 million ethanol, Oregon)
Virent (US)	Bioforming platform (catalytic chemistry)	Fuels and chemicals	Demonstration
Gevo (US)	Synthetic biology and chemistry	Isobutanol	Commercial plant (isobutanol and ethanol, Minnesota)
Avantium (NL)	High-throughput catalyst development	Furanics	Pilot
Aemetis (US)	Consolidated bioprocessing	Chemicals (isoprene) and fuels	Pilot
Amyris (US)	Yeast metabolic engineering/metabolic engineering	Farnesene (muconic acid from Draths)	Commercial plant (BZ)
Verdezyne (US)	Yeast metabolic engineering	Fuels and chemicals (adipic acid)	Pilot

Table 8.3 Selected companies using biochemical technologies for converting lignocellulosic biomass into lignin and lignin derivatives as primary products. [225-226]

Company (country)	Technology	Primary Products	Technology scale (capacity, location)
DOE (US)	Base-catalyzed depolymerization	Chemicals from lignin	Pilot
CIMV (FR)	Organosolv	Lignin, paper pulp and C5-sugars	Pilot
Lignol (CA)	Organosolv/ biochemical	Ethanol and lignin	Pilot
Wageningen UR (NL)	Fractionation and conversion	Lignin and derivatives	Pilot
Borregaard (NO)	Acidic pulping, enzyme hydrolysis, catalytic oxidation	Cellulose, ethanol, lignosulfonates, vanillin	Commercial (5.3 million gallons/year ethanol; ~150,000 tonnes/ÿear lignosulfonates ; NO)

8.7 References

[1] CatchBio consortium in http://www.catchbio.com/about_catchbio

[2] J.M. Bergthorson, M. Ngadi, S.D. Salusbury, B. Fishbein, M. Bakari and V. Toepel, *2nd Generation Biomass Conversion Efficiency*,. Technical Report, submitted to International Air Transport Association, 2009 in http://www.iata.org/SiteCollectionDocuments/Documents/IATAConversionTechnologiesFinalv2.pdf

[3] S.P.S. Chundawat, G.T. Beckham, M.E. Himmel and B.E. Dale, *Ann. Rev. Chem. Biomol. Eng. 2*, 212, 2011 in http://www.annualreviews.org/doi/abs/10.1146/annurev-chembioeng-061010-114205

[4] J.D. Stephen, W.E. Mabee and J.N. Saddler, *Biofuels Bioprod. Bioref. 4*, 503, 2010 in http://onlinelibrary.wiley.com/doi/10.1002/bbb.239/full

[5] A. Demirbas, *Biorefineries*, Springer, 2010 in http://www.springerlink.com/content/xl21819684u157l4

[6] Y. Zheng, Z. Pan and R. Zhang, *Int. J. Agric. Biol. Eng. 2*, 51, 2009 in http://www.ijabe.org/index.php/ijabe/article/viewFile/168/83

[7] L.R. Lynd, W.H. Van Zyl, J.E. McBride and M. Laser, *Curr. Opin. Biotechnol. 16*, 577, 2005 in http://agwest.sk.ca/bioproducts/documents/CurrentOpinioninBiotechnology_2.pdf

[8] Purdue University , 2009 in http://www.purdue.edu/uns/x/2009a/090507McCannEFRC.html

[9] M. L. Y. Liu, *Fermentation of hemicellulose rich liquid fraction derived from steam pretreated softwoods*, Thesis, Simon Fraser University, 2010 in http://www.swst.org/wfs/PDF%20Files/ubc_2010_fall_liu_michael.pdf

[10] U.S. Deparment of Energy, *Energy Efficiency & Renewable Energy, Biomass program*, 2008 in http://www1.eere.energy.gov/biomass/biochemical_processes.html

[11] O.J. Sanchez, E.S. Fraga and C.A. Cardonna, *Process synthesis for fuel ethanol production from lignocellulosic biomass using an optimization-based strategy*, World Renewable Energy Congress IX and Exhibition, 2006 in http://biotecnologia.ucaldas.edu.co/publicaciones/EtOH-ProcSynthesis%20full%20text.pdf

[12] K. Ojeda, E Sanchez and V. Kafarov, *Energy 36*, 2119, 2011 in http://www.sciencedirect.com/science/article/pii/S0360544210004536

[13] Y. Sun and J. Cheng, *Bioresour. Technol. 83*, 1, 2002 in http://www.sciencedirect.com/science/article/pii/S0960852401002127

[14] Molecular-Plant-Biotechnology.info, *Production of Enzymes by Microorganisms*, 2011 in http://www.molecular-plant-biotechnology.info/applied-biotechnology/microbial-biotechnology/production-of-enzymes-by-microorganisms.html

[15] http://www.mapsenzymes.com/enzymes.asp

[16] G.E. Nedwin, T. Schaefer and P. Falholt, *Enzyme Discovery: Screening, Cloning, Evolving, Chemical Engineering Progress 101*, 48, 2005 in http://www.novozymes.com/en/innovation/publishing-our-scienceour-publications/Pages/reviews-.aspx

[17] H. Dalboge and L. Lange, *Tibtech 16*, 265, 1998 in http://www.sciencedirect.com/science/article/pii/S0167779998012025

[18] R.W. King, K.D. Lustig, P.T. Stukenberg, T.J. McGarry and M.W. Kirschner, *Science 277*, 973, 1997 in http://www.sciencemag.org/content/277/5328/973.full

[19] H. Dalboge, FEMS *Microbiol. Rev. 21*, 29, 1997 in http://onlinelibrary.wiley.com/doi/10.1111/j.1574-6976.1997.tb00343.x/full

[20] H.S. Olsen, *Enzymes from research to commercial production at Novozymes*, 2008 in http://www.biotechacademy.dk/upload/institutter/bic/sites/biotech%20academy/bioweek/enzymes.pdf

[21] D.B. Wilson, *Curr. Opin. Biotechnol. 20* 295, 2009 in http://www.sciencedirect.com/science/article/pii/S0958166909000676

[22] Reuters, *Novozymes says Italy plant heralds biofuel future*, 2011 in http://www.reuters.com/article/2011/04/12/us-novozymes-biofuels-plant-idUSTRE73B2KR20110412

[23] Novozymes, *Cellulosic ethanol*, 2011 in http://www.novozymes.com/en/solutions/bioenergy/cellulosic-ethanol/Pages/default.aspx

[24] Novozymes in http://bioenergy.novozymes.com/en/cellulosic-ethanol/cellic/Documents/Cellic%20brochure.pdf

[25] Novozymes, *New Enzymes Turn Waste Into Fuel*, 2010 in http://www.novozymes.com/en/news/news-archive/Pages/45713.aspx

[26] Genencor, *Genencor Launches a Market-Leading Enzyme to Advance Cellulosic Biofuel Production Worldwide*, 2011 in http://www.businesswire.com/news/home/20110622005037/en

[27] M. Stiklen, *Curr. Opin. Biotechnol. 17*, 315, 2006 in https://www.msu.edu/~stickle1/Sticklen_Curr%20Opinion_paper.pdf

[28] U.S. Department of Energy, Energy Efficiency and Renewable Energy, Biomass Program, *Dilute Acid Hydrolysis*, 2007 in http://www1.eere.energy.gov/biomass/printable_versions/dilute_acid.html

[29] P. C. Badger, *Ethanol from Cellulose: A General Review*, p. 17, in J. Janick and A. Whipkey (eds.), *Trends in new crops and new uses*, ASHS Press, Alexandria, VA, 2002 (paper available in www.hort.purdue.edu/newcrop/ncnu02/v5-017.html)

[30] D. I. Bransby, *Cellulosic Biofuels Technologies*, Auburn University, 2007 in www.adeca.alabama.gov/C17/ATF/Document%20Library/Cellulosic%20Biofuel%20Technologies.pdf

[31] A.K.CHANDEL et al., *Economics and environmental impact of bioethanol production technologies: an appraisal, Biotechnol. Mol. Biol. Rev. 2*, 14, 2007

[32] U.S. Department of Energy, Energy Efficiency and Renewable Energy, Biomass Program, *Concentrated Acid Hydrolysis*, 2006 in http://www1.eere.energy.gov/biomass/printable_versions/concentrated_acid.html

[33] Arkenol Process simplified Flow-Chart in http://www.arkenol.com/Arkenol%20Inc/images/ahflow.gif

[34] BlueFire Renewables, *About* in http://bfreinc.com/about-test

[35] BlueFire Renewables, *Our Technology* in http://bfreinc.com/our-technology

[36] Chemicals-Technology.com, *BlueFire Renewables Cellulosic Ethanol Facility*, 2011 in http://www.chemicals-technology.com/projects/bluefire-facility

[37] Masada, Technology Development in http://masadaonline.com/id2.html

[38] Masada, *CES OxyNol Process* in http://www.masadaonline.com/id3.html

[39] P.F. STANBURY, A. WHITAKER and S.J. HALL, *Principles of Fermentation Technology*, 1999 in http://www.rsc.org/ebooks/archive/free/BK9780854046065/BK9780854046065-00001.pdf

[40] B. PUMPHREY and C. JULIEN, Fermentation Basics, 1996 in http://www.biocompare.com/Articles/ApplicationNote/1542/FERMENTATION-BASICS.html

[41] University of North Carolina at Pembroke, Biotechnology Research and Training Center, 2011 in http://www.uncp.edu/biotech/facility

[42] L. DU PLESSIS, *Co-expression of cellulase genes in Saccharomyces cerevisiae for cellulose degradation*, Thesis, Stellenbosch University, 2008 in http://scholar.sun.ac.za/bitstream/handle/10019.1/1577/Du%20Plessis,%20L_colour.pdf?sequence=2

[43] M.J. TAHERZADEH and K. KARIMI, BioRes. **2**, 707, 2007 in http://www.ncsu.edu/bioresources/BioRes_02/BioRes_02_4_707_738_Taherzadeh_K_EnzymeBased_Ethanol_Review.pdf

[44] C.N. HAMELINCK, *Outlook for advanced fuels*, Thesis, Universiteit Utrecht, 2004 in http://igitur-archive.library.uu.nl/dissertations/2005-0209-113022/full.pdf

[45] M. BALAT, H. BALAT and C. OZ, *Progr. Energy Comb. Sci. 34*, 551, 2008 in http://www.biotek.lu.se/fileadmin/bioteknik/Utbildning/Kurser/KKKA05/ProgressETOHprocess.pdf

[46] A.K. CHANDEL, C. ES, R. RUDRAVARAM, M.L. NARASU, L.V. RAO and P. RAVINDRA, Biotechnol. *Mol. Biol. Rev. 2*, 14, 2007 in http://www.academicjournals.org/bmbr/PDF/pdf2007/Feb/Chandel%20et%20al.pdf

[47] Oasisnakhoil, *Economics and environmental impact of bioethanol production technologies: an appraisal*, 2008 in http://www.oasisnakhoil.com/oct08.html

[48] J. AXELSSON, Separate Hydrolysis and Fermentation of Pretreated Spruce, Thesis, Linköpings Universitet, 2011 in http://www.essays.se/essay/2f10e10561

[49] J.D. MCMILLAN, *Biotechnological Routes to Biomass Conversion*, NREL, 2004 in http://www.nrel.gov/docs/gen/fy04/36831g.pdf

[50] R. HARUN, B. LIU and M.K. DANQUAH, *Progress in Biomass and Bioenergy Production*, InTech, 2011 in http://www.intechopen.com/source/pdfs/16655/InTech-Analysis_of_process_configurations_for_bioethanol_production_from_microalgal_biomass.pdf

[51] C.N. HAMELINCK, G. VAN HOOIJDONK and A.P.C. FAAIJ, *Biomass Bioenergy 28*, 384, 2005 in http://www.star-int.info/pdf/bdocs/transport/Bioethanol_Biomass.pdf

[52] L.X. LAY, *Bioproducts from sulfite pulping: Bioconversion of sugar streams from pulp, sludge, and spent sulfite liquor*, Thesis, University of Washington, 2010 in http://depts.washington.edu/sfrbbl/Files/Lisa%20Lai%20thesis%20final.pdf

[53] K. OLFSSON, D. RUNQUIST, B. HAHN-HAGERDAL and G. LIDEN, *AMB Express 1:4*, 2001 in http://www.amb-express.com/content/1/1/4

[54] K. OLOFSSON, M. BERTILSSON and G. LIDEN, Biotechnol. *Biofuels 1*, 7, 2008 in http://www.biotechnologyforbiofuels.com/content/1/1/7

[55] A.A. VERTES, N. QURESHI and H. YUKAWA, *Biomass to biofuels: strategies for global industries*, John Wiley & Sons, Chichester, 2009 in http://www.mendeley.com/research/biomass-to-biofuels-strategies-for-global-industries

[56] B. HAHN-HAGERDAL, K. KARHUMAA, C. FONSECA, I. SPENCER-MARTINS and M.F. GORWA-GRAUSLUND, *Appl. Microbiol. Biotechnol. 74*, 937, 2007 in http://www.springerlink.com/content/17jl577258xw1204

[57] K. OLOFSSON, B. PALMQVIST and G. LIDEN, *Biotechnol. Biofuels 3*, 17, 2010 in http://www.biotechnologyforbiofuels.com/content/3/1/17

[58] L.R. LYND, W.H.VAN ZYL, J.E. MCBRIDE and M. LASER, *Curr. Opin. Biotechnol. 16*, 577, 2005 in http://agwest.sk.ca/bioproducts/documents/CurrentOpinioninBiotechnology_2.pdf

[59] M. JIN, V. BALAN, C. GUNAWAN, B.E. DALE, *Biotechnol. Bioeng. 108*, 1290, 2011 in http://onlinelibrary.wiley.com/doi/10.1002/bit.23059/abstract

[60] MEHROTRA and SUMBALI, *Principles of Microbiology*, Tata McGraw-Hill Education, 2009 in http://www.tmhshop.com/9780070141209

[61] http://en.wikipedia.org/wiki/Fermentation

[62] D.R. SHONNARD, *Major Metabolic Pathways*, Michigan Technological University, http://www.chem.mtu.edu/~drshonna/cm4710f07/lectures/chapter5.pdf

[63] G. KIRCHNER in http://homepages.ius.edu/GKIRCHNE/Glycolysis.htm

[64] http://www.bioinfo.org.cn/book/biochemistry/chapt14/bio8.htm

[65] http://www.biotopics.co.uk/edexcel/biotechnol/ferm.html

[66] Wikipedia Commons, *Ethanol Fermentation,* in http://en.wikipedia.org/wiki/Ethanol_fermentation

[67] *Fates of Pyruvate under Aerobic and Anaerobic Conditions* in http://www.bioinfo.org.cn/book/biochemistry/chapt14/sim3.htm

[68] B. HAHN-HAGERDAL, K. KARHUMAA, M. JEPPSSON and M.F. GORWA-GRAUSLUND, *Adv. Biochem. Engin/Biotechnol. 108*, 147, 2007 in http://www.springerlink.com/content/rh497k37g2748790

[69] K. TODAR, *Lactocccus lactis: nominated as the Wisconsin State Microbe* in http://textbookofbacteriology.net/featured_microbe.html

[70] K. TODAR, Lactic Acid Bacteria, 2011 in http://textbookofbacteriology.net/lactics_2.html

[71] H. SONG and S.Y. LEE, *Enz. Microb. Technol. 39*, 352, 2006 in http://www.sciencedirect.com/science/article/pii/S0141022906001190

[72] S.J. LEE, H. SONG and S.Y. LEE, *Appl. Environ. Microbiol. 72*, 1939, 2006 in http://www.ncbi.nlm.nih.gov/pmc/articles/PMC1393240

[73] InnovativeIndustry.net, *Biosuccinic Acid: Rising Bioplastic and Feedstock for Green Chemicals*, 2011 in http://www.innovativeindustry.net/biosuccinic-acid-rising-bioplastic-and-feedstock-for-green-chemicals

[74] BioAmber Company site in http://www.bio-amber.com/corporate_info.html

[75] European Biotechnology Science & Industry News, *Manufacturing New Platform Chemicals*, 2010 in http://www.eurobiotechnews.eu/western-europe/france/articledetail/?tx_ttnews%5Bpointer%5D=8&tx_ttnews%5Btt_news%5D=12843&tx_ttnews%5BbackPid%5D=39&cHash=c086312e9c

[76] BioAmber, *BioAmber and Cargill Partner to Develop Succinic Acid Producing Microorganism* Press release, 2011 in http://www.bio-amber.com/img/pdf/BioAmber_Cargill_Press_Release_May9.pdf

[77] BioAmber, *BioAmber partners with Mitsubishi Chemical in Succinic Acid*, Press release, 2011 in http://www.bio-amber.com/img/pdf/BioAmber_Mitsubishi_press_release_May9.pdf

[78] DSM, *DSM and Roquette to open commercial scale bio-based succinic acid plant by 2012*, Press release, 2011 in http://www.dsm.com/en_US/cworld/public/media/pages/press-releases/36_11_dsm_and_roquette_to_open_commercial_scale_bio_based_succinic_acid_plant_2012.jsp

[79] A. UIHLEIN and L. SCHEBEK, *Biomass Bioenerg. 33*, 793, 2009 in http://www.sciencedirect.com/science/article/pii/S0961953408002572

[80] Argonne National Laboratory, Baseline and Projected Water Demand Data for Energy and Competing Water Use Sectors, 2008 in http://www.ead.anl.gov/pub/doc/ANL-EVS-TM08-8_water_demand_report.pdf

[81] DuPont in http://www.ddce.com/about/sustainability.html

[82] EPA, 2010 in http://www.epa.gov/otaq/renewablefuels/420r10006.pdf

[83] F. Cherubini and G. Jungmeier, *Int. J. Life Cycle Assess. 15*, 53, 2010 in http://www.springerlink.com/content/964384p486257268

[84] F. Cherubini and S. Ulgiati, *Appl. Energy 87*, 47, 2010 in http://www.sciencedirect.com/science/article/pii/S0306261909003596

[85] L.Haspeslagh, TOTAL Company, 2007, Internal report

[86] Environment News Service, *Cellulosic Ethanol Production Far Behind Renewable Fuel Standard,* in http://www.ens-newswire.com/ens/oct2011/2011-10-11-093.html

[87] R.Crowe, "RAND Report Sparks Debate about Advanced Biofuels" in http://www.renewableenergyworld.com/rea/news/article/2011/01/rand-report-sparks-debate-about-advanced-biofuels

[88] Abengoa Locations and Sites in http://www.abengoabioenergy.com/corp/web/en/acerca_de/oficinas_e_instalaciones/bioetanol/

[89] Abengoa, *Biomass Enzymatic Hydrolysis* in http://www.abengoabioenergy.com/corp/web/en/nuevas_tecnologias/tecnologias/ruta_bioquimica/enzimas/

[90] *Cellulosic Snapshot,* EPM reviews the status of U.S. cellulosic ethanol projects in http://www.ethanolproducer.com/articles/6695/cellulosic-snapshot

[91] Biopact, *An in-depth look at biorefinery concepts*, 2007 in http://news.mongabay.com/bioenergy/2007/07/in-depth-look-at-biorefinery-concepts.html

[92] Biopact, *Abengoa Bioenergy enters Brazilian market via acquisition of Dedini Agro*, 2007 in http://news.mongabay.com/bioenergy/2007/08/abengoa-bioenergy-enters-brazilian.html

[93] BiofuelsDigest, *Abengoa closes $132M DOE loan guarantee, path cleared for cellulosic ethanol project*, 2011 in http://biofuelsdigest.com/bdigest/2011/09/30/abengoa-closes-132m-doe-loan-guarantee-path-cleared-for-cellulosic-ethanol-project/

[94] M. Sapp, *BiofuelsDigest*, 2012 in http://www.biofuelsdigest.com/bdigest/2012/11/07/televent-signs-it-deal-for-abengoas-kansas-cellulosic-ethanol-plant/

[95] J. Lane, *10 TOP Strategic Investors in Biofuels and Material*", BiofuelsDigest, 2011, in http://biofuelsdigest.com/bdigest/2011/06/06/10-top-strategic-investors-in-biofuels-materials/

[96] The Economist, *Generation* game, 2010 in http://www.economist.com/node/15464489

[97] Shell, News and Media Releases, 2011 in http://www.shell.com/global/aboutshell/media/news-and-media-releases/2011/shell-cosan-raizen-biofuels-02062011.html

[98] Energyboom, *E boom capital: Shell-Cosan Brazilian ethanol JV will double output*, 2011 in http://www.energyboom.com/biofuels/e%E2%80%A2boom-capital-shell-and-cosan-brazilian-ethanol-jv-will-double-output

[99] *Cosan and Shell announce Rayzen*, 2009, in http://www.cosan.com.br/cosan2009/web/conteudo_eni.asp?idioma=1&tipo=31249&conta=46&id=125044

[100] J. Lane, *Shell and its biofuel babies in review*, *BiofuelsDigest*, 2010 in http://biofuelsdigest.com/bdigest/2010/06/08/shell-re-ups-with-iogen-through-2012-shell-and-its-biofuels-babies-in-review/

[101] Iogen, Iogen's Cellulosic Ethanol Demonstration Plant in http://www.iogen.ca/company/demo_plant/index.html

[102] S. Retka-schill, *Ethanol Producer Magazine*, 2010, in http://www.ethanolproducer.com/articles/7043/1000-keys

[103] Iogen Process in http://www.iogen.ca/cellulosic_ethanol/what_is_ethanol/process.html

[104] http://www.biofuelsdigest.com/bdigest/2012/05/01/shell-iogen-and-cancellation-in-manitoba-answers-to-your-questions

[105] http://www.greencarcongress.com/2012/04/shell-and-iogen-abandon-project-to-build-large-scale-cellulosic-ethanol-facility-in-manitoba.html

[106] Iogen, News Release, 2013 in http://www.iogen.ca/news_events/press_releases/2013_01_31_novozymes.pdf

[107] Genencor, *Genencor Launches First Ever Commercial Enzyme Product for Cellulose Ethanol*, 2007 in www.genencor.com/cms/connect/genencor/media_relations/news/frontpage/gen_businessupdate_393_en.htm

[108] Danisco, *Genencor*, 2005 in www.danisco.com/cms/connect/corporate/about%20danisco/organisation/divisions/genencor/genencor_en.htm

[109] Dupont Danisco Cellulosic Ethanol LLC in http://www.ddce.com/

[110] Frontiers of Engineering, *Reports on Leading-Edge Engineering from the 2007 Symposium*, 2008, in http://www.ddce.com/technology/AllesJenkins_SustainableBiorefineriesinNAE.pdf

[111] USDA, Farm Service Agency, *Testimony for Judith Canales, Administrator, Rural Business Service and Juan Garcia, Deputy Administrator, in* http://agriculture.house.gov/pdf/hearings/GarciaandCanales110720.pdf

[112] DuPont, News Releases, 2011 in http://www2.dupont.com/media/en-us/news-events/corporate-news-releases.html

[113] http://biofuels.dupont.com/cellulosic-ethanol/conversion/

[114] Verenium, *Verenium Begins Commissioning of Nation's First Cellulosic Ethanol Demonstration-Scale Plant*, 2008, in http://ir.verenium.com/releasedetail.cfm?releaseid=412789

[115] Investors-Hub, *Fundamental and technical discussions about Verenium Corp.* http://investorshub.advfn.com/boards/board.aspx?board_id=11154

[116] Alternativenergy, *BP and Verenium announce pivotal biofuels agreement*, 2010, in http://www.bp.com/genericarticle.do?categoryId=9024973&contentId=7063769

[117] J. Fahey, *BP's Biofuel play : plan B or PR?* , Forbes, 2010 in http://www.forbes.com/sites/energysource/2010/07/15/bp-verenium-biofuel-ethanol/

[118] BP, Press release, 2012 in http://www.bp.com/genericarticle.do?categoryId=2012968&contentId=7079431

[119] D.E. Ray, *Has ethanol's boost to corn prices run its course?* in http://www.agpolicy.org/weekdoc/444.doc

[120] J. Mick, Bailytech, *POET Cuts Its Cellulosic Ethanol Costs from $4.13/Gallon to $2.35/Gallon* in http://www.dailytech.com/POET+Cuts+Its+Cellulosic+Ethanol+Costs+from+413Gallon+to+235Gallon/article16870.htm

[121] http://www.dsm.com/en_US/cworld/public/media/pages/press-releases/01_12_dsm_and_poet_to_make_advanced_biofuels_a_reality_by_2013.jsp

[122] POET-DSM, News release, 2012 in http://www.poetdsm.com/news/andritz.aspx

[123] Dong Company Website, *Inbicon elected as bioethanol company of the year*, 2011 in http://www.dongenergy.com/EN/Media/Newsroom/News/Pages/Inbiconelectedasbioethanolcorporationoftheyear.aspx

[124] Inbicon Company Website, *Hydrothermal Pretreatment*, in http://www.inbicon.com/Technologies/Hydrothermal_pretreatment/Pages/Hydrothermal%20pretreatment.aspx

[125] Inbicon Company Website, *Enzymatic Hydrolysis*, in http://www.inbicon.com/Technologies/Enzymatic_hrydolysis/Pages/Enzymatic%20Hydrolysis.aspx

[126] European Biofuels Technology Platform, *Inbicon Biomass Refinery for production of Cellulosic Ethanol* http://www.biofuelstp.eu/cell_ethanol.html#inbicon

[127] Inbicon Company Website, *Kalundborg Demonstration Plant* in http://www.inbicon.com/Projects/Kalundborg%20Demonstration%20plant/Pages/Kalundborg%20Demonstration%20plant.aspx

[128] M. McDermott, Treehugger Science and Technology, *Danish Cellulosic Ethanol Technology (Hopefully) Coming Soon to North America*, 2008, in http://www.treehugger.com/files/2008/08/danish-cellulosic-ethanol-technology-coming-to-north-america.php

[129] S. Writers, *Statoil Now Blending Inbicon's Cellulosic Ethanol For Danish Drivers*, Bio Fuel Daily, 2010, in http://www.biofueldaily.com/reports/Statoil_Now_Blending_Inbicon_Cellulosic_Ethanol_For_Danish_Drivers_999.html

[130] Inbicon, Press Release, 2012 in http://us1.forward-to-friend1.com/forward/show?u=df7293807f22176b27268cf3a&id=2637779208

[131] M. KANELLOS, *Valero Invests in Mascoma: Have Biofuels Hit the Tipping Point?*, Greentechmedia, 2011, in http://www.greentechmedia.com/articles/read/valero-invests-in-mascoma-have-biofuels-hit-the-tipping-point/

[132] *Major breakthroughs toward cellulosic ethanol,* 2009 in http://nextbigfuture.com/2009/05/major-breakthroughs-towards-cellulostic.html

[133] Mascoma Company Website, *What is Consolidated Bioprocessing (CBP)?*, in http://mascoma.com/pages/sub_cellethanol04.php

[134] C.B. CORNELL, *Mascoma Announces Major Research Advance for Cellulosic Ethanol,* Gas2, 2009, in http://gas2.org/2009/05/07/mascoma-announces-major-research-advance-in-cellulosic-ethanol/

[135] *Mascoma and Valero Sign Deal to Develop Commercial Cellulosic Ethanol Biorefinery,* American Fuels, 2011 in http://americanfuels.blogspot.com/2011/01/mascoma-and-valero-sign-deal-to-develop.html?utm_source=feedburner&utm_medium=feed&utm_campaign=Feed%3A+AmericanFuels+%28American+Fuels%29

[136] Mascoma, Press Release, 2012 in http://www.mascoma.com/download/TransFerm%20press%20release%20FEW-v7.pdf

[137] BlueFire Renewables, News Release, 2012 in http://bfreinc.com/2012/02/sucresource-signs-gs-caltex-first-cellulose-to-sugar-plant-sale-and-partnership

[138] J.J. BOZELL and G.R. PETERSON, Green Chem. **12**, 539, 2010 in http://onlinelibrary.wiley.com/doi/10.1002/chin.201028264/full

[139] US Department of Energy, *Top Value Added Chemicals from Biomass*, 2004 in http://www1.eere.energy.gov/biomass/pdfs/35523.pdf

[140] ZeaChem in http://www.zeachem.com/index.php

[141] J.M. LANE, *The 50 Hottest Companies in Bioenergy 2010-2011, Selectors Data Book*, 2010 in http://www.ascension-publishing.com/BIZ/Hot50.pdf

[142] Zeachem, 2011 in http://www.zeachem.com/press/pressrelease060111.php

[143] Zeachem, Press Release, 2012 in http://www.zeachem.com/press/pressrelease102912.php

[144] US Department of Energy, Energy Efficiency and Renewable Energy, 2012 in http://www1.eere.energy.gov/biomass/pdfs/ibr_arra_zeachem.pdf

[145] Virent, 2007 in http://www.virent.com/MeetVirent/our_story.html

[146] Virent, BioForming, Technology, 2007 in http://www.virent.com/BioForming/technology.html

[147] P.G. BLOMMEL and R.D. CORTRIGHT, *Production of Conventional Liquid Fuels from Sugars*, White paper, 2008 in http://www.biofuelstp.eu/downloads/Virent_Technology_Whitepaper.pdf

[148] Virent, Press release, 2011 in http://www.virent.com/News/press/06-06-11_Virent_Makes_Paraxylene.pdf

[149] SpecialChem, 2011 in http://www.specialchem4polymers.com/resources/latest/displaynews.aspx?id=6110

[150] Greener package, 2011 in http://www.greenerpackage.com/bioplastics/virent_produces_plant-based_paraxylene

[151] Consumer Energy Report, 2011 in http://www.consumerenergyreport.com/2011/06/14/qa-with-virent-cto-randy-cortright

[152] http://www.virent.com/partners/corporate

[153] Coca-Cola, 2011 in http://www.thecoca-colacompany.com/dynamic/press_center/2011/12/plantbottle-partnerships.html

[154] P&G, 2012 in http://news.pg.com/press-release/pg-corporate-announcements/coca-cola-ford-heinz-nike-and-procter-gamble-form-collabora

[155] Virent, Press Release, 2012 in http://www.virent.com/news/shell-using-technology-licensed-from-virent/

[156] Virent, Press Release, 2012 in http://www.virent.com/news/virent-makes-jet-fuel-from-virdias-cellulosic-biomass/

[157] Gevo in http://www.gevo.com/about/company-overview/

[158] Gevo, Our Science and Technology in http://www.gevo.com/our-business/our-science-and-technology/

[159] Gevo, 2009 in http://ir.gevo.com/phoenix.zhtml?c=238618&p=irol-newsArticle&ID=1491575&highlight=

[160] Total, 2009 in http://www.total.com/en/press/press-releases/consultation-200524.html&idActu=2021

[161] Gevo in http://www.gevo.com/our-business/isobutanol

[162] Green Car Congress, 2011 in http://www.greencarcongress.com/biobutanol

[163] Gevo, News Release, 2011 in http://ir.gevo.com/phoenix.zhtml?c=238618&p=irol-newsArticle&id=1640177

[164] Gevo, News Release, 2012 in http://ir.gevo.com/phoenix.zhtml?c=238618&p=irol-newsArticle&ID=1699401

[165] Gevo, News Release, 2012 in http://ir.gevo.com/phoenix.zhtml?c=238618&p=irol-newsArticle&id=1737809

[166] Avantium in http://www.avantium.com

[167] Avantium in http://www.yxy.com/company-profile

[168] Avantium in http://avantium.com/catalysis/R-D-solutions/A-look-inside-our-R-D/technologies.html

[169] E. DE JONG, *Furanics : Versatile Molecules Applicable for Biopolymers Applications*, Avantium in http://www.soci.org/News/~/media/Files/Conference Downloads/Bioplastic Processing Apr 09/Jong.ashx and http://160.36.150.12/OBP/FIB2010/Resources/fib-de%20Jong,%20Ed.pdf

[170] Avantium, Press Release, 2011 in http://www.avantium.com/news-events/press-releases/Avantium-and-The-Coca-Cola-Company-sign-partnership-agreement-to-develop-next-generation-100-plant-based-plastic-PEF/

[171] Avantium, Press Release, 2012 in http://avantium.com/news/Avantium-and-Danone-sign-development-partnership-for-Next-Generation-bio-based-plastic-PEF.html

[172] The Street, 2011 in http://www.thestreet.com/story/11310708/1/ae-biofuels-announces-company-name-change-to-aemetis.html

[173] Aemetis, Press Releases, 2012 in http://www.aemetis.com/aemetis-announces-acquisition-of-cilion-inc-and-california-ethanol-plant/

[174] Aemetis, Business Update Conference, 2012 in http://www.aemetis.com/wp-content/uploads/2012/12/Aemetis-Business-Update-Conference-Call-2012-12-20.pdf

[175] http://www.businesswire.com/news/home/20111115005749/en/AE-Biofuels-Announces-Company-Change-Aemetis

[176] http://www.icis.com/blogs/green-chemicals/2011/11/ae-biofuels-enters-renewable-c.html

[177] Aemetis, Technology in http://www.aebiofuels.com/?section=technology

[178] http://ethanolproducer.com/articles/7946/ae-biofuels-acquires-biotech-company-zymetis

[179] Aemetis, Z-Microbe in http://www.aebiofuels.com/?section=technology&subsection=z-microbe

[180] Aemetis, ATSCH in http://www.aebiofuels.com/?section=technology&subsection=ATSCH

[181] http://www.liviakis.com/comp_profiles/Aemetis-Investor-Presentation-Website.pdf

[182] Aemetis, 2012 in http://www.aemetis.com/media/press-releases

[183] Aemetis, 2012 in http://www.aemetis.com/media/press-releases

[184] Verdezyne, 2011 in http://verdezyne.com/verdezyne/index.cfm

[185] Verdezyne, News, 2011 in http://verdezyne.com/verdezyne/News/documents/VerdezynePilotPlantReleaseBusinessFINAL.pdf

[186] Verdezyne, Technology, 2011 in http://verdezyne.com/verdezyne/Technology/index.cfm

[187] Verdezyne, Products, 2011 in http://verdezyne.com/verdezyne/Products/index.cfm

[188] Textile World, 2011 in http://www.textileworld.com/Articles/2011/December/Verdezyne_Opens_Biobased_Adipic_Acid_Pilot_Plant_In_California.html

[189] Verdezyne, Press Release, 2012 in http://verdezyne.com/verdezyne/News/documents/Verdezyne_UF%20Press%20Release%20-%20FINAL.pdf

[190] Verdezyne, Recent News, 2012 in http://verdezyne.com/verdezyne/News/documents/Verdezyne-DuPont_release%20FINAL.pdf

[191] Amyris in http://www.amyris.com/en/about-amyris/overview

[192] Amyris in http://www.amyris.com/en/markets

[193] EMBL-EBI in http://www.ebi.ac.uk/chebi/searchId.do?chebiId=CHEBI:10418

[194] NABC, 2012 in http://www.nabcprojects.org/fermentation_lignocellulosic_sugar.html

[195] NABC, 2011 in http://www.nabcprojects.org/about.html

[196] NABC, 2011 in http://www.nabcprojects.org/process_strategies.html

[197] NABC, 2011 in http://www.nabcprojects.org/news_releases.html

[198] http://www.icis.com/blogs/green-chemicals/2011/10/amyris-buys-draths.html

[199] Bioplastek 2011 in http://myemail.constantcontact.com/Is-the-Time-Right-for--Drop-in-Bioplastics-.html?soid=1101885321533&aid=QqQ4sS4qDUY

[200] Amyris, Press release, 2011 in http://www.amyris.com/News/122/Amyris-ETH-Bioenergia-Sign-Agreement-to-Produce-Biofene-in-Brazil

[201] Amyris in http://www.amyris.com/en/newsroom/218-total-and-amyris-partner-to-produce-renewable-fuels

[202] Amyris, Press Release, 2013 in http://www.amyris.com/News/266/Amyris-Ships-First-Truckload-of-Biofene-from-its-New-Plant-in-Brazil

[203] US Department of Energy, Energy Efficiency and Renewable Energy, Biomass Program, *Lignin-Derived Co-products,* 2005 in http://www1.eere.energy.gov/biomass/printable_versions/lignin_derived.html

[204] CIMV in http://www.cimv.fr/cimv/20-.html

[205] CIMV in http://www.cimv.fr/cimv-technology/cimv-technology/5-.html

[206] L. Kham, Y. Le Bigot, M. Delmas and G. Avignon, *Ind. Crops Prod. 21*, 9, 2005 in http://www.cimv.fr/uploads/delignification-wheat-straw-using-mixture-carboxylic-acids-and-peroxoacids-industrial-crops-and-products-2005.pdf

[207] M. Delmas, *Chem. Eng. Technol. 31*, 792, 2008 in http://www.cimv.fr/uploads/vegetal-refining-and-agrichemistry-chemical-engineering-and-technology-2008.pdf

[208] A. LEPONIEMI, *Fibres and energy from wheat straw by simple practice*, Dissertation, VTT Publications 767, 2011 in http://www.vtt.fi/inf/pdf/publications/2011/P767.pdf

[209] http://ec.europa.eu/research/index.cfm?lg=en&pg=newsalert&year=2010&na=na-010310

[210] Biocore, Biocore in brief in http://www.biocore-europe.org/page.php?optim=biocore-in-brief

[211] Technip, Press Releases, 2012 in http://www.technip.com/en/press/technip-and-cimv-sign-green-chemistry-cooperation-agreement

[212] Lignol in http://www.lignol.ca

[213] Lignol, Press Release, 2011 in http://www.lignol.ca/news/News-2011/Lignol-High-Performance-Lignin-Press-April-28.pdf

[214] X. PAN, C. ARATO, N. GILKES, D. GREGG, W. MABEE, K. PYE, Z. XIAO, X. ZHANG and J. SADDLER, Biotechnol. Bioeng. **90**, 473, 2005 in http://144.92.31.11/profiles/faculty/Pan-publications/No%206.pdf

[215] Lignol, Press Release, 2013 in http://www.lignol.ca/news/News-2013/News%20Release%2031Jan3013%20final.pdf

[216] Wageningen UR, 2011 in http://www.wur.nl/UK/about

[217] Wageningen UR, 2011 in http://www.wur.nl/UK/newsagenda/news/Launch_of_Wageningen_UR_Lignin_Platform_.htm

[218] Food & Biobased Research, Wageningen UR, 2011 in http://www.fbr.wur.nl/UK/newsagenda/archive/agenda/2011/Workshop_Wageningen_UR_Lignin_Platform.htm

[219] Biofuels & BiobasedProducts, *LignoValue*, 2010 in http://www.biobased.nl/lignovalue

[220] http://www.borregaard.com/About-us/Organisation

[221] http://www.borregaard.com/content/view/full/10227

[222] P.A. SORLIE, Borregaard, Corporate presentation, 2011 in http://hugin.info/111/R/1565694/486007.pdf

[223] M. LERSCH, Creating value from wood. The Borregaard biorefinery, in http://www.bioref-integ.eu/fileadmin/bioref-integ/user/documents/Martin_Lersch__Borregaard_-_Creating_value_from_wood_-_The_Borregaard_biorefinery.pdf

[224] http://en.wikipedia.org/wiki/Lignosulfonates

[225] US Department of Energy, Energy Efficiency & Renewable Resources, *Biomass Program, Research & Development*, 2011 in http://www1.eere.energy.gov/biomass/integrated_biorefineries_text_only.html

[226] W.T. COYLE, USDA, Next generation Biofuels : *Near Term Challenges and Implications for Agriculture*, 2010 in http://www.ers.usda.gov/publications/BIO0101/BIO0101.pdf

Chapter 9

Thermochemical Conversion of Lignocellulosic Biomass

9.1 Introduction

The thermochemical platform for conversion of lignocellulosic biomass aims to efficiently produce biofuels and chemicals via processes that use heat and chemistry.

Primary routes for biomass thermal conversion are schematized in **Figure 9.1**. [1-2]

Combustion is the thermal conversion of organic matter with an oxidant (generally oxygen) to produce primarily carbon dioxide and water. The oxidant is in stoichiometric excess, i.e. complete oxidation. Biomass combustion mainly leads to heat energy which can be utilized for steam and/or electricity production. [3]

Gasification is the thermal conversion of organic matter at elevated temperatures and in oxygen-deficient conditions (reducing conditions) to produce primarily permanent gases, with char, water, and condensibles as minor products. The main permanent gases resulting from biomass gasification are the mixture of carbon monoxide and hydrogen known as ***synthesis gas*** or ***syngas***.

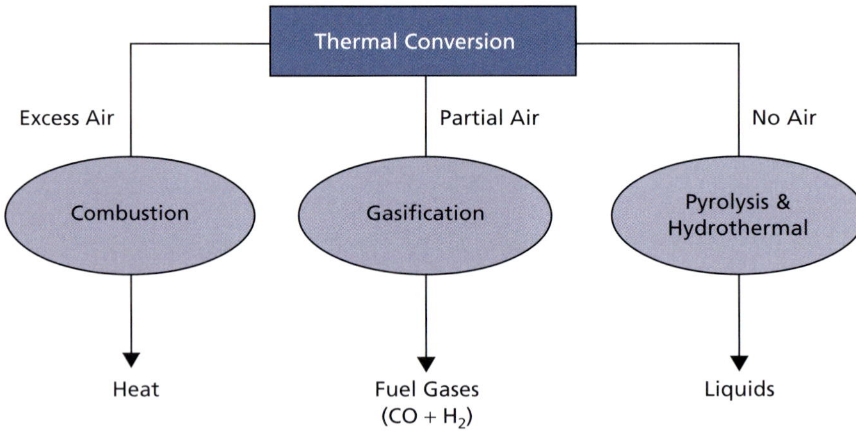

Figure 9.1 Primary routes for biomass thermal conversion. [1]

Pyrolysis is the thermal conversion of organic matter in the absence of oxygen. Biomass pyrolysis commonly refers to lower temperature thermal processes producing primarily liquids, with the possibility of chemical and food byproducts. Pyrolysis produces liquids, known as ***bio-oil***, charcoal (also known as char), and gas. ***Hydrothermal treatments,*** which involve the processing of biomass in subcritical and supercritical water medium, are very similar to pyrolysis, except that it occurs at much higher feedstock moisture contents. [4] They include biomass liquefaction, gasification, and carbonization.

The ***term Biomass-to-Liquids*** (BTL) is applied to synthetic fuels made from biomass through a thermochemical route. [5]

9.2 Processes

9.2.1 Combustion

Use of wood and other forms of biomass as fuels for generating electricity and heat has become a focus of renewed interest in many parts of the world. [6] Biomass, in contrast to fossil fuels, has a unique potential for making a positive environmental impact. Biomass can be burned without emitting large amounts of nitrogen oxides. Moreover, because the sulfur content is very low, emissions of sulfur dioxide will also be low. In addition, using biomass as a fuel would mobilize agricultural residues and forestry residues.

Each type of biomass has specific properties that determine its performance as a fuel in combustion and/or gasification devices. [6] The most important properties relating to the thermal conversion of biomass are:

- moisture content,
- ash content,
- volatile matter content. Elemental composition,
- heating value, and
- bulk density.

Volatile matter refers to the part of biomass that is released when the biomass is heated (up to 400 to 500 °C). [6] During this heating process the biomass decomposes into volatile gases and solid char. Biomass typically has a high volatile matter content (up to 80%), whereas coal has a low volatile matter content (less than 20%). The heating value of a fuel is an indication of the energy chemically bound in the fuel with reference to standardized equipment. In defining the thermal properties of biomass, it is important to note that it consists of water, ash, and ash-free matter and that the proportion of each is critical in evaluating the suitability of biomass as a fuel.

The design of a combustion or gasification system depends heavily on the specific biomass material – its morphology, moisture content, and mix of contaminants. The last will determine which flue-gas cleaning systems will be applied.

Combustion systems based on steam cycles to generate electricity are technically mature and commercially available. [6] The concepts based on fluidized-bed combustion are technically proven.

9.2.2 Gasification

General description

Biomass gasification is a high-temperature process (600–1000 °C) to decompose the complex components of biomass into simpler gaseous molecules, primarily hydrogen, carbon monoxide, and carbon dioxide. [7] In most cases, some ***char*** and ***tars*** are also formed, along with methane, water and other constituents. An overview of the gasification technology is given in **Figure 9.2**. [8]

Pretreatment of the biomass feedstock is generally the first step in gasification. [9] Pretreatment involves drying, pulverizing, and screening or torrefaction (see § 7.4.5). Optimal gasification requires dry fuels of uniform size, with a moisture content not higher than 15–20%. [9]

Hydrogen and carbon monoxide (syngas) are the desired product gases, because, unlike combustion gases, they can be directly fired into a gas turbine for power generation or used in chemical synthesis. An advantage of biomass gasification over biomass combustion is that the power generation efficiency of a gas turbine combined with a steam turbine (***combined cycle gas turbine***) can be as much as twice the efficiency of biomass combustion processes, which use a steam cycle alone. The reason is that by allowing a higher temperature of

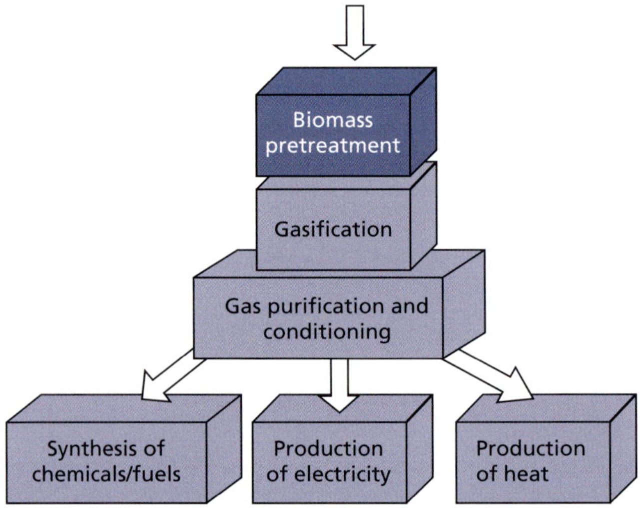

Figure 9.2 Overview of the biomass gasification technology. [8]

the gas in a gas turbine than of the steam in a steam turbine, due to material constraints, the Carnot efficiency is improved. [10]

Biomass gasification can be described as a two-stage process. [11] In the first stage called *pyrolysis*, heat vaporizes the volatile components of biomass in the absence of oxygen at temperatures ranging from 450 to 600 °C. Pyrolysis vapors consist of carbon monoxide, hydrogen, methane, carbon dioxide, volatile tars, and water. The residue, about 10–25% of the original fuel mass is charcoal. [11] The second stage of gasification is called char conversion. This occurs at temperatures of 700–1200 °C. [11] The charcoal residue from the pyrolysis stage reacts with oxygen, producing carbon monoxide.

Gasification of biomass has been achieved through a wide range of designs for the gasification chamber. [7] The design greatly influences the relative proportions of product gases and contaminants, and, therefore, the energy content and potential uses. Gasification with direct heating in the gasification chamber allows a limited supply of air, oxygen, steam, or a combination to serve as a partial oxidizing agent for heat generation. Indirect heating provides gasification through an external heat source, with or without the addition of steam, with minimal oxygen. The addition of steam causes a steam reforming reaction that increases the amount of hydrogen in the product gases, raising the energy content of the product gas.

Biomass gasification produces a combustible mixture of raw gases that vary according to the feedstock and gasification approach. [7] Steam reformed, indirect, and oxygen-fired direct gasification systems produce biobased syngas, a medium energy combustible and reactive mixture rich in hydrogen and carbon monoxide with a heating value in the range of 10–20 MJ/Nm^3. [7, 12] Syngas can be used as auxiliary fuel in boilers, as a fuel for electricity and steam generation via gas turbines or fuel cells, or as a feedstock in the production of fuels and chemicals. Air-fired direct gasification systems produce a low energy, biobased fuel gas with a heating value of 2.5–8.0 MJ/Nm^3. This low energy gas is generally suitable for combustion to produce thermal energy.

The gasification products cannot be easily stored, so the gasification process is integrated with other conversion processes to utilize the output. [7] The product gas must be cleaned of solids, tars, and other contaminants as a function of the intended use. The successful removal of alkali metals from the gasified biomass is the most critical step towards demonstration of pressurized gasifiers, since alkali may condense on the rotor blades of the gas turbine leading to corrosion. [10] For atmospheric gasifiers, the removal of tars is also critical.

Gasification followed by Fischer-Tropsch process

Biomass gasification is typically followed by a ***Fischer-Tropsch*** process or synthesis, which is a set of chemical reactions that convert a mixture of hydrogen and carbon monoxide (syngas) into liquid hydrocarbons. [13] The process, a key component of gas to liquids technology, produces a petroleum substitute typically from coal, natural gas, or biomass for use as fuels, such as gasoline and diesel, and chemicals.

The diagram of the Fischer-Tropsch process is shown in **Figure 9.3**.

The process usually includes the following steps:

- cleaning up syngas following gasification;
- Fischer-Tropsch liquid synthesis using catalysts such as cobalt, zinc, chromium, iron, copper, and molybdenum. Each catalyst is specifically selective for certain liquid fuels and sensitive to different contaminants and poisons; and
- processing that liquid into fuels and chemicals. [15]

Several reactions are required to adjust the H_2/CO ratio of the syngas. Most important is the water-gas-shift reaction (WGS), which provides a source of hydrogen at the expense of carbon monoxide (**Equation 9.1**): [16]

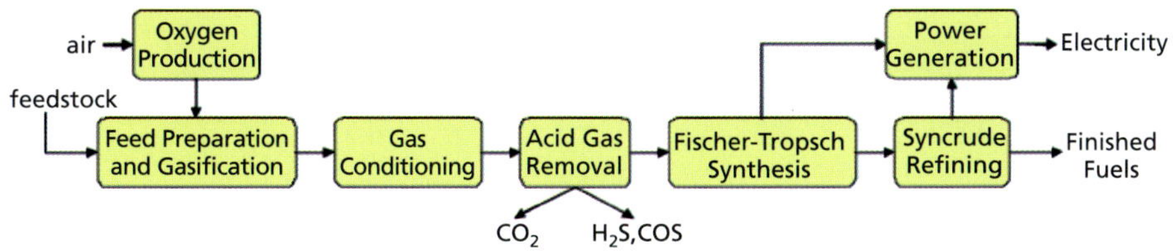

Figure 9.3 Generic Fischer-Tropsch process diagram. [14]

$$H_2O + CO \rightarrow H_2 + CO_2 \tag{9.1}$$

The following chemical equation describes the original Fischer-Tropsch process (**Equation 9.2**):

$$(2n + 1)\ H_2 + n\ CO \rightarrow C_nH_{2n+2} + n\ H_2O \tag{9.2}$$

where n is a positive integer which can be adjusted in function of the conditions of reaction (temperature, pressure, catalyst). [17] In addition to alkane formation, competing reactions result in the formation of alkenes, alcohols, and other oxygenated hydrocarbons.

Mainly straight-chain saturated hydrocarbons from CH_4 to heavy waxes may be produced from the Fischer-Tropsch process. [10] Some olefins are also produced. The reaction mechanism is a polymerization process in which the building block is the CH_2 unit. Due to the ***stepwise*** growth mechanism, the Fischer-Tropsch product distribution generally follows an Anderson-Schulz-Flory distribution (**Figure 9.4**).

The distribution can be expressed as (**Equation 9.3**):

$$W_n/n = (1-\alpha)^2\,\alpha^{n-1} \tag{9.3}$$

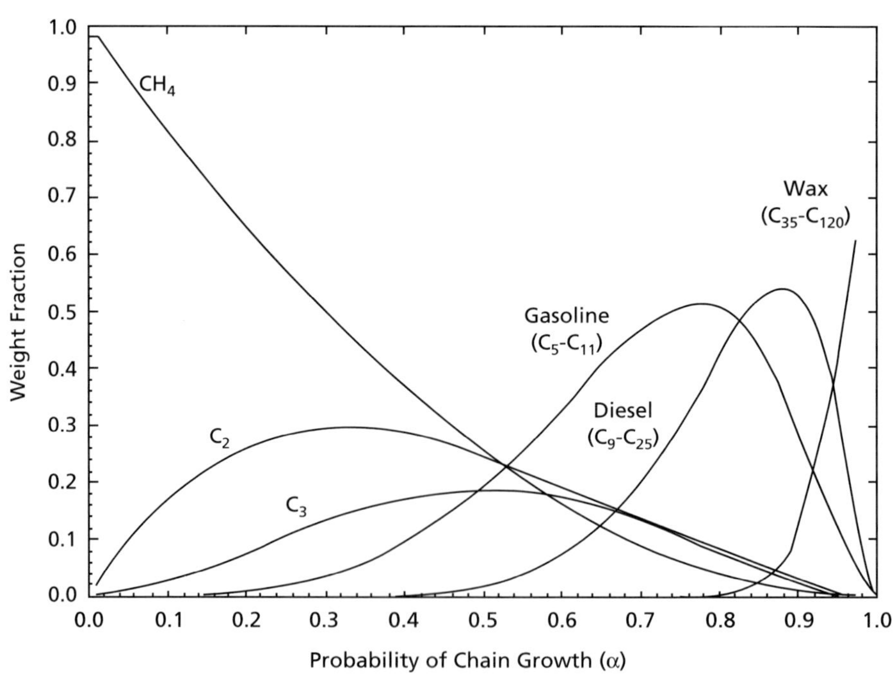

Figure 9.4 The Anderson-Schulz-Flory distribution followed by the Fischer-Tropsch process. The chain growth probability varies with the processing conditions. [18-19]

where α is the chain growth probability, i.e. the probability that a molecule will continue reacting to form a longer chain, and W_n is the weight fraction of hydrocarbon molecules containing n C atoms. [20]

A higher chain growth probability means that the product contains more long-chain hydrocarbons, and hence less CH_4. In general, the chain growth probability is largely determined by the catalyst and the specific process conditions. It is known that the higher the pressure, and the lower the temperature and inlet H_2/CO ratio, the higher the chain growth probability. It is commonly accepted that the highest yield to diesel is achieved by first making waxes, which is then hydrocracked into the diesel fraction (C_{12}-C_{20}). In this way, the formation of shorter by-products is minimized. Consequently, the worldwide Fischer-Tropsch research is currently focused on how to prepare catalysts that give high chain growth probability. Cobalt catalysts seem to be the way forward. [21] Most of the gasification and Fischer-Tropsch research is centered in the United States and with multinational petroleum companies. However, cobalt is a scarce resource and the United States do not have a domestic source available.

To reduce the loss of syngas to the purge, hydrogen can be recovered by pressure-swing adsorption (PSA) before the final gas purge. [22] The PSA technology is based on a physical binding of gas molecules to adsorbent material. [23] An interesting development in the field of hydrogen recovery is the use of industrial membranes. [24] Most hydrogen separation membranes operate on the principle that only hydrogen can penetrate through the membrane because of the inherent properties of the material. [25]

Gasification followed by fermentation

Synthesis gases, consisting of H_2, CO and very often some CO_2, may be produced from biomass according to the approximate **Equation 9.4**: [26]

$$8\ CH_2O\ (\text{biomass}) + O_2 \rightarrow 6\ CO + 2\ CO_2 + 8\ H_2 \qquad (9.4)$$

The components of synthesis gas may be converted into ethanol by certain anaerobic bacteria [27] according to **Equations 9.5** and **9.6**:

$$6\ CO + 3\ H_2O \rightarrow CH_3CH_2OH + 4\ CO_2 \qquad (9.5)$$

$$6\ H_2 + 2\ CO_2 \rightarrow CH_3CH_2OH + 3\ H_2O \qquad (9.6)$$

Overall theoretical (no side-products) **Equation 9.7** may be obtained by combining **Equations 9.5** and **9.6** with **Equation 9.4**:

$$8\ CH_2O + O_2 \rightarrow 7/3\ CH_3CH_2OH + 10/3\ CO_2 + H_2O \qquad (9.7)$$

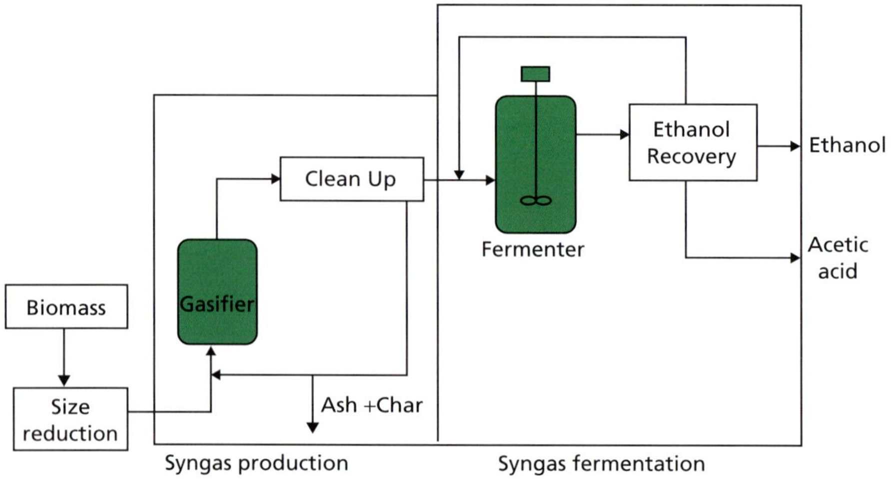

Figure 9.5 Schematic drawing of the process for the production of ethanol from biomass gasification followed by fermentation. [28]

Since nearly all the biomass (including the lignin) can be converted into gas by **Equation 9.4**, yields of ethanol of ~50% of the total biomass are possible (135 gallon/ton). Furthermore, overall combustion energy efficiency in the theoretical case is close to 85%. [28]

The gasification/fermentation process for producing ethanol from biomass is shown in **Figure 9.5**.

After gasification and gas purification, the next process step is fermentation. In this case the cleaned syngas is anaerobically converted into ethanol. Biological conversion of chemicals from syngas offers several advantages over catalytic techniques:

- Biological conversion occurs under mild temperatures and pressures, whereas catalytic reactors are operated at high temperatures and pressures.
- The reaction specificity of enzymes is typically higher than that of inorganic catalysts.
- Most biological catalysts are tolerant to sulphur gases, reducing the cost of gas cleanup prior to the conversion step.
- Biological conversion does not require a set CO/H_2 ratio.

Several acetate-producing ***acetogenic*** bacteria (often called homoacetogenic bacteria [29-30]) are capable of metabolizing syngas into ethanol. Two of the most promising strains are *Butyribacterium methylotrophicum* and *Clostridium ljungdahlii*. [31-32]

Both are gram-positive bacteria, meaning that they have a high amount of peptidoglycan in their cell wall. The *Clostridium* species has currently the most potential for the development of ethanol.

In acetogenic bacteria like *C. ljungdahlii*, the Wood-Ljungdahl pathway, [33] also called the acetyl-CoA pathway, is used to fix CO_2 or CO and convert it into acetyl-CoA. [27] Most of these organisms are able to grow at the expense of H_2 plus CO_2 as the sole energy source. [30] H_2 then serves as the electron donor for CO_2 reduction to acetate. The methyl group of acetate is formed from CO_2 via formate and reduced C1 intermediates bound to tetrahydrofolate. The carboxyl group is derived from CO, which is synthesized from CO_2 by carbon monoxide dehydrogenase. The latter enzyme also catalyzes the formation of acetyl-CoA from the methyl group plus CO. Acetyl-CoA is then converted either to acetate or to other products.

In 2010, Köpke et al. [27] detected genes encoding all respective enzymes in *C. ljungdahlii. C. ljungdahlii* has several systems to either produce ethanol or to use it for growth. Utilization and generation of ethanol can proceed via acetyl-CoA and acetaldehyde using NAD^+-dependent acetaldehyde and ethanol dehydrogenases. *C. ljungdahlii* might use one mechanism for ethanol consumption and the other for ethanol production.

In 2011, Kopke et al. [34] presented evidence of 2,3-butanediol production by three non-pathogenic acetogenic *Clostridium* species – Clostridium autoethanogenum, *C. ljungdahlii* and *C. ragsdalei* – using carbon monoxide-containing industrial waste gases or syngas as sole source of energy.

9.2.3 Pyrolysis

General description

Pyrolysis is the thermal decomposition of organic materials in the absence of oxygen. [35-36] It is important to differentiate pyrolysis from gasification. Gasification decomposes lignocellulosic biomass to syngas by carefully controlling the amount of oxygen present. Biomass pyrolysis gives rise to the production of solid (charcoal), liquid (condensable ***bio-oil***) and gaseous products. [37-38] The older literature generally equates pyrolysis to carbonization, in which the main product is a solid char. Today, the term pyrolysis often refers to processes in which oils are preferred products. [35]

The general changes that occur during pyrolysis can be described as follows: [35]

- Heat transfer from a heat source, to increase the temperature inside the fuel.
- The initiation of primary pyrolysis reactions at this higher temperature releases volatiles and forms char.
- The flow of hot volatiles toward cooler solids results in heat transfer between hot volatiles and cooler unpyrolyzed fuel.

- Condensation of some volatiles in the cooler parts of the fuel, followed by secondary reactions, can produce tar.
- Autocatalytic secondary pyrolysis reactions proceed while primary pyrolytic reactions simultaneously occur.
- Further thermal decomposition, reforming, water gas shift reactions, radical recombination, and dehydrations can also occur, which vary with the process residence time, temperature and pressure.

Fundamental research on ***fast pyrolysis*** (vapor residence time up to 2 s [39]) has shown that high yields of the lower-value solid char from traditional ***slow pyrolysis*** (vapor residence time from 5 to 30 minutes [40]) can be replaced by higher-value fuel gas, fuel oil, or chemicals from fast pyrolysis. [35] Fast pyrolysis is a technology capable of converting lignocellulosic biomass into combustible bio-oil with product yields typically above 70 wt%. [38] Product yields and characteristics depend not only on the properties of the biomass feedstock but also on processing conditions. [38] Production of high bio-oil yields requires very rapid thermal degradation of the biomass feedstock followed by fast removal of the volatile products from the reaction chamber in order to minimize secondary reactions. These conditions are achieved in reactors that allow very efficient heat transfer conditions, very low gas residence times, and a precise control of the reaction temperature. [38]

The biomass feedstock needs to be conditioned in order to reduce moisture (up to 15%) and particle size (~2–3 mm for fluidized bed reactors [41]). [38]

Degradation of biomass components

Biomass pyrolysis products are a complex mixture of the products from the individual pyrolysis of its components.

Cellulose degradation occurs at 240–350 °C to produce levoglucosan (**Figure 9.6**) and anhydrocellulose. [35] Levoglucosan is produced when the glucosan radical is generated without the bridging oxygen from the preceding monomer unit. The C6 hydroxyl moiety transfers a proton to the C1 cation during formation of a 1–6 oxygen bridge.

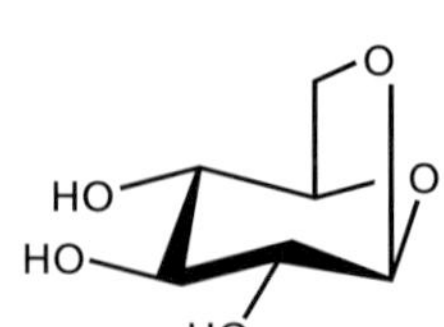

Figure 9.6 Levoglucosan (1,6-anhydro β-D-glucopyranose). [42]

Hemicelluloses decompose at temperatures of 200-260 °C, resulting in more volatiles, less tars and less chars than cellulose. [35] Most hemicelluloses do not yield significant amounts of levoglucosan. Much of the acetic acid liberated from wood pyrolysis is attributed to deacetylation of the hemicelluloses.

Lignin decomposes when heated at 280–500 °C. [35] Lignin pyrolysis yields phenols vie the cleavage of ether and carbon-carbon linkages. Lignin is more difficult to dehydrate than cellulose and hemicelluloses. Lignin produces more residual char than does the pyrolysis of cellulose.

Fast pyrolysis

Pyrolysis processes may be slow (conventional) pyrolysis or fast pyrolysis, depending on the operating conditions used (**Table 9.1**).

The aim of slow pyrolysis is to produce mainly charcoal, whereas fast pyrolysis is intended to convert biomass into a maximum quantity of liquid. [45] Both processes have in common that the energy in the biomass feedstock is concentrated in a smaller volume by which transport costs and storage space can be reduced. Also beneficial is that a more uniform, stable, and cleaner-burning product is obtained that could be used as an intermediate.

The essential characteristics of fast pyrolysis for maximum oil production include very rapid heating of the biomass, operating temperature ~500 °C, and rapid quenching of the vapors produced. [45] Crucial is high rates of heating of the solid biomass particles. The energy required to pyrolyze biomass depends upon the feedstock and the yield of bio-oil. Values from the literature are in the range 0.8–1.5 MJ/kg.

Fast pyrolysis produces energy, liquids, gases and char. [46] Small particles, less than a quarter inch in size, are delivered to a high-heat reactor. The fuel must be small in size to assure high heat transfer rates during the process. Around 500 °C, the material is transformed into a vapor, which in turn is cooled, condensed, and recollected as a liquid bio-oil. The produced bio-oil has a heating value that is about half of conventional fuel oil. [35] Fast pyrolysis processes produce 60–75 wt% of liquid bio-oil, 15–25 wt% of solid char and 10–20 wt% of non condensable gases, depending on the feedstock used. [35] No waste is generated because the bio-oil and char can be used as fuels and the gas can be recycled.

Table 9.1 Main types of pyrolysis. [35-36, 43-44]

Pyrolysis type	Residence time	Heating rate	Temperature (°C)	Main products
Slow	5–30 min	low	400–500	Oil, gas, char
Fast	0.5–5 s	very high	450–650	Bio-oil

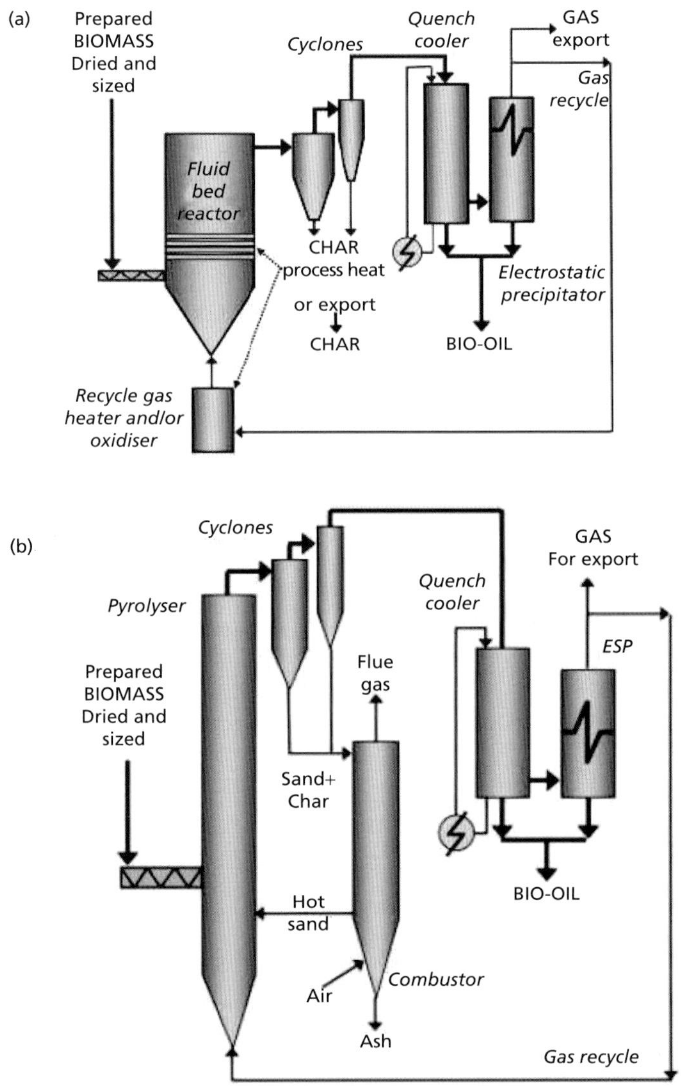

Figure 9.7 Schematics of fluidized beds. (a) Bubbling fluidized beds. (b) Circulating fluidized beds. [41]

Several reactor configurations can be used for fast pyrolysis: fluidized bed, rotating cone and vacuum, among others. [41] Bubbling fluidized bed reactors are the most common design and use sand to provide a large thermal mass and even heat distribution (**Figure 9.7** (a)). Since the rate-limiting step in the process is heat transfer to the biomass particles, small particles are necessary (~2–3 mm). Circulating fluidized beds operate similarly, however the sand and char are circulated, separated, and the clean sand is recycled to the reactor (**Figure 9.7** (b)).

Among the promising developments, catalytic pyrolysis of biomass is a thermochemical conversion process where a heterogeneous catalyst is used as a heat carrier in the pyrolysis reactor for the in-situ upgrading of the quality of the bio-oil. [47] The technology could result in a marketable bio-oil to be finally used as transportation fuel.

Bio-oil

Bio-oils are dark, brown, free-flowing organic liquids that are comprised of highly oxygenated compounds. [35] Synonyms for bio-oil include namely pyrolysis oils, pyrolysis liquids and bio-crude oil.

Bio-oil is formed by rapid heating of biomass and rapid cooling of the vapor fraction. This quenching traps many products that would further react if the residence time was higher. The resulting liquid product is highly variable depending on the type of feedstock and can be made of hundreds of decomposition products of cellulose, hemicelluloses, and lignin. [41] Bio-oil is typically characterized by a high water content (<35%) and a low pH (~2). This oil can be upgraded (i.e. deoxygenated and stabilized) to a more desirable product for use as a transportation fuel.

Bio-oil contains many reactive compounds, which contribute to unusual attributes. [35] Oxygen-containing compounds in bio-oil include water, hydroxyaldehydes, hydroxyketones, sugars, carboxylic acids, and phenolics. [48] Bio-oil can be considered a microemulsion in which the continuous phase is an aqueous solution of holocellulose decomposition products and small molecules from lignin decomposition. [35] The continuous liquid phase stabilizes a discontinuous phase that is largely composed of pyrolytic lignin polymers.

It is possible to manipulate the chemistry of bio-oils by changing the thermal conditions in which they are produced or carrying out the pyrolysis in the presence of catalysts. [49]

Catalytic upgrading of bio-oil and tall-oil

Bio-oil

Untreated bio-oil is a liquid with ~25% water that cannot be easily separated. [50] It is a complex mixture of oxygenated compounds that is unstable in long-term storage and is not miscible with any conventional hydrocarbon-based fuel. Bio-oil can be stabilized and converted to a conventional hydrocarbon fuel by removing the oxygen through hydrotreating typically followed by hydrocracking (**Figure 9.8**). [51-52] Hydrotreating and hydrocracking (hydroprocessing) are commonly employed in the petroleum industry to remove undesired compounds from crude oil and to break large hydrocarbon molecules.

Hydrotreating is an exothermic process commonly used to selectively remove impurities such as sulfur and nitrogen that could affect downstream equipment. Hydrotreating takes place in a hydrogen-rich environment. [52] Typical proc-

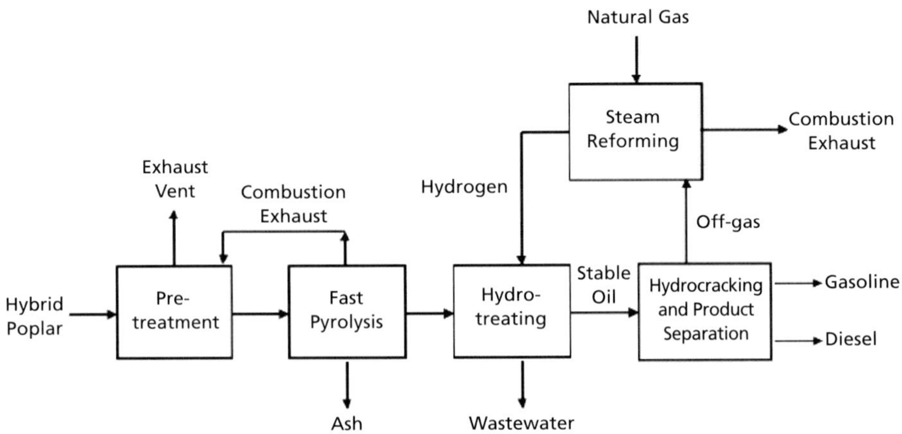

Figure 9.8 Fast pyrolysis followed by hydrotreating/hydrocracking upgrading. [50]

ess conditions for hydrotreating are pressures of 7–10 MPa and temperatures of 300–400 °C using a cobalt-molybdenum catalyst.

Single-stage hydrotreating has proved to be difficult, producing a heavy tar-like product. [50] Dual-stage processing, where mild hydrotreating is followed by more severe hydrotreating has been found to overcome the reactivity of the bio-oil.

Hydrocracking breaks down heavy molecules into shorter chains. [52] For example, heavy hydrocarbons with 30 or more carbon atoms can be split into chains with diesel (C12) or gasoline (C8) range. Process conditions are slightly more severe than for hydrotreating, with pressures of 10–14 MPa and temperatures of 400–450 °C using a nickel-molybdenum catalyst.

Tall-oil

Tall oil, also called liquid rosin or tallol, is a viscous yellow-black odorous liquid obtained as a by-product of the Kraft process of wood pulp manufacture when pulping mainly coniferous trees. [53] Tall oil is the third largest chemical by-product in a Kraft mill after lignin and hemicelluloses. The yield of crude tall oil from the process is in the range of 30–50 kg/tonne pulp. Crude tall oil has three major components: resin acids, fatty acids and unsaponifiables (also called neutral elements). [54] Tall oil, like pyrolytic oils, can be catalytically upgraded through hydrotreatment to transportation fuels and useful chemicals. [55]

In February 2012, the Finnish paper company UPM announced that it will invest in a biorefinery producing biofuels from crude tall oil in Lappeenranta, Finland. [56] The industrial scale investment is the first of its kind globally. The biorefinery will produce annually ~100,000 tonnes of advanced second generation biodiesel for transport. Construction of the biorefinery will begin

Figure 9.9 UPM hydrotreatment process. Source: UPM. [57]

in the summer of 2012 and be completed in 2014. UPM's total investment will amount to approximately EUR 150 million.

UPM's wood-based process includes five major steps (**Figure 9.9**):

- Making crude tall oil: separation from black liquor, and acidulation and decanting.
- Pretreating the crude tall oil: removal of metals, solids and water.
- Hydrotreatment process: through catalysts, heat and pressure, removal of sulfur, nitrogen and oxygen, as well as isomerization and cracking.
- Recycle gas purification.
- Fractionation: distillation of heavier and lighter components, as well as obtention of biodiesel and biogasoline. [57]

Fermentation of bio-oil to ethanol

In the absence of inorganic contaminants, especially alkali and alkaline earth metals, thermal depolymerization of cellulose and hemicelluloses by fast pyrolysis yields both fermentable sugars and anhydrosugars, which can be hydrolyzed to fermentable sugars (**Figure 9.10**). [58]

In many respects, bio-oil fermentation is similar to conventional biochemical production of cellulosic ethanol, except that acid or enzymatic hydrolysis is replaced by pyrolysis for the purpose of depolymerizing biomass carbohydrates. [60]

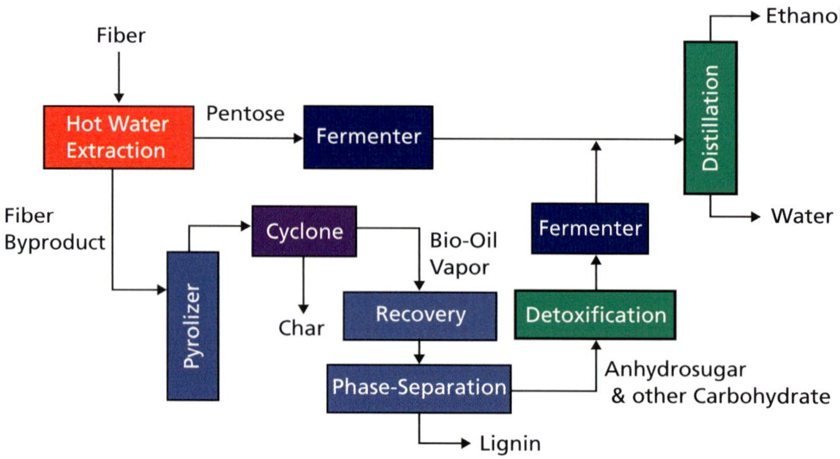

Figure 9.10 Bio-oil fermentation process. [59]

Fermentative utilization of bio-oil faces two formidable challenges. [61] First is the fact that most bio-oil-associated sugars are present in the anhydrous form. Metabolic engineering has enabled utilization of the main anhydrosugar, levoglucosan, in workhorse biocatalysts. The second challenge is the fact that bio-oil is rich in microbial inhibitors. Collection of bio-oil in distinct fractions, detoxification of bio-oil prior to fermentation, and increased robustness of the biocatalyst are effective methods for addressing this inhibition. [62]

9.2.4 Hydrothermal treatments

In nature, it takes millions of years to convert liquid slurry of organic material into hydrocarbon oils, and for this, heat and pressure are required.

Replicating this hydrothermal process in an industrial process means typically: [63]

- Water evaporation is not needed.
- Temperature is high (e.g. 250–400 °C) and pressure is high to maintain water in liquid state (e.g. 30–300 bars) or in a supercritical state (374 °C–221 bars).
- Duration of treatment can be variable depending on the type of fuel desired, such as biocrude oil, gas and coal (e.g. 5 min to several hours).
- Catalysts may eventually be employed to favor the obtention of specific fuel type (e.g. transforming crude oil into methane gas).

In the conditions described above, a lot of different chemical reactions are involved leading to complex pathways and structures. The major reactions involved during hydrothermal processes are hydrolysis, pyrolysis, depolymerization, and decomposition leading to low weight molecules, but also dehydration, polymerization, and condensation leading to aggregates and high weight structures. In these conditions, water plays the role of solvent, catalyst and reactant, and hydrolysis reactions are auto-catalyzed by release of acidic monomers in the reaction media. [64]

All in all, there are several interesting facts showing that hydrothermal technology can be used as an efficient method to treat different types of biomass sources: [65-66]

- Hydrothermal treatment can treat continuously large volumes of organic wastes.
- The process avoids drying/evaporation steps (e.g. compared to flash pyrolysis), giving a much higher energy efficiency.
- Hydrothermal liquefaction produces bio-oils with low residual humidity and high heating value.

Hydrothermal liquefaction

Starting in the 1970s during the first oil crises, research on hydrothermal decomposition and liquefaction of biomass has lead to the development of several pilot plants and industrial initiatives:

- BOM (U.S. Bureau of Mines) process or wood-to-oil process: original process developed in the Albany facility, involving dried and ground wood chips slurried in heavy oil medium. Wood wastes are submitted to liquefaction at 370 °C/ 276 bars with addition of sodium carbonate and carbon monoxide. [67]
- LBL (Lawrence Berkeley Laboratory) process: the second step of improvement of the BOM process, involving equipment changes of Albany facility and drastic change in the process conditions because wood chips are not dried anymore and are broken down into a pumpable water slurry by acid hydrolysis.

The hydrothermal upgrading (HTU) process is the industrial result of the previous developments described above. It is a thermochemical process for the conversion of biomass to biofuel. [68] Originally investigated by Shell Research, it is now being developed by a consortium including Shell Netherlands, Stork Engineers & Contractors B.V., Biofuel B.V., TNO-MEP, and Biomass Technology Group B.V. [69] A demonstration plant due to start in 2007-2008 has never been built. A similar process called TCP (Thermal Conversion Process) has been developed by Changing World Technologies (CWT) to produce renewable diesel from wastes. [70-71]

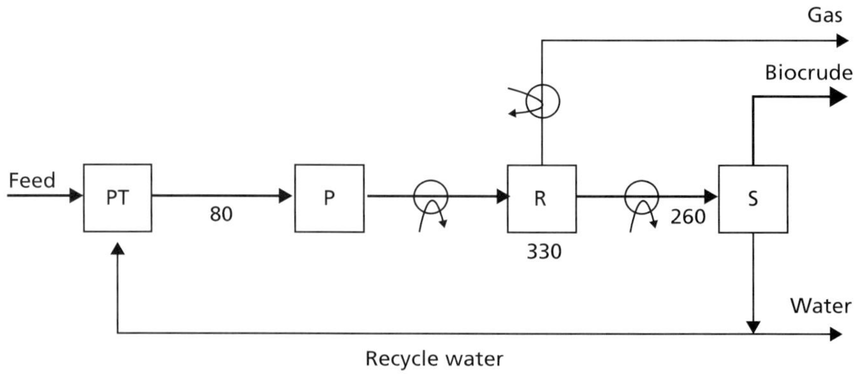

Figure 9.11 Simplified scheme of the hydrothermal upgrading (HTU) process. PT, pretreatment, P, pump; R, reactor; S, separator. Figure: temperatures in °C. [72]

In the HTU process (**Figure 9.11**), biomass is reacted in liquid water at a temperature of 300–350 °C, a pressure of 120–180 bars, and a residence time of 10–30 minutes. [68]

The process chemistry involves deoxygenation by formation of carbon dioxide. The resulting organic product is mainly a liquid referred to as ***biocrude***. The biocrude is a mixture with a wide molecular weight distribution and consists of various kinds of molecules. It does not mix with water and has a high calorific value. It solidifies at ~80 °C. Its heating value and other properties make it attractive for co-firing in coal-fired power stations. Catalytic hydrodeoxygenation (upgrading) has been shown to convert biocrude into hydrocarbons of which the diesel fraction has excellent ignition properties.

The HTU process is capable of converting all types of biomass and especially wet feedstocks such as municipal organic waste, sugar beet pulp, verge grass, with moisture contents of typically 80%.

Hydrothermal gasification

Industries and municipalities generate substantial amounts of biomass as high-moisture waste streams, such as manure, food processing sludges, stillage from ethanol production, and municipal waste-water sludge. [73] Due to the high moisture content, common thermochemical processes are not cost-effective at converting these streams to valuable fuels, chemicals and power.

Gas can be obtained from wet biomass: mainly methane at low temperature range (300–500 °C) and hydrogen at higher ranges (400–600 °C). To get reasonable yields, development of performing solid catalysts is needed, and until now all attempts remain at the laboratory stage. [64]

In the 1990s, the U.S. Department of Energy's Pacific Northwest National Laboratory (PNNL) tried to improve the so-called Thermochemical Environment Energy System (TEES) based on different metals (Ru, Rh, Ni). [74] In 2008, Elliot from PNNL reviewed the laboratory research on catalytic hydrothermal gasification of biomass. [75] Research included liquid-phase, subcritical processing as well as supercritical processing. The U.S. Department of Energy has also financed a project on low-temperature, catalytic hydrothermal gasification (LTCHG) in 2006. [73] Currently research is underway, and the Laboratory for Energy and Materials Cycles at the Paul Scherrer Institut in Switzerland has claimed 70% process efficiency for Ni based catalysts. [76]

Hydrothermal carbonization

In high water content conditions, hydrothermal degradation of biomass is initiated by hydrolysis, which exhibits a lower activation energy than most of the pyrolytic decomposition reactions. Hemicelluloses, lignin, and cellulose decompose at temperatures ranging from 180 to 220 °C. But as time goes on, polysaccharide oligomers and monomers, together with aromatic lignin compounds tend to recombine/aggregate/condensate to form carbonaceous structures, also identified as ***biochar***. The process is mentioned in literature as HTC (hydrothermal ***carbonization***). [63]

Recently, Antonietti and his team [77-78] from the Max Planck Institute of Colloids and Interfaces (Germany) claimed conversion of biomass into carbon and water with 100% carbon efficiency under specific hydrothermal degradation conditions (200 °C during 12 hours with formic acid as catalyst) (**Figure 9.12**).

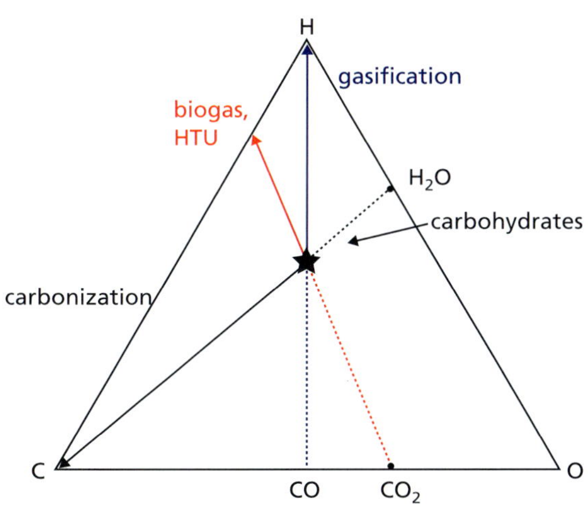

Figure 9.12 Comparison of hydrothermal carbonization (HTC) with other biomass conversion techniques (carbonization, black; gasification, blue; biogas and HTU (hydrothermal upgrading), red). Source: Max Planck Institute. [79]

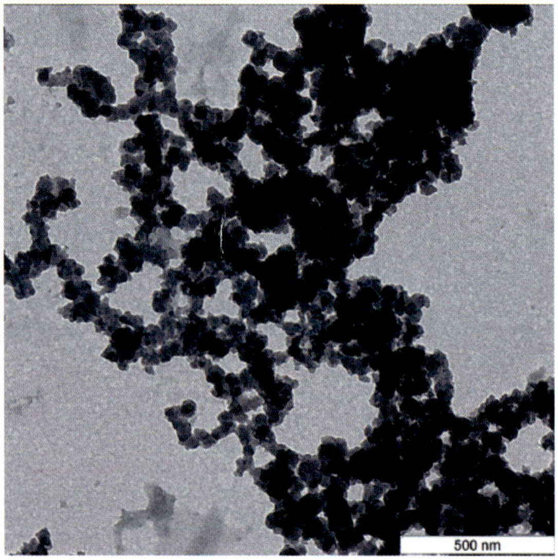

Figure 9.13 Nanospherical carbon particles obtained by HTC from biomass. Source: Max Planck Institute. [79]

For Antonietti and his team, as the reaction does not produce CO_2, while being exothermic, it is an adequate way to generate energy and to create a negative CO_2 balance. Furthermore, in these conditions, were produced specific nanospherical carbon structures potentially useful for many applications such as a new kind of fuel cell (**Figure 9.13**).

Production of biochar already exists as an industrial process for treatment of waste sludge. [80-81]

9.3 Life cycle assessments

9.3.1 Environmental performance of lignocellulosic ethanol production through biochemical or thermochemical conversion

Lignocellulosic biomass can be converted into ethanol through biochemical or thermochemical process. In 2010, Mu et al. [82] completed a life cycle assessment study comparing the environmental performances of the two lignocellulosic ethanol production routes. They compiled material and energy flow data from process simulation models to develop life cycle inventory and compared the fossil fuel consumption, greenhouse gas emissions, and water consumption of both biomass-to-ethanol processes. The biochemical conversion involved hydrolysis and fermentation while the thermochemical conversion involved gasification and catalytic synthesis (**Figure 9.14**).

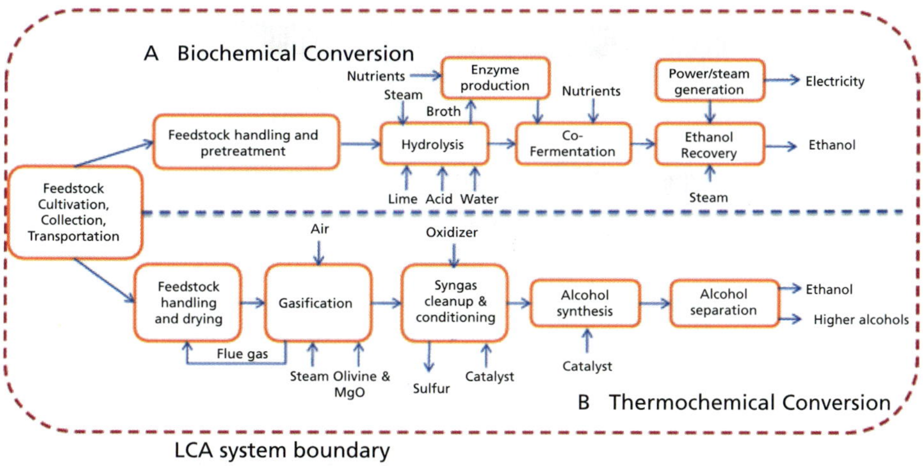

Figure 9.14 Schematic diagram of lignocellulosic ethanol production through biochemical conversion and thermochemical conversion. Copyright © 2010, Springer Science+Business Media, LLC. [82]

Process simulation models suggested that biochemical conversion has slightly better performance on greenhouse gas emission and fossil fuel consumption than thermochemical conversion, but that thermochemical conversion has significantly less direct, indirect, and life cycle water consumption. [82] Also, if the thermochemical plant operates as a biorefinery with mixed alcohol co-products separated for chemicals, it has the potential to achieve better performance than biochemical pathway across all environmental impact categories considered due to higher co-product credits associated with chemicals being displaced.

9.3.2 Economic and environmental comparison of a biochemical and a thermochemical lignocellulosic ethanol processes

In 2009, Foust et al. [83] performed an economic and environmental comparison of a biochemical and a thermochemical lignocellulosic ethanol conversion processes (**Figures 9.15** and **9.16**).

There is now a great interest in how yields, economics, and environmental impacts of the various conversion processes of lignocellulosic biomass to transportation fuels compare. [83] The information regarding these conversion processes is typically very difficult to use on a comparison basis because different underlying assumptions, such as feedstock costs, plant size, co-product credits, or assumed state of technology, have been utilized. In this study, a rigorous comparison of different lignocellulosic

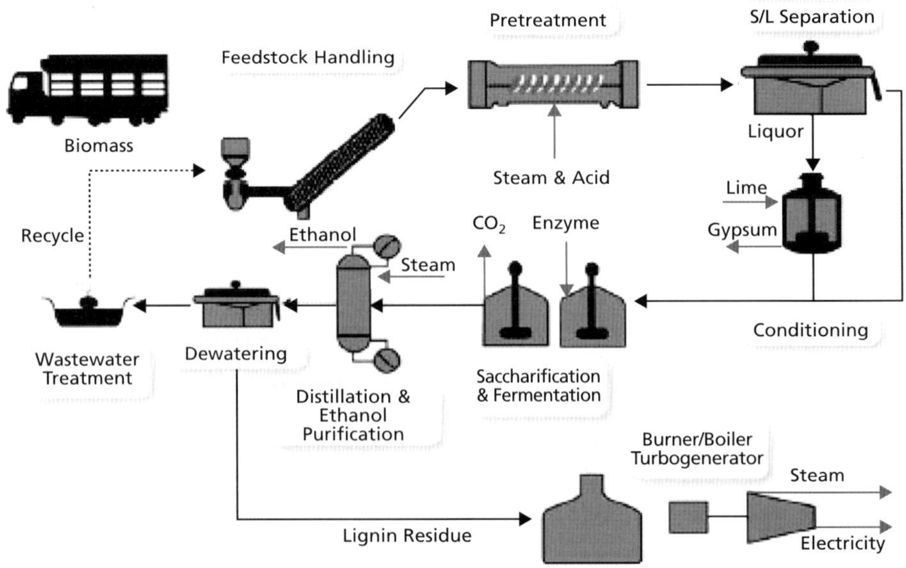

Figure 9.15 Biochemical conversion process. Copyright © 2009, Springer Science+Business Media B.V. [83]

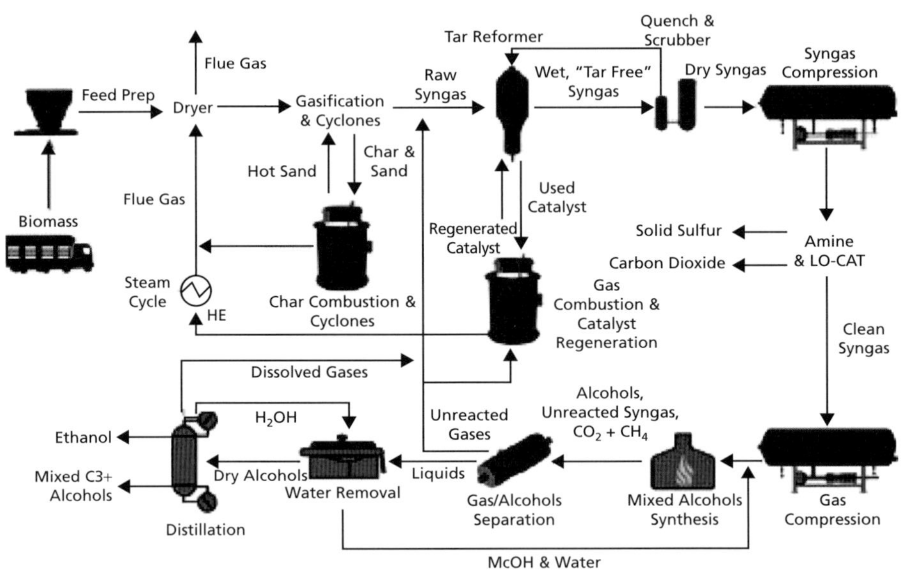

Figure 9.16 Thermochemical conversion process. Copyright © 2009, Springer Science+Business Media B.V. [83]

ethanol conversion processes was performed with standard underlying economic and environmental assumptions so that exact comparisons can be made. The key finding of this study is that although the biochemical and thermochemical processes analyzed have their individual strenghs and weaknesses, the two processes have very comparable yields, economics and environmental impacts.

9.4 Production of fuels and/or chemicals: case studies

9.4.1 Range Fuels

The Wall Street Journal recently published an editorial called "The Range Fuels Fiasco," [84] which gave rise to a political affair in the USA. [85] Range Fuels used a so-called innovative and proprietary two-step thermochemical process to convert biomass, such as wood chips, switchgrass, corn stover, sugarcane bagasse, and olive pits to cellulosic biofuels and clean, renewable power. Promising production of 100 million gallons of ethanol from wood chips each year, the company won the 2008 North American Fuels Technology Innovation Green Excellence of the Year Award. It also got several hundred million dollars from private and public investors (e.g. \$76 million grant from the US Department of Energy in 2007, as well as an \$80 million loan guarantee from the U.S. Department of Agriculture in 2010). Facing "technical problems" and requiring more money, the plant closed at the beginning of 2011 after barely producing one batch of cellulosic ethanol, and the DOE suspended federal support for the company's operations and construction. [86]

Even if failures and drawbacks are inevitable in the history of a nascent technology, experiences like the "The Range Fuels Fiasco" have been shaking the basement of the lignocellulosic ethanol and second-generation biofuels development in the USA.

The Range Fuels process was a thermochemical conversion process, being able to treat different types of biomass. In the first step of the process, heat, pressure, and steam are used to convert the non-food biomass to syngas. Excess energy in this step is recovered and used to generate clean, renewable power. In the second step, the cleaned syngas is passed over a proprietary catalyst and transformed into cellulosic biofuels, which can then be separated and processed to yield a variety of low carbon biofuels, including cellulosic ethanol and methanol.

9.4.2 Choren

Recently, the petroleum company Shell sold stake in the German company Choren. In 2007, Choren was one of the world's leading gasification technology companies for lignocellulosic biomass feedstock. Its gasification/Fischer-Tropsch technology for production of second-generation fuels was announced to be one of the first to be industrialized. Abandoned by Shell in 2009, Choren has begun bankruptcy proceedings in 2011. [87]

Since 2003, Choren was operating a Shell Fischer-Tropsch synthesis plant in connection with a biomass gasification unit for the production of diesel. [88] In this plant, the diesel fraction was obtained by distillation of the Fischer-Tropsch raw product. Carbo-V is the name of Choren's patented biomass gasification process that is used to produce their diesel under the brand name Sundiesel®. The Carbo-V process is a three-stage gasification process involving a low temperature gasification, a high temperature gasification, and an endothermic entrained bed gasification. [89-90]

9.4.3. Coskata

Coskata, based in Illinois, USA, is a biology-based renewable energy company whose low-cost platform technology allows for the production of fuels and chemicals from a variety of input material (including biomass, agricultural and municipal wastes, and other carbonaceous material). [91]

Like all companies involved in second-generation biofuels, Coskata is eagerly looking for public loans to help scaling-up technology and raise capital for its projects. In that aspect, 2011 was definitely a good year for the company that was notified by the US Department of Agriculture (USDA) of their intent to provide a $250 million loan guarantee in support of a commercial biorefinery that will utilize Coskata's technology. [92] In fact, Coskata grabs biggest slice of USDA's $405M in loan guarantees for biofuel biorefineries before Enerkem ($80M) and Ineos Bio ($75M). [93]

Coskata's conversion process produces FlexEthanol™, or flexible feedstock ethanol. It includes three major steps:

- Incoming material converted to synthesis gas.
- Fermentation of synthesis gas into ethanol.
- Separation and recovery of ethanol. [94]

Coskata has already demonstrated the efficiency of their pilot plant located in Madison (Pensylvannia) by successfully converting wood biomass and municipal solid waste into fuel-grade ethanol. The process leverages proprietary microorganisms and patented bioreactor designs in a unique conversion process that can produce fuel-grade ethanol from virtually any carbon-based feedstock and producing ~100 gallons of fuel from each dry ton of biomass. [95] The objective targeted by Coskata was to produce commercial grade of bioethanol for less than $1.00 per gallon, and the planned capacity was 55 million gallons per year. [93, 96]

In July 2012, Coskata announced that they were abandoning their $100 million IPO (initial public offering) along with a fundamental shift in company strategy. No longer will their immediate plans involve the conversion of wood into ethanol, but instead they will focus on natural gas to ethanol. [97]

9.4.4 Shell

Shell first became involved in gasification technology when the Shell gasification Process (SGP) units were commissioned in the 1950s to produce syngas using gaseous and liquid feedstocks, such as refinery residual oil. [98] In 1972, Shell began developing a gasification process for coal (Shell Coal Gasification Process – SCGP).

Shell coal gasification technology uses a dry-feed, pressurized, entrained-flow, slagging gasifier that operates on a wide variety of feedstocks. Dried, pulverized coal is fed to the gasifier through pressurized lock hoppers using a transport gas (syngas or nitrogen). Preheated 95% pure oxygen and steam (as a moderator) are mixed and fed to the injector. The coal reacts with oxygen at typical conditions of 1480–1600 °C and 24–41 bars to produce syngas consisting of only small amounts of CO_2 and no hydrocarbon liquids or gases.

The Shell technology uses a refractory-lined reactor vessel, equipped with an inner membrane wall consisting of circulating water/steam-filled tubes (**Figure 9.17**).

Entrained flow gasification, which is commercially available on large scale (mainly for coal and liquid fuels), can also reach the highest efficiency from biomass to syngas. [99] The benefit of this process is that tars, that are formed when biomass is gasified at temperatures around 900 °C, do not form, since the process operates ~400 °C higher, where tars are broken down into CO, H_2, CO_2 and H_2O. This avoids an expensive gas cleaning section.

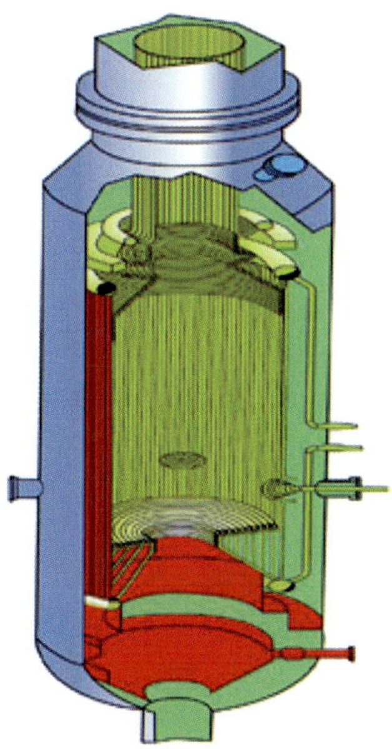

Figure 9.17 Shell entrained flow gasifier. [98]

9.4.5 INEOS

Switzerland-based INEOS is a leading petrochemical company comprising 15 businesses each with major chemical company heritage. Its production network spans 60 manufacturing facilities in 13 countries throughout the world. [100] INEOS Bio is one of the global businesses in INEOS. INEOS Bio has developed a highly innovative thermochemical and biochemical technology for the production of advanced biofuel and renewable energy from a wide range of low-cost carbon materials. [101-102]

The INEOS Bio Ethanol technology converts a wide range of low cost organic materials, including household and commercial wastes, into ethanol for use as a renewable road transport fuel or petrochemical intermediate. [103] The process is a combined thermochemical and biochemical process and includes as steps: gasification, fermentation and purification (**Figure 9.18**).

At the heart of the process are INEOS Bio's patented bacteria. The bacteria are highly selective to ethanol, are inexpensive compared to conventional catalysts, continually regenerate themselves, are tolerant to most impurities, and operate at low temperatures and pressures.

In October 2012, INEOS Bio announced that its joint venture project, INEOS New Planet Energy (INPE) is now producing renewable power using INEOS Bio's technology. The facility located in Florida is producing renewable power for the facility and for export to the local community. [105] At full production, the Center is expected to produce 8 million gallons of cellulosic ethanol and 6 megawatts of renewable power using biomass including yard, vegetative and agricultural wastes.

9.4.6 Fulcrum Bioenergy

Fulcrum Bioenergy, based in California, USA, converts municipal solid waste (MSW) into ethanol. [106] The company utilizes sorted, post-recycled MSW and converts it into ethanol using a two-step process that consists of an innovative gasification system followed by Fulcrum's proprietary alcohol synthesis process. In the fist step, the MSW is converted into syngas. In the second step, the syngas is catalytically converted into ethanol (**Figure 9.19**).

For the gasification step, Fulcrum worked with InEnTech to combine two gasification technologies into a single energy-efficient process. For the alcohol

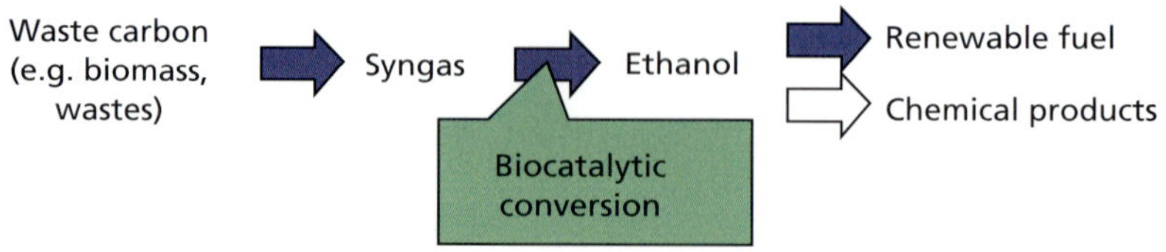

Figure 9.18 Simplified INEOS Bio Ethanol technology process. [104]

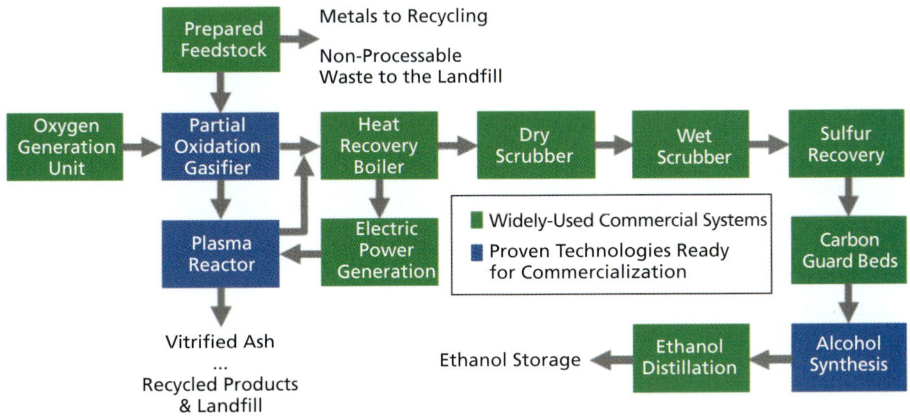

Figure 9.19 Fulcrum's process overview. [107]

synthesis step, the company worked with Saskachewan Council Research (SRC) and Nipawin Biomass Ethanol New Generation Cooperative Ltd (Nipawin) to integrate their thermochemical catalyst into its proprietary process. [108] Fulcrum plans to construct its first commercial-scale ethanol production facility in Nevada, the Sierra BioFuels Plant. At this facility, they expect to produce ~10 million gallons of ethanol per year at an unsubsidized cash operating cost of less than $1.3 per gallon, net of the sale of co-products such as renewable energy credits. [109-110] In August 2012, the US Department of Agriculture awarded Fulcrum's Sierra BioFuels plant a conditional commitment for a $105 million loan guarantee. The plant will produce ethanol as well as renewable energy to be used at the plant utilizing processed municipal waste – household garbage – as a feedstock.

9.4.7 Changing World Technologies

Changing World Technologies (CWT), based in New York, USA, has developed a hydrous pyrolysis process, called Thermal Conversion Process (TCP), to produce diesel from wastes. [70] TCP mimics the earth natural geothermal by using water, heat and pressure to transform organic and inorganic wastes into oil, gases, carbons, metals and ashes. Even heavy metals are transformed into harmless oxides. [111] CWT uses a two-stage process: at first a hydrous pyrolysis at 200-300 °C under pressure followed by a cracking of the pyrolysis oil into light hydrocarbons at ~500 °C. [112-113]

Renewable Environmental Solutions (RES), a CWT's subsidiary, was formed in 2000 to commercialize the TCP in the agricultural and food processing industries. [114-115] The first TCP plant, located in Carthage, Missouri, is now since 2004 an operational facility producing a domestic supply of renewable diesel from agricultural and livestock waste streams. The RES plant can now process up to 300 tons/day of waste producing 15 million gallons/year of Renewable Diesel Fuel Oil (RDO). [116-117]

In 2011, the US Environmental Protection Agency (EPA) has designated CWT's Renewable Diesel Oil, currently produced in its Missouri facility, as a biomass-based diesel and Advanced Biofuel under the EPA's Renewable Fuel Standards Program (RFS). [116, 118]

9.4.8 InEnTec

InEnTec, based in Oregon, USA, is the world leader in advanced systems for waste gasification. [119] Its patented Plasma Enhanced Melter® (PEM®) technology can transform virtually any waste into clean energy products for transportation fuels, electricity generation, and industrial products. PEM® has its origin in many decades of work in the development of two technologies – plasma technology and glass melter technology. [120]

The PEM® system (**Figure 9.20**) uses heating from electrically conducting gas (a plasma) to convert waste feeds into valuable products including energy (ultra-clean power generation or transportation fuels production, using the syngas) and industrial materials (chemicals from the syngas, roofing tiles, insulating panels, sand-blasting media, and other construction-related from the glass; and recoverable metals). [121]

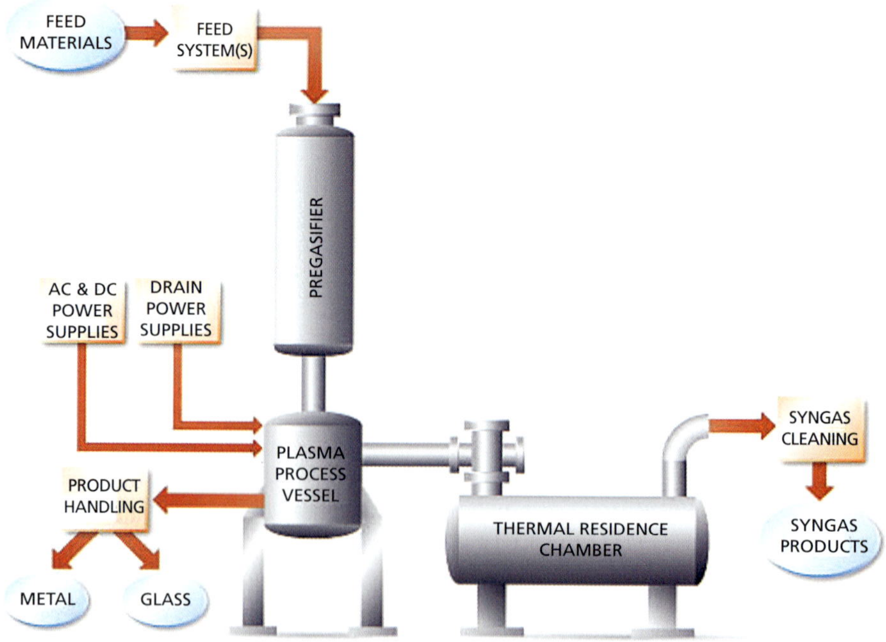

Figure 9.20 InEnTec's Plasma Enhanced Melter® technology. [121]

The PEM® process includes: (1) a feed system; (2) a pregasifier in which a majority of the organic fraction of the feedstock is converted into syngas; (3) a plasma process vessel in which the remaining feed materials are dropped onto a molten glass surface near the plasma-arc zone, resulting in the gasification of the remaining organic materials and in the incorporation of the inorganic components into the molten glass bath; and (4) a thermal residence chamber in which syngas streams from the pregasifier and the plasma process chamber are ducted. [121]

In March 2010, InEnTec formed a joint venture with Waste Management Inc. called S4 Energy solutions to build and operate plasma gasification facilities using InEntec technology including a project at the Columbia Ridge Landfill in Oregon. [122] The Oregon facility now converts municipal solid waste into syngas. [123] In addition, other higher-value materials may be processed in the system. InEntec also constructed a facility on Dow Corning's property in Michigan. The facility processes industrial byproducts from Dow Corning operations and converts them into hydrochloric acid and syngas. The facility has been in commercial operation since 2010.

9.4.9 Green Power

Green Power Inc. is based in Washington, USA. Its process is based on a simple single step process known as catalytic depolymerization (CDP). [124] The shredded and dried feed waste material is filled together with a catalyst and a neutralizing agent into a reaction vessel. [125] Heating to ~350 °C and heavy agitation starts the cracking process resulting in a mixture of oily hydrocarbons which, after reaching the required molecular size, turn into vapor which is conveyed into the distillation or rather condensation column. By fractional distillation, the various hydrocarbon fractions are separated into naphta, kerosene, diesel, and heavy oil.

In 2008, Green Power completed its first prototype plant that is capable of capable of processing 100 tons of municipal and other waste per day in a low heat and low pressure, proprietary catalytic system (**Figure 9.21**). [125-127]

9.4.10 AVA-CO_2

In October 2010, the Swiss company AVA-CO_2 launched the world's first industrial-size HTC (HydroThermal Carbonization) plant with an annual capacity to process 8,400 tonnes of biomass. [81, 128] Organic (plant) matter in an aqueous suspension is transformed into a lignite-like product at temperatures of 180–220° C and under increased pressure. The resulting CO_2-neutral biocoal can be burnt or be employed in the standard industrial applications for lignite. The AVA cleancoal® has high ash melting point and excellent combustion characteristics. The process transforms all carbon from the vegetable

Figure 9.21 Picture of Green Power's plant being utilized for processing. Source: Green Power [125]

waste to coal (theoretical carbon efficiency of 100% but practical carbon efficiency of 90% expected in industrial applications) (**Figure 9.22**). No methane is released in the process but there is a small amount of CO_2 that is lost in the process (2–3%). [129] In addition to cleancoal®, the process can also produce biochar. As a component in soil improvement products, biochar contributes to the sustainable improvement of arid and depleted soils. Furthermore, AVA biochar binds CO_2 in the soil, ensuring long-term removal of greenhouse gases from the atmosphere. Hydrothermal carbonization transforms biomass into CO_2-neutral coal in an anaerobic chemical process that splits hydroxide anions (HO^-) and hydrogen cations (H^+) off the organic molecules, releasing them as water (H_2O). The dehydrated organic molecules combine to a multitude of different carbon compounds. This process is very similar to that of natural carbonization. As water is a highly stable molecule, the removal of each water molecule from the organic molecules releases energy. This means that HTC is a highly efficient exothermic process. [130]

As a simplification, biomass rich in cellulose and hemicelluloses and poor in lignin is the ideal feedstock. [131] Biomass types such as grassy materials, spent grains, waste coffee grounds, sugar beets and cereal husks are considered very well suited. Wood chips, bark, saw dust, and other woody waste make difficult feedstock because of their high lignin content.

AVA-CO_2's HTC technology uses a multi-batch process. [132] In comparison to continuous procedures, this system is simpler and thus shows greater reliability. The plants are built in a modular way that enables maxi-

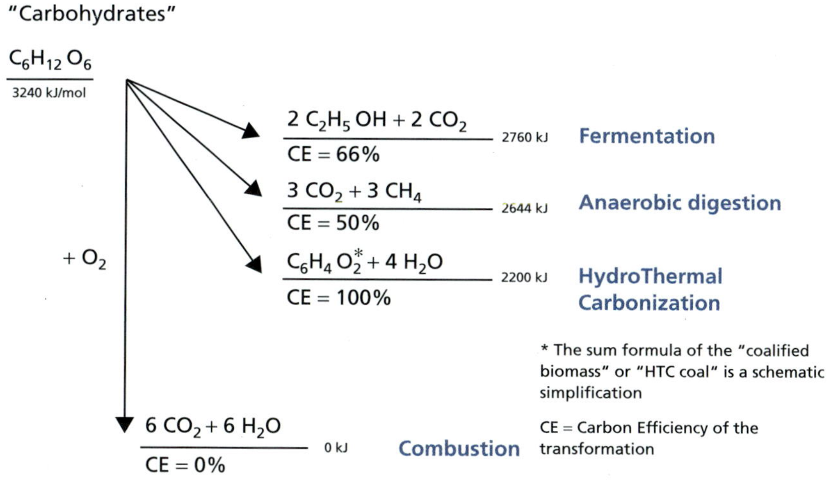

Figure 9.22 Theoretical carbon efficiency for fermentation, anaerobic digestion, HTC and combustion. [130]

mum flexibility. The modular system also allows projects to be implemented into existing infrastructures.

In October 2012, the first commercial HTC-Plant started test operations in Relzow, Germany. [133-134]

9.4.11. Enerkem

Enerkem is a Montreal, Canada-based company producing fuels and chemicals from waste. [135] Based on syngas production, cleaning and conditioning, the Enerkem process uses catalytic reactors to convert syngas into alcohol. In Enerkem's process, the production of methanol is required for obtaining final cellulosic ethanol. Therefore, Enerkem can either sell its methanol as an end-product, or use it as an intermediate to make other products. [136]

The process can use municipal solid waste (MSW) from different sources as well as a broad range of residues. It uses relatively low temperatures and pressures, which reduces energy requirements and costs. [136]

Enerkem has two main projects of waste-to-ethanol commercial plants. The first one in Edmonton (Canada) will initially produce 9.5 million gallons of ethanol per year from 100,000 tons of sorted municipal solid waste. [137] The second project is located in Pontotoc (Mississippi, USA). Funded by USDA, it plans to produce 10 million gallons of ethanol per year. [138]

Enerkem operates two facilities dedicated to innovation in Sherbrooke (Quebec) and Edmonton (Alberta, Canada). Their mission is to develop renewable chemical intermediates, such as alcohols and acrylates, which Enerkem plans later to commercialize (**Figure 9.23**).

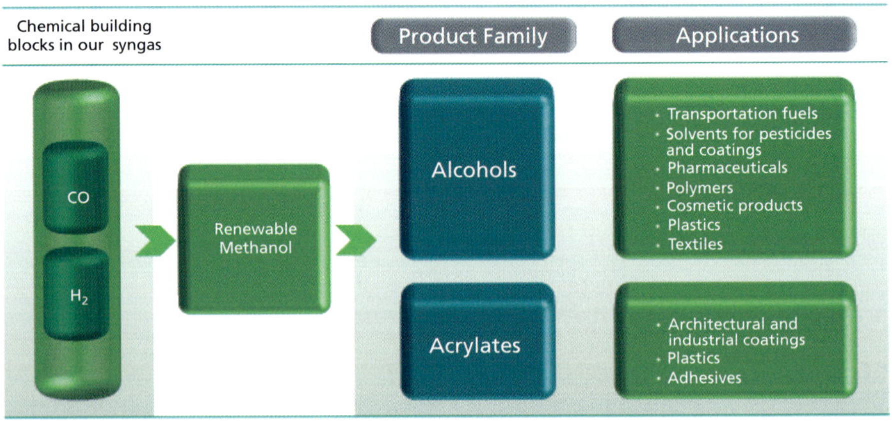

Figure 9.23 Enerkem's technology platform. Source: Enerkem [139]

With its unique technology platform, Enerkem has the ability to expand its product offering beyond cellulosic ethanol. By converting waste into chemical grade synthesis gas, and then methanol, Enerkem has two chemical building blocks from which it can produce its current and future products.

In February 2012, Enerkem and Ethanol Greenfields (Canada) announced Quebec's first waste-to-biofuels production facility. [140] The full-scale commercial cellulosic ethanol plant will be located in Varennes (Quebec) and will use Enerkem's proprietary technology to convert non-recyclable municipal solid waste into biofuels. The anticipated annual production capacity of this plant is ~38 million liters (10 million gallons). With a full-scale waste-to-fuels facility under construction in Edmonton (Canada) and another one under development in Mississippi, the Varennes facility represents Enerkem's third full-scale commercial project. In June 2012, Enerkem began ethanol production from waste at its demonstration facility in Westbury, Quebec. [141]

9.4.12 Ensyn

Ensyn Corporation, based in Delaware, USA, is a producer of renewable liquid fuels and renewable chemical products. [142] Ensyn's RTP (Rapid Thermal Process) technology is a processing system that uses heat to thermally crack carbon-based feedstocks such as wood biomass into high yields of a higher-value liquid product (**Figure 9.24**). [143] With RTP, conversion typically takes place in less than two seconds.

The RTP pyrolysis process is based on the application of a hot transported bed (typically sand) that is circulating between two vessels. Feedstocks, such as wood residues, are subjected to fast, thorough contact with the hot sand for under a few seconds, resulting in the thermal cracking of the feedstock to gases

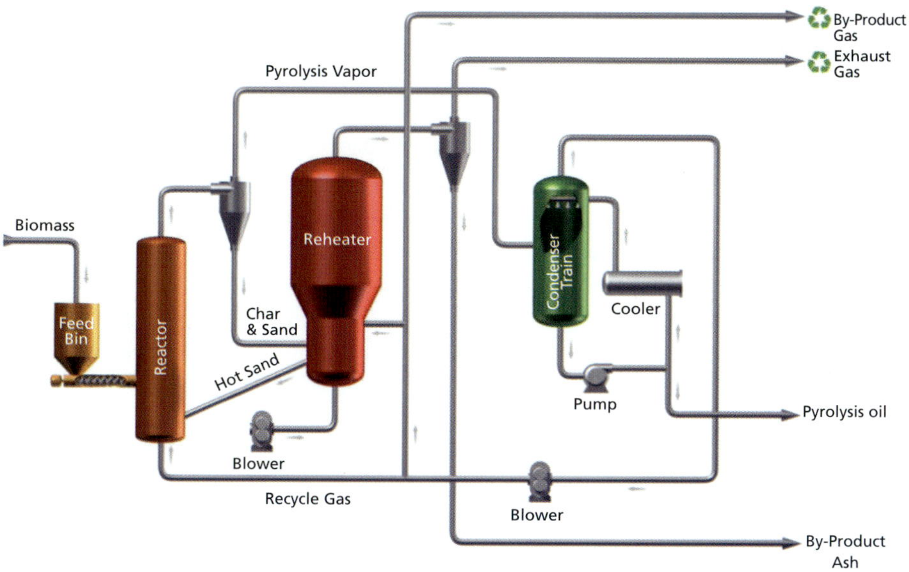

Figure 9.24 Ensyn's RTP flow diagram (from left to right, colored devices: feed bin, reactor, reheater, condenser train). Source: Ensyn [144]

and vapors. Product vapors are rapidly quenched, or cooled, and recovered as a light liquid product. RTP represents an analogue to Fluid Catalytic Cracking, a very common process used in most refineries.

Using RTP Renewable Fuel Oil (RFO) for heat generation represents the most basic application of Ensyn's energy and fuels business. [145] Ensyn is now in the process of development of stand-alone RFO facilities to satisfy the growing demand for renewable liquid heating fuels. Ensyn's RTP also represents significant advantages for the generation of electricity from solid biomass. [146]

Ensyn's RTP liquids can be converted to transportation fuels, such as gasoline, diesel and jet fuel that are indistinguishable from their petroleum-based counterparts. [147] Ensyn is carrying out its transportation fuels program with two strategic relationships, UOP LLC, a Honeywell company, and Chevron Technologies Ventures, a division of Chevron USA, Inc. The approach selected by Ensyn and its strategic collaborators for their transport fuels initiatives with Ensyn is a two-step process:

- First, non-food biomass is converted to high yields of liquids using Ensyn's non-catalytic RTP process.
- Second, RTP liquids are upgraded to drop-in transportation fuels using conventional technologies used in petroleum refining.

Upgrading of Ensyn's RTP liquids has been demonstrated in pilot facilities, and the focus has been to select the optimal technologies for full-scale commercial operations. The Renfrew RTP facility, 100% owned by Ensyn, is now Ensyn's Commercial Merchant Facility for fuels production. [148] The Rentrew facility, which was commissioned in 2007, is located in Ontario, Canada. Ensyn produces now commercial quantities of RTP fuels at this facility. The principal RTP unit at Renfrew has a nominal processing capacity of 75 dry tonnes (150 tonnes per day of green material) per day. However, it routinely operates at capacities in excess of its nominal design rate, with a maximum demonstrated processing capacity of 100 dry tonnes per day.

Ensyn initiated commercial RTP operations in the food industry in 1989, making RTP liquids for the production of food and flavoring ingredients in conjunction with Red Arrows Products Company, LLC, of Wisconsin. [149] Since 1989, Ensyn has designed, built and delivered multiple commercial RTP facilities to Red Arrow, four of which are still in operation.

In the 1990s, following the commercial success of RTP food ingredients, Ensyn initiated a development program to identify, develop and commercialize numerous additional chemical products that could be sourced from its RTP liquids. [150] Ensyn identified a number of resins and crosslinkers in RTP liquids that were effective natural replacements for resins manufactured from fossil fuels. Its first commercial resin was a natural resin (NR), a highly reactive ligninic product that is a substitute for phenol in some resins, and is used especially as an adhesive in the manufacture of man-made wood products. [151] The natural resin is obtained from a ligninic fraction of the RTP liquid. Additional natural chemical products have been identified and tested by third parties.

9.4.13 Avello Bioenergy

Based in Iowa, USA, Avello Bioenergy is commercializing proprietary technology in the evolving field of biomass fast pyrolysis. [152] Avello brings a novel and innovative approach to pyrolysis oil fractionation to the industry. The company has transformed fast pyrolysis into a versatile technology platform producing a range of products from biomass (**Figure 9.25**). The platform provides multiple petroleum replacement products for use in existing and emerging markets.

Avello™FRAC is the Avello process/biorefinery that converts biomass into several unique pyrolysis oil fractions and biochar using innovative pyrolysis technology. The process includes proprietary technology licensed from Iowa State University to produce pyrolysis oil fractions with superior qualities when compared to conventional bio-oil. [154]

Avello™FRAC produces petroleum replacement products for material, energy, soil and chemical applications: [152]

- Bioasphalt™, which is a renewable additive or replacement for petroleum based asphalt binders in paving and roofing material applications.

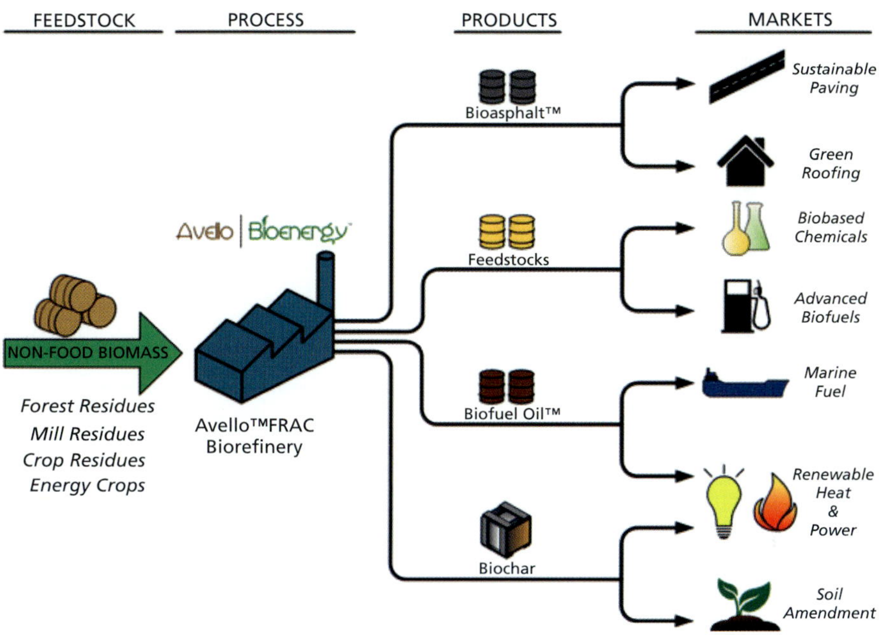

Figure 9.25 Avello Bioenergy technology platform. Source: Avello Bioenergy. [153]

- Chemical feedstocks, of which potential markets include bulk and specialty chemicals, lignin-based chemicals, phenolic resins, adhesives, antioxidants, flavour and flagrances. The properties of Avello™FRAC fractions show potential for easier upgrading to drop-in biofuels and fuel additives.
- Biofuel oil, which is a stable, low carbon, renewable liquid fuel for direct fossil fuel replacement or blending applications.
- Biochar, which will be marketed as a soil amendment, fuel source for heat and power and as a potential carbon sequestration agent.

In 2012, Avello Bioenergy operates a pilot plant at Iowa State University. [155] The facility can process up to a quarter ton of biomass per day. Avello has also been approved by the Iowa power fund for $2.5 million grant to begin building a 2.5 ton/day (2.3 tonne/day) demonstration unit early in 2012.

9.4.14 Anellotech

Anellotech, Inc., based in New York City, USA, has developed a clean technology platform called catalytic pyrolysis for inexpensively producing petrochemicals and transportation fuels from renewable non-food biomass. [156] The inexpensive production is achieved by performing all chemical conversions in one

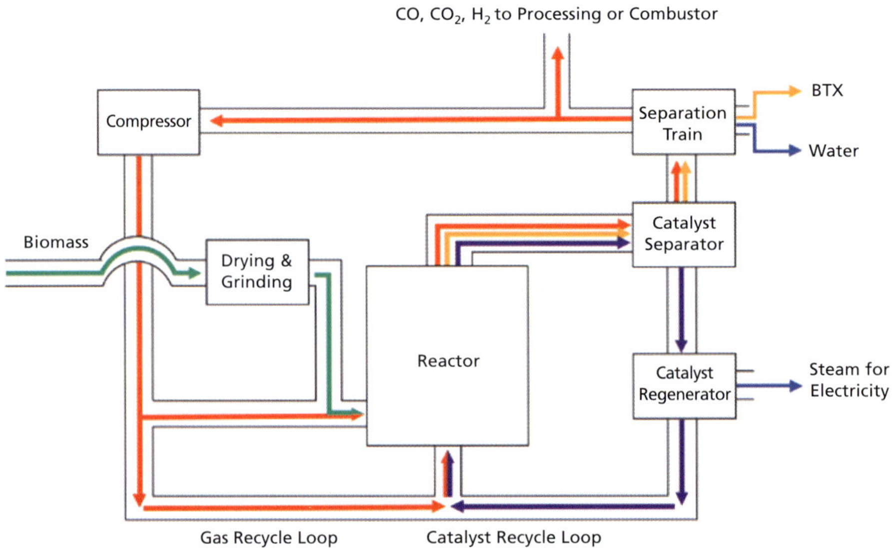

Figure 9.26 Anellotech process. Source: Anellotech. [157]

reactor, using an inexpensive catalyst. The first application of the technology, Biomass to Aromatics™ (BTA) will produce green benzene, toluene and xylenes (BTX). Anellotech will own and operate its own plants, license technology and sell proprietary catalysts to licensees.

Biomass to Aromatics is based on basic science research by Prof. George Huber's research lab at the University of Massachusetts, Amherst. [157] A basic schematic of this process is shown in **Figure 9.26**.

The biomass, such as wood waste, corn stover and sugar cane bagasse, is first dried and ground before injection into a fluidized bed reactor in the presence of an inexpensive zeolite catalyst. In this one brief step, biomass is rapidly heated without oxygen and the resulting gases are immediately catalytically converted into aromatic hydrocarbons. The resulting BTX mixture can be sold to petrochemical companies for processing in existing separation units, or distilled by Anellotech and sold directly into the market. Other than the reactor, regenerator and the catalyst, the process equipment is standard items. [158-159]

The patented process can currently create 50 gallons of BTX per tonne of biomass. [160] Anellotech aims to produce 85 gal/tonne in the future. In 2010, a pilot plant on the University of Massachusetts Amherst campus was operational. [161]

BTX is a $100 billion global market historically growing at rates above global GDP. These are three of the seven primary chemical feedstocks used by the chemical industry to make an array of products such as automobile tires, nylon, laundry detergents, polyester clothing, and packaging. In addition, equally significant volumes of toluene and xylenes are blended into gasoline to boost octane ratings. Beyond Biomass to Aromatics, Anellotech can apply the technology to produce green transportation fuels, olefins and BTX derivatives.

In November 2012, Anellotech signed licensing agreement with University of Massachusetts-Amherst for a new technology tripling the yield of renewable p-xylene from non-food biomass using Anellotech's core, catalytic fast pyrolysis technology. [162]

9.4.15 Dynamotive

Dynamotive Energy Systems Corporation, based in Vancouver, Canada is an energy solutions provider. [163] Its carbon/greenhouse gas neutral fast pyrolysis technology uses medium temperatures and oxygen-free conditions to turn dry waste biomass and energy crops into BioOil® for power and heat generation. BioOil® can be further converted into vehicle fuels and chemicals.

Dynamotive has developed two innovative technologies: fast pyrolysis and pyrolysis oil upgrading. [164-165]

Dynamotive's fast pyrolysis process was developed by Dr. Desmond Radlein at Resource Transform International (RTI). [166] Dynamotive licensed the technology from RTI in 1996 and acquired all patents in 2000 (**Figure 9.27**).

Prepared feedstock (<10% moisture and 1-2 mm particle size) is fed into the bubbling fluid-bed reactor which is heated to 450–500 °C in the absence

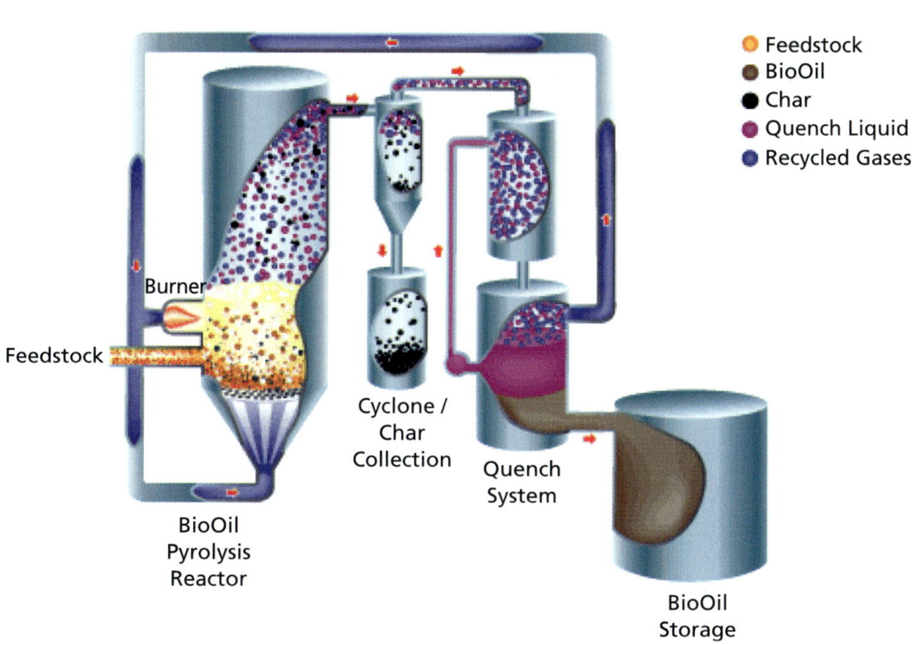

Figure 9.27 Dynamotive's fast pyrolysis. Source: Dynamotive. [166]

of oxygen. The feedstock flashes and vaporizes. The resulting gases pass into a cyclone where solid particles, char, are extracted. The gases enter a quench tower where they are quickly cooled using BioOil already made in the process. The BioOil condenses and falls into the product tank, while non-condensable gases are returned to the reactor to maintain process heating. The entire reaction from injection to quenching takes only two seconds. Three products are produced: BioOil (60–75 wt%), char (15-20 wt%) and non-condensable gases (10–20 wt%). A fourth product, BioOil Plus, can be produced by adding back the separated char into the BioOil, in a finely ground form. Char has emerging value for soil productivity enhancement and carbon sequestration.

Dynamotive's pyrolysis upgrading is a two stage process involving hydroreforming and hydrotreating. [165] In the hydroreforming step, hydrogen is added to the pyrolysis oil in a reactor that incorporates an industrial catalyst. Water, methanol and acetic acid are removed from the pyrolysis oil and the energy density of the remaining oil is substantially increased. The resulting hydrocarbon liquid is completely miscible with other hydrocarbon fuels and has an energy content that is almost 90% of diesel. The hydrotreating step further upgrades this fuel. Additional hydrogen is added and additional oxygen is removed. The general outcome of this second step is to produce a hydrocarbon liquid with characteristics the same as a mixture of jet fuel, petrol and diesel. The various fuels may then be separated.

Dynamotive has developed and commercialized its patented fast pyrolysis process, and has built and operated two commercial plants. [167] The commercial plants are located in West Lorne, Ontario (130 tonnes/day capacity), and Guelph, Ontario (200 tonnes/day capacity).

9.4.16 BioTFuel

"BioTFuel" [168] is the name of a project that involves Axens (France), the CEA (French Atomic Energy Commission), the IFP Energies nouvelles (previously French Institute of Petroleum), Sofiproteol (France), ThyssenKrupp Uhde (Germany), and Total (France) as partners. The overall budget is EUR 112.7 million of which EUR 33 million from public funds. The project will use Uhde's proprietary gasification process. [169] It includes the construction and operation of two pilot plants in France to produce biodiesel and biokerosene based on biomass gasification. The plants are scheduled to go into operation in 2012. [170]

The Uhde's Prenflo™ PDQ (Prenflo with Direct Quench) [171] process was selected for its suitability in processing a variety of feedstocks and in generating hydrogen-rich synthesis gases, such as for Fischer-Tropsch synthesis applications, by which diesel and kerosene can be produced. [172] The global project includes four main steps: pretreatment, gasification, gas treatment and synthesis (**Figure 9.28**). [173-174]

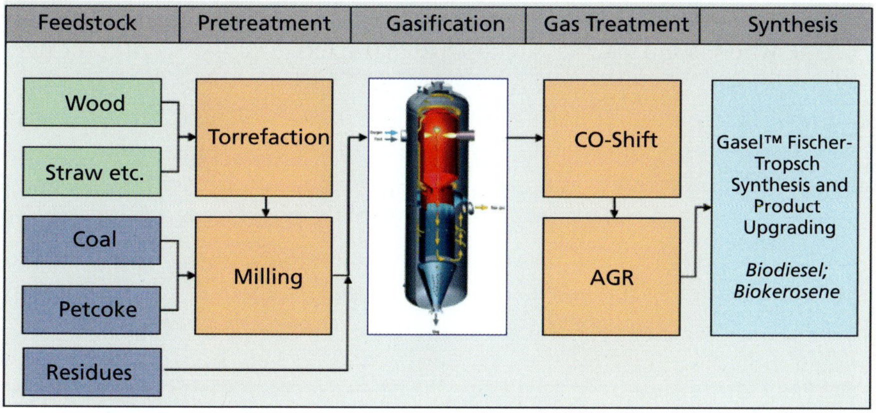

Figure 9.28 BioTFuel integrated process chain. [175]

Sofiproteol is in charge of the first step of the process, which consists in activation of the raw biomass. Sofiproteol is a financial and industrial company and promotes the activities of researching, marketing, processing, and adding value to oilseed and protein crop productions. [176] The chosen technology is torrefaction (see section 7.4.5) in a multiple hearth furnace. [177] The torrefaction process will be located on the Sofiproteol site in Venette (Picardie, France).

The Prenflo PDQ gasifier used in the second step of the process operates at pressures of 40 bar and higher and at temperatures above 2000 °C. Gas temperature at the outlet of the gasifier/quench is 200–250 °C. Carbon conversion is greater than 99%, and typical composition of the raw syngas is more than 85 vol% CO + H_2, 6–8 vol% CO_2 and less than 0.1 vol% CH_4. [172] Already used for fossil feedstocks, the technology has to be adapted to biomass. The raw gas produced from the gasifier needs to be treated and purified (step 3) before going to the Fisher-Tropsch step of the process. [168]

The final step is based on already existing Gasel™ process developed by IFP Energies nouvelles with Axens and Eni. It comprises both Fischer-Tropsch conversion of the syngas and products upgrading by hydrocracking and hydroisomerization. [178] Total will implement gasification, gas treatment and Fischer-Tropsch synthesis on its industrial site of Dunkerque, France.

9.4.17 Rentech

Rentech, based in Los Angeles, California, stands for Renewable Energies Technologies. [179] The company provides clean energy solutions. [180-181] The patented, commercially proven, Rentech-SilvaGas Biomass Gasification Process produces syngas from a wide variety of biomass feedstocks (**Figure 9.29**). [182]

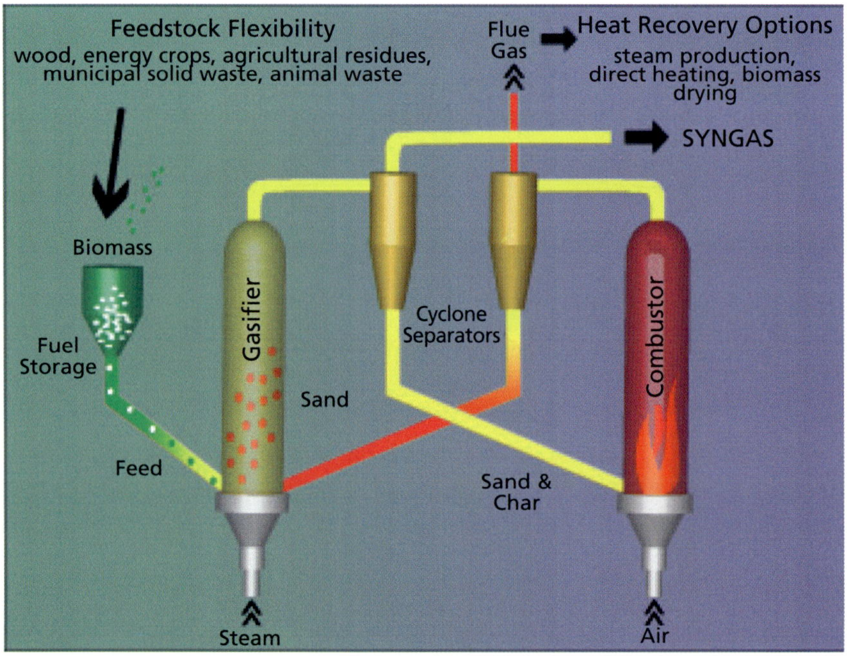

Figure 9.29 Rentech-SilvaGas gasification process. Source: Rentech [182]

With approximately $100 million in funding, SilvaGas Corporation developed biomass gasification technology that has operated successfully on a commercial scale. A gasifier in Burlington, Vermont, USA, converted 400 dry tons per day of wood-based biomass into synthesis gas used for power production. The SilvaGas gasifier operated on this scale between 1998 and 2001 during a series of operating campaigns. Rentech acquired SilvaGas in 2009.

In 2011, Rentech acquired majority ownership of ClearFuels Technology Inc., a project development company with proprietary reformation technology for the gasification of biomass. [181] In 2010, Rentech and ClearFuels jointly had received a conditional $22.6 million grant from the U.S. Department of Energy for a project to construct a biomass gasifier at Rentech's Energy Technology Center (RETC) in Denver. [183] The 20 ton-per-day ClearFuels biomass gasifier is being integrated with Rentech's Product Demonstration Unit for the production of renewable synthetic fuels from biomass. [184]

The Rentech-ClearFuels high-efficiency biomass gasifier is a flexible versatile process for the conversion of multiple biomass feedstocks to renewable biofuels and other renewable energy products such as clean, renewable syngas, hydrogen, steam and power. [184] The gasification technology produces hydrogen as well as syngas from cellulosic feedstocks through the use of a High Efficiency Hydrothermal Reformer (HEHTR). [185]

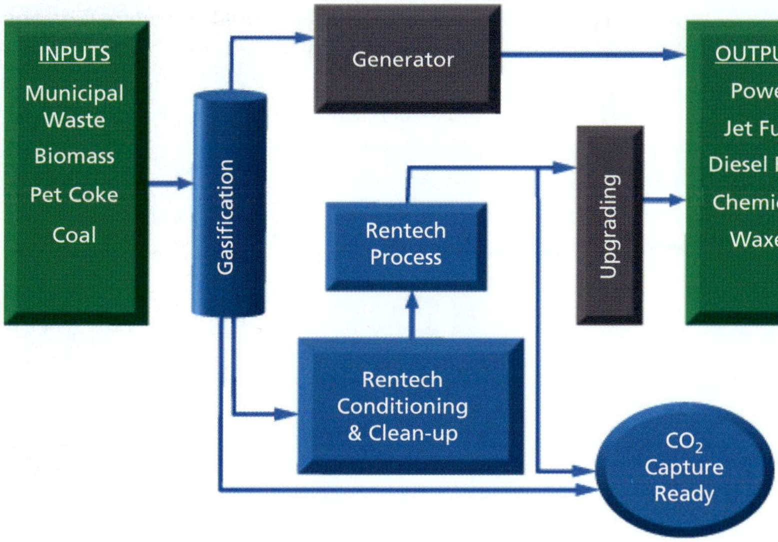

Figure 9.30 Rentech process. [186]

The Rentech Process is a proprietary technology that, together with gasification and upgrading technologies, converts synthesis gas from biomass and fossil resources into hydrocarbons that are subsequently processed and upgraded into ultra-clean synthetic fuels and specialty waxes and chemicals (**Figure 9.30**). [186-187] The Rentech Process is based on Fischer-Tropsch chemistry. The Rentech process can also be combined with the Rentech-SilvaGas biomass gasifier to produce renewable power.

Rentech's product demonstration unit is located in Colorado, USA at the Rentech Energy Technology Center. [188] This facility is designed to produce over 400 gallons per day of ultra-clean synthetic jet fuel, aviation fuel, ultra-low sulfur diesel, and specialty waxes and chemicals, and is scalable for greater output.

9.4.18 EnerTech

EnerTech Environmental, Inc., based in Georgia, USA is a renewable energy company dedicated to protecting public health and the environment through the development and commercialization of clean energy technologies for biosolids (sewage sludge) and other organic wastes. [189]

EnerTech's patented SlurryCarb™ process converts high-moisture wastes into a renewable solid fuel. [80] The process includes the following features:

- Biosolids are subjected to pressure and heat.
- Upon reaching the desired reaction temperature, the biosolids break down into carbon (carbonized product) and light gases.

- The result is a slurry with molecules that are much smaller than the original biosolids and very high in energy.
- Excess moisture is then extracted from the slurry. [190]

The dried product forms a fuel (called eFuel) that has a heating value of ~6,500 to 7,000 Btu/lb (15,100 - 16,200 kJ/kg). [80]

Located in Rialto, California, the Rialto SlurryCarb™ facility is the world's first commercial SlurryCarb™ facility. The 270,000 tons per year facility is providing a long-term biosolids recycling plan for five municipalities in the Los Angeles region. [191]

9.5 Overview of selected companies

An overview of selected companies using gasification to convert lignocellulosic biomass into fuels and/or chemicals is shown in **Table 9.2**.

Important observations from **Table 9.2** are:

- INEOS and InEnTec operate at commercial scale whereas Fulcrum Bioenergy and Enerkem have commercial projects.
- Gasification plant closures occurred in Europe and the USA.
- Several demonstration facilities have been constructed, especially in North America.
- There are many gasification technologies available. Various technologies are now being subjected to scaling-up.
- Most companies involved in biomass gasification develop fuels as primary products.

An overview of selected companies using pyrolysis to convert lignocellulosic biomass into fuels and/or chemicals is shown in **Table 9.3**.

Some important observations from **Table 9.3** are:

- Fast pyrolysis for biorefining has reached the commercial scale: Changing World Technologies produces diesel through its Thermal Conversion Process (hydrous pyrolysis/cracking), Ensyn produces fuels and chemicals through its Rapid Thermal Process (fast pyrolysis), and Dynamotive produces fuels and chemicals through its fast pyrolysis/upgrading process.
- Organic waste materials are used as feedstocks by Changing World Technologies. Woody biomass is the most popular feedstock among Ensyn and Dynamotive. [194]
- Most companies involved in biomass pyrolysis try to develop both fuels and chemicals from lignocellulosic biomass.
- There are many pyrolysis technologies available.

Table 9.2 Overview of selected companies using gasification for converting lignocellulosic biomass into fuels and/or chemicals. [192-193]

Company (country of the parent company)	Technology	Primary product	Technology scale (capacity in gallons per year, location)
Range Fuels (US)	Gasification/catalytic alcohol synthesis	Ethanol	Plant closed
Choren (DE)	Gasification/Fischer-Tropsch	Diesel	Plant closed
Coskata (US)	Gasification/fermentation	Ethanol	Shift to natural gas instead of biomass
Shell (NL-UK)	Gasification	Syngas	Commercial for coal
INEOS (CH)	Gasification/fermentation	Ethanol	Commercial (8 million ethanol; 6 MW power, Florida)
Fulcrum Bioenergy (US)	Gasification/catalytic alcohol synthesis	Ethanol	Commercial project (10 million, Nevada)
InEnTec (US)	Gasification (Plasma Enhanced Melter)	Fuels and chemicals	Commercial
Enerkem (CA)	Gasification/catalytic synthesis of methanol	Ethanol and chemicals (alcohols and acrylates)	Demonstration (Quebec) Commercial projects (10 million, Mississippi; 10 million, Quebec)
BioTFuel (EU)	Torrefaction/Gasification/ Fischer-Tropsch	Fuels	R&D project
Rentech (US)	Gasification/Fischer-Tropsch	Fuels	Demonstration (~0.15 million, Colorado)

An overview of selected companies using catalytic depolymerization; hydrothermal carbonization and tall oil hydrotreating to convert lignocellulosic biomass into fuels is shown in **Table 9.4**.

Some important observations from **Table 9.4** are:

Table 9.3 Overview of selected companies using pyrolysis for converting lignocellulosic biomass into fuels and/or chemicals. [192-193]

Company (country of the parent company)	Technology	Primary product	Technology scale (capacity, location)
Changing World Technologies (US)	Hydrous pyrolysis/ cracking (Thermal Conversion Process)	Diesel	Commercial (15 million gallons/year, Missouri)
Ensyn (US)	Fast pyrolysis (Rapid Thermal Process)	Fuels and chemicals	Commercial (100 dry tons biomass/ day, Ontario)
Avello Bioenergy (US)	Fast pyrolysis	Fuels and chemicals	Demonstration plant project
Anellotech (US)	Catalytic pyrolysis (Biomass to Aromatics)	Chemicals (aromatics) and fuels	Pilot
Dynamotive (CA)	Fast pyrolysis/hydro-reforming/hydrotreating	Fuels and chemicals	Commercial (130 tonnes biomass/ day, Ontario; 200 tonnes biomass / day, Ontario)

Table 9.4 Overview of selected companies using catalytic depolymerization, hydrothermal treatment and tall oil hydrotreating for converting lignocellulosic biomass into fuels. [192-193]

Company (country of the parent company)	Technology	Primary product	Technology scale (capacity, location)
Green Power (US)	Catalytic depolymerization (CDP)	Fuels	Prototype plant (100 tons/day and 120 gallons diesel/ton of input, [126] Washington)
AVA-CO_2 (CH)	Hydrothermal carbonization (HTC)	Coal	Commercial
EnerTech (USA)	Hydrothermal carbonization (HTC)	Coal	Commercial (60,000 tons coal/year, California)
UPM (FI)	Tall oil hydrotreating	Diesel	Commercial project (28 million gallons/year, FI)

- UPM is to invest in a biorefinery producing biofuels from tall oil through a hydrotreating process. The industrial-scale investment is the first of its kind in the world. [195-196] Construction of the facility should be completed in 2014. The forest biorefinery may well offer a unique opportunity for the world pulp and paper industry to improve its business performance. [197]
- Hydrothermal carbonization has become a commercial technology.
- Companies using catalytic depolymerization, hydrothermal carbonization or tall oil hydrogenation produce fuels as primary products.

9.6 References

[1] R.L. Bain, *An Introduction to Biomass Thermochemical Conversion*, NREL, 2004, in www.nrel.gov/docs/gen/fy04/36831e.pdf

[2] R.L. Bain, *Biomass gasification*, NREL, 2006 in http://www.ars.usda.gov/sp2UserFiles/Program/307/biomasstoDiesel/RichBainpresentationslides.pdf

[3] A. Pandey, C. Larroche, S.C. Ricke, C.G. Dussap, E. Gnansounou *Biofuels: Alternative feedstocks and Conversion Processes*, Elsevier Science and Technology, 2011

[4] S. Kumar, *Hydrothermal Treatment for Biofuels: Lignocellulosic Biomass to Bioethanol, Biocrude and Biochar*, University of Auburn, Auburn University, 2011 in http://etd.auburn.edu/etd/handle/10415/2055

[5] European Biofuels, *Biomass to Liquids*, 2011 in http://www.biofuelstp.eu/btl.html

[6] P. Quaak, H. Knoef and H. Stassen, *Energy from Biomass, A Review from Combustion and Gasification Technologies*, World Bank Technical Paper N° 422, 1999 in http://www.burner.su/gasifier/3119957-Energy-from-Biomass-A-Review-of-Combustion-and-Gasification-Technologies.pdf

[7] Wisconsin Biorefining Development Initiative, *Biomass gasification*, http://www.biorefine.org/proc/biomassgas.pdf

[8] N. Sundac, *Catalytic Cracking of Tars from Biomass Gasification*, Lund University, 2007 in http://www.docstoc.com/docs/93900154/Catalytic-cracking-of-tar-from-biomass-gasification

[9] Oregon Department of Energy, *Biomass Energy Technology*, 2010 in http://www.oregon.gov/ENERGY/RENEW/Biomass/bioenergy.shtml

[10] S. Logdberg, *Development of Fischer-Tropsch Catalysts for Gasified Biomass*, Thesis, KTH Chemical Science and Engineering, 2007 in http://kth.diva-portal.org/smash/record.jsf;jsessionid=2b378d58adb34605e89f3f63ce4a?parentRecord=diva2:12316&pid=diva2:12318

[11] M. NIAOUNAKIS and C.P. HALVADAKIS, *Olive Processing Waste Management: Literature Review and Patent Survey*, Pergamon, 2011 in http://www.amazon.com/Olive-Processing-Waste-Management-Second/dp/0080448518

[12] Biorefinery Euroview Project, *Current technology and future technological perspectives*, 2008 in http://www.iea-bioenergy.task42-biorefineries.com/publications/?eID=dam_frontend_push&docID=638.

[13] H. BOERRIGTER and B. VAN DER DRIFT, *"Biosyngas" Key Intermediate in Production of Renewable Transportation Fuels, Chemicals and Electricity: Optimum Scale and Economic Prospects of Fischer-Tropsch Plants*, 2005 in http://www.ecn.nl/docs/library/report/2005/rx05181.pdf

[14] T.G. KREUTZ, E.D. LARSON, G. LIU and R.H. WILLIAMS, Princeton University, 25th Annual International Pittsburgh Coal Conference, 2008 in http://mitei.mit.edu/system/files/kreutz-fischer-tropsch.pdf

[15] D.I. BRANSBY, *Cellulosic Biofuels Technologies*, Auburn University, 2007 in www.adeca.alabama.gov/C17/ATF/Document%20Library/Cellulosic%20Biofuel%20Technologies.pdf

[16] Purdue University, Center for Catalyst Design, *Water-gas-shift reaction* in https://engineering.purdue.edu/CCD/index.php?page=wgs

[17] B.H. BOWEN and M.W. IRWIN, *Coal-to-Gas and Coal-to-Liquids*, Purdue University, 2007 in http://www.purdue.edu/discoverypark/energy/assets/pdfs/cctr/outreach/Basics3-CTG-CTL-Aug07.pdf

[18] Ø. VESSIA, *Fischer-Tropsch reactor fed by syngas*, 2006 in http://www.zero.no/transport/biodrivstoff/fischer-tropsch-reactor-fed-by-syngas

[19] P.L. SPATH and D.C. DAYTON, *Preliminary Screening-Technical and Economic Assessment of Synthesis Gas to Fuels and Chemicals with Emphasis on the Potential of Biomass-Derived Syngas*, 2003 in http://www.nrel.gov/docs/fy04osti/34929.pdf

[20] WIKIPEDIA COMMONS, *Ficher-Tropsch Process* in http://en.wikipedia.org/wiki/Fischer%E2%80%93Tropsch_process

[21] I. CHORKENDORFF and J.W. NIEMANTSVERDRIET, *Concepts of Modern Catalysis and Kinetics*, Wiley-VCH, Weinheim, 2003 in http://books.google.be/books/about/Concepts_of_modern_catalysis_and_Kinetic.html?id=oT1RAAAAMAAJ&redir_esc=y

[22] A. DE KLERK, *Fischer-Tropsch Refining*, Wiley-VCH, Weinheim, 2011 in http://www.wiley-vch.de/publish/en/books/ISBN978-3-527-32605-1

[23] LINDE, *Hydrogen Recovery by Pressure Swing Adsorption* in http://www.linde-india.com/userfiles/image/File/Hydrogen%20Recovery%20by%20PSA.pdf

[24] National Energy Technology Laboratory, *Increasing Security and Reducing Carbon Emissions of the U.S. Transportation Sector: A Transformational Role for Coal with Biomass*, 2007 in http://catalogue.nla.gov.au/Record/4326918

[25] Energy & Environmental Research Center, *Hydrogen Separation Membranes*, 2010 in http://www.undeerc.org/ncht/pdf/eercmh36028.pdf

[26] E.C. Clausen and J.L. Gaddy, *Ethanol from Biomass by Gasification/Fermentation*, University of Arkansas, 1992 in http://www.anl.gov/PCS/acsfuel/preprint%20archive/Files/38_3_CHICAGO_08-93_0855.pdf

[27] M. Kopke, C. Held, S. Hujer, H. Liesegang, A. Wiezer, A. Wollherr, A. Ehrenreich, W. Liebl, G. Gottschalk and P. Durre, *Proc. Natl. Acad. Sci. 107*, 13087, 2010 in http://www.pnas.org/content/107/29/13087.full

[28] Eindhoven University of Technology, Telos Ingenia Consultants & Engineers, *Bio-ethanol from Bio-syngas*, 2005 in http://www.telos.nl/Telos/Medewerkers/MKasteren/downloads_getfilem.aspx?id=161201

[29] V. Muller, *Appl. Environ. Microbiol. 69*, 6345, 2003 in http://aem.asm.org/cgi/content/full/69/11/6345

[30] G. Diekert and G. Wohlfarth, *Antonie Van Leewenhoek* 66, 209, 1984 in http://www.springerlink.com/content/l4329p8127pt8724

[31] D. Biello, *Can fermenting microbes save us from climate change?*, American Scientist, 2010 in http://blogs.scientificamerican.com/observations/2010/06/29/can-fermenting-microbes-save-us-from-climate-change/

[32] R.S. Tanner, L.M. Miller and D. Yang, *Int. J. Syst. Bacteriol. 43*, 232, 1993 in http://www.ncbi.nlm.nih.gov/pubmed/7684239

[33] L.G. Ljungdahl, *Ann Rev. Microbiol. 40*, 415, 1986 in http://www.annualreviews.org/doi/abs/10.1146/annurev.mi.40.100186.002215?journalCode=micro

[34] M. Kopke, C. Mihalcea, F. Liew, J.H. Tizard, M.S. Ali, J.J.Conolly, B. Al-Sinawi and S.D. Simpson, *Appl. Environ. Microbiol. 77*, 5467, 2011 in http://www.ncbi.nlm.nih.gov/pmc/articles/PMC3147483/

[35] D. Mohan, C.U. Pittman and P.H. Steele, *Energy & Fuels 20*, 848, 2006 in http://pubs.acs.org/doi/abs/10.1021/ef0502397

[36] European Biomass Industry Association, *Pyrolysis*, http://www.eubia.org/211.0.html

[37] B.V. Babu, *Biomass pyrolysis: a state-of-the-art review*, Biofuels, Bioprod. Bioref. **2**, 393, 2008 in http://onlinelibrary.wiley.com/doi/10.1002/bbb.92/abstract

[38] G. San Miguel, J. Makibar and A.R. Fernandez-Akarregi, *New Advances in the Fast Pyrolysis of Biomass*, Proceedings of the 12th International Conference of Environmental Science and Technology, Rhodes, Greece, 2011 in http://www.srcosmos.gr/srcosmos/showpub.aspx?aa=15168

[39] A.V. Bridgwater, *Chem. Eng. J. 91*, 87, 2003 in http://144.206.159.178/ft/157/561134/12519301.pdf

[40] Q. Yan, H. Toghiani, F. Yu, Z. Cai and J. Zhang, *Forest Prod. J. 61*, 367, 2011 in http://www.fpl.fs.fed.us/documnts/pdf2011/fpl_2011_yan001.pdf

[41] C.M. Balonek, *Autothermal Oxidative Pyrolysis of Biomass Feedstocks over Noble Metal Catalysts to Liquid Products*, Dissertation, University of Minnesota, 2011 in http://conservancy.umn.edu/bitstream/115660/1/Balonek_umn_0130E_12142.pdf

[42] Toronto Research Chemicals Inc. in http://www.trc-canada.com/detail.php?CatNum=A648100&CAS=498-07-7&Chemical_Name=1,6-Anhydro-%CE%B2-D-glucopyranose&Mol_Formula=C6H10O5&Synonym=1,6-Anhydro-D-glucose%3B%20Levoglucosan%3B%201,6-Anhydro-D-glucopyranoside%3B%20 1,6-Anhydroglucose%3B%20Leucoglucosan%3B%20NSC%2046243%3B

[43] Samoa Fiber Holdings, *Fast Pyrolysis*, 2010, http://www.sfiber.com/bio-oil/fast-pyrolisis.html

[44] R.P. Overend, *Thermochemical conversion of Biomass, Renewable Energy Sources Charged with Energy from the Sun and Originated from Earth-Moon Interaction*, Encyclopedia of Life Support Systems in http://www.eolss.net/Sample-Chapters/C08/E3-08-01-05.pdf

[45] R.C. Brown Ed., *Thermochemical Processing of Biomass, Conversion into Fuels, Chemicals and Power*, John Wiley & Sons Ltd., Chichester, UK, 2011 in http://onlinelibrary.wiley.com/doi/10.1002/9781119990840.ch5/summary

[46] D. Cassidy, *Fast Pyrolysis* Forest Encyclopedia Network, 2008 in http://new.forestencyclopedia.net/p/p1206/view

[47] A.A. Lappas, K.G. Kalogiannis, E.F. Iliopoulou, K.S. Triantafyllidis and S.D. Stefanidis, *WIREsEnergy Environ. 1*, 285, 2012 doi: 10.1002/wene.16 in

[48] J. Frederick and K. Lisa, *Biomass Pyrolysis*, LUT/Environmental Engineering/Summer 2012 in https://noppa.lut.fi/noppa/opintojakso/bh50a2000/materiaali/bio_lecture_4_pyrolysis_slides_for_lut.pdf

[49] J.G. Speight Ed., *The Biofuels Handbook*, RSC Energy Series, 2011 in http://www.rsc.org/shop/books/2011/9781849730266.asp

[50] S.B. Jones, C. Valkenburg, C.W. Walton, D.C. Elliott, J.E. Holladay, D.J. Stevens, C. Kinchin and S. Czernik, *Production of Gasoline and Diesel from Biomass via Fast Pyrolysis, Hydrotreating and Hydrocracking: A Design Case*, US Department of Energy, 2009 in http://www.pnl.gov/main/publications/external/technical_reports/pnnl-18284.pdf

[51] Z. Qi, C. Jie, W. Tiejun and X. Ying, *Energy Convers. Manage. 48*, 87, 2007 in http://www.che.ncsu.edu/ILEET/CHE596web_Spr2010/resources/biomass-biofuels/Review-of-Biomass-Pyrolysis.pdf

[52] M.W. Wright, J.A. Satrio, R.C. Brown, D.E. Daugaard and D.D. Hsu, *Techno-Economic Analysis of Biomass Fast pyrolysis to Transportation Fuels*, 2010 in http://www.nrel.gov/docs/fy11osti/46586.pdf

[53] http://en.wikipedia.org/wiki/Tall_oil

[54] H. WANSBROUGH, *Tall Oil Production and Processing*, Grant and Hockh's Chemical Dictionary (5th ed.) McGraw-Hill Book Company, 1987 in http://nzic.org.nz/ChemProcesses/forestry/4G.pdf

[55] R.K. SHARMA and N.N. BAKHSHI, *Can. J. Chem. Eng. 69*, 1071, 1991 in http://onlinelibrary.wiley.com/doi/10.1002/cjce.5450690505/abstract

[56] UPM, 2012 in http://www.upmpaper.com/en/Papers/paper-news/Pages/UPM-to-build-the-world's-first-biorefinery-producing-wood-based-biodiesel.aspx

[57] UPM, 2012 in http://www.upm.com/EN/ABOUT-UPM/Businesses/Biofuels/biorefinery-investment/process-and-raw-material/Pages/default.aspx

[58] R.C. BROWN, *Thermochemical Processing of Biomass: Conversion into Fuels, Chemicals, and Power*, 2011 in http://onlinelibrary.wiley.com/doi/10.1002/9781119990840.ch1/summary

[59] R.C. BROWN and J. HOLMGREN, *Fast Pyrolysis and Bio-Oil Upgrading*, Iowa Stat University and UOP in http://www.ascension-publishing.com/BIZ/HD50.pdf

[60] IOWA State University, *Bio-Oil Fermentation* in http://www.biorenew.iastate.edu/research/thermochemicals/hybrid-thermal-conversion.html

[61] L.R. JARBOE, Z. WEN, D. CHOI and R.C. BROWN, *Appl. Microbiol. Biotechnol. 91*, 1519, 2011 in http://link.springer.com/article/10.1007%2Fs00253-011-3495-9?LI=true

[62] T. JIN, Y. LIANG, D.S. LAYTON, M.M. DEATON, Z. CHI, R.C. BROWN, Z. WEN and L.R. JARBOE, *Hybrid Thermochemical Processing: Fermentation of Pyrolytic Substrates*, American Institute of Chemical Engineers Annual Meeting, 2012 in https://aiche.confex.com/aiche/2012/webprogram/Paper271041.html

[63] M. CROCKER, *Thermochemical Conversion of Biomass to Liquid Fuels and Chemicals*, (Royal Society of Chemistry), 2010 in http://www.springer.com/energy/renewable+and+green+energy/book/978-1-84973-035-8

[64] D. KNEZEVIC *Hydrothermal conversion of biomass*, Thesis, 2009 in http://doc.utwente.nl/67359/1/thesis_D_Knezevic.pdf

[65] J.A. LIBRA et al., *Biofuels 2*, 89, (2011), in http://www.karlheinzemmerich.de/Dokumente/HTC_Review_Biofuels_2011.pdf

[66] S. XIU et al., *Biofuel's Engineering Process Technology*, 167 Conversion to Liquid Fuel http://www.intechopen.com/source/pdfs/17481/InTech-Biorefinery_processes_for_biomass_conversion_to_liquid_fuel.pdf

[67] D.C. ELLIOTT, *Process Development for Biomass Liquefaction*, Pacific Northwest laboratory, 1977, in http://www.anl.gov/PCS/acsfuel/preprint%20archive/Files/25_4_SAN%20FRANCISCO_08-80_0257.pdf

[68] Joanneum Research, *HTU technology for the production of a transportation fuel from biomass*, 2007 in http://energy-in-industry.joanneum.at/energy-in-industry/index.php/HTU_technology_for_the_production_of_a_transportation_fuel_from_biomass

[69] R.L. BAIN, *Thermochemical Technologies for Conversion of Biomass to Fuels and Chemicals*, NREL, 2006 in http://www.nrel.gov/docs/gen/fy04/36831e.pdf

[70] Changing World Technologies, 2011 in http://www.changingworldtech.com

[71] Pure Energy Systems Wiki, *Directory: Waste to Energy, Waste to Energy Technologies and Resources*, 2011 in http://peswiki.com/index.php/Directory:Waste_to_Energy

[72] F. GOUDRIAAN, J. NABER and E. VAN DEN BERG, *Conversion of Biomass Residues to Transportation Fuels With The HTU® Process* in http://www.adktroutguide.com/files/HTU_Process.pdf

[73] U.S. Department of Energy; Energy Efficiency and Renewable Energy, *Biomass Program, Catalytic Hydrothermal Gasification of Wet Biomass Feedstocks*, 2006 in http://www1.eere.energy.gov/biomass/pdfs/hydrothermal_gasification.pdf

[74] Pacific Northwest National Laboratory, PNNL Awards and News in Bio-Based Products in http://www.pnl.gov/biobased/awards.stm

[75] D.C. ELLIOTT, *Biofuels, Bioprod.Bioref. 2*, 254, 2008 in http://www.adktroutguide.com/files/Elliott_hydrothermal_gasification_of_biomass.pdf

[76] P. Scherrer Institut, Laboratory for Energy and Materials Cycles, *Hydrothermal Gasification of Woody Biomass* in http://cpe.web.psi.ch/CPE/posters/Rom%20Poster%20Maurice%20Waldner%20(Handout).pdf

[77] M.M. TITIRICI, A.THOMAS and M. ANTONIETTI., *New J. Chem., 31*, 787, 2007 in http://pubs.rsc.org/en/Content/ArticleLanding/2007/NJ/b616045j

[78] M. ANTONIETTI, *Magic Coal from the Steam Cooker*, Max Planck Research **3**, 20, 2006 in http://www.mpg.de/974909/F001_Focus_020_025.pdf

[79] M. ANTONIETTI, *Hydrothermal carbonization: basics and recent progress*, Max Planck Institute of Colloids and Interfaces in http://xa.yimg.com/kq/groups/22438052/1352627346/name/HTC+(antonietti).pdf

[80] EnerTech; 2010 in http://enertech.com/technology/slurrycarb.html

[81] AVA-CO_2 Company Website in http://www.ava-co2.com/web/pages/en/about-ava-co2.php

[82] D. MU, T. SEAGER, P.S. RAO and F. ZHAO, *Environ. Manage. 46*, 565, 2010 in http://www.springerlink.com/content/9339428031848561

[83] T.D. FOUST, A. ADEN, A. DUTTA and S. PHILLIPS, *Cellulose 16*, 547, 2009 in http://link.springer.com/article/10.1007%2Fs10570-009-9317-x

[84] "*The Range Fuels Fiasco: a case study in the folly of politically directed investment*" in http://online.wsj.com/article/SB10001424052748704364004576132453701004530.html

[85] T.P. CARNEY, "Green-energy plant sucks up subsidies, then goes bust" in http://washingtonexaminer.com/politics/2011/02/green-energy-plant-sucks-subsidies-then-goes-bust

[86] *"Range Fuels hit financials technical roadblocks"* http://coloradoenergynews.com/2011/01/range-fuels-hits-financial-technical-roadblocks/

[87] R. RAPIER, 2011 in http://www.consumerenergyreport.com/2011/07/08/what-happened-at-choren/

[88] M. RUDLOFF, *Transport Fuel Synthesis – Fischer-Tropsch Diesel*, CHOREN Industries, Birkenfeld, 2005 in www.baum-consult.de/Renew/documents/14_M_Rudloff%5CM.Rudloff%20-%20Fischer-Tropsch-Diesel%20synthesis%20-%20paper.pdf

[89] M. RUDLOFF, First commercial BTL production facility-The CHOREN beta plant Freiberg, 15th European Biomass Conference, Berlin, 2007 in superseded website www.choren.com

[90] Choren 3 steps gasification process in superseded website http://www.choren.com/en/carbo-v/carbo-v/

[91] Coskata Company website in http://www.coskata.com/

[92] Business Wire, 2011, in http://www.businesswire.com/news/home/20110120006425/en/Coskata-Supported-Biorefinery-Selected-Receive-250-Million

[93] I. KUO, GreenBeat, 2011, in http://venturebeat.com/2011/01/20/usda-awards-405m-in-biofuels-loan-guarantees-coskata-biggest-slice/

[94] Costaka Flexthanol Process in http://www.coskata.com/process/index.asp?source=7E352957-657F-44D4-8CEC-3FCA8BBB2D7C

[95] B. AMERSON, Green County Democrat, 2011, in http://greenecountydemocrat.wordpress.com/tag/coskata

[96] X-conomy, 2011, in http://www.xconomy.com/detroit/2011/08/22/coskata-lands-series-d-funding

[97] R. RAPIER, Consumer Energy Report, *Analyzing Kostaka's Major Strategy Shift*, 2012 in http://www.consumerenergyreport.com/2012/07/30/analyzing-coskatas-major-strategy-shift

[98] National Energy Technology Laboratory, *Shell Gasifiers* in http://www.netl.doe.gov/technologies/coalpower/gasification/gasifipedia/4-gasifiers/4-1-2-3_shell.html

[99] A. VAN DER DRIFT, H. BOERRIGTER, B. CODA, M.K. CIEPKIK and K. HEMMES, Energy Research Centre of the Netherlands (ECN), *Entrained Flow Gasification for Biomass,* 2004 in http://www.ecn.nl/docs/library/report/2004/c04039.pdf

[100] INEOS in http://www.ineos.com/abo_pro.html

[101] INEOS Bio in http://www.ineosbio.com/57-Welcome_to_INEOS_Bio.htm

[102] http://www.biofuelswiki.org/pub/Home/PresentationTheStateOfTheArtOfCellulosicEthanol/Propp_Alan.pdf

[103] INEOS *Bio Ethanol Technology* in http://www.ineosbio.com/61-INEOS_Bio_Ethanol_technology.htm

[104] Florida Department of Environmental Protection, *Technical Evaluation and Preliminary Determination*, Draft, Applicant Ineos New Planet Bioenergy, 2010 in http://www.dep.state.fl.us/air/emission/bioenergy/indian_river/INEOSTECH.pdf

[105] Ineos Bio, Press Release, 2012 in http://www.ineosbio.com/76-Press_releases-44.htm

[106] Fulcrum in http://www.fulcrum-bioenergy.com/process.html

[107] Fulcrum Bioenergy, Inc., Nevada Legislative Commission, Exhibit C - Energy, 2009 in http://www.leg.state.nv.us/Interim/75th2009/Exhibits/Energy/E121509C.pdf

[108] BiofuelsDigest, 2011 in http://biofuelsdigest.com/bdigest/2011/09/26/fulcrum-bioenergys-115m-ipo-the-10-minute-version

[109] J. LANE, *Fulcrum Bioenergy's $115M IPO: The 10-Minute Version*, Altenergystocks, 2011 in http://www.altenergystocks.com/archives/biofuels/ethanol

[110] Fulcrum Bioenergy Facility, in http://www.fulcrum-bioenergy.com/facilities.html

[111] Changing World Technologies, 2011 in http://www.changingworldtech.com/what/index.asp

[112] http://en.wikipedia.org/wiki/Hydrous_pyrolysis

[113] Patent publication number: US 2004/0192980 A1 in http://www.google.com/patents?id=VrGfAAAAEBAJ

[114] http://www.changingworldtech.com/who/index.asp

[115] http://www.changingworldtech.com/when/index.asp

[116] http://en.wikipedia.org/wiki/Changing_World_Technologies

[117] Global Emerging Markets, 2012 in http://www.gemny.com/investments/ri.shtml

[118] http://www.changingworldtech.com/

[119] InEnTec, 2011 in http://www.inentec.com

[120] InEnTec, *About InEnTec, History of the PEM®*, 2011 in http://www.inentec.com/about-inentec/history-of-inentec-and-the-pemtm.html

[121] InEnTec, *PEM® Technology, Process Details*, 2011 in http://www.inentec.com/pemtm-technology/process-details.html

[122] Waste Business Journal, 2010 in http://www.wastebusinessjournal.com/news/wbj20100928C.htm

[123] InEnTec, PEM Facilities, 2012 in http://www.inentec.com/pem-facilities.html

[124] Green Power Inc. *How does it work*, 2011 in http://www.cleanenergyprojects.com/How-does-it-work.html

[125] Green Power Inc. *About us*, 2012 in http://www.cleanenergyprojects.com/About-Us.html

[126] Pure Energy Systems Wiki, 2011 in http://peswiki.com/index.php/Directory:Green_Power_Inc's_NanoDiesel:Catalytic_Pressureless_Depolymerization_(Oiling)

[127] http://www.pureenergysystems.com/NEC/conferences/2008/EnvironmentalHallofFame--Chicago/displays/GreenPowerInc.pdf

[128] http://www.businesswire.com/news/home/20101026006679/en/AVA-CO2-Introduces-Industrial-Size-Hydrothermal-Carbonisation-HTC-Plant

[129] AVA-CO2, personal communication, 2012

[130] AVA-CO2, *Hydrothermal carbonization* in http://www.ava-co2.com/web/pages/en/technology/hydrothermal-carbonization.php

[131] AVA-CO2, *AVA Biomass*, in http://www.ava-co2.com/web/media/downloads_EN/datenblaetter/Factsheet_AVA_Biomass_EN.pdf

[132] AVA-CO2, *AVA HTC Module* in http://www.ava-co2.com/web/media/downloads_EN/datenblaetter/Factsheet_AVA_HTC-Module_EN.pdf

[133] Eurosolid, Hydrothermale Carbonisierung in http://www.eurosolid.de/lang_DE/htc.html

[134] AVA-CO2, Press Release, *HTC-1: The First Industrial Plant for hydrothermal carbonization Wordwide by AVA-CO2 Begins Trial Operations*, 2012 in http://www.ava-co2.com/web/media/downloads_EN/medienmitteilungen/PressRelease_EN_HTC-1_Relzow.pdf

[135] Enerkem, *Profile* in http://enerkem.com/en/about-us/profile.html

[136] Enerkem, Process in http://enerkem.com/en/technology-platform/process.html

[137] C.D. Bader, *Where Will the Green Barrels Lead Us*? MSW Management, 2010 in http://www.mswmanagement.com/web-articles/greenwaste-to-energy-5.aspx

[138] PS Newswire in http://www.prnewswire.com/news-releases/enerkem-awarded-loan-guarantee-from-us-department-of-agriculture-for-its-mississippi-waste-to-fuel-biorefinery-114292734.html

[139] Enerkem, Technology Platform, Innovation in http://www.enerkem.com/en/technology-platform/innovation.html

[140] Canada Newswire, 2012 in http://www.newswire.ca/en/story/916875/enerkem-and-greenfield-ethanol-announce-quebec-s-first-waste-to-biofuels-production-facility

[141] http://www.prnewswire.com/news-releases/enerkem-begins-ethanol-production-from-waste-at-its-demonstration-facility-in-westbury-quebec-157810175.html

[142] Ensyn in http://www.ensyn.com/about-ensyn/overview

[143] Ensyn, Technology Overview in http://www.ensyn.com/technology/overview

[144] ENSYN, *The RTP Pyrolysis Pathway Maximizing Value of Biomass Residues*, 2011 in http://www.ensyn.com/wp-content/uploads/rich-widget/file/EC%20Corp%20PPT%20April%202011NEW.pdf

[145] Ensyn, Renewable Fuel Oil in http://www.ensyn.com/products/fuel-products/pyrolysis-heating-oil

[146] Ensyn, Renewable Electricity in http://www.ensyn.com/products/fuel-products/power-generation

[147] Ensyn, Renewable Transport Fuels in http://www.ensyn.com/products/fuel-products/transport-fuel

[148] Ensyn, *Key RTP Facilities* in http://www.ensyn.com/technology/key-rtp-facilities/

[149] Ensyn, *Food Chemicals* in http://www.ensyn.com/products/other-product/food

[150] Ensyn, *Other Renewable Chemicals* in http://www.ensyn.com/products/other-product/chemicals

[151] US Patent 6326461 in http://www.freepatentsonline.com/6326461.html

[152] http://www.avellobioenergy.com/

[153] Avello Bioenergy, Technology in http://www.avellobioenergy.com/index.cfm?nodeID=22131

[154] Avello Bioenergy, Technology, *Avello™FRAC*, 2012 in http://www.avellobioenergy.com/en/technology/avellofrac

[155] http://www.canadianbiomassmagazine.ca/index.php?option=com_content&task=view&id=2553&Itemid=132

[156] Anellotech in http://www.anellotech.com

[157] Anellotech, Technology in http://www.anellotech.com/tech.html

[158] Slideshare, Pepsi PET announcement analysis in http://www.slideshare.net/jimlunt/pepsi-co-tpa-annoucement-analysis

[159] D. DE GUZMAN, *Battle of the Bioplastic Bottle*, ICIS Green Chemicals, 2011 in http://www.icis.com/blogs/green-chemicals/2011/03/battle-of-the-bioplastic-bottl.html

[160] http://www.icis.com/cgi-bin/mt/mt-search.cgi?IncludeBlogs=148&search=anellotech

[161] http://www.biofuelsdigest.com/biotech/2010/11/26/umass-amherst-anellotech-pioneer-conversion-of-bio-oil-to-chemical-intermediates/

[162] Anellotech, Press Release, 2012 in http://www.anellotech.com/press.html

[163] Dynamotive in http://www.dynamotive.com/about

[164] http://www.cleantechdatabase.org/networks/250/index.html

[165] Dynamotive, Technology in http://www.dynamotive.com/technology/

[166] Dynamotive, Fast Pyrolysis in http://www.dynamotive.com/technology

[167] http://www.dynamotive.com/assets/resources/2012/08/Dynamotive_Commercial_Case.pdf

[168] IFP Energies Nouvelles , *Biogazole et biokérosène de 2e génération - IFPEN partenaire du projet BioTfueL* in http://www.ifpenergiesnouvelles.fr/actualites/dossiers/biogazole-et-biokerosene-de-2e-generation-ifpen-partenaire-du-projet-biotfuel

[169] European Biofuels Technology Platform, *Biomass to Liquids (BtL)*, in http://www.biofuelstp.eu/btl.html

[170] The Engineer, *Uhde process for Biofuel r&D project, 2010* in http://www.theengineer.co.uk/uhde-process-for-biofuel-rd-project/1001236.article

[171] Uhde. *Prenflo Gasification* in http://www.uhde.eu/fileadmin/documents/brochures/uhde_brochures_pdf_en_11.pdf

[172] Green Car Congress. *Uhde's PRENFLO Gasification Process to be Part of €113M BioTfueL BTL Project in France, 2010* in http://www.greencarcongress.com/2010/03/uhde-20100328.html

[173] Uhde, *Gasification : what does the market want*, 10th European Gasification Conference, Amsterdam, 2010 in http://www1.icheme.org/gasification2010/pdfs/workshopgasi10gasificationwhatdoesthemarketwantkarstenradke.pdf

[174] Danish Methanol Association, *BioTFuel* in http://www.starch.dk/methanol/energy/straw.asp

[175] S. Jensen and N. Ulrich, *Uhde's activities in the development of a BTL process chain*, 10th European Gasification Conference, Amsterdam; 2010 in http://www1.icheme.org/gasification2010/pdfs/uhdeactivitiesinthedevelopmentofabtlprocesschain.pdf

[176] Sofiproteol Group in http://www.sofiproteol.com/en/group/

[177] J.F. Rous, Sofiproteol, La *Picardie, des projets d'envergure pour le développement des bioraffineries du futur*, Amiens, 2010 in http://www.lille.inra.fr/var/lille/storage/htmlarea/9_%20JF%20Rous%20-%20SOFIPROTEOL%2010032011.pdf

[178] Axens (IFP Group Technologies), Fischer *Tropsch and Upgrading* in http://www.axens.net/our-offer/by-market/alternative-fuels/natural-gas/44/fischer-tropsch-a-upgrading.html

[179] http://www.rentechinc.com/companyOverview.php

[180] http://www.rentechinc.com/index.php

[181] http://www.rentechinc.com/pdfs/Rentech_ClearFuels_Transaction_Press_Release%20FINAL%204%2019%2011.pdf

[182] http://www.rentechinc.com/silvaGas.php

[183] http://biomassmagazine.com/articles/3457/rentech-clearfuels-receive-doe-grant/

[184] http://www.rentechinc.com/clearFuels.php

[185] http://www.rentechinc.com/gasifier.php

[186] http://www.rentechinc.com/rentechProcess.php

[187] http://www.rentechinc.com/upgrader.php

[188] http://www.rentechinc.com/pdu.php

[189] http://enertech.com/about/index.html

[190] http://enertech.com/downloads/SlurryCarbOverview.pdf

[191] http://enertech.com/facilities/sitedevelopments/rcrf.html

[192] US Department of Energy, Energy Efficiency & Renewable Resources, *Biomass Program, Research & Development*, 2011 in http://www1.eere.energy.gov/biomass/integrated_biorefineries_text_only.html

[193] W.T. Coyle, USDA, Next generation Biofuels : *Near Term Challenges and Implications for Agriculture*, 2010 in http://www.ers.usda.gov/publications/BIO0101/BIO0101.pdf

[194] Canadian Biomass in http://www.canadianbiomassmagazine.ca/content/view/1779/38/

[195] http://www.goodnewsfinland.com/archive/themes/bioeconomy/41e1f423

[196] European Biofuels in http://www.biofuelstp.eu/talloil.html

[197] V. Chambost, R. Eamer and P.R. Stuart, *Pulp & Paper Canada 108*, 30, 2007 in http://www.pulpandpapercanada.com/paptac/PDFs/June07/strategicplanning.pdf

Chapter 10

Perspectives

10.1 Introduction

Simultaneously resolving energy security and environmental concerns is a key challenge for policy makers today. [1] The world is facing a fast-growing human population and the consequent growing demand for food, water and energy. [2-3] Furthermore, anthropogenic climate change is a severe threat to mankind and requires a significant reduction of our current greenhouse gas emissions to avoid detrimental consequences for the planet. Only the use of new technologies will allow us to progressively bridge the gap between economic growth and environmental sustainability. These technologies will transform the world economy from one based on fossil resources to one based on renewable resources such as biomass, while improving the sustainable production of energy, fuels, chemicals, and materials. [4]

Biorefineries were identified as a potential solution to mitigate the threat of climate change and to meet the growing demand for energy and non-energy products. [2] Generally, one can make a distinction between first- and second generation biorefineries. [5] First-generation biorefineries, which are already well implemented around the world, convert edible biomass into energy and non-energy products. They have generated the "food versus fuel" debate about increased biofuel production and rising food prices, as well as the displacement of food crops. [6] Furthermore, direct and indirect land-use change affects the biodiversity as well as releasing carbon stored in the land. [7-8] As a consequence, the sustainability of many first-generation biorefineries has been increasingly questioned. Second-generation biorefineries use lignocellulosic feedstocks, such as organic wastes and residues, and convert them into energy and non-energy products. The advantages are multiple:

- No competition with food production.
- Much higher amounts available.
- Use of waste and residue materials that do not find applications today.
- Use of abandoned land.
- Diversion of waste materials from landfills.
- In general, lower environmental impacts. [8]

In this way, the new biorefineries can offer considerable potential to promote rural development and improve economic conditions worldwide. However, while second-generation technologies are more efficient, they could become unsustainable if they compete with food crops for available land. Thus, their sustainability will depend on whether producers comply with criteria like minimum lifecycle greenhouse gas (GHG) reductions, including land-use change and social standards.

Most lignocellulosic biorefineries are expected to be ready for large-scale commercial production in a few years. The landscape of active players is rather scattered and fragmented with many relatively small technology players, but there is an ever increasing number of large players starting to invest.

Two of the main industry drivers, in addition to energy security and environmental concerns, are mandates and policies. [2] Fuel regulations are examples of potential industry drivers.

Generally, the markets for biobased products are expected to grow very strongly globally due to four irreversible trends. [2] First, the economics of fossil-based products are deteriorating since crude oil resources are getting scarce. Second is the growing need for national energy security and geopolitical security. Third, public pressure for environmental sustainability is increasing. Last, rapid democratic growth will drive demand. These fundamental trends triggered a vast interest in biobased products and placed high on the strategic agenda of most players in a variety of industries.

This chapter will start with a presentation of the U.S. Biomass Program and key European Union (EU) research projects. It will then deal with perspectives on:

- Lignocellulosic and algal feedstocks for biorefineries including biomass residues and wastes, energy crops, and microalgae, all originating from photosynthesis.

- Biochemical and thermochemical conversion processes.
- Energy and non-energy products from lignocellulosic biomass.
- Sustainability in terms of environmental impacts (LCA), economic criteria, and social standards.

10.2 Representative programs in the U.S. and EU

10.2.1 U.S. Biomass Program

The Biomass Program is one of the nine technology development programs within the Office of Energy Efficiency and Renewable Energy (EERE) at the US Department of Energy (DOE). The Biomass Program works with industry, academia, and U.S. national laboratory partners on a balanced portfolio of research in biomass feedstocks and conversion technologies. [9] Through research, development and demonstration efforts geared toward the development of integrated biorefineries, the Biomass Program is helping transform the U.S.'s renewable biomass resources into commercially viable, high-performance energy and non-energy products.

The Biomass Program primarily focuses on research and development efforts to ensure that cellulosic ethanol is commercially viable by 2012 and that biobased aviation fuel, diesel fuel and gasoline are price competitive by 2017. [9]

The Biomass Program's research, development, demonstration and deployment (RDD&D) efforts are organized around five key technical and three cross-cutting elements (**Figure 10.1**). [10]

The first two technical program elements – feedstock supply and conversion – primarily focus on research and development. [10] The next two

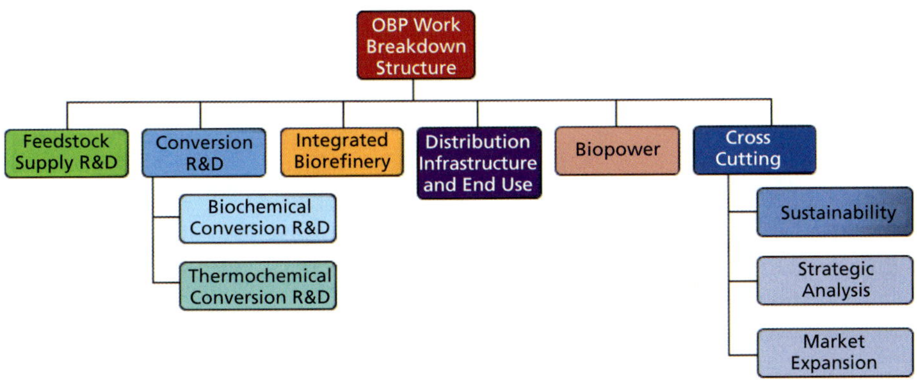

Figure 10.1 Office of the Biomass Program work breakdown structure. Source: EERE. [10]

technical areas, integrated biorefineries and distribution infrastructure, primarily focus on demonstration and deployment. The fifth technical element, biopower, includes R&D activities to develop improved technologies facilitating the use of biomass as a feedstock for power generation. The cross-cutting elements, sustainability, strategic analysis and market expansion, focus on addressing barriers that could impede the development of biomass technologies. The organization of the work allows the Program to allocate resources for pre-commercial, enabling technology development, as well as for demonstration and deployment of technologies across the biomass-to-bioenergy supply chain.

The Biomass Program works through public-private, cost-sharing partnerships to address critical challenges in the deployment of technologies for integrated biorefineries. [11] These biorefinery projects prove the viability of various feedstock and conversion pathways and reduce technical and financial risks by following a progression from pilot to demonstration scale and from demonstration to commercial scale. Each step in this progression enables validation of production performance at scale, paving the way for commercial production.

Strategically locating biorefinery projects in different areas of the U.S. promotes local and regional economic development and conversion technologies

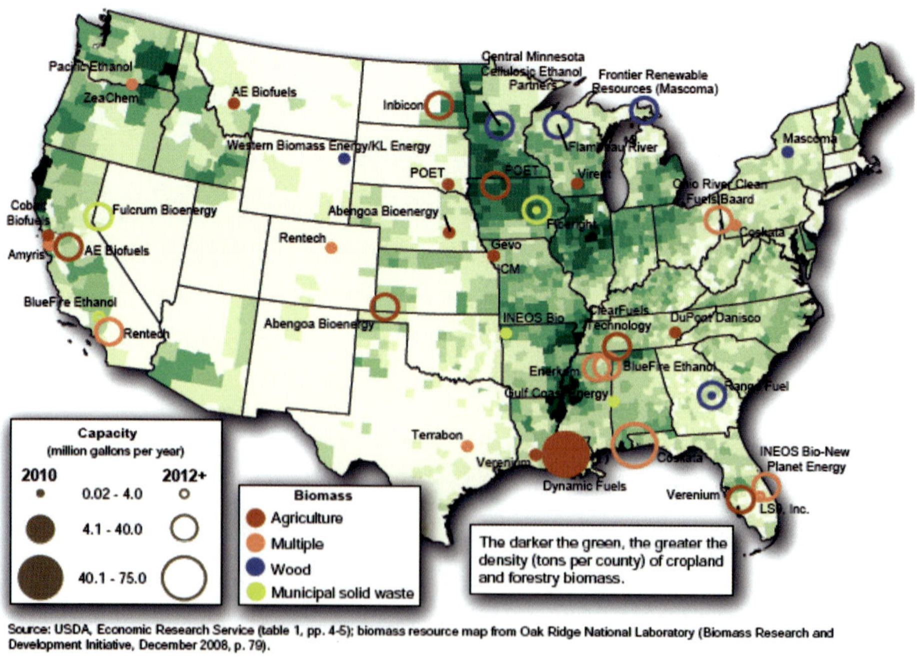

Source: USDA, Economic Research Service (table 1, pp. 4-5); biomass resource map from Oak Ridge National Laboratory (Biomass Research and Development Initiative, December 2008, p. 79).

Figure 10.2 Lignocellulosic biorefineries located across the United States near biomass supplies (2010). Source: USDA. [12]

optimized for the biomass feedstocks in each region (**Figure 10.2**). [11] The overall impact from the expansion of lignocellulosic biorefineries on U.S. agriculture initially will be small because production will be very limited, requiring only modest amounts of biomass from all sources in the next several years. [12] Some companies will exploit already existing streams of forestry waste and municipal solid waste. Other companies are in the process of developing supply arrangements for agricultural biomass (particularly crop residues and energy crops). Eventually, agriculture could play a significant role as lignocellulosic fuel production expands.

10.2.2 EU research initiative

The Directive on renewable energy (Directive 2009/28/EC) establishes a common framework for the promotion of energy from renewable sources. It sets national targets for all Member States, such that the EU will reach a 20% share of energy from renewable sources by 2020 and a 10% share of renewable energy specifically in the transport sector. The new Directive on renewable energy also establishes the sustainability criteria for biofuels. [13]

A major research initiative of the European Commission about the sustainable use of biomass has started in 2010. [14] Researchers and industry are going to develop new ways to convert biological feedstock into energy and valuable material using biorefinery technology. The Commission will fund the program with €52 million for 4 years. 81 partners from universities, research institutes and industry in 20 countries will invest an additional €28 million. The programme will contribute to the European Lead Market Initiative on non-energy biobased products. It aims to facilitate the translation of technological and non-technological innovation into commercial products and services. Three large collaborative projects will address the entire value chain and one project will provide coordination of ongoing biorefinery research projects (**Table 10.1**).

1. The EuroBioRef Project (European Multilevel Integrated Biorefinery Design for Sustainable Biomass Processing) deals with the entire process of transformation of biomass, from fields to final commercial products. It

Table 10.1 The 4 main projects funded by the European Commission. [13]

Project Acronym	Number of partners	Project cost (€million)
SupraBio	17	19.0
EuroBioRef	28	37.4
Biocore	24	20.3
Star-Colibri	12	2.41

involves 28 partners from 14 different countries and is coordinated by the Centre National de la Recherche Scientifique (CNRS), France. [14-15]

The EuroBioRef basic concept uses a new flexible approach to combine "virtual integration" with proximity to both feedstock sources and product markets (**Figure 10.3**). [16]

2. The project Biocore (see § 8.5.3) will process agricultural residues such as wheat and rice straws and different sorts of woods. The products will be second generation biofuels, bulk chemicals, polymers, speciality molecules, heat, and power. Biocore involves 24 partners from 13 different countries and is coordinated by the Institut National de la Recherche Agronomique (INRA), France. [14, 17] Biocore will use the patented organosolv technology developed by the French company CIMV.

3. The Star-Colibri project (Strategic Targets for 2020 – Collaboration Initiative on Biorefineries) promoted coordination to overcome fragmentation in the field of biorefineries research. The project consortium was led by the European Federation of Woodworking Industries and comprised 12 partners from 6 countries. [14, 18] Star-Colibri's latest publications and deliverables included the European Biorefinery Joint Strategic Research Roadmap for 2020 and the Joint European Biorefinery Vision for 2030. [19] The project successfully ended in 2011 after 2 years of activity.

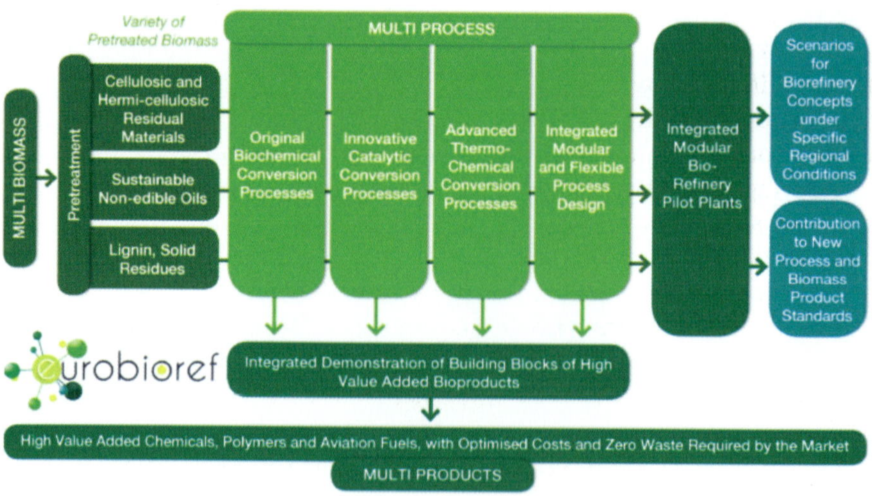

Figure 10.3 Flexibility, adaptability and multidimensional integration of the EuroBioRef project that has received funding from the European Union Seventh Framework Programme [FP7/2007-2013] under grant agreement No. 241718 (www.eurobioref.org). [16]

4. SupraBio aims at sustainable products from economic processing of biomass. Economic and sustainable production of fuels, chemicals, and materials from biomass requires capture of the maximum energy and monetary value from sustainable feedstock. SupraBio achieves this by focusing on innovative research and development of critical unit operations, by using process intensification to match economic production to the scale of available feedstock and by process integration that provides energy from process waste, optimizes utilities to minimize environmental impact and maximizes value from product mix (**Figure 10.4**). The project is coordinated by Brunel University London. [14, 20]

Other projects funded by the European Commission should also be mentioned, like the SUNLIBB project with one cooperative partner in Brazil [22] or the CaneBioFuel project, including Novozymes and Lund University. [23]

10.3 Cellulosic and lignocellulosic feedstocks

Lignocellulosic biorefineries utilize biomass consisting of the residual non-food parts of current crops or other non-food sources, such as forestry residues, perennial grasses or algae, the goal being to maximize the value of the products obtained from the whole lignocellulosic content of the plant. [2, 24]

Despite the strategic relevance of biobased products in general for many industries, numerous technical and strategic challenges still hamper commercial industrialization. [2] Some of these challenges are related to feedstock processing and logistics and feedstock yield.

It is needed to optimize feedstock processing and logistics. [2] This includes different areas: (1) developing densification techniques, allowing the transport of originally low-density feedstocks at low cost, and (2) establishing preservation techniques to control physical and chemical modification of biomass during pre-conversion processing. Developing highly coupled feedstock logistics systems that can deal with the seasonal nature of feedstock production economically is also vital.

Specific research is needed to improve feedstock yield and composition of biomass for optimal conversion efficiency. [2, 25] This involves plant genomics, breeding programs and the chemical engineering of desirable traits (e.g. drought resistance, photo-cycle intensity, cold-tolerance, sugar composition C5/C6).

Therefore, investment is essential to (1) support the development of global biomass supply chain, and (2) develop and support a reliable upstream supply chain able to mobilize a sufficient level of feedstock available for conversion, but not at the expense of food/land use. [2]

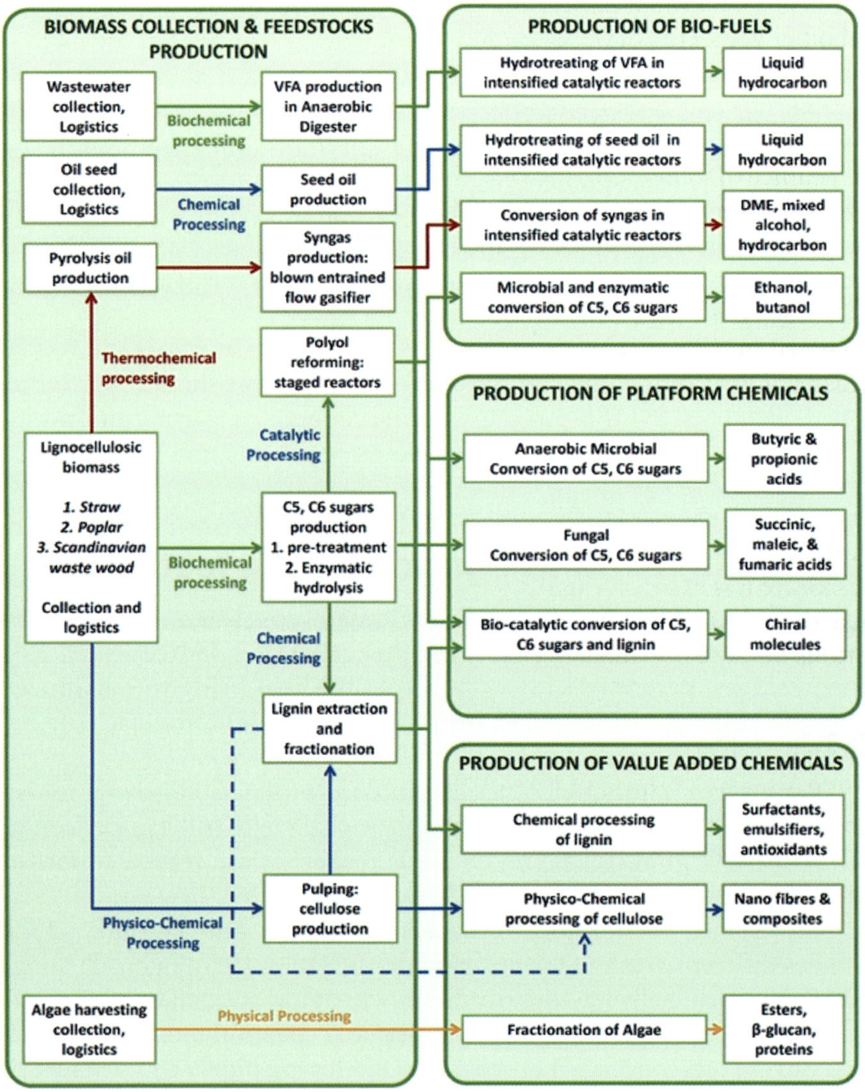

Figure 10.4 Organization and scope of SupraBio; VFA, volatile fatty acids. Source: SupraBio. [21]

10.3.1 Biomass residues and wastes

The constraints related to the availability of additional land suggest that lignocellulosic biorefineries should focus on currently available feedstock sources in the initial phase of the industry's development. [5] Agricultural and forestry residues form a readily available source of biomass and can provide feedstock from current harvesting activities without need for additional land cultivation. Biomass residues in general include agricultural residues (corn stover, straws from wheat, rice, and other grain crops), wood residues resulting from lumber, furniture and fiber production, forest residues (tops and limps from harvest for wood products, material from fuel reduction treatments), black liquors from pulp production, animal wastes, and clean wood from trimmings and construction demolition. [26]

Municipal solid waste (MSW) constitutes an alternative promising source of lignocellulosic biomass. As an example, Fulcrum Bioenergy, Inc. has announced partnerships with two waste companies to supply feedstock for their new production facility. [27] Waste Management of Nevada and Waste Connections will supply pre-sorted MSW to Fulcrum's Sierra Biofuels plant, where it will be converted into ethanol and renewable electricity using a two-step process that consists of gasification followed by alcohol synthesis (see § 9.4.6).

10.3.2 Energy crops

Miscanthus, switchgrass (**Figure 10.5**), and other perennial grasses are also considered to be promising sources of lignocellulosic biomass. [28] Perennial grasses are less expensive to produce because they do not have to be replanted each year. Fast-growing trees, such as willow and poplar, are also attractive options because of harvesting and storage advantages.

10.3.3 Algae

Would algae be a solution to overcome some or most of the problems encountered with lignocellulosic biomass exploitation? Several reasons are in favor of algal feedstocks:

- Algae grow in variable climates on non-arable land with non-potable water, so we can reasonably consider there is no competition with food crops.

Figure 10.5 *Miscanthus giganteus* (left: courtesy of Jean-Michel DEPLANQUE, Tournai, Belgium) and switchgrass (right). [29]

- Algal biomass is rich in lipids that can be extracted and converted to fuels and chemicals.
- Algae do not contain lignin, and therefore they should be better adapted to biochemical valorization. [30]

From a chemical point of view, algae convert CO_2 and water, using solar energy, into biomass and O_2. Like for all plants, yields for this transformation are very low.

Algae can be converted through different process routes starting from pond and bioreactor cultivation methods up to the production of methane, hydrogen, fuels, and various molecules. [31-32]

Algae type, climate, cultivation system and process control are many factors influencing yields of biomass production from water and CO_2. Today ponds are operated as pilot units (1–50 ha) by different companies (LiveFuels, SeamBiotic, PetroSun, etc.) and industrial plants based on pond technology are under construction (~800 ha). [30] Yields obtained are 20 g/m²/day of dry algal biomass, i.e. 70 T/ha/year. Investments are close to $100,000 /ha with operating costs close to $53,000 /ha/year.

Much higher yields can be reached in confined photo-bioreactors (~50–60 g/m²/day, i.e. 175–210 T/ha/year), by capturing solar light via optical fibers. [30] Demonstration units are being built around the world by companies such as BioFuel Systems (Spain), AlgaeLink (The Netherlands) and Valcent Vertigro (USA).

Most of the oil major companies have projects in algal research: Shell with HR Biopetroleum, Chevron with Solazyme, Exxon with Synthetic Genomics, etc. Today, exploitation costs depend largely on the production technology used but they are generally still too high to compete with traditional oils. [33] Development of a competitive industrial process seems still far away.

10.3.4 Research

It has been shown that genetic modification of lignin reduces recalcitrance and improves ethanol production from switchgrass. [34] This apparent reduction in the recalcitrance of transgenic switchgrass has the potential to lower processing costs for biomass fermentation-derived fuels and chemicals. Utilization of the transgenic approach probably is one of the most effective and economical ways for feedstock improvement.

Basic plant science research should also be conducted to increase photosynthesis efficiency from ~3% to 4%. [35] Basic organic chemistry research is needed on proteins, carbohydrates, and oils. In 2011, the U.S. National Science Foundation (NSF) and the U.K. Biotechnology and Biological Sciences Research Council (BBSRC) collaborated in issuing new awards that will investigate more efficient ways to harvest sunlight to make new biofuels and biomaterials. [36]

10.3.5 Supply

According to a report from the U.S. Idaho National Laboratory (INL), one of the principal challenges to establish lignocellulosic biorefineries as a self-sustaining enterprise is organizing the logistics of the biomass feedstock supply system in a way that maintains the economic and ecological viability of supply system infrastructures while providing the needed quantities of resources. [37]

The U.S. Department of Energy aims to displace 30% of the 2004 gasoline use with biofuels (60 billion gal/yr) by 2030 as outlined in the Energy Independence and Security Act of 2007. [37] Fifteen billion gal/yr are projected to come from grains, and the remaining 45 billion gal/yr will be produced from lignocellulosic biomass resources, requiring ~530 million tons of lignocellulosic biomass to be sustainably delivered to biorefineries annually. In order for the biofuels industry to be an economically viable enterprise, the feedstock supply system (i.e. moving the biomass from the field to the biorefinery) cannot consume more that 25% of the total cost of the biofuel production. [37]

The INL report provides supportive designs for an evolutionary progression from exiting biomass supply systems to a uniform-format, bulk solid supply system. [37] This type of approach has been demonstrated and proven very successful for feed grains. U.S. DOE/EERE has emphasized four critical feedstock supply barriers in its Biomass Multi-Year Program Plan of November 2011: [38]

- Incorporate sustainability and feedstock supply risk into assessment of current and future biomass resource quantities, prices, and characteristics.
- Develop baseline productivity for major feedstocks on a regional and sustainable basis.
- Develop feedstock materials to meet stability, density, flowability, and quality targets associated with a uniform-format feedstock supply.
- Develop commercial-scale biomass supply systems by increasing capacity and efficiency of associated unit operations.

10.4 Lignocellulosic processes

A variety of biochemical and thermochemical technologies exist for the production of biobased energy, fuels, chemicals, and materials. There is currently no clear commercial or technical advantage between the two pathways, even after many years of R&D and the development of pilot or demonstration plants. [39] Both sets of technologies remain largely unproven at the fully commercial scale, are under continual development, and have significant barriers yet to be overcome.

10.4.1 Biochemical platform

For the biochemical route, many challenges remain, including:

- understanding how pretreatments affect the physicochemical nature of heterogeneous cell walls.
- determining how enzymes deconstruct the cell wall effectively with the aim of designing superior catalysts.
- resolving issues associated with the co-optimization of pretreatment, enzymatic hydrolysis and fermentation. [40]

The US DOE in its 2011 Biomass Multi-Year Program Plan identified eleven biochemical conversion challenges and barriers, including biomass fractionation, biomass variability, biomass recalcitrance, pretreatment chemistry, pretreatment costs, cellulase production cost, cellulase loading, cleanup/separation, catalyst development, biological process integration and biochemical/thermochemical interface. [38] However, the following two issues were estimated as critical:

- Lowering/stabilizing enzyme costs and understanding enzyme companies' marketing strategies.
- Moving beyond fermentation-to-ethanol technologies by developing fermentative organisms, catalysts and other hybrid bio/chemical conversion routes.

Figure 10.6 illustrates the evolution of the cost and its components in the biochemical conversion of corn stover to ethanol. [38] The figure shows that the largest expected reduction in the cost will be obtained in the areas of pretreatment, enzymes and fermentation organisms.

10.4.2 Thermochemical platform

The U.S. DOE in its 2011 Biomass Multi-Year Program Plan identified nine thermochemical conversion challenges and barriers, including feeding dry biomass, feeding or drying wet biomass, gasification of biomass, pyrolysis of biomass and bio-oil stabilization, syngas cleanup and conditioning, fuel synthesis and upgrading, validation of syngas quality, sensors and controls, and thermochemical integration. [38] However, the following three high-impact research areas were estimated as critical:

- quality of biomass intermediates (syngas or bio-oil),
- fuels synthesis from bio-oil and from syngas, and
- reactor process optimization.

Thermochemical conversion R&D has prioritized its efforts in overcoming technical barriers based on techno-economic analysis. The analysis results are illustrated for the gasification route in **Figure 10.7**, and for the pyrolysis route in **Figure 10.8**.

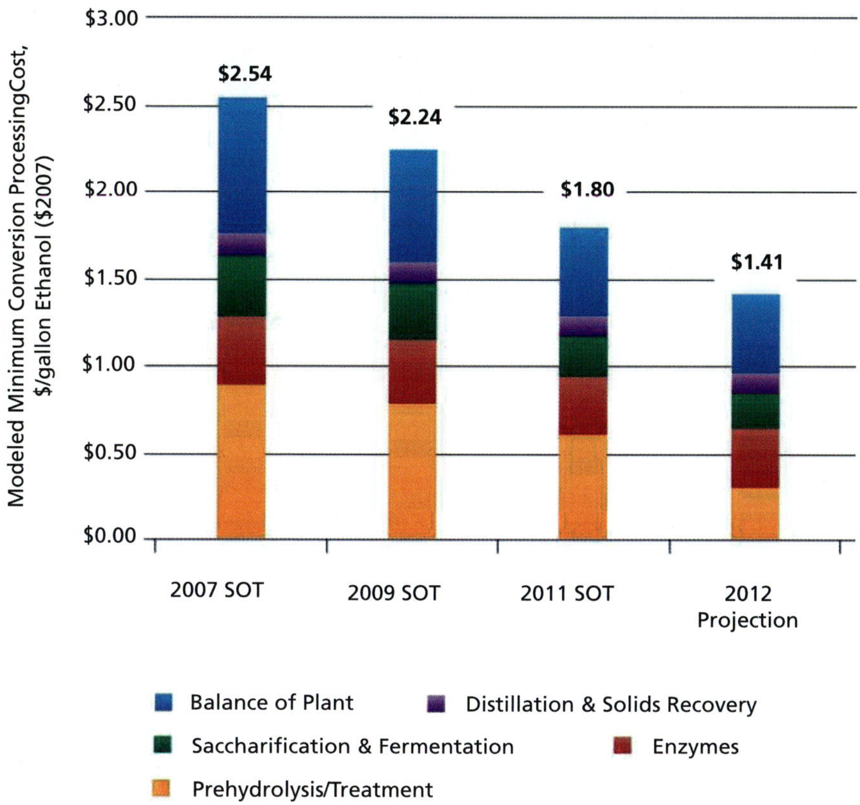

Figure 10.6 Biochemical conversion of corn stover to ethanol ($/gal in 2007 dollars); SOT, state of technology. [38]

Figure 10.7 shows that a total potential reduction in conversion cost of 61% between 2007 and projected 2012 can be achieved via improvements in all six areas.

Figure 10.8 shows that a total potential reduction of 75% between 2009 and projected 2017 can be achieved via improvements in all four areas.

10.5 Fuels and chemicals from lignocellulosic biomass

10.5.1 Fuels

Types of lignocellulosic fuels

While the production of first-generation biofuels is in an advanced stage regarding both processing and infrastructure, the production of second-generation biofuels is mainly in a pilot or demonstration stage. The main obstacle for lignocellulosic fuels is both high initial investment costs and well as higher

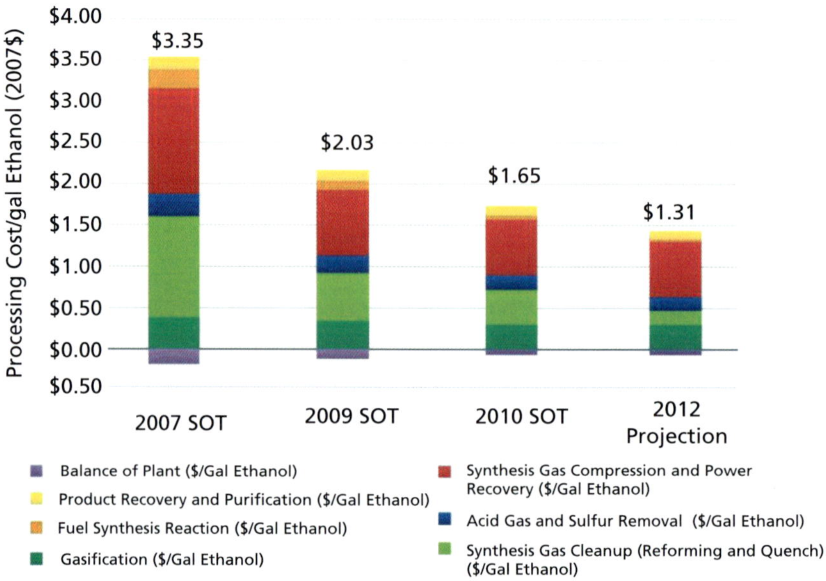

Figure 10.7 Thermochemical conversion of woody feedstocks to ethanol ($:gal in 2007 dollars) via gasification; SOT, state of technology. [38]

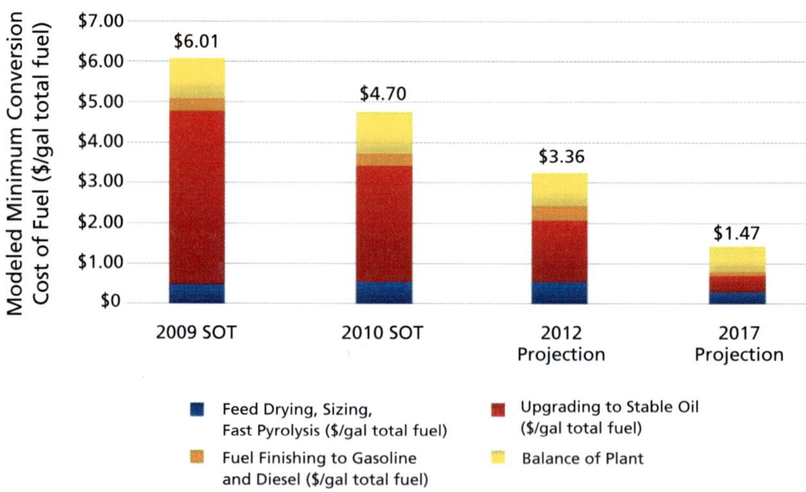

Figure 10.8 Thermochemical conversion of woody feedstocks to renewable gasoline and diesel blend stocks ($/gal in 2007 dollars) via pyrolysis; SOT, state of technology. [38]

costs for the end-product compared to fossil fuels or many first-generation biofuels. Though R&D investments are significant worldwide, it remains uncertain when second-generation will become commercially competitive.

An overview of the lignocellulosic biofuels is given in **Table 10.2**.

Table 10.2 Classification of lignocellulosic biolfuels. [41]

Biofuel group	Specific biofuel	Production process
Bioethanol	Cellulosic ethanol Cellulosic isobutanol	Hydrolysis and fermentation
Synthetic biofuels (Biomass-to-Liquids, or BTL)	Fischer-Tropsch (FT) hydrocarbons (diesel, kerosene) Methanol Heavier alcohols (butanol and mixed) Dimethyl ether (DME)	Gasification and synthesis
Methane	Synthetic natural gas (SNG)	Gasification and synthesis
Bio-hydrogen	Hydrogen	Gasification and synthesis or biological processes

Fischer-Tropsch diesel and cellulosic ethanol are the two most common second-generation biofuels. [41] Both fuels can be blended with conventional diesel and gasoline, or used pure. Another promising lignocellulosic fuel is bio-SNG, a synthetic gas similar to natural gas. The gas can be produced from a wide variety of biomass feedstocks and can be compressed or liquefied for use as transport fuel in modified vehicles.

IEA view on support policies

In 2009, the International Energy Agency (IEA) wrote as a conclusion of its Report on 1st- to 2nd-Generation Biofuel Technologies: "Even at high oil prices, second-generation biofuels will probably not become fully commercial nor enter the market for several years to come without significant additional government support." [39]

Promotion of second-generation biofuels can help provide solutions to energy security and diversification, rural economic development, GHG mitigation and help reduce other environmental impacts. [39] Policies designed to support the promotion of second-generation biofuels must be carefully developed if they are to avoid unwanted consequences and potentially delay commercialization. The relatively high cost of support currently offered for many first-generation biofuels can be an impediment to the development of second-generation biofuels, as the goal of some current policies that support the industry are not always in alignment with policies that foster innovation.

Advances in technology should enable second-generation biofuels to build on infrastructure and markets established by first-generation biofuels. [39] This assumes that future policy support should be carefully designed in order to foster the transition from first- to second-generation and to take into account the specificities of first- and second-generation biofuels, the production of sustainable feedstocks, and other related policy goals being considered.

Key policy points, as outlined in the IEA Report, are that:

- policies to support first- and second-generation biofuels should be part of a comprehensive strategy to reduce CO_2 emissions;
- enhanced R&D investment in second-generation biofuels is needed;
- accelerating the demonstration of commercial-scale second-generation biofuels in different regions is required;
- deployment policies of second-generation biofuels are either blending targets or tax credits; and
- environmental performance and certification schemes need to be developed. [39]

10.5.2 Chemicals

Building blocks

In 2004, the US DOE identified twelve promising building blocks obtainable from biomass sugars. [42] This group of platform molecules included four-carbon 1,4-diacids (succinic, fumaric and malic), 2,5 furan dicarboxylic acid, 3-hydroxy propionic acid, aspartic acid, glucaric acid, glutamic acid, itaconic acid, levulinic acid, 3-hydroxybutyrolactone, glycerol, sorbitol and xylitol/arabinitol.

To the DOE list, Bozell et al. [43] in 2010 added a new group of attractive sugar-derived building blocks, including ethanol, furans, hydrocarbons, and lactic acid. Isoprene and farnesene are examples of biohydrocarbons developed using engineered microbes.

As shown in our case studies related to the biochemical route, isobutanol, adipic acid, ethyl acetate and paraxylene are additional promising building blocks obtainable from biomass. Furthermore, as shown in our case studies related to the thermochemical route, aromatic intermediates can be produced via catalytic pyrolysis, and alcohols and acrylates (two chemical building blocks) via gasification and catalysis.

The lignin fraction of biomass is more and more considered as an attractive source for aromatic compounds, including benzene, toluene, and xylenes as platform chemicals.

Major chemical, oil and consumer goods companies are heavily investing in the biochemical sector. An already mentioned example of this investment is the partnership of Coca-Cola with Virent, Gevo and Avantium to accelerate the commercialization of a 100% biobased PET or PEF for its next-generation plant-based bottles. [44]

U.S. industry perspective

In the U.S., biobased chemicals and plastics are considered to be a new driver of U.S. economic development. [45] Historically, U.S. chemicals and plastics industry was responsible for 5 million U.S. jobs and $20 billion positive trade balance. Over last two decades, competitive advantage for chemicals and plastics manufacturing has moved to the Middle East and Asia.

Biobased chemicals and plastics represent an historical opportunity to reverse these trends through the creation of a new generation of renewable, sustainable products developed and produced in the U.S. Recent advances in biotechnology are now making it possible to manufacture many traditional chemicals from renewable biomass instead of petroleum.

The U.S. has a substantial potential competitive advantage in the manufacture of biobased chemicals and plastics:

- They are home to much of the world's leading industrial biotechnology, which enables the creation of a wide range of biobased products.
- They have a strong agricultural sector.
- They are the largest chemicals and plastics market in the world.
- Most of the downstream industry for biobased products is already in place. [46]

A growing number of early stage successes indicate this sector has tremendous potential. [46] Even in its nascent state, the biobased products industry accounts for over 5,700 direct jobs, and is likely responsible for over 40,000 economy wide. Achieving the industry's full potential could create tens of thousands of high-paying green jobs in the U.S. within the next five years.

The biobased products should be nurtured. [46] To foster growth of the sector, federal policy should provide strong support for research, development, and commercialization of innovative biobased products, including grants and loans for construction of biorefineries, a strong biobased markets program, and tax incentives for pioneering commercial production.

EU Lead Market Initiative

In Europe, non-energy biobased products constitute one of Europe's six lead markets that are supported by actions to lower barriers to bring new products or services onto the market in the framework of the Lead Market Initiative. [47]

To be successful, non-energy biobased products have to satisfy various end-use requirements at a competitive cost and to show that their entire life cycle is environmentally friendly. Perceived uncertainty about product properties and weak market transparency however hinder the fast take-up of those products. [48] The European Commission's action plan for non-energy biobased products integrates all necessary actions to favor the innovation of the new products. The actions range from improving standardization, labelling, and certification to ensure the quality and consumer information on the new products to harnessing the purchases of public authorities. [49]

There are important interlinks between some non-energy biobased products and bioenergy which influence the degree and timing of introduction of non-energy biobased products. [48] The large interdependencies and complex value chains across a wide range of products characterize the lead market approach in this domain, which calls for a coherent and coordinated approach in particular across the different policy areas affecting biobased lead markets (Common Agriculture Policy, enterprise policy, environmental policy, etc.).

10.6 The sustainability challenge

In 1989, the World Commission on Environment and Development (Brundtman Commission) has defined sustainable development as "development which meets the needs of current generations without compromising the ability of future generations to meet their own needs." [50] The concept supports strong economic and social development, in particular for people with a low standard of living. At the same time, it underlines the importance of protecting the natural resource base and the environment (**Figure 10.9**). Economic and social well being cannot be improved with measures that destroy the environment. Intergenerational solidarity is also crucial: all development has to take into account its impact on the opportunities for future generations.

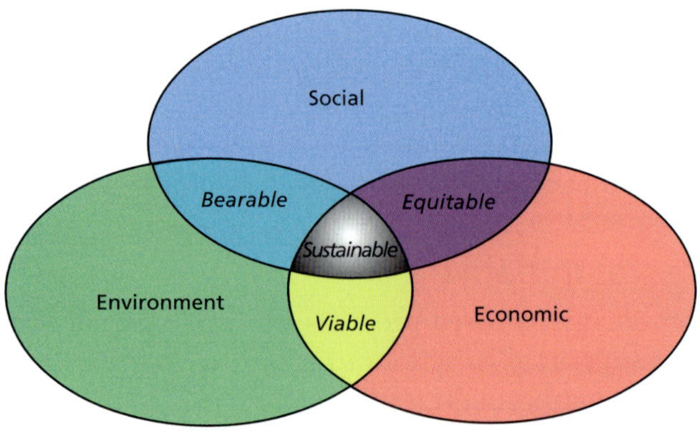

Figure 10.9 The three pillars of sustainability. [51]

Most biobased production routes to fuels, chemicals, and power can save energy in production processes and significantly reduce CO_2 emissions. [52-54] Second-generation technologies have typically a lower environmental impact than first-generation ones. Vegetable solvents emit few or no volatile organic compounds, and biobased chemicals offer potential to reduce the generation of toxic wastes. [55] The social impacts primarily relate to employment creation and the potential for rural development. The economic benefits derive from a growing biobased chemicals market and the macroeconomic savings of biobased chemicals, when compared to petrochemical based chemicals.

While one of the original intentions of switching to biobased production is the conservation of resources, caution is advised to ensure that the implementation of first- and second-generation biorefineries does not jeopardize the environment. [2] The following issues are of most concern:

- Land use change and its effect on GHG emissions. Land use change affects the GHG balance by conversion of native systems – often by slash and burn – and release of soil carbon. This key factor is characterized into two main types:
    1. Direct land-use change occurs if feedstock for biorefinery purposes displaces a prior land-use (e.g. forest), thereby generating possible changes in the carbon stock of that land.
    2. Indirect land-use change occurs if pressure on agriculture due to the displacement of previous activity or use of the biomass induces land-use changes on other lands. [56] The displacement of current land-use for biobased production can trigger direct land-use changes elsewhere.
- Use of fertilizers, pesticides, and water. These sustainability issues are associated with biomass production.
    1. Fertilizers are essential for the production of adequate quantities of food, feed, chemicals, and bioenergy. Sustainability of fertilizer use depends on its benefits to soil quality and health. [57] It also includes ensuring adequate and timely availability of fertilizers to the farmers at affordable prices. [58] The future will involve more efficient capture of nutrients from animal manures, crop residues, and other organic materials.
    2. Pesticides are considered essential to protect crops from insects, rodents, and fungi. [59] However, they can also cause environmental damage such as water pollution and present risks to human health. Rules for sustainable use of pesticides have been set by the EU to reduce the risks and impacts of pesticide use on people's health and the environment. [60] Actions for sustainable use of pesticides include training of professional pesticide users, distributors, and advisors; gathering of information about poisoning; prohibition of aerial

spraying; minimization or banning of the use of pesticides in critical areas; and promotion of low-pesticide-input management.

3. Water, which is already scarce in many regions, is a critical input to biorefineries, and is used in large volumes for feedstock production (irrigation) and processing. [61] Other concerns include the impacts associated with nutrient-rich agricultural runoff and wastewater from processing.

- Link between commodity prices and biorefineries. It is being reported that the increase in biobased production is one of many factors in food price increases. As global demand for food is expected to double within the coming 50 years and global demand for transportation fuels is expected to increase even more rapidly, there is a great need for renewable energy supplies that do not cause significant environmental harm and do not compete with food supply. Lignocellulosic fuels, which typically do not compete with food crops for land may put less pressure on the link between food prices and fuel.

- Reputational risks. Concerns are that biorefineries may pose a threat to biodiversity, damage rural communities through large multinational corporations, adversely affect labor conditions, make excessive use of water resources, or damage the food supply. All of these factors represent a reputational and commercial risk for corporations investing in biorefineries.
- Legislation-driven deforestation. Deforestation due to demand for biofuels is closely linked to land-use change. Although current legislative targets for biofuels favor biorefineries and biobased economy, they do not sufficiently address sustainability issues. In many circumstances, incentives exist to cut down forests in the name of emissions reduction. [2]

Sustainability risks relating to domestic biomass production originating from wastes and agricultural and forestry residues, where no land-use change occurs, are currently considered to be low. [55] However, biomass certification can serve as an instrument to guarantee the sustainability of biomass. [62]

Biomass-based energy plays an increasingly important role among renewable energies. [63] Biobased chemicals are generally seen as the most promising segment of industrial biotechnology. [64] Thus, there is a foreseeable considerable competition between material and energetic biomass use. Cascading can provide one alternative to mitigate these competition conflicts by aiming at a combined material and energetic use of biogenic resources and thus using synergies in terms of sustainable development. The concept of cascading describes the strategy of utilizing products or their components by a sequence of multiple material and, at the end of their life cycle, energetic uses. Using biomass to produce material products first, and then to recover its energy content from the waste can provide multiple benefits and ensures more efficient use of biomass. [65] Reutilization of biomass may reduce demand for land and maximize greenhouse gas mitigation potential.

10.7 References

[1] International Energy Agency, *Energy Security and Climate Policy*, 2007 in http://www.iea.org/publications/free_new_Desc.asp?PUBS_ID=1883

[2] World Economic Forum, *The Future of Industrial Biorefineries*, 2010 in http://www3.weforum.org/docs/WEF_FutureIndustrialBiorefineries_Report_2010.pdf

[3] OECD Green Groth Studies, Energy, 2011 in http://www.oecd.org/dataoecd/37/42/49157219.pdf

[4] Wageningen UR, *Biobased Economy across the board*, 2011 in http://www.themabiobasedeconomy.wur.nl/UK/newsagenda/news/Biobased_Economy_across_the_board.htm

[5] International Energy Agency, *Sustainable production of second-generation biofuels*, 2010 in http://www.iea.org/papers/2010/second_generation_biofuels.pdf

[6] International Centre for Trade and Sustainable Development, *Biofuel Production, Trade and Sustainable Development*, 2008

[7] BirdLife International, *Environmental impacts of current biofuels*, 2012 in http://www.birdlife.org/eu/EU_policy/Biofuels/eu_biofuels2.html

[8] R. Zah, TA-Swiss, Future Perspectives of 2nd generation Biofuels, 2010

[9] Department of Energy, Energy Efficiency and Renewable Energy, *Biomass Program*, 2011 in http://www1.eere.energy.gov/biomass/about.html

[10] Department of Energy, Energy Efficiency and Renewable Energy, *Biomass Program, Research, Development, Demonstration, and Deployment*, 2011 in http://www1.eere.energy.gov/biomass/research_development.html

[11] Department of Energy, Energy Efficiency and Renewable Energy, *Biomass Program, Biofuels, Biopower and Bioproducts: Integrated Biorefineries*, 2010 in http://www1.eere.energy.gov/biomass/pdfs/ibr_portfolio_overview.pdf

[12] W.T. Coyle, U.S. Department of Agriculture, Economic Research Service, Next generation Biofuels : *Near Term Challenges and Implications for Agriculture*, 2010 in http://www.ers.usda.gov/media/122837/bio0101.pdf

[13] SupraBio, *Biorefineries*, 2010 in http://www.suprabio.eu/biorefineries/

[14] European Commission, Research and Innovation, *Commission steps up biomass use - Nearly € 80 million for biorefinery research*, 2010 in http://ec.europa.eu/research/index.cfm?lg=en&pg=newsalert&year=2010&na=na-010310

[15] EuroBioRef, *European Multilevel Integrated Biorefinery Design for Sustainable Biomass Processing* in http://www.eurobioref.org

[16] http://www.eurobioref.org/index.php/concept-innovation

[17] Biocore, *The Biocore project* in http://www.biocore-europe.org/

[18] Star-COLIBRI, *Star-COLIBRI project: biorefinery clustering* in http://www.star-colibri.eu

[19] Star-COLIBRI, *Latest news* in http://www.star-colibri.eu/news/latest-news/star-colibri-project-2-years-of-succesful-collaborations-come-to-an-end-2.html

[20] SupraBio, *Welcome*, 2010 in http://www.suprabio.eu/

[21] SupraBio, *Suprabio at a glance*, 2010 in http://www.suprabio.eu/suprabio-at-a-glance

[22] SUNLIBB, *Sustainable Liquid Biofuels from Biomass Biorefining* in http://www.york.ac.uk/org/cnap/SUNLIBB/index.html

[23] Cane BioFuel in http://www.canebiofuel.com/

[24] SupraBio, Newsletter, January 2011 in http://www.suprabio.eu/klanten/supra/media/documenten/SUPRABIO%20newsletter%20number%201.pdf

[25] OECD, *Industrial Biotechnology and Climate Change, Opportunities and Challenges*, 2011 in http://www.oecd.org/sti/biotechnologypolicies/49024032.pdf

[26] US Climate Change Technology Program, *Biomass Residues*, 2003 in http://www.climatetechnology.gov/library/2003/tech-options/tech-options-2-3-7.pdf

[27] Energy Business Daily, 2011 in http://energybusinessdaily.com/renewables/fulcrum-bioenergy-signs-two-feedstock-agreements-for-new-biorefinery

[28] Energy Future Coalition , 2007 in http://www.energyfuturecoalition.org/biofuels/fact_ethanol_cellulose.htm

[29] http://en.wikipedia.com/wiki/Switchgrass

[30] L. Haspeslagh, *Aquatic phototrophs for the production of fuels and green chemicals*, Total, http://www.biofuelstp.eu/spm3/pdf/TOTAL_Microalgae.pdf

[31] http://en.wikipedia.org/wiki/Algaculture

[32] Kema, *Micro-algae to biofuel* in http://www.kema.com/Images/Micro-algae%20to%20biofuel.pdf

[33] A.O. Alabi, M. Tampier and E. Bibeau, *Microalgae Technologies & Processes for Biofuels/Bioenergy Production in British Columbia: Current Technology, Suitability & Barriers to Implementation*, Final Report submitted to the British Columbia Innovation Council, Seed Science, 2009 in http://www.globalbioenergy.org/uploads/media/0901_Seed_Science_-_Microalgae_technologies_and_processes_for_biofuelsbioenergy_production_in_British_Columbia.pdf

[34] C. Fu, J.R. Mielenz, X. Xiao, Y. Ge, C.Y. Hamilton, M. Rodriguez Jr., F. Chen, M. Foston, A. Ragauskas, J. Bouton, R.A. Dixon and Z.Y. Wang, *Proc. Natl. Acad. Sci USA 108*, 3803, 2011 in http://www.pnas.org/content/early/2011/02/04/1100310108.full.pdf+html

[35] Biomass Research and Development Technical Advisory Committee, Biomass Research and Development Initiative, *Roadmap for Bioenergy and Biobased Products in the United States*, 2007 in http://www.usbiomassboard.gov/pdfs/obp_roadmapv2_webkw.pdf

[36] National Science Foundation, 2011 in http://www.nsf.gov/news/news_summ.jsp?cntn_id=119017

[37] J.R. Hess, C.T. Wright, K.L. Kenney and E.M. Searcy, *Uniform-Format Solid Feedstock Supply System*, Idaho National Laboratory, INL/EXT-09-15423, 2009 in http://www.inl.gov/portal-files/solid-feedstock-supply-system.pdf

[38] U.S. Department of Energy, Energy Efficiency and Renewable Energy, Biomass Multi-Year Program Plan, 2011 in http://www1.eere.energy.gov/biomass/pdfs/mypp_november_2011.pdf

[39] R. Sims, M. Taylor, J. Saddler and W. Mabee, *IEA's Report on 1st- to 2nd-Generation Biofuel Technologies*, RenewableEnergyWorld.com, 2009 in http://www.renewableenergyworld.com/rea/news/article/2009/03/ieas-report-on-1st-to-2nd-generation-biofuel-technologies

[40] S.P.S. Chundawat, G.T. Beckham, M.E. Himmel and B.E. Dale, *Annu Rev. Chem. Biomol. Eng. 2*, 121, 2011 in http://www.annualreviews.org/doi/full/10.1146/annurev-chembioeng-061010-114205

[41] A. Eisentraut, Sustainable Production of Second-Generation Biofuels, International Energy Agency, 2010 in http://www.oecd.org/dataoecd/17/12/44567743.pdf

[42] US Department of Energy, *Top Value Added Chemicals from Biomass*, 2004 in http://www1.eere.energy.gov/biomass/pdfs/35523.pdf

[43] J.J. Bozell and G.R. Peterson, *Green Chem. 12*, 539, 2010 in http://pubs.rsc.org/en/content/articlelanding/2010/gc/b922014c

[44] Biofuelsdigest, 2011 in http://www.biofuelsdigest.com/bdigest/2011/12/16/the-jesus-molecule

[45] R. Singh, *The U.S. Industry Perspective on Assessing Sustainability of Bio-based Products*, Biotechnology Industry Organization, 2010 in http://www.oecd.org/dataoecd/21/8/45699411.pdf

[46] Biotechnology Industry Organization, *Biobased Chemicals and Products: A New Driver for Green Jobs*, 2010 in http://www.bio.org/articles/biobased-chemicals-and-products-new-driver-green-jobs

[47] European Commission, Enterprise and Industry, 2012 in http://ec.europa.eu/enterprise/policies/innovation/policy/lead-market-initiative

[48] European Commission, Enterprise and Industry, 2012 in http://ec.europa.eu/enterprise/policies/innovation/policy/lead-market-initiative/biobased-products/index_en.htm

[49] http://ec.europa.eu/enterprise/policies/innovation/policy/lead-market-initiative/files/lmi_brochure_en.pdf

[50] United Nations Economic Commission for Europe, UNECE in 2004-2005, *Sustainable development-concept and action* in http://www.unece.org/oes/nutshell/2004-2005/focus_sustainable_development.html

[51] Green Planet Ethics in http://greenplanetethics.com/wordpress/sustainable-development-can-we-balance-sustainable-development-with-growth-to-help-protect-us-from-ourselves

[52] Organisation for Economic Co-operation and Development, *Industrial Biotechnology and Climate Change*, 2011 in http://www.oecd.org/dataoecd/15/54/49024032.pdf

[53] Organisation for Economic Co-operation and Development, The Application of Biotechnology to Sustainability, 2002 in http://www.oecd.org/dataoecd/61/13/1947629.pdf

[54] EuropaBio, How industrial biotech can tackle climate change, 2009 in http://www.europabio.org/industrial/positions/how-industrial-biotechnology-can-tackle-climate-change

[55] Europe Innova, Eco-Innovation, Biochem, 2010 in http://www.europe-innova.eu/c/document_library/get_file?folderId=177014&name=DLFE-11072.pdf

[56] E. GNANSOUNOU, L. PANICHELLI, A. DAURIAT and J.D. VILLEGAS, Accounting for indirect land-use changes in GHG balances of biofuels, EPFL, 2008 in http://infoscience.epfl.ch/record/121496/files/Accounting_for_ILUC_in_biofuels_production.pdf

[57] Agri-Briefs, Agronomic News Items, Is Fertilizer Use Sustainable?, 2001 in https://www.ipni.net/ppiweb/agbrief.nsf/5a4b8be72a35cd46852568d9001a18da/77ae1690b210379985256a030067956e!OpenDocument

[58] Food and Fertilizer Technology Center, *Fertilizer Policy for ensuring sustainable food production in the Asian and Pacific region*, 2008 in http://www.agnet.org/library.php?func=view&id=20120103111630&type_id=1

[59] EurActiv.com, *Pesticides: Striking the right balance?* 2008, in http://www.euractiv.com/sustainability/pesticides-striking-right-balanc-linksdossier-188479

[60] European Commission, DG Health and Consumers, 2012 in http://ec.europa.eu/food/plant/plant_protection_products/sustainable_use_pesticides/index_en.htm

[61] C. EASTON and S. HERBERT, Biomass As Energy - Avoiding Risks and Promoting Sustainability, 2010 in http://terrachoice.com/wp-content/uploads/2011/10/Easton-Herbert_Biomass-as-Energy.pdf

[62] Biomass Technology Group, *Sustainability Criteria & Certification Systems for Biomass Production*, 2008 in http://ec.europa.eu/energy/renewables/bioenergy/doc/sustainability_criteria_and_certification_systems.pdf

[63] Wuppertal Institute for Climate, Environment and Energy, *Biomass Cascading towards Climate Protection and an Optimised Expansion of Renewable Energies*, 2012 in http://www.wupperinst.org/en/projects/proj/index.html?projekt_id=254&bid=164

[64] BioConomy, Biobased Chemicals Europe-Building the Value Chain: Infocast in Milan *Summary and Trends*, 2011 in http://bioconomy.net/2011/03/16/biobased-chemicals-europe-building-the-value-chain-infocast-in-milan-summary-and-trends

[65] International Panel for Sustainable Resource Management, *Toward sustainable production and use of resources: Assessing Biofuels*, 2009 in http://www.unep.fr/scp/rpanel/pdf/Assessing_Biofuels_Presentation.pdf

Glossary

A

$(\alpha/\alpha)_6$ barrel Fold consisting of six inner and six outer α helices forming a barrel-like structure.

α helix Right- or left-handed coil conformation, resembling a spring, in which every backbone group donates a hydrogen bond to the backbone CO group four residues earlier.

Acetogen Organism capable to reduce CO_2 (homoacetogenesis) to acetate via the acetyl coenzyme A (acetyl-CoA) or Wood-Ljungdahl pathway; acetogens are strictly anaerobic bacteria that can grow by the conversion of C1 compounds such as H_2-CO_2, CO, or formate to acetate.

Actin Globular, roughly 42-kDa protein found in most eukaryotic cells.

Aerobic organism Organism that can survive and grow in an oxygenated environment.

Afterglow Oxidation of the char.

Aglycone Non-sugar group of a glycoside.

Allomorph Any of alternative crystalline forms of a substance, such as native cellulose.

Amoeba Genus of protozoa, well known as a representative unicellular organism.

Amorphogenesis Physical changes, such as swelling, segmentation, and destratification of cellulose, that enhance enzymatic hydrolysis.

Amylopectine Branched α-linked-D-glucan.

Amylose Linear α-1,4-D-glucan.

Anaerobic digestion Series of processes in which microorganisms break down biodegradable material in the absence of oxygen.

Anaerobic organism Organism that does not require oxygen for growth and may even die in its presence.

Angiosperm Flowering plant in which the seeds are enclosed in an ovary.

Anhydroglucopyranose unit Cyclic six-membered glucose residue that results from the reaction of an alcohol and a hemiacetal to form an acetal and water.

Anomeric center (anomeric carbon) Asymmetric carbon atom at C1 generated by hemiacetal ring formation.

Anomer one of two diastereoisomers (designated α and β) generated by hemiacetal ring formation.

Antiparallel Parallel but oriented in opposite directions; a chain packing is antiparallel when the chains running side by side alternate in direction.

Apiogalacturonan Pectic polysaccharide consisting of an α-1,4-linked D-galacturonic acid backbone with apiose and apiobiose side chains via O2 or O3 links.

Aquasolv Hot water extraction method.

Aqueous phase reforming Production of hydrogen from biomass-derived oxygenated compounds such as sugars, sugar alcohols, and glycerol.

Arabinoxylans Water-soluble hemicelluloses consisting of β-1,4-xylan backbone with arabinose side-chains.

Asymmetric unit (or repeating unit) Smallest portion of a crystal structure to which crystallographic symmetry can be applied to generate one unit cell.

ATP synthase Enzyme that provides energy for the cell to use through the synthesis of adenosine triphosphate (ATP).

Autohydrolysis Process of converting lignocellulosic biomass into fermentable sugars by exposure to high-temperature steam.

Autotroph Organism that produces organic compounds from inorganic molecules by photosynthesis or inorganic chemical reactions.

Axial Perpendicular to the mean plane of the ring.

B

β barrel Large β sheet that twists and coils to form a closed structure in which the first strand is hydrogen bonded to the last.

$(\beta/\alpha)_8$ barrel Fold consisting of 8 repeating units of β/α module, in which the 8 β strands form an inner parallel β sheet arranged in a barrel structure, which is surrounded by the 8 α helices.

β-1,3;1,4 glucans Hemicellulose polysaccharides consisting of β-1,4-linked glucans with interspersed single β-1,3-linkages.

β-glucosidase Glucosidase enzyme which acts upon β-1,4-bond linking two glucose or glucose-substituted molecules.

β jelly roll Variant of a Greek key topology with both ends of a sandwich or a barrel fold being crossed by two interstrand connections.

β sandwich Fold consisting of two β sheets that pack together, face-to-face, in a layered arrangement.

β sheet Assembly of β strands that are hydrogen-bonded to each other.

β strand A stretch of amino acids typically 5-10 amino acids long whose peptide backbones are almost fully extended.

Bacteria Group of unicellular microorganisms.

Biochar Charcoal created by pyrolysis of biomass.

Biochemical conversion Technological platform involving essentially three basic steps: biomass pretreatment, conversion of pretreated biomass into sugars, and sugar fermentation.

Biocrude Crude-oil substitute made from biomass.

Biofuel Solid, liquid, or gas fuel derived from biomass.

Biological carbon cycle One of the major biogeochemical cycles in which carbon moves from the atmosphere, through the Earth's ecosystems (the biosphere), and back to the atmosphere.

Biomass Material of biological origin excluding material embedded in geologic formation and/or fossilized.

Biomass-to-Liquids (BTL) Multi-step process to produce liquid fuels (such as gasoline and diesel) from biomass through a thermochemical route; the term BTL is also applied to the fuels made through the BTL process.

Bio-oil A liquid fuel produced by the pyrolysis of biomass.

Biorefinery Facility that integrates biomass conversion processes and equipment to produce energy (fuels, power, and heat), and chemicals and materials from biomass.

Biosphere Global sum of all ecosystems.

Bryophyte Embryophyte (land plant) that is nonvascular.

Building block A basic element or part of something.

Bundle-sheath cells Photosynthetic cells arranged into tightly packed sheaths around the veins of a leaf.

C

C3 plant Plants that survive solely on C3 fixation; C3 fixation is named for the 3 carbon atoms present in the first product of carbon fixation in these plants.

C4 plant Plants that use C4 fixation; C4 fixation is named for the 4 carbon atoms present in the first product of carbon fixation in these plants.

Callose Plant polysaccharide comprised of glucose residues linked together by β-1,3-linkages.

Capnophilic Ability to thrive in conjunction with carbon dioxide.

Carbohydrate-active enzyme Enzyme that degrades, modifies or creates glycosidic bonds.

Carbonization Conversion of an organic substance into carbon or a carbon-containing residue.

Carotenoids A class of more than 600 naturally occurring pigments synthesized by plants, algae, and photosynthetic bacteria.

Carbohydrate-binding module Contiguous amino acid sequence within a carbohydrate-active enzyme with a discreet fold having carbohydrate-binding activity.

Catalytic subunit (module) Contiguous amino acid sequence within a carbohydrate-active enzyme with a discreet fold having catalytic activity.

Cellobiohydrolase Exoglucanase, which cleaves the β-1,4-glucosidic bond at one of the ends of the polysaccharide.

Cellulase Class of enzymes produced mainly by fungi, bacteria, and protozoans that catalyze the hydrolysis of cellulose.

Cellulose synthase Glycosyltransferase that catalyzes the chemical reaction:

UDP-glucose + $(\beta\text{-1,4-D-glucosyl})_n \rightarrow$ UDP + $(\beta\text{-1,4-D-glucosyl})_{n+1}$.

Cellulose synthase complex (CSC) Cellulose synthesizing complex also called terminal complex.

Cellulose synthase-like genes Superfamily of plant genes whose amino acid sequences are related to the CesA (cellulose synthase catalytic subunit) genes.

Cellulosome Multienzyme complex produced by many cellulolytic microorganisms.

Center chain Chain passing through the center of the unit cell.

Center sheet Sheet formed by hydrogen bonding between adjacent center chains.

Char Solid material forming during the initial stage of combustion of a carbonaceous material.

Chemiosmosis Movement of ions across a selectively-permeable membrane down their electrochemical gradient.

Chiral (or handed) A object or system is called chiral if it differs from its mirror image, and its mirror image cannot be superposed on the original object; a chiral molecule and its mirror image are called enantiomers.

Chlorophyll Green pigment that absorbs sunlight and uses its energy to synthesize carbohydrates from CO2 and water.

Choroplasts Organelles found in plant cells and other eukaryotic organisms that conduct photosynthesis.

Cholesteric Describing a type of liquid crystal with a helical structure and which is therefore chiral.

Cinnamic acid 3-Phenylacrylic acid ($C_9H_8O_2$).

Cinnamyl alcohol dehydrogenase (CAD) Enzyme responsible for the last step in lignin monomer biosynthesis, catalyzing the conversion of coniferaldehyde, sinapaldehyde, and *p*-coumaraldehyde to coniferyl alcohol, sinapyl alcohol, or *p*-coumaryl alcohol, respectively.

Cladogram Tree-like diagram showing evolutionary relashionships.

Cohesin Reiterated domain on scaffoldin which interacts with the dockerin domain on each enzymatic subunit.

Cladogram Tree-like diagram showing evolutionary relationships.

Combinatorial synthesis A process to prepare large sets of organic compounds by combining sets of building blocks.

Combined cycle gas turbine An energy generation technology that uses a gas turbine to create electricity and then captures the resulting waste heat to create steam, which in turn drives a steam turbine, increasing the system's power output without an increase in fuel.

Combustion or burning is the sequence of exothermic chemical reactions between a fuel and an oxidant accompanied by the release of energy (mainly heat).

Commelinoid (or **commelinid) monocots** Monophyletic group comprising the palms, the spiderworts, the graminaceous plants, and the gingers.

Condensation reaction Chemical reaction in which two molecules combine to form one single molecule, with loss of a small molecule such as water.

Conformation Spatial arrangement of a molecule.

Conformation of the hydroxymethyl group (cellulose) Conformation expressed by two letters, the first referring to the torsion angle χ (O5-C5-C6-O6), i.e. the relative orientation of the C6-O6 bond to the C5-O5 bond about the C5-C6 bond, and the second to the torsion angle χ' (C4-C5-C6-O6).

Coniferin Glucoside of coniferyl alcohol.

Conserved domain Recurring unit in polypeptide chains (sequence and structure motif), determined by comparative analysis; molecular evolution uses such domains as building blocks, and these may be recombined in different arrangements to make different proteins with different functions.

Conserved plant-specific region One of the two plant-specific regions that is relatively conserved.

Convergent evolution Acquisition of the same biological trait in unrelated lineages.

Crystal planes All lattice planes and lattice directions are described by Miller indices h, k, l; the crystal plane (hkl) is a plane defined by the points **a**/h, **b**/k, and **c**/k (**a, b, c**: unit-cell vectors); crystal directions [hkl] are specified as the coordinates of the lattice point the closest to the origin along the desired direction; crystal planes (hkl) in the real-space (or direct lattice) are characterized by the normal vector and the interplanar spacing.

Cyanobacteria (blue-green algae) Phylum of Bacteria that obtain their energy through photosynthesis, using water as an electron donor and producing oxygen.

Cytochromes Membrane-bound hemoproteins that contain heme groups and carry out electron transport.

Cytochrome b_6f Enzyme found in the thylakoid membrane, catalyzing the transfer of electrons from plastoquinol to plastocyanin.

D

Designer cellulosome Artificial cellulosome comprised of recombinant chimeric scaffoldin constructs and selected dockerin-containing enzyme hybrids.

Dicotyledon (Dicot) Flowering plant whose seed has two embryonic leaves or cotyledons.

Dinoflagellate Single-celled aquatic protist.

Directed evolution Protein engineering strategy in which random mutagenesis is applied to a protein, and a selection regime is used to pick out variants that have the desired properties.

Dockerin Protein domain on each cellulosomal enzyme subunit that interacts with the cohesin domain on scaffoldin.

E

EF-hand motif Helix-loop-helix structural domain found in a large family of calcium-binding proteins.

Electron transport chain An electron transport chain couples a reaction between an electron donor (such as NADH) and an electron acceptor (such as O_2) to the transfer of H^+ ions across a membrane, through a set of mediating biochemical reactions.

Embden Meyerhof Parnas (EMP) pathway A chain of ten enzyme-catalyzed reactions that oxidize glucose into pyruvate.

Endosperm Tissue produced under the seeds of most flowering plants around the time of fertilization.

Endoglucanase Endo-β-1,4-D-glucan glucanohydrolase that usually causes a random breach of β-glucosidic bonds.

Endwise degradation Peeling-off reaction.

Endwise polymerization Polymerization in which, after a certain initial period, the monomer precursors are joined to the ends of the growing polymer instead of combining with each other.

Equatorial (sugar) In the mean plane of the ring.

Equatorial (lattice plane) Parallel to the chain axis.

Euryarchaeota Phylum of the Archaea.

Exoglucanase β-1,4-D-glucan cellobiohydrolase, which cleaves the β-glucosidic bond at one of the ends of the polysaccharide.

Exon DNA region in a gene that codes for the protein.

Expression cloning A technique in DNA cloning that uses expression vectors to generate a library of clones, with each clone expressing one protein. This expression library is then screened for the property of interest.

F

FAME Fatty acid methyl esters that constitute biodiesel.

Fast pyrolysis Pyrolysis characterized by moderate temperatures (400-600 °C) and rapid heating rates.

Fermentation Process of extracting energy from the oxidation of organic compounds, such as carbohydrates, using an endogeneous electron acceptor which is usually an organic compound; fermentation is caused by enzymes that are produced by microorganisms.

Ferulate 5-hydroxylase (F5H) Enzyme responsible for the last hydroxylation of the syringyl-type lignin precursors.

Ferulic acid or coniferic acid 4-Hydroxy-3-methoxycinnamic acid, which is an abundant phenolic phytochemical found in plant cell wall components, such as arabinoxylans as side chains. It is related to trans-cinnamic acid. As a component of lignin, ferulic acid is a precursor.

Fibrobacteres Bacterial phylum that includes many of the major rumen bacteria.

Firmicute Division of Bacteria, most of which have Gram-positive cell wall structure.

First-generation biobased product Product that has been derived from edible biomass.

First-generation biorefinery Biorefinery that uses edible biomass as feedstock.

Fischer-Tropsch Catalyzed chemical reaction in which synthesis gas is converted into liquid hydrocarbons of various forms.

Flaming combustion Gas phase oxidation process.

Forcing (radiative) A measure of the influence a factor has in altering the balance of incoming and outgoing energy in the Earth-atmosphere system and an index of the importance of the factor as a potential climate change mechanism.

Fungus Eukaryotic organism that is a member of the kingdom Fungi.

Furanose Simple sugar that contains a five-membered ring consisting of four carbon atoms and one oxygen atom.

Furfural Aromatic aldehyde (OC_4H_3CHO), also named furan-2-carbaldehyde.

G

Galactoglucomannans Hemicellulose polysaccharides consisting of a β-1,4-linked backbone of D-glucose and D-mannose residues, with α-1,6-linked D-galactose side groups.

Galactomannans Hemicellulose polysaccharides consisting of a β-1,4-linked D-mannose backbone with α-1,6-linked D-galactose side groups.

Gasification Process that converts organic or fossil-based carbonaceous materials into carbon monoxide, hydrogen, carbon dioxide, and methane. This is achieved by reacting the material at high temperatures (>700 °C), without combustion, with a controlled amount of oxygen and/or steam.

Geologic carbon cycle One of the major biogeochemical cycles in which carbon moves between rocks and minerals, the oceans, and the atmosphere over large time scales.

Glowing combustion Flameless form of combustion deriving its heat from heterogeneous reactions occurring on the surface of a solid fuel when heated in an oxidizing environment.

Glucan Polyglucose.

Glucanase Any enzyme that breaks down a glucan.

Glucomannans Water-soluble hemicelluloses consisting of β-1,4-linked D-glucose and D-mannose; they are mainly straight-chain polymers, with a small amount of branching; they are found in the roots of the Asian konjac plant and also as hemicelluloses in the wood of conifers and cotyledons.

Glucosidase Enzyme of the hydrolase class that hydrolyzes glucose residues from glucosides.

Glucosidic bond Chemical linkage formed between the hemiacetal group of glucose and an alcohol group.

Glucuronoarabinoxylans Hemicellulose polysaccharides having a β-1,4-xylan backbone with glucuronic acid and arabinose side-chains.

Glucuronoxylans Hemicellulose polysaccharides having a β-1,4-xylan backbone with glucuronic acid side-chains.

Glycan Polysaccharide.

Glycan synthase Glycosyltransferase that catalyzes the synthesis of glycans.

Glycolysis Metabolic pathway that converts glucose into pyruvate; glycolysis takes place in the cytoplasm of the cell.

Glycone Sugar group of a glycoside.

Glycosidase Glycoside hydrolase.

Glycoside Any molecule formed from the reaction of sugars and alcohols at the anomeric carbon.

Glycoside hydrolase (also called **glycosidase**) Enzyme that hydrolyzes the glycosidic bond between two or more carbohydrates or between a carbohydrate and a non-carbohydrate moiety.

Glycosidic bond Chemical linkage formed between a hemiacetal group of a sugar molecule and an alcohol group.

Glycosyltransferase (GT) Enzyme that catalyzes the transfer of a monosaccharide residue from an activated donor substrate to an acceptor molecule, forming glycosidic bonds.

Golgi apparatus Organelle found in most eukaryotic cells whose primary function is to process and package the macromolecules that are synthesized by the cell.

Greek key Topology in which typically three antiparallel β strands connected by hairpins are followed by a longer connection to the fourth strand, which lies adjacent to the first.

Green diesel A form of diesel fuel which is derived from renewable feedstock by using biomass to liquid or vegetable oil refining technologies to convert the oils into paraffinic hydrocarbons; green diesel feedstock can be sourced from a variety of oils including canola, algae, jatropha, and salicornia, in addition to tallow.

Green fluorescent protein Protein composed of 238 amino acids originally isolated from the jellyfish *Aequorea Victoria* that fluoresces green when exposed to blue light.

Greenhouse gases (GHG) Greenhouse gases are gases in the atmosphere that absorb and emit radiations within the thermal infrared range.

Guaiacyl (G) 4-Hydroxy-3-methoxyphenyl.

Gymnosperm Plant whose seeds are not enclosed in an ovule.

H

Heme Prosthetic group (i.e. tightly bound cofactor) that consists of an iron atom contained in the center of a large heterocyclic organic ring called a porphyrin.

Hemicellulases Group of enzymes that hydrolyze hemicelluloses.

Hemicelluloses Branched glycans composed of 5-carbon and 6-carbon sugars. In plant cell walls, hemicellulose binds to cellulose to form a matrix.

Heterotroph Organism that cannot fix carbon, but uses organic carbon for growth.

Homogalacturonan Linear pectin chain of α-1,4-linked-D-galacturonic acid residues in which some of the carboxyl groups are methyl esterified.

Homolog A gene related to a second gene by descent from a common ancestral DNA sequence; the term may be applied to the relationship between genes separated by the event of speciation (ortholog) or to the relationship between genes separated by the event of genetic duplication (paralog).

Hydrocellulose Product with lower degree of polymerization and higher crystallinity obtained from dilute acid hydrolysis of cellulose.

Hydrogenation Chemical reaction that typically constitutes the addition of pairs of hydrogen atoms to a molecule.

Hydrogenolysis Chemical reaction whereby a carbon-carbon or a carbon-heteroatom single bond is cleaved by hydrogen.

Hydrothermal treatment Processing of biomass in sub- and supercritical water medium.

Hydrothermolysis Hydrolysis of lignocellulosic materials using pure water at elevated temperatures.

Hypervariable region One of the two plant-specific regions that represents highly divergent DNA sequences.

I

Inversion (mechanism) Single nucleophilic displacement that leads to an inversion of the anomeric configuration.

Inverting glycosyltransferase Enzyme that catalyzes the transfer of sugar moieties with an inversion of the anomeric configuration.

Ionic liquid Liquid that contains, essentially, only ions.

Isoform (protein) Any of several different forms of the same protein formed because of single nucleotide polymorphisms.

J

Jelly roll Variant of Greek key topology with both ends of a sandwich or a barrel fold being crossed by two interstrand connections.

K

Krebs cycle (or citric acid cycle) Series of enzyme-catalyzed reactions, which is of central importance in all living cells that use oxygen as part of cellular respiration.

L

Laccases Multi-copper oxidases of wide specificity that carry out one-electron oxidation of phenolic and related compounds, and reduce O_2 to water.

Levelling-off degree of polymerization (LODP) When the degree of polymerization of a cellulose is plotted against time of hydrolysis, the degree of polymerization decreases sharply initially and then levels off at a value characteristic of the cellulose called LODP.

Light-dependent reactions or **light reactions** Reactions of photosynthesis that convert light energy into chemical energy, producing ATP and NADH.

Light-harvesting complex (green plants) An array of protein and pigment molecules embedded in the thylakoid membrane which transfer light energy to one chlorophyll *a* molecule at the reaction center of a photosystem.

Light-independent reactions or **dark reactions** Reactions of photosynthesis that use the ATP and NADPH synthesized during the light-dependent reactions to provide the energy for the synthesis of glucose and other organic molecules from inorganic CO_2 and water.

Lignification The process that includes the biosynthesis of monolignols, their transport to the cell wall, and polymerization into the final molecule.

Lignin Generic term for a large group of aromatic polymers resulting from the oxidative combinatorial coupling of 4-hydroxyphenylpropanoids.

Lignin peroxidase (LiP) Hemoprotein that catalyzes the oxidative cleavage of C-C bonds in a number of model compounds, and oxidizes benzyl alcohols to aldehydes or ketones.

Ligninase Lignin-degrading enzyme.

Lignocellulosic biomass (or simply biomass) Plant biomass that is composed of cellulose, hemicelluloses, and lignin.

Lithosphere Rigid outermost shell of a rocky planet; it comprises the crust and the uppermost mantle.

Lycophyte Oldest living vascular plant division, including some of the most primitive extant species.

Lyocell fiber Regenerated cellulose fiber produced via the direct dissolution of cellulose in organic solvents such as N-methylmorpholine-N-oxide.

Lyotropic A material is called lyotropic if it forms liquid crystal phases because of the addition of a solvent.

M

Manganese peroxidase (MnP) Hemoprotein involved in the oxidative degradation of lignin in white-rot basidiomycetes; it catalyzes the oxidation of Mn(II) to Mn(III) by hydrogen peroxide.

Mannan Polymannose.

Mercerization Treatment for cotton fabric and thread that gives fabric a lustrous appearance.

Mesophase (or liquid crystal phase) Intermediate phase in ordering between the crystalline solid and isotropic liquid phases.

Mesophyll Tissue between the upper and lower epidermis of a leaf blade, consisting of parenchyma-like cells containing numerous chloroplasts.

Metabolism Set of enzyme-catalyzed reactions that convert substrates that are external to the cell into various internal products, i.e. the set of reactions that happen in living organisms to maintain life.

Metabolite Metabolites are the intermediates or products of metabolism. A primary metabolite is directly involved in normal growth, development and reproduction. A secondary metabolite is not directly involved in those processes, but usually has an important ecologic function. Ethanol and lactic acid are primary metabolites produced by industrial microbiology.

Microfibril (cellulose) Fibrous crystalline aggregate of cellulose molecules.

Microtubule One of the components of the cytoskeleton; they have a diameter of 25 nm and length varying from 200 nm and 25 μm.

Module (enzymes) Independently folding, structurally and functionally discrete unit of modular enzymes.

Molecular screening (high throughput) Automated, simultaneous testing of thousands of distinct chemical compounds in models of biological mechanisms.

Monoclinic One of the seven lattice point groups; in the monoclinic system, the crystal is described by three vectors of unequal length that form a rectangular prism with a parallelogram as its base.

Monocotyledon (monocot) Flowering plant whose seed has a single embryonic leaf or cotyledon.

Monolignols (or hydroxycinnamyl alcohols) Lignin monomers, the most abundant of which are *p*-coumaryl (4-hydroxycinnamyl) alcohol, coniferyl (3-methoxy 4-hydroxycinnamyl) alcohol, and sinapyl (3,5-dimethoxy 4-hydroxycinnamyl) alcohol.

Monooxygenase Enzyme that incorporates one hydroxyl group into substrates in many metabolic pathways.

Morphology Study of form, comprising shape, size, and structure.

N

Nematic Describing the structure of some liquid crystals whose molecules align in loose parallel lines.

Non-commelinoid monocot APG (Angiosperm Phylogeny Group) clade comprising orchids and lilies.

Non-processive glycosyltransferase Enzyme that catalyzes the addition of only one sugar residue.

Nucleotide Structural unit of DNA (deoxyribonucleic acid) and RNA (ribonucleic acid) that consists of a nitrogenous base, a sugar, and a phosphate group.

Nucleotide sugar (or sugar nucleotide) Activated monosaccharides that act as donors of sugar residues in nucleotide sugars metabolism; they are substrates for glycosyltransferases.

O

Olefin metathesis or **transalkylidenation** A chemical reaction where the bonds between different atoms are broken and new bonds are formed; in olefin metathesis, the carbon bonds change places.

Open reading frame Portion of an organism's genome which contains a sequence of bases that could potentially encode a protein.

Operon Functioning unit of key nucleotide sequences including an operator, a common promoter, and one or more structural genes, which is controlled as a unit to produce messenger RNA (mRNA) in the process of protein transcription.

Organelle A differentiated structure within a cell that performs a specific function.

Organosolv Pulping technique that uses an organic solvent to solubilize lignin and hemicelluloses.

Origin chain Chain passing through the origin of the unit cell.

Origin sheet Sheet formed by hydrogen bonding between adjacent origin chains.

Orthologs Genes in different species that are similar to each other because they originated from a common ancestor; orthologs retain the same function in the course of evolution.

Oxycellulose Solid, and often fibrous, water-insoluble products obtained from oxidation of cellulose.

Oxidative combinatorial coupling (lignin polymerization) Oxidative radicalization of phenols, followed by combinatorial radical coupling (see **combinatorial synthesis**).

Oxygen evolving complex (***OEC***) Water oxidizing enzyme involved in the photooxidation of water during the light reactions of photosynthesis.

P

Parallel down Sense of the polymer chain when the *z* coordinate of O5 is smaller than that of C5.

Parallel up Sense of the polymer chain when the z coordinate of O5 is greater than that of C5.

Paralogs Homologs whose evolution reflects gene duplication events; paralogs evolve new functions.

Peak oil The point in time when the maximum rate of global petroleum extraction is reached, after which the rate of production enters terminal decline.

Pectin Any of several heterogeneous polysaccharides that are composed primarily of α-1,4-D-galacturonic acid units or their methyl esters; in plant primary cell walls, pectins form a gel phase in which the cellulose-hemicelluloses network is embedded.

Pentose phosphate pathway A process that generates NADPH and pentoses from glucose 6-phosphate.

Peroxidase Hemoprotein catalyzing the oxidation by hydrogen peroxide of a number of substrates.

Phenylpropanoids A class of organic compounds that are biosynthesized from the amino acid phenylalanine.

Photoautotroph Autotrophic organism that carries out photosynthesis to acquire energy and fix carbon.

Photoheterotroph Heterotrophic organism that uses light for energy, but cannot use CO_2 as their sole carbon source; consequently, they use organic compounds to satisfy their carbon requirements.

Photolyze Photodissociate.

Photophosphorylation Production of ATP using the energy of sunlight.

Photorespiration A process in plant metabolism by which RuDP has oxygen added to it, instead of CO_2 during normal photosynthesis.

Photosynthesis Metabolic pathway that converts light energy into chemical energy.

Photosystem I The second photosystem (protein complex) in the photosynthetic light reactions of algae, plants, and some bacteria.

Photosystem II The fist photosystem (protein complex) in the photosynthetic light reactions.

Phototendering Process by which organic fibers and textiles lose strength and flexibility as a result of exposure to sunlight.

Plasma membrane Selectively permeable lipid bilayer found in all cells. A plasma membrane has two surfaces, one toward the outside of the cell and termed the *exoplasmic* (outer) *surface* (ES) and one adjacent to the cytoplasm and termed the *protoplasmic* (inner) *surface* (PS). When the bilayer is split during the freeze-fracture-etch process, two additional surfaces known as *fracture faces* are revealed. The fracture faces are termed the *EF* and the *PF faces* according to their association with the exoplasmic or protoplasmic leaflets of the membrane, respectively.

Plastids Major organelles found in the cells of plants and algae; they are responsible for photosynthesis, for the storage of products such as starch, and for the synthesis of many classes of molecules.

Polarity (cellulose) Cellulose polarity refers to the chemical difference of the two ends of the molecule, inducing possible parallel and antiparallel chain packings.

Polyhydric alcohol Hydrogenated form of carbohydrate, whose carbonyl group has been reduced to a hydroxyl group. Sugar alcohols have the general formula $H(HCHO)_{n+1}H$, whereas sugars have $H(HCHO)_nHCO$.

Polymorph Any of different crystalline forms of the same chemical compound.

Porphyrin Heterocyclic macrocycles composed of four modified pyrrole subunits.

Primary cell wall (plant) Cell wall of growing cells, which are relatively thin, and only semi-rigid to accommodate future cell growth.

Primary scaffoldin Enzyme-binding scaffoldin.

Primitive lattice A crystal lattice in which there are lattice points only on the cell corners.

Processive glycosyltransferase Enzyme that catalyzes the addition of multiple sugar residues.

Processivity Ability of an enzyme to repetitively continue its catalytic function without dissociating from its substrate.

Protein engineering The design and construction of new proteins with novel or desired functions, through the modification of amino acid sequences using recombinant DNA technology.

Protists Group of eukaryotic microorganisms.

Purple bacteria Proteobacteria that are capable of producing energy through photosynthesis; they contain a single type of reaction center (type I or II) and do not produce oxygen because the electron donor involved in photosynthesis is not water.

Pyranose Simple sugar that contains a six-membered ring consisting of five carbon atoms and one oxygen atom.

Pyrolysis Thermochemical decomposition of organic material at elevated temperatures in the absence of oxygen. Pyrolysis typically occurs under pressure and at operating temperatures above 430 °C.

Q

Quantum requirement (light reactions of photosynthesis) Number of light photons absorbed for the transfer of one electron.

Quinone methides A class of conjugated organic compounds that contain a cyclohexadiene with a carbonyl and an exocyclic double bond carbon (methylene).

R

Raceway pond Shallow artificial pond used in the cultivation of algae.

Rational design Protein engineering strategy based on detailed knowledge of the structure and function of the protein.

Reaction center (*RC*) (photosynthesis) Complex of several proteins, pigments, and other cofactors assembled together to execute the primary energy conversion reactions of photosynthesis.

Recombinant DNA A form of synthetic DNA that is engineered through the combination or insertion of one or more DNA strands, thereby combining DNA sequences that would not normally occur together.

Red fluorescent protein Protein originally isolated from the coral *Discosoma striata* that has an optimal absorption at 558 nm and emits light at 583 nm.

Reducing end (cellulose) Chain end containing an unsubstituted hemiacetal, the nonreducing end containing an additional hydroxyl group at C4.

Regeneration Precipitation of dissolved cellulose.

Repeating unit (or **asymmetric unit)** Smallest portion of a crystal unit to which crystallographic symmetry can be applied to generate one unit cell.

Respiration Set of the metabolic reactions and processes that take place in organisms' cells to convert biochemical energy from nutrients into adenosine triphosphate (ATP), and then release waste products. The reactions involved in respiration are catabolic reactions that involve the oxidation of one molecule and the reduction of another.

Retention (mechanism) Double displacement mechanism that leads to the retention of the anomeric configuration.

Rhamnogalacturonan I (RG I) Pectic polysaccharide containing a backbone of the repeating galacturonic acid-rhamnose disaccharide, in which the rhamnosyl residues can be substituted with arabinan, galactan and/or arabinogalactans.

Rhamnogalacturonan II (RG II) Highly branched pectic polysaccharide containing an α-1,4-linked galacturonic acid backbone with side chains comprising a wide and unusual range of sugar residues.

Rieske proteins Iron-sulfur protein (ISP) components of cytochrome b_6f complexes and *bc*1 complexes which were first discovered by J.S. Rieske and co-workers in 1964; they contain a high potential 2Fe-2S cluster; the pair of iron atoms in the cluster are bound to two cysteine and two histidine residues, in addition to two sulfur atoms.

Rosette A structure of cellulose-synthesizing terminal complexes.

RuBisCO (Ribulose-1,5-bisphosphate carboxylase/oxygenase) Enzyme involved in the Calvin Cycle, that catalyzes the fist major step of carbon fixation.

S

Scaffoldin Multifunctional integrating subunit of the cellulosome.

Second generation biobased product Product that has been derived from non-edible biomass.

Second-generation biorefinery Biorefinery that uses non-edible biomass as feedstock.

Secondary cell wall (plant) Structure found in many plant cells located between the primary cell wall and the plasma membrane; the cell starts producing the secondary cell wall only after the primary cell wall is completed and the cell has stopped growing; it is thicker and stronger than the primary cell wall.

Shikimic acid An important biochemical intermediate in plants and microorganisms, better known under its anionic form shikimate (3,4,5-trihydroxycyclohex-1-ene-1-carboxylate); the shikimate pathway is the biosynthetic sequence employed by plants and bacteria to generate the aromatic amino acids.

Sinapyl alcohol 4-(3-hydroxy-1-propenyl)-2,6-dimethoxyplenol; a unit in the lignin polymer.

Solanoceae Family of flowering plants that contains a number of important agricultural crops such as potato, tomato, and aubergine, as well as many toxic plants.

Solvolysis Special type of nucleophilic substitution or elimination where the nucleophile is a solvent molecule.

Slow pyrolysis Pyrolysis characterized by slow biomass heating rates, lengthy gas and solids residence times, and typically low temperature.

Stepwise polymerization Polymerization in which any two monomers present in the reaction mixture can link together at any time, and growth of the polymer is not confined to chains that are already forming.

Stomata Openings in the epidermis that regulates gas exchange in the plant; they are formed by two specialized epidermal cells called guard cells.

Sugar nucleotide (or nucleotide sugar) Activated monosaccharides that act as donors of sugar residues in nucleotide sugars metabolism; they are substrates for glycosyltransferases.

Supercritical (fluid) Substance above its critical temperature and critical pressure, where distinct liquid and gas phases do not exist.

Supramolecular assembly Well defined complex of molecules held together by noncovalent bonds.

Synthesis gas (or **syngas**) Gas mixture that contains varying amounts of carbon monoxide and hydrogen.

Syringyl (S) 4-Hydroxy-3,5-dimethoxyphenyl.

Syringin Glucoside of sinapyl alcohol.

T

Tars Condensable organic compounds with boiling point between 80 and 350 °C that contain a high degree of aromatic rings.

Tautomerization Chemical reaction resulting in the migration of a proton and the shifting of bonding electrons; the concept of tautomers that are interconvertible by tautomerizations is called tautomerism.

Terminal complex Cellulose-synthesizing complex that contains a number of cellulose synthases organized in spinnerets at the cell membrane.

Thermochemical conversion Technological platform involving generally biomass gasification or pyrolysis processes.

Thermotropic Exhibiting different phases at different temperatures; a liquid crystal is thermotropic if the order of its components is determined or changed by temperature.

Thylakoid lumen Chloroplast compartment enclosed by the thylakoid membrane.

Thylakoid membranes (chloroplast) Internal system of interconnected membranes that carry out the light reactions of photosynthesis.

Transcription factor Protein that binds to specific DNA sequences, thereby controlling the transfer (or transcription) of genetic information from DNA to mRNA.

Transglycosidase Glycosidase that is also able to form glycosidic bonds.

Transverse microfibril Microfibril that is transverse to the main cell axis.

Triclinic One of the seven lattice point groups; in the triclinic system, the crystal is described by three vectors of unequal length that are not mutually orthogonal.

Tubulin One of several members of a family of globular proteins; the most common members of the tubulin family are α-tubulin and β-tubulin, the proteins that make up microtubules.

Tunicate Subphylum of a group of underwater saclike filter feeders with siphons that are members of the phylum Chordata.

Turgor pressure Main pressure of the cell contents against the cell wall in plant cells and bacteria cells.

Type I cohesin Domain of the scaffoldin that binds the cognate dockerin domain of each enzymatic component.

Type II cohesin Domain on cell-surface anchoring proteins that interacts with a type II dockerin domain.

Type I dockerin Domain on the enzymes that interacts with the type-I cohesin domains.

Type II dockerin Domain on scaffoldin that interacts with the type II cohesin domains contained on cell-surface anchoring proteins.

U

Uniplanar orientation Preferential orientation of cellulose microfibrils to the cell wall surface (film surface).

Unit cell The smallest repeating unit that can generate the whole crystal with only translation operations.

Uridine diphospho-glucose (UDPGlc) A compound in which α-glucopyranose is esterified at C1 with the terminal phosphate group of uridine-5'-pyrophosphate.

Uronic acid Sugar acid with both a carbonyl and a carboxylic acid function; it can be thought of as a sugar in which the terminal carbon's hydroxyl function has been oxidized to a carboxylic acid; the names of uronic acids are generally based on their parent sugars (e.g. glucuronic acid).

V

Vascular plant Green plant having a vascular tissue system (e.g. ferns, gymnosperms, angiosperms).

Versatile peroxidase (VP) Hemoprotein, combining catalytic properties of manganese peroxidase (oxidation of Mn(II)), lignin peroxidase (Mn-independent oxidation of non-phenolic aromatic compounds).

Viscose fiber Regenerated cellulose fiber produced via the dissolution of a cellulose dithiocarbonate ester called xanthate in an aqueous solution of caustic soda.

W

Whisker Short single crystal fiber or filament used as a reinforcement in a matrix.

X

Xylan Polyxylose.

Xylanase Class of enzymes that hydrolyze the β-1,4 linkage in the xylan backbone, yielding short xylooligomers. Most known xylanases belong to families GH 10 and 11.

Xylogalacturonans Pectic polysaccharide consisting of an α-1,4-linked D-galacturonic acid backbone with β-1,3-linked xylose side chains.

Xyloglucan endotransglycosylase/hydrolases (XTH) Enzymes that are able to cleave and to reattach xyloglucans polymers in plant cell walls.

Xyloglucans Hemicellulose polysaccharides with a linear β-1,4-glucan backbone substituted at O6 with mono-, di-, and triglycosyl side chains.

Z

Zinc finger Superfamily of protein domains that can bind to DNA; a zinc finger consists of two antiparallel β strands and an α helix.

Z-scheme Sawtooth shifts in redox potential as electrons pass from P680 (reaction center chlorophyll of photosystem II) to $NADP^+$.

ZSM-5 (Zeolite Sieve of Molecular porosity) Aluminosilicate zeolite mineral belonging to the pentasil (a pentasil unit consists of eight five-membered rings) family of zeolites; it is widely used in the petroleum industry as a heterogeneous catalyst for hydrocarbon isomerization reactions such as the isomerization of metaxylene to paraxylene.

Index

Symbols

A

B

C

D

E

F

T

U

V

W

X

Z